THE HANDBOOK OF OPERATOR FATIGUE

In memory of Roy Davies
"The Dean of Vigilance."

The Handbook of Operator Fatigue

Edited by

GERALD MATTHEWS
University of Cincinnati, Ohio, USA

PAULA A. DESMOND
Dell Inc., USA

CATHERINE NEUBAUER
University of Cincinnati, Ohio, USA

and

P.A. HANCOCK
University of Central Florida, USA

CRC Press
Taylor & Francis Group
Boca Raton London New York

CRC Press is an imprint of the
Taylor & Francis Group, an **informa** business

CRC Press
Taylor & Francis Group
6000 Broken Sound Parkway NW, Suite 300
Boca Raton, FL 33487-2742

First issued in paperback 2017

© 2012 by Taylor & Francis Group, LLC
CRC Press is an imprint of Taylor & Francis Group, an Informa business

No claim to original U.S. Government works
Version Date: 20160616

ISBN 13: 978-1-138-07781-2 (pbk)
ISBN 13: 978-0-7546-7537-2 (hbk)

Visit the Taylor & Francis Web site at
http://www.taylorandfrancis.com

and the CRC Press Web site at
http://www.crcpress.com

Contents

PART VIII   OPERATIONAL COUNTERMEASURES

List of Figures

List of Tables

The Editors

Gerald Matthews, PhD, Professor of Psychology, University of Cincinnati, USA

Mailing address:
Department of Psychology, University of Cincinnati, Cincinnati, OH 45221
Email:
Gerald.Matthews@uc.edu

Paula A. Desmond, PhD, Usability Senior Engineer, Enterprise Experience Design Group, Dell Inc., USA

Mailing address:
Dell Inc., One Dell Way, Round Rock, TX 78682
Email:
Paula_Desmond@Dell.com

Catherine Neubauer, MA, University of Cincinnati, USA

Mailing address:
Department of Psychology, University of Cincinnati, Cincinnati, OH 45221
Email:
neubauce@mail.uc.edu

P.A. Hancock, PhD, Provost's Distinguished Research Professor and Pegasus Professor, University of Central Florida, USA

Mailing address:
Department of Psychology, University of Central Florida, Orlando, FL 32816
Email:
Peter.Hancock@ucf.edu

The Contributors

Phillip L. Ackerman, School of Psychology, Georgia Institute of Technology, Atlanta, GA

Siobhan Banks, Centre for Sleep Research, University of South Australia, Adelaide, South Australia, Australia

Gregory Belenky, Sleep and Performance Research Center, Washington State University, Spokane, WA

Abigail Brown, Center for Community Research, DePaul University, Chicago, IL

Molly Brown, Center for Community Research, DePaul University, Chicago, IL

Charles Calderwood, School of Psychology, Georgia Institute of Technology, Atlanta, GA

John A. Caldwell, Fatigue Science and Archinoetics, LLC, Honolulu, HI

Christopher Christodoulou, Department of Neurology, SUNY at Stony Brook, Stony Brook, NY

Erin Marie Conklin, School of Psychology, Georgia Institute of Technology, Atlanta, GA

Diane L. Cox, Faculty of Health and Wellbeing, University of Cumbria, Carlisle, UK

Ashley Craig, Royal Rehabilitation Centre, The University of Sydney, Sydney, New South Wales, Australia

Paula A. Desmond, Enterprise Experience Design Group, Dell Inc., Round Rock, TX

Meredyth Evans, Center for Community Research, DePaul University, Chicago, IL

Simon Folkard, Department of Psychology, Swansea University, Swansea, UK

Jonathan French, Human Factors and Systems Department, Embry-Riddle Aeronautical University, Daytona Beach, FL

Valerie J. Gawron, The MITRE Corporation, McLean, VA

Andrew Graham, New South Wales Centre for Road Safety, North Sydney, New South Wales, Australia

P.A. Hancock, Department of Psychology, University of Central Florida, Orlando, FL

Julie Hatfield, Injury Risk Management Research Centre, University of New South Wales, Sydney, New South Wales, Australia

William S. Helton, Department of Psychology, University of Canterbury, Christchurch, New Zealand

Edward M. Hitchcock, Organizational Science and Human Factors Branch, National Institute for Occupational Safety and Health, Cincinnati, OH

G. Robert J. Hockey, Department of Psychology, University of Sheffield, Sheffield, UK

Melinda L. Jackson, Sleep and Performance Research Center, Washington State University, Spokane, WA

Leonard A. Jason, Center for Community Research, DePaul University, Chicago, IL

R.F. Soames Job, New South Wales Centre for Road Safety, North Sydney, New South Wales, Australia

Osami Kajimoto, Osaka City University Graduate School of Medicine, Osaka, Japan

William D.S. Killgore, Neuroimaging Center, Mclean Hospital, Belmont, MA

Gerald P. Krueger, Krueger Ergonomics Consultants, Alexandria, VA

Hirohiko Kuratsune, Osaka City University Graduate School of Medicine, Osaka, Japan

Gerald Matthews, Department of Psychology, University of Cincinnati, Cincinnati, OH

James C. Miller, Miller Ergonomics, San Antonio, TX

Catherine Neubauer, Department of Psychology, University of Cincinnati, Cincinnati, OH

Kelly J. Neville, Human Factors and Systems Department, Embry-Riddle Aeronautical University, Daytona Beach, FL

Timothy David Noakes, UCT/MRC Research Unit for Exercise Science & Sports Medicine, Sports Science Institute, Cape Town, South Africa

Alan T. Pope, NASA Langley Research Center, Hampton, VA

Roger R. Rosa, National Institute for Occupational Safety and Health, Washington, DC

Chika Sakashita, Injury Division, George Institute for Global Health, Sydney, New South Wales, Australia

Dyani J. Saxby, Department of Psychology, University of Cincinnati, Cincinnati, OH

Mark W. Scerbo, Department of Psychology, Old Dominion University, Norfolk, VA

Andrew P. Smith, School of Psychology, Cardiff University, Cardiff, UK

Chad L. Stephens, NASA Langley Research Center, Hampton, VA

James L. Szalma, Department of Psychology, University of Central Florida, Orlando, FL

Yvonne Tran, Royal Rehabilitation Centre, The University of Sydney, Sydney, New South Wales, Australia

Lloyd D. Tripp, Air Force Research Laboratory, Wright Patterson Air Force Base, Dayton, OH

Philip Tucker, Department of Psychology, Swansea University, Swansea, UK

Hans P.A. Van Dongen, Sleep and Performance Research Center, Washington State University, Spokane, WA

Joel S. Warm, Air Force Research Laboratory, Wright Patterson Air Force Base, Dayton, OH

Yasuyoshi Watanabe, RIKEN Center for Molecular Imaging Science (CMIS), Osaka City University Graduate School of Medicine, Osaka, Japan

Jim Waterhouse, Research Institute for Sport and Exercise Sciences, Liverpool John Moores University, Liverpool, UK

Ann Williamson, School of Aviation, University of New South Wales, Sydney, New South Wales, Australia

PART I

Introduction to Operator Fatigue

1
An Overview of Operator Fatigue

Gerald Matthews, Paula A. Desmond, Catherine Neubauer and P.A. Hancock

INTRODUCTION TO OPERATOR FATIGUE

Fatigue is one of the most puzzling enigmas in all of psychology. We all know what it feels like, and yet a rigorous definition remains elusive. Many experimental studies show the detrimental effects of fatigue (Matthews et al., 2000), but sometimes individuals who appear to be highly fatigued continue to show normal levels of performance (Holding, 1983). The earliest systematic investigations showed that subjective feelings of fatigue do not necessarily correspond to objective performance loss (Thorndike, 1914). It is also challenging to identify the neural and psychological processes that mediate the impact of fatigue on performance. Fatigue, in part, reflects fundamental changes in neural function (Saper, Scammell & Lu, 2005), but also depends critically on an operator's interest in the task and the high-level cognitive processes that regulate motivation (Hockey, 1997).

There is ample evidence that fatigue impairs operator performance within a range of applied domains, especially industry and transportation. Indeed, operator fatigue has been implicated in any number of high-profile disasters, including those at the Three Mile Island nuclear plant, the Bhopal chemical works, and on board the *Exxon Valdez* tanker and the Space Shuttle Challenger (Rosekind et al., 1995). Furthermore, fatigue is a "hidden killer" through its impact on health, as well as safety.

At the time of writing, the US Federal Aviation Authority (FAA) is investigating instances of air traffic controllers having fallen asleep at three separate airports between January and April 2011 (FAA Press Release, April 13, 2011).[1] The fatigue issues that controllers face are well known (Gregory, Oyung & Rosekind, 1999) and national aviation authorities are tasked with ensuring the wakefulness of controllers. In one sense, it is extraordinary that such problems still occur, although they doubtless reflect social, political and financial influences. The Associated Press (AP) reported a suggestion, based on extensive evidence, from Gregory Belenky of Washington State University that controllers working night shifts should be allowed to take short recuperative naps.[2] The same report also quotes a political response from a US senator: "I think that is totally bogus. There are so many professions that have to work long hours." Clearly, science and society do not always see the problem in the same way.

---

1 See http://www.faa.gov/news/press_releases/news_story.cfm?newsId=12664 (accessed April 14, 2011).
2 See http://news.yahoo.com/s/ap/20110415/ap_on_re_us/us_sleeping_air_traffic_controllers (accessed April 14, 2011).

Despite politicians' tenuous grasp on the scientific method, research has in fact made great strides in developing a range of practical tools for assessing fatigue and for implementing countermeasures (e.g., Caldwell & Caldwell, 2004; Hancock & Desmond, 2001). Work scheduling, naps and rests, judicious use of stimulant drugs such as caffeine, and adaptive task design all provide interventions that are effective in specific circumstances. This handbook is intended to provide an accessible survey of the spectrum of research on operator fatigue, and a resource for those seeking answers to fatigue problems in operational practice. In this introduction, we seek to present the main issues in operator fatigue research, organized in relation to the sections of the book, and to highlight the specific contribution made by each contributory chapter.

This chapter is joined by two companions. From a historical perspective, the need for sleep has been recognized from the earliest times and accounts from antiquity provide the starting point for Miller's survey of the history of fatigue in Chapter 2. The chapter traces the key events that accelerated fatigue research, including concerns about the lengthy shifts of early twentieth-century factory workers, and the performance decrements evident in World War II aviators and radar operators, culminating in the contemporary differentiation of multiple fields of fatigue research relevant to industrial, transportation and military operations. As Miller discusses, the historical trend is towards a more systems-based perspective on the impact of multiple components of fatigue on the operator's interactions with technology.

Hockey's chapter, which follows, focuses on the challenges of fatigue and performance research. He points out four key issues that become evident fairly early in the development of the field: the link between fatigue and output, the relationship between fatigue and brain energy, the importance of operator motivation, and the division of fatigue in multiple types. All of these issues still remain unresolved despite extensive research. Indeed, questionable assumptions persist that work causes fatigue, and fatigue represents a depletion of brain energy. Hockey sets out a research agenda for finally answering the defining questions of fatigue research, based on a model of the adaptive management of goal-oriented human activity.

THE NATURE OF FATIGUE

Fatigue may be succinctly defined as the likelihood of falling asleep, but there is rather more to the condition than sleepiness alone. There are actually several different ways in which we can parse the multiple domains of fatigue research, including discriminating sources of fatigue, types of fatigue response, and underlying processes (Matthews et al., 2000).

Sources of Fatigue

What makes us tired? Answers that come to mind include lack of sleep, late nights, hard work, boredom and illness. In fact, fatigue research has crystallized into a variety of subdomains associated with different sources of fatigue. One source is an insufficiency of sleep (or poor quality of sleep). A second source derives from the 24-hour circadian cycle in wakefulness and alertness. While sleep loss and circadian rhythms are distinct influences on fatigue, they are increasingly studied together. A major advance has been the development of "two-process" models of fatigue (Borbély, 1982) that allow fatigue to be specified from the combined impact of a homeostatic need for sleep, increasing with hours of wakefulness, and circadian rhythms driven by an internal clock. Models of this kind provide the basis for evaluating shift systems: night workers face the dual penalty of sleep deprivation alongside the loss of alertness driven by the circadian phase during which they are forced to be active.

Further sources of fatigue are derived from the task itself. Occupational studies suggest that rather long work shifts exceeding eight hours are necessary to observe loss of productivity (Knauth, 2007). In transportation, fatigue may be an issue even for drivers working normal, daytime shifts (Friswell & Williamson, 2008), pointing to the hazards for the "wakeful-but-fatigued" driver (Matthews et al., 2011). In the laboratory, studies of vigilance have been especially important for investigating task-induced fatigue (Davies & Parasuraman, 1982). In specific circumstances, even short-duration tasks requiring sustained attention can show rapid and operationally significant performance decrements (Temple et al., 2000). More recently, the problem of cognitive fatigue has surfaced in a range of performance contexts other than traditional vigilance paradigms (Ackerman, 2011). A variety of task factors (as well as operator characteristics) may influence the speed of onset of fatigue (Ackerman, Calderwood & Conklin, Chapter 6, this volume). High workload and monotonous tasks that afford the operator little scope for implementing compensatory strategies appear to be most vulnerable to fatigue effects (Matthews, Warm et al., 2010a). Conversely, tasks that offer high levels of challenge and intrinsic interest can be highly fatigue-resistant (Holding, 1983).

Sleep loss, time-of-day effects and task workload thus define three major areas of fatigue research, but there remain yet other sources of fatigue. Various conditions associated with compromised neural functioning, including infection and other health problems, use of sedative drugs and nutritional factors may all impact on fatigue (Matthews, Davies et al., 2000; Smith, 2011). Environmental stressors may produce fatigue as one of a range of psychological symptoms (Hancock, 1984). These stressors may include, for example, loud noise, high temperatures and computer displays that elicit visual fatigue. Stress and fatigue should be discriminated, although, as in cases of burn-out, stress and fatigue may appear as a common syndrome.

Fatigue Responses

Fatigue may be differentiated not just in relation to the events and stimuli that precipitate the condition but also to the responses that signal the fatigue state. Modern research continues to be shaped by the tripartite division between physiological outcomes proposed by Bartley & Chute (1947), such as muscular fatigue, subjective feelings of tiredness and discomfort, and decrements in performance.

All types of response have their complexities, limiting development of valid metrics for fatigue. Seemingly the most obvious form of physiological fatigue is muscular fatigue resulting from prolonged physical exertion, which is especially relevant to occupations requiring physical work and to sports psychology. Muscle fatigue is often attributed to progressive impairments in cellular metabolism associated with failure of oxygenation and depletion of chemical sources of energy, as well as the harmful effects of the accumulating metabolites produced by prolonged muscular activity. However, as Noakes discusses in Chapter 7 of this volume, it appears that the neuropsychological regulation of physical activity is more important than any localized cellular process. Just as problematic is the extent to which psychological fatigue corresponds to an impairment in brain metabolism. Work continues on the role in fatigue of intracellular metabolic processes, stress hormones, immune system functioning and neurological pathways, as Chapter 14 by Watanabe, Kuratsune & Kajimoto indicates. There are also intriguing parallels between loss of vigilance and declining cerebral blood-flow velocity, as Warm, Tripp, Matthews & Helton discuss in Chapter 13. Nonetheless, robust neurophysiological correlates of fatigue remain somewhat elusive (see Hockey, Chapter 3).

Subjective fatigue may include awareness of bodily discomfort, but it refers also to experience of mental states including tiredness, sleepiness, apathy and mind-wandering. The salience of subjective symptoms raises the difficult issue of how to interpret conscious experience within a psychological science of fatigue, as is discussed in detail in Chapter 4 by Hancock, Desmond & Matthews. A persistent difficulty is that while subjective states are clearly influenced by brain processes, correspondences between states and brain activations appear to be imperfect, even when a broad statistical association can be established (Barrett & Wager, 2006). One perspective is simply to reject subjective states in favor of purely neurological investigations. Another is to see subjective measures as providing more valid indices of broad-based biocognitive systems than single psychophysiological measures (Thayer, 1989). The validity of subjective measures in general is addressed further in Part III of this volume, "Assessment of Fatigue."

Traditional arousal theory assumed that performance variation reflected a unitary dimension of cortical arousal. When fatigue is associated with loss of arousal, a global deficit across the range of performance tasks should be seen. The classical Yerkes-Dodson Law permits only task difficulty to moderate this behavioral fatigue response: easy tasks should show a greater performance decrement. Critiques of the traditional arousal theory may be found elsewhere (e.g., Hancock & Ganey, 2003; Hockey & Hamilton, 1983). Most relevant here is that modern cognitive neuroscience identifies numerous distinct modules or components for information-processing that may be differentially sensitive to fatigue.

Studies of fatigue have confirmed that choice of performance index is critical. Especially in the complex, multi-component tasks of real life, fatigue may influence some performance indices but not others. To give just one example, Neubauer et al. (in press) found that fatigue elicited by vehicle automation did not impair subsequent vehicle control, indexed by variability of lateral position on the road (a standard driving performance metric) – but fatigued drivers were slow to react to an unexpected event requiring an evasive maneuver. Furthermore, as the Cambridge Cockpit studies of the 1940s suggested (Davis, 1948), fatigue may be more evident in the breakdown in coordination of multiple subtasks than in any single behavioral response.

Processes for Fatigue

The complexities of both the input and output functions for fatigue suggest that it is futile to try to capture its threats to operator performance with any single function or characteristic, as applications of the Yerkes-Dodson Law have vainly attempted to do. Instead, a range of fatigue effects needs to be distinguished and placed within an appropriate cognitive and/or neural architecture. Different process models may also be appropriate for different sources of fatigue.

Process models for fatigue and stress may be broadly divided into two types (Matthews, 2001). Biocognitive models specify (in this context) fatigue effects in terms of changes in parameters of an explicit architecture for the task domain of interest. The architecture may be defined either neurologically or in terms of virtual information-processing operations, such as resource allocation. The cognitive demands of sustained attention may deplete a general resource that leads to vigilance decrement (Hancock & Warm, 1989; Warm, Parasuraman & Matthews, 2008). Formal models based on this principle may be developed by utilizing structural equation models relating resource indices to performance (Matthews, Warm et al., 2010b). Fatigue may also be linked to parameters of more elaborated general architectures for cognition (Gunzelmann et al., 2011).

Matthews (2001) distinguished cognitive-adaptive models from biocognitive models. Even if there are no changes in component process efficiency, fatigue may influence how the functionality afforded by the cognitive architecture is used. The operator may become more concerned with avoiding excessive personal discomfort than with striving for excellence. This perspective is formalized in cybernetic models that link fatigue to control mechanisms for effort regulation (Hockey, 1997; Hockey, Chapter 3, this volume). At the empirical level, the focus is on strategy changes that may accompany fatigue, such as a shift from a reactive to a proactive style of operation (Hockey, Wastell & Sauer, 1998).

There is no escape from the complexities of fatigue effects on operator performance. Depending on the source of fatigue, the nature of the task and the operator's goals, skills and personal qualities, we may see changes in a variety of underlying neurological and psychological processes. Superficially, the range of processes that may be sensitive (or insensitive) to fatigue can produce an inconsistent picture. Strong process-based models provide a method for better-informed prediction of performance deficits, provided that we can make the choice from the various models available that appropriately match the operational context (Matthews, Warm et al., in press).

In this volume, three chapters address the fundamental nature of fatigue. In differentiating multiple processes for fatigue effects, we may lose sight of the forest for the trees. Hancock, Desmond & Matthews (Chapter 4) counter the trend towards fractionation of fatigue effects by returning to evolutionary principles. Adaptation depends critically on successful energy management. As a rejoinder to Hockey's questioning (in Chapter 3) of the link between fatigue and brain energy, in Chapter 4, Hancock, Desmond & Matthews point out that physiological metabolic resources limit the steady-state supply of energy necessary for adaptation. The psychological concept of attentional resources may correspond more directly to metabolic sources of energy and brain glucose levels than has hitherto been realized.

Ackerman, Calderwood & Conklin (Chapter 6, this volume) return us to the theme of the complexity of cognitive fatigue effects in identifying multiple task characteristics that moderate the development and impact of fatigue. Task characteristics fall into three major categories that influence available cognitive effort, the allocation of effort to the task, and the level of off-task distraction. In practice, cognitive tasks differ in the extent to which their characteristics tax the operator in these different ways. In particular, the extent to which the task environment engages the operator's motivation is a practically significant influence on operator fatigue. Thus, accomplishing a good person-environment fit at work is at least a partial defense against fatigue.

Ackerman, Calderwood & Conklin also introduce the importance of individual differences in fatigue vulnerability. Szalma (Chapter 5, this volume) provides a conceptual analysis of individual differences in fatigue and stress that may afford more effective countermeasures tailored to the individual operator. He points out that several relevant theories express common themes of individual differences in adaptation to demanding task environments, variously expressed in terms of compensatory control (Hockey, 1997), dynamic resource allocation (Hancock & Warm, 1989), and transactional stress processes (Matthews, 2001). A theoretical integration of these different models may support understanding of how individual difference factors such as extraversion–introversion moderate adaptive processes.

The final chapter in this section, by Noakes, introduces another core issue – the relationship between physical and mental fatigue – from the standpoint of sports psychology. As Noakes points out in Chapter 7, it is often assumed that prolonged muscular exertion leads to some catastrophic failure of biological homeostasis, causing the athlete to cease the activity. However, studies show that athletes are able to anticipate and regulate physical fatigue to maintain performance. Noakes' Central Governor Model (CGM) specifies the psychological factors that regulate physical exertion, accommodating the roles of mental toughness and motivation in sports performance.

MEASUREMENT OF FATIGUE

Hancock, Desmond & Matthews (Chapter 4, this volume) draw attention to "Muscio's paradox." Tests for fatigue exist, but, because we have no solid definition of fatigue, we cannot tell if these tests provide valid measurements and we cannot develop better tests from first principles.

One solution to the measurement problem is simply not to measure. In the classic experimental tradition we can treat fatigue as a construct that is inferred to explain findings from controlled studies of fatigue manipulations. For example, interactive effects of sleep loss and other stressors can be explained in terms of sleep deprivation lowering cortical arousal, without the need to take measurements (Broadbent, 1971). However, with multiple components of fatigue and multiple components of performance or information processing, the need to assess which elements of fatigue are elicited by a given experimental manipulation becomes more acute (Hockey & Hamilton, 1983; Matthews, 2001). The issue is especially pressing for studies of task-induced fatigue, in which time on-task may have a variety of effects on operator functioning, in addition to increasing fatigue. In principle, the vigilance decrement can be explained satisfactorily in relation to the observer's changing statistical model of the task (i.e., expectancy theory: Frankmann & Adams, 1962), without any need to refer to fatigue or any other general state construct. Invoking fatigue as a factor that mediates vigilance decrement requires measurement of fatigue, so that its role as an intervening variable can be directly tested (e.g., Helton, Matthews & Warm, 2009).

In fact, as in other controversial areas of psychological assessment such as intelligence-testing, progress can be made by beginning with an operational definition. That is, we may start with a somewhat ad hoc fatigue scale. Provided the scale passes basic tests of reliability and criterion validity, its use in research allows it to be refined over time. Studies of its properties allow a theory of the construct measured by the scale to be developed and tested, establishing construct validity. Theory development, in turn, leads to advances in measurement. Thus, intelligence test development has led to measures of valid, practically useful tests of general cognitive ability to which psychological theories may be attached, although "IQ" does not cover all the facets of intelligence that might be encompassed by a broader definition (Sternberg, 2000). Similarly, any given scale for fatigue will have its limitations, but exploring alternative operationalizations of fatigue through research will provide both informative measures and a basis for theory construction.

There is now quite a wide variety of fatigue scales available, which relate systematically to related constructs such as personality traits (Christodoulou, Chapter 8, this volume; Matthews, 2011). However, scales should be used with caution. The criteria for workload assessment established by O'Donnell & Eggemeier (1986) apply also to fatigue measures. The fatigue index should be *sensitive* to changes in operator state. Ideally, it should also be *diagnostic* of the source of fatigue, for example, differentiating sleepiness from task-induced fatigue. It should also be *selective*, in changing only in response to changes in fatigue, and not other influences such as stress and workload. It has been challenging to find indices that satisfy all these criteria.

Fatigue may be measured through behavioral, psychophysiological, and/or subjective means. The attractions of objective measurement are obvious. For example, the Psychomotor Vigilance Test (PVT; see Balkin et al., 2004) provides a convenient, widely used measure of sleepiness based on reaction time assessment. It has been widely used in research on sleep and performance. However, in much performance research, the use of behavioral measures is limited by a potential circularity. (Put differently, a performance index may fail to diagnose the source of fatigue.) If we want to show that fatigue affects performance, a measure of fatigue that is independent of performance is surely needed.

As Miller here notes, use of psychophysiological measures in research on sleep and time of day effects dates to the early decades of the twentieth century. This followed on Hans Berger's first recording of EEG in 1924. Increasing slow-wave activity in the EEG has become one of the primary indices of fatigue in research on both sleep deprivation and sustained performance (Craig & Tran, Chapter 12, this volume). A range of additional response systems has been explored, including eye movements, cerebral bloodflow and the biochemical changes associated with chronic fatigue states. These are discussed in detail in Part IV, in this book, "The Neuroscience of Fatigue." There may be issues regarding the sensitivity of psychophysiological measures, if we assume that fatigue should be distinguished from low cortical arousal. It is not clear that measures such as the EEG are optimal for detecting the changes in motivational control that some accounts (Hockey, Chapter 3, this volume) place at the core of the fatigue process.

Subjective measures are deceptively easy to construct and use; some research uses only a single rating scale, such as the widely used Stanford Sleepiness Scale (SSS; see Hoddes et al., 1973). Measures may be developed for use in specific contexts such as health (Christodoulou, Chapter 8, this volume) and vehicle operation (Matthews & Desmond, 1998). Subjective scales also allow discrimination of the general disposition to become fatigued (i.e. "trait" fatigue) from the immediate experience of tiredness and other symptoms (i.e. "state" fatigue). Unitary fatigue scales with good psychometric properties and practical utility do exist (Michielsen et al., 2004). However, subjective assessments support multidimensional models that discriminate different symptoms of fatigue such as tiredness, emotional disturbance and off-task thoughts (Matthews, Campbell et al., 2002). Since different sources of fatigue may provoke different constellations of symptoms, multidimensional assessments may improve the diagnosticity of measurement. For example, active and passive fatigue states differentiated conceptually by Desmond & Hancock (2001) correspond to different patterns of subjective state, and have differing impacts on safety and performance (Neubauer, Matthews & Saxby, Chapter 23, this volume). Of course, subjective measures are not well suited for assessment of "implicit" or unconscious fatigue responses. Ideally, a range of modes of fatigue assessment is needed to optimize measurement and prediction of performance (Matthews, Warm et al., 2010b).

In Chapter 8, Christodoulou surveys the different means for behavioral, psychophysiological and subjective assessment of fatigue. Recent innovations in assessment, such as the use of experience-sampling methods, allow fatigue to be tracked over time in naturalistic settings. Other topical issues in assessment include explaining dissociations between subjective fatigue and objective performance deficits, and attempts to distinguish sleepiness from fatigue.

Two linked chapters in this volume (Chapter 9, Matthews, Desmond & Hitchcock; Chapter 10, Matthews, Hancock & Desmond) explore multidimensional assessment strategies for performance environments. Matthews, Desmond & Hitchcock set out a taxonomy for organizing different aspects of trait and state fatigue to guide multidimensional assessments. In addition to discriminating objective and subjective fatigue constructs, the taxonomy also fractionates subjective fatigue into multiple components, including awareness of physical symptoms, sleepiness, and emotional, cognitive and motivational aspects of mental fatigue. Matthews, Hancock & Desmond address the application of multidimensional models to understanding and predicting fatigue effects. Countermeasures for operator fatigue may require both a focus on building resilience (lessening the impact of external stress factors on the fatigue state) and a focus on protecting performance when fatigue does develop (decreasing the operator's sensitivity to fatigue states).

THE NEUROSCIENCE OF FATIGUE

There are several critical reasons to look beyond performance assessment alone and explore underlying neuropsychological aspects of fatigue. There are limits to what a purely performance-based theory of fatigue can accomplish. Broadbent (1971) speculated that multiple brain arousal systems might control fatigue and stressor effects, a view confirmed by contemporary cognitive neuroscience. However, as systems multiply, it becomes increasingly difficult to infer which systems control performance in any given experimental study from performance data alone. Theory needs to incorporate data from the increasingly varied range of tools available for studying neural processes, notably brain-imaging. Indeed, advances in neuroscience may counter the limitations of self-report assessments in refining existing psychophysiological measures such as EEG (Craig & Tran, Chapter 12, this volume).

Studies of human neurophysiology have made great advances in understanding the bases for sleep and circadian variation in physiological functions such as maintaining body temperature (Czeisler & Gooley, 2007). Studies showing that sleep–wake behavior is influenced by an endogenous, clock-like process have proved particularly important. Light plays a critical role in synchronizing circadian rhythms to the 24-hour day. Wever (1979) placed participants in "temporal isolation," with no natural light, clocks or other cues to the time, and studied variation in cycles of sleep and body temperature. (According to Czeisler & Gooley (2007), Kleitman pioneered this method by placing two volunteers in Mammoth Cave, Kentucky, for a month). Studies also provided false time cues, such as a clock that ran fast, in order to study internal rhythms in a state of forced oscillation. Performance may be studied within the broader context of the sleep biology of the neural circadian pacemakers responsible.

Therefore, neuroscience may point us directly towards more effective countermeasures for fatigue, including a better understanding of stimulant drugs such as caffeine and modafinil. More subtly, improved neuropsychological assessments of fatigue may provide a basis for adaptive automation. If we can measure fatigue on a continuous basis through psychophysiology, technology can intervene to support the human operator when fatigue reaches some predetermined criterion level (Stephens, Scerbo & Pope, Chapter 26, this volume). In the occupational context, understanding the dynamics of the brain processes that regulate sleep and wakefulness, as expressed in two-process models (Borbély, 1982), affords improvement in work scheduling. These applications are discussed in several chapters in Part VIII of this volume, "Operational Countermeasures" (French & Neville, Chapter 29; Tucker & Folkard, Chapter 28; Van Dongen & Belenky, Chapter 30).

While recognizing the importance of neuroscience, we should also avoid the lure of excessive "explanatory neurophilia" (Weisberg et al., 2008). There are important elements of fatigue, including its motivational and strategic expressions, which do not correspond to brain processes in any simple manner (although they may have psychophysiological correlates), and should therefore be understood predominantly in psychological terms. At a practical level, a part of the solution to the problems of fatigue management is thus instruction and training in strategies for self-management.

In Part IV of this volume, Banks, Jackson & Van Dongen introduce "The Neuroscience of Fatigue." As they point out (Chapter 11), the neurology of circadian rhythms is better understood than that of sleep. A central circadian pacemaker located within the hypothalamus interacts with additional peripheral pacemakers distributed across the body and brain. The neurobiology of sleep remains somewhat mysterious, but evidence is accumulating that sleep is a property of discrete neural assemblies rather than the brain as a whole. Advances in neuroscience, coupled with use of EEG and fMRI, support understanding of performance deficits in fatigue states, and of sleep effects on memory. The reader is also referred to the later Chapter 30 by Van Dongen & Belenky that sets out the utility of neuroscience models for developing countermeasures.

Turning to psychophysiology, Craig & Tran (Chapter 12, this volume) review the use of EEG for assessment of fatigue. Typically, the EEG is analyzed to obtain measures of the activity or power within specified frequency bands. A systematic review of studies using a variety of experimental paradigms showed two consistent effects associated with reduced cortical arousal. Power reliably increased in two frequency bands: theta (4–7.5 Hz) and alpha (8–13 Hz). Theta is linked to drowsiness and loss of alertness; alpha with a relaxed but wakeful state. Changes in the EEG appear to precede performance impairment, so that EEG may be used to provide an early warning of the onset of fatigue.

Warm, Tripp, Matthews & Helton (Chapter 13, this volume) introduce fatigue-monitoring techniques based on measurement of cerebral hemodynamics, including Transcranial Doppler Sonography (TCD) and Functional Near-Infrared Spectroscopy (fNIRS). TCD permits assessment of cerebral bloodflow velocity (CBFV), and it appears especially promising for measuring task-induced cognitive fatigue during vigilance. Declining CBFV accompanies the vigilance decrement and, especially in the right hemisphere, it appears to be sensitive to workload parameters. The CBFV decline also seems to be distinct from loss of cortical arousal.

Chronic fatigue may be generated by the biochemical pathways of the body. Watanabe, Kuratsune & Kajimoto (Chapter 14, this volume) survey research on possible biomarkers of fatigue. In fact, multiple biomarkers may exist reflecting different mechanisms. They include metabolites of cellular energy production, hormones produced by the hypothalamo–pituitary–adrenal axis (linking fatigue to stress), and immunological abnormalities. The physiological model advanced by Watanabe et al. is consistent with evidence from brain-imaging studies that implicate serotonergic regulation of the prefrontal cortex and anterior cingulate cortex in fatigue. Correspondingly, Banks, Jackson & Van Dongen (Chapter 11) suggest that the high metabolic rate of the prefrontal cortex confers on it the greatest need for recovery and metabolic slowing during sleep.

PERFORMANCE EFFECTS OF SLEEP LOSS AND CIRCADIAN RHYTHMS

Sleep loss typically leads to performance deficits across a wide range of different tasks (Matthews, Davies et al., 2000; Miller, Matsangas & Shattuck, 2008). Indeed, the interest in such studies is as much concerned with cognitive functions that are preserved under sleep deprivation as in those that are impaired (e.g., Tucker et al., 2010). Research that goes beyond simply demonstrating adverse effects of sleep loss has typically focused on one or more additional issues: the moderating effects of task demands, integration of theory with neuroscience, and methodological variants of the sleep deprivation paradigm.

The cognitive revolution of the 1960s brought with it a new focus on the specific information-processing routines that might be sensitive to sleep loss. Cognitively informed research brought out some subtleties, such as possible effects of sleep deprivation on attentional selectivity, as well as on overall attentional efficiency (Hockey & Hamilton, 1983). Recent work has also emphasized compensatory processes that may operate to counteract sleep loss, as well as the operator's continued "metacognitive" capacity to monitor declining performance (Baranski, 2011). However, in contrast to some other areas of stress research (Matthews, Davies et al., 2000), findings tend to emphasize the ubiquity of performance deficits across multiple processing components. Where performance is maintained, the key factor may be operator motivation and the intrinsic interest of the task, consistent with the theoretical analyses of fatigue presented by some of the present authors (Ackerman, Calderwood & Conklin, Chapter 6, this volume; Hockey, Chapter 3). Thus, it is hard to take cognitive theory beyond rather general notions, including loss of attentional resources, reduced task-directed effort, and more frequent "microsleeps" and lapses in attention (Matthews, Saxby et al., 2011).

Advances in neuroscience may provide a path towards a more fine-grained account of sleep deprivation effects. We saw previously that converging lines of evidence suggest that the prefrontal cortex and allied brain structures for cognitive control may be especially vulnerable. A focus on impairments in executive control of more specialized neurocognitive processing systems may thus be productive (Killgore, Chapter 15, this volume). The role of prefrontal structures in emotion might also provide some insight into how task interest protects against fatigue deficits. Application of contemporary models that differentiate multiple components of executive processing may also enhance theory development (Matthews, Gruszka & Szymura, 2010; Tucker et al., 2010).

A third research issue derives from real-life contexts for sleep loss. Often, operators experience partial rather than total sleep loss, perhaps sleeping five to six hours per night, and building up a progressive "sleep debt" (Van Dongen et al., 2003). Operators may also experience sleep disturbance, sleep at varying phases of the circadian cycle, and even extended sleep durations. Investigations of these more subtle forms of sleep disruption are needed.

Circadian rhythm research has also followed a trajectory from concerns with overall alertness and performance, to the more differentiated view of performance change associated with cognitive psychology (Minors & Waterhouse, 1981), and, to some extent, back to the more unitary perspective suggested by the concept of a central circadian pacemaker (Banks, Jackson & Van Dongen, Chapter 11, this volume). Research from the 1970s and 1980s demonstrated some rather intriguing changes in task strategy across the waking day (e.g., Oakhill, 1986), but contemporary research is increasingly driven by the neuroscience models described in several contributions to this volume.

In Chapter 15 Killgore highlights the effects of sleep loss on metabolic activity within the prefrontal and parietal cortices, and their consequences for cognitive performance. In addition to loss of alertness, Killgore also reviews some less well-known effects of sleep loss, for example, on psychomotor coordination, emotional processing and high-level decision-making and risk assessment. Sleep loss may also impair social cognition and moral judgment, especially in highly emotive contexts. Models of cognitive and emotional regulation may serve to explain these findings.

Waterhouse (Chapter 16) reviews the classic studies of circadian rhythms on performance, as well as contemporary perspectives and research directions. Periodicity reflects both the endogenous body clock, entrained to the 24-hour day, and exogenous components derived from the individual's lifestyle. Two-process models allow the effects of time of day to be distinguished from "time awake." These models are essential for understanding performance effects associated with work scheduling and with the "jet lag" induced by rapid time-zone transitions. Waterhouse also reviews the individual differences in circadian function that may cause people to be either morning types or evening types.

Gawron (Chapter 17) provides a systematic review of sleep loss effects that emphasizes their significance for operational environments. She tabulates the various studies that demonstrate degradations in visual performance, cognitive function, routine task completion, vigilance and physical performance. Both accuracy and speed of performance are affected by sleep loss. The review demonstrates the relevance of sleep deprivation research to the defense, public safety, and industry domains. The review confirms the diversity of tasks vulnerable to impairment, but also indicates functions, such as spatial cognition, that may hold up relatively well under sleep deprivation.

FATIGUE AND HEALTH

Fatigue is a common symptom of a range of physical illnesses and psychiatric conditions. In a volume on operator fatigue, the primary interest is in the performance concerns of occupational health psychology (Matthews, Panganiban & Gilliland, in press). In keeping with a transactional perspective (Szalma, Chapter 5, this volume), maintaining occupational health requires a focus on both external and internal fatigue factors. Adverse working conditions may provoke fatigue or burn-out that is sufficiently severe and prolonged to threaten health, with the Japanese notion of death through overwork ("karoshi") being an extreme example. Tasks and work environments require to be designed to minimize chronic fatigue. Attention to work scheduling is also necessary, as is discussed in chapters in the subsequent parts: "Applied Contexts for Operator Fatigue" and "Countermeasures to Fatigue." The impact of fatigue on worker health and well-being was noted by the

National Academy of Science (Colten & Altevog, 2006). Long hours and/or night work can present a significant public health problem.

Turning to the characteristics of individuals, a substantial part of the workforce may be vulnerable to fatigue because of pre-existing medical, neurological or psychological conditions and disorders. Chronic fatigue syndrome (CFS) is a particular concern because it can severely impact the person's ability to undertake even light work (Cox, Chapter 19, this volume) and it produces a range of deficits in information processing (Smith, 2011). Strategies for optimizing the contribution of the fatigue-prone worker are also needed, especially as Western nations attempt to transition individuals from living off disability benefits to productive employment. Methods are needed for matching work to the capabilities of the individual, potentially requiring both task redesign and detailed assessment of job applicants.

Many otherwise healthy workers may experience short-lived fatigue because of acute infections, such as the common cold or through taking medications that cause drowsiness (Smith, 2011). People are often inclined to "work through" minor illnesses. In routine work settings, one issue is to determine the point at which fatigue and other symptoms are severe enough that productivity is substantially impaired. Guidelines are needed to determine when absence from work is necessary to aid recovery from the infection, and to avoid infecting others. In safety-critical work, such as vehicle operation, there is the added imperative to prohibit individuals from working when their performance is compromised; fitness-for-duty tests (Miller, Chapter 2, this volume) may be useful in this regard.

Jason, Brown, Evans & Brown (Chapter 18, this volume) review the psychiatric context for chronic fatigue. Fatigue is commonly reported in primary care settings. Diagnosis of an underlying disorder may be challenging; it may even be hard to determine whether fatigue reflects medical or psychiatric sources. There is also comorbidity between CFS and psychiatric conditions, including mood disorders, which may produce fatigue. CFS remains enigmatic, but Jason et al. discuss diagnostic strategies for differentiating the condition from other disorders.

Cox (Chapter 19) provides a complementary chapter on CFS that focuses on its impact on everyday life and work, and the range of therapies available for it. She points out that individuals with moderate or severe chronic fatigue will typically have ceased working; the most severely affected may be bedridden much of the time. Fortunately, there are a variety of psychological therapies that can support a return to work. These include cognitive behavior therapy, graded exercise therapy, and activity management programs that enhance and support coping with symptoms.

Smith (Chapter 20) reviews the effects of upper respiratory tract infections (URTIs), which include the common cold. Studies show that URTIs are associated with both fatigue and operationally significant performance decrements across a range of tasks, including safety-critical tasks such as driving vehicles. Colds and influenza may produce different patterns of cognitive impairment. Smith also reviews possible mechanisms, including immunological and central nervous system concomitants of infection, and the use of medication to counter fatigue.

APPLIED CONTEXTS FOR OPERATOR FATIGUE

Folk memories persist of the grim, exhausting work conditions of the Industrial Revolution, recorded in Victorian literature by writers, most notably Charles Dickens. Fortunately, legislation and statutory regulations, changes in social norms, and awareness of the costs of fatigue have mitigated the problem in modern industrial nations, although sweatshop conditions remain in the developing world. Nonetheless, fatigue at work is commonplace (Maslach, Schaufeli & Leiter, 2001; Rosa, Chapter 21, this volume). It may even be exacerbated by social trends, including the work schedules of the 24/7 society, work–family conflicts, the increasing need for operators to monitor automated systems, and the need for workers to prove their value by working long hours in economically trying times.

Occupational fatigue is a concern for both individuals and organizations. Fatigue may lead to a loss of productivity, but, more dramatically, to an increased risk of accidents to workers, and to the general public. The effects of fatigue on social functioning noted by Killgore (Chapter 15, this volume) also suggest its potential to disrupt what is sometimes called contextual performance; i.e., those informal but essential elements of the job such as communication and teamworking.

The fatigue experienced by factory workers provided an initial impetus to applied research (Miller, Chapter 2, this volume). However, fatigue may be even more salient in other occupational contexts. In particular, transportation requires extended periods of vehicle operation, and specialized literatures for air, surface and maritime operations have developed. Perhaps the most extreme fatigue is experienced by military personnel, due to the exigencies of combat. Miller et al. (2008) describe the sleeplessness of troops racing north to Baghdad during the 2003 Operation Iraqi Freedom.

A recent review (Williamson et al., 2011) documents systematically the threat to safety posed by the various forms of fatigue. Evidence comes from studies using self-report, objective performance measurement and accident analysis methodologies. Williamson et al. conclude that there is strong evidence for detrimental effects of sleep loss (sleep homeostasis) and task-induced fatigue on safety, in both industrial and transportation contexts. They note that a limited number of studies have investigated task effects on safety outcomes. Evidence for circadian effects was weaker, probably because these effects interact with sleep loss. Further research is needed especially to identify populations vulnerable to sleep deprivation, such as persons with sleep disorders; to determine how much recovery sleep is needed following an episode of sleep loss; to investigate circadian effects more systematically; and to determine the maximum safe durations for continuous work in various safety-critical occupations.

In Chapter 21 of this volume, the topic of occupational fatigue is introduced by Rosa's review of the impact of long work hours. In the U.S., substantial numbers of people work hours longer than the standard 40-hour week. Although long hours may be beneficial, e.g. in boosting wages, studies conducted by the National Institute of Occupational Safety and Health (NIOSH) demonstrate increasing risks of fatigue from shifts exceeding eight hours, expressed in both safety- and health-related outcomes. Rosa notes that poor sleep constitutes an independent risk factor, but the combined effects of working long hours and poor sleep can be especially harmful.

Two chapters address driver fatigue. In Chapter 22, Job, Graham, Sakashita & Hatfield present a systems approach to understanding and remediating the dangers of fatigue. They point out that enhancing public awareness of fatigue is hindered by difficulties in defining fatigue, in pinpointing the role that fatigue plays in crashes, and in various limitations of existing countermeasures. Within the safe systems approach, attempts at modifying individual drivers' behaviors are complemented by more general attempts at improving safety through stronger enforcement of speed limits, and safety-enhancing engineering of vehicles and roadways.

Neubauer, Matthews & Saxby (Chapter 23, this volume) discuss fatigue from the more traditional perspective of the individual driver. After reviewing applied studies, they set out a transactional model that emphasizes fatigue as a product of how the individual driver appraises and copes with the demands of prolonged, sometimes monotonous driving. They emphasize the value of assessing driver fatigue as a multidimensional state (see Matthews, Desmond & Hitchcock, Chapter 9, this volume), in order to distinguish theoretically distinct forms of fatigue and stress, including active and passive fatigue (Desmond & Hancock, 2001).

Caldwell (Chapter 24, this volume) reviews fatigue in another relevant transportation domain, aviation. Aircrew may deal with time zone transitions as well as extended duty periods, variable work–sleep schedules, and night operations. The two-process models we have cited throughout this introductory chapter again provide the basis for countermeasures. These include both measures taken during the pre-flight period, directed towards ensuring that the pilot is adequately rested, and measures taken in-flight (e.g., rest periods, caffeine) to mitigate the immediate fatigue state.

In the final chapter in Part VII: "Applied Contexts for Operator Fatigue", Krueger surveys soldier fatigue and performance effectiveness. He points out that the introduction of technologies for night-fighting has served to promote military doctrines of continuous and sustained operations. Soldiers may be expected to work and fight for long periods without relief, until a mission objective is accomplished. Chapter 25 goes on to review the operational consequences of soldier fatigue, and strategies for developing countermeasures, including the greater need to educate both soldiers and military leaders about fatigue.

COUNTERMEASURES

Folk wisdom served to provide the first countermeasures to fatigue, based on adequate sleep and rest. The human factors challenge is to improve on these commonsense remedies. The principal elements of countermeasures research are as follows. First, operator fatigue must be assessed and monitored, so that interventions may be delivered accordingly. Second, in cases of acute on-the-job fatigue, the operator must either be relieved of work, or performance must be boosted in some fashion. Third, risks of chronic fatigue must be managed in order to reduce the operator's subsequent vulnerability to acute fatigue.

We have discussed the various assessment issues in the preceding sections. The only further point to make is that the optimal form of assessment depends on both the time-span for fatigue, and the practical constraints of the performance domain. For acute fatigue, there is typically a need for continuous performance monitoring, which is effected most readily by behavioral or psychophysiological indices. For example, monitoring drivers' eye movements for eye closures may provide a practical fatigue index, which can be used to trigger an intervention (Lal & Craig, 2001). Advances in technology for ambient psychophysiological monitoring support such assessments. For chronic fatigue, a wider range of measures is potentially of value, including validated self-report scales and biochemical indices.

Short-term interventions most often take the form of providing short rest breaks, or countering decreasing cortical arousal with stimulant drugs, typically caffeine but also synthetic substances such as modafinil (Minzenberg & Carter, 2008). Light plays a special role in maintaining synchrony of circadian rhythms (Czeisler & Gooley, 2007). There is also some potential for human factors design solutions, given the importance of operator motivation and task interest in resisting fatigue (cf., Hancock, 1997, 2009). Technology might also be used for performance enhancement in the form of adaptive automation that takes over operator functions when fatigue is detected (Stephens, Scerbo & Pope, Chapter 26, this volume). Such efforts are part of a larger field of augmented cognition which seeks to integrate psychophysiological monitoring and artificial intelligence to enhance operator capabilities in real time (Schmorrow & Reeves, 2007).

Interventions for chronic fatigue are best known from the fatigue management programs developed for occupational settings. Consistent with a transactional perspective (Szalma, Chapter 5, this volume), programs might include changes to the external working environment and interventions that enhance the performance of the individual operator. Recommendations for such programs made by Knauth & Hornberger (2003) include suggestions for work practices, including optimizing shift systems, an issue discussed by several contributors to the countermeasure section in this book. Other relevant aspects of job design include workload, job rotation, and provisions for maintaining work–life balance (Rosa, 1995). It is important also to educate workers and management in the various steps that can be taken to manage alertness, and to maintain health (overlapping with the health issues previously discussed). Similar, multi-pronged interventions can be applied to a variety of operational domains, including vehicle driving (Job, Graham, Sakashita & Hatfield, Chapter 22, this volume) and aviation (Caldwell, Chapter 24, this volume).

The countermeasures section in this book begins with a survey of adaptive automation solutions to acute operator fatigue, especially in the aviation context (Stephens, Scerbo & Pope, Chapter 26, this volume). Adaptive automation requires identification of a physiological response system that indicates fatigue, as well as signal-processing algorithms that identify the nature of cognitive deficits as precisely as possible. Stephens et al. discuss recent developments in EEG recording and neural network methods of analysis that might support practically useful methods for enhancing operator performance following detection of fatigue.

For driver fatigue, the problem is rather different. Automated aids to performance are becoming more prevalent, but the priority is generally to encourage the driver to take a rest break rather than to continue while fatigued. Williamson (Chapter 27, this volume) reviews the various countermeasures to driver fatigue, divided into those that discourage drivers from commencing the trip while fatigued and those directed towards the alleviation of acute fatigue states. There are various means of detecting fatigue, notably those based on eye movements, but operational difficulties in their use remain. Williamson also discusses the issues raised by drivers' difficulties in detecting their own sleepiness, and resistance to ceasing driving.

The remaining chapters address the key issue of work scheduling, which, as the current travails of the FAA suggest, is an essential component of fatigue management. Tucker & Folkard (Chapter 28, this volume) provide a general review of strategies for designing of work shifts. The majority of studies indicate that permanent night shifts should be avoided, so that the two-process model may be applied to optimizing rotating shifts. Other key issues include the shift durations and the scheduling of rest periods within and between shifts.

The two-process model is also the basis for formal quantitative modeling of fatigue according to biomathematical principles. French & Neville (Chapter 29, this volume) trace the rapid developments in fatigue modeling that have ensued since the 1990s. They focus especially on criteria for model validation, and means for improving operational validity. They conclude that models provide a valuable decision aid in scheduling duty periods and detecting potential fatigue issues in schedules, but the final decision must always be made by the human manager.

Van Dongen & Belenky (Chapter 30, this volume) provide a complementary account of model-based fatigue risk management that emphasizes the neurobiological basis for modeling. They also stress that cognitive performance instability is the main operational risk factor related to fatigue, potentially leading to disasters when the consequences of human error are severe. Studies of railroad and airline operations illustrate how models may establish a "bright line," or threshold level, to distinguish safe from unsafe work schedules. Like French & Neville, Van Dongen & Belenky state the need for human judgment in application of models, comparing such model predictions to weather forecasts.

CONCLUSIONS: ISSUES FOR OPERATIONAL PRACTICE

To conclude our introduction, we identify six general issues and themes that cut across different theoretical perspectives and domains of application. These are issues that practitioners should reflect on carefully prior to designing and evaluating any countermeasures to operator fatigue.

Clarity in theory
We have referred to a variety of different theoretical perspectives on fatigue throughout this introduction. Theories are expressed at different levels of analysis, including brain physiology, virtual information processing (e.g., resource theory), and the high-level goals and motivations of the operator. All these theories have the potential for application, but there is obvious scope for confusion. Seemingly simple concepts such as "energy" can be elusive and there are differing views on whether physical energy is a useful metaphor for any psychological processes. Psychological theories are better seen as tools for research than expressions of absolute truth. The challenge for the practitioner is to choose the theory that is most appropriate for their domain of application. For example, neurobiological models may be especially well suited to contexts in which operators may actually fall asleep; motivational theories may work better for managing task-induced fatigue.

Dynamic and regulative perspectives
Sleep homeostasis, circadian rhythms and vigilance decrement are all time-bound processes. Thus, understanding of fatigue must necessarily be dynamic, as expressed in general theories of stress and adaptation (Hancock & Warm, 1989). The classic problem of circadian rhythm research is specifying endogenous oscillators together with the forcing function provided by external events. By contrast, studies of task-induced fatigue increasingly emphasize cognitive control, and the operator's attempts at coping with monotony and perceptions of impairment. Research into the impact of fatigue on executive

control may provide a means for integrating neurobiological and cognitive–psychological perspectives. In any case, the practitioner should be prepared to think about operator strategy and how efforts to compensate for fatigue may be supported.

The importance of measurement

The ambiguities of the term "fatigue" have been recognized from the beginnings of systematic study in the area. In operational settings, fatigue may co-exist with a host of potentially overlapping constructs including boredom, physical discomfort, distraction, emotional distress and social disconnection. Previously, we emphasized the value of multidimensional models for defining operational fatigue states with the precision necessary for development and assessment of countermeasures. Both subjective and objective fatigue indices may contribute to such assessments.

The case for complexity

Much of the ground we have covered in this introduction has emphasized the complexity of fatigue. There are contrasting definitions, conceptualizations and theories of fatigue, supporting a multiplicity of assessments. The cognitive science perspective suggests the need to evaluate fatigue effects across numerous, distinct information-processing components. The key fatiguing agents – sleep homeostasis, circadian rhythms, task demands, health issues – may influence performance through somewhat different sets of mechanism. Practical interventions must also accommodate features unique to the domain of application. Thus, we can reasonably recommend in-depth, fine-grained analyses of fatigue in operational settings as the strongest basis for intervention.

The case for simplicity

A contrasting case can also be made for use of simplifying principles in practical contexts. Although a compelling argument can be made for multidimensionality of fatigue responses, simple unitary scales have proved their worth in applied settings (e.g., Michielsen et al., 2004). We have seen too that two-process models (Borbély, 1982) are a valuable aid to designing work schedules, although they do not make the distinctions between different fatigue outcomes and cognitive processes that are suggested by experimental studies. One of the key skills for the practitioner may be in determining when relatively simple, unitary fatigue models are sufficient for the practical problem at hand, and when a finer level of granularity is required.

Multidisciplinarity in research and practice

Finally, the diversity of contributions to this volume illustrates the pressing need for multidisciplinary research. Fatigue is traditionally a province of psychology, but the critical contributions of neuroscience and other biosciences need no further justification. Recent evidence for the role of gene expression in chronic fatigue (Smith, 2011) suggests that future research teams may need a molecular geneticist. Another critical discipline is engineering, given that human interactions with technology have sharpened the focus on fatigue (cf., Hancock, 1997, 2007). The potential of augmented cognition to enhance countermeasures suggests a role for specialists in AI to optimize algorithms for tailoring interventions to operator state. In addition, expertise in industrial, transportation, military and other application domains is critical. Prospects for furthering our understanding and management of operator fatigue clearly depend on such multidisciplinary efforts.

REFERENCES

Ackerman, P.L. (2011). 100 years without resting, in P.L. Ackerman (ed.), *Cognitive Fatigue: The Current Status and Future for Research and Application*. Washington, DC: American Psychological Association, 11–37.

Balkin, T.J., Bliese, P.D., Belenky, G., Sing, H., Thorne, D.R., Thomas, M., Redmond, D.P., Russo, M. & Wesensten, N.J. (2004). Comparative utility of instruments for monitoring sleepiness-related performance decrements in the operational environment. *Journal of Sleep Research*, 13, 219–27.

Baranski, J.V. (2011). Sleep loss and the ability to self-monitor cognitive performance, in P.L. Ackerman (ed.) *Cognitive Fatigue: Multidisciplinary Perspectives on Current Research and Future Applications*. Washington, DC: American Psychological Association, 67–82.

Barrett, L.F. & Wager, T.D. (2006). The structure of emotion: Evidence from neuroimaging studies. *Current Directions in Psychological Science*, 15, 79–83.

Bartley, S. & Chute, E. (1947). *Fatigue and Impairment in Man*. New York: McGraw-Hill.

Borbély, A.A. (1982). A two process model of sleep regulation. *Human Neurobiology*, 1, 195–204.

Broadbent, D.E. (1971). *Decision and Stress*. London: Academic Press.

Caldwell, J.A. & Caldwell, J.L. (2004). *Fatigue in Aviation: A Guide to Staying Awake at the Stick*. Williston, VT: Ashgate Publishing.

Colten, H.R. & Altevog, B.M. (eds) (2006). *Sleep Disorders and Sleep Deprivation: An Unmet Public Health Problem*. Washington, DC: National Academy of Sciences.

Czeisler, C.A. & Gooley, J.J. (2007). Sleep and circadian rhythms in humans. *Cold Spring Harbor Symposia on Quantitative Biology*, 72, 579–97.

Davies, D.R. & Parasuraman, R. (1982). *The Psychology of Vigilance*. London: Academic Press.

Davis, D.R. (1948). *Pilot Error: Some Laboratory Experiments*. London: TSO.

Desmond, P.A. & Hancock, P.A. (2001). Active and passive fatigue states, in P.A. Hancock and P.A. Desmond (eds), *Stress, Workload, and Fatigue*. Mahwah, NJ: Lawrence Erlbaum, 455–65.

Frankmann, J.P. & Adams, J. (1962). Theories of vigilance. *Psychological Bulletin*, 59, 257–72.

Friswell, R. & Williamson, A. (2008). Exploratory study of fatigue in light and short haul transport drivers in NSW, Australia. *Accident Analysis and Prevention*, 40, 410–17.

Gregory, K., Oyung, R. & Rosekind, M. (1999). Managing alertness and performance in air traffic control operations. *FAA Office of Aviation Medicine Reports*, DOT-FAA-AM-99-2, Jan.

Gunzelmann, G., Moore, L.R., Gluck, K.A., Van Dongen, H.P.A. & Dinges, D.F. (2011). Fatigue in sustained attention: Generalizing mechanisms for time awake to time on task, in P.L. Ackerman (ed.), *Cognitive Fatigue: Multidisciplinary Perspectives on Current Research and Future Applications*. Washington, DC: American Psychological Association, 83–101.

Hancock, P.A. (1984). Environmental stressors, in J.S. Warm (ed.), *Sustained Attention in Human Performance*. New York: Wiley, 103–42.

Hancock, P.A. (1997). *Essays on the Future of Human–Machine Systems*. Eden Prairie, MN: BANTA.

Hancock, P.A. (2009). *Mind, Machine, and Morality*. Chichester, UK: Ashgate.

Hancock, P.A. & Desmond, P.A. (eds) (2001). *Stress, Workload, and Fatigue*. Mahwah, NJ: Lawrence Erlbaum.

Hancock, P.A. & Ganey, H.C.N. (2003). From the inverted-U to the extended-U: The evolution of a law of psychology. *Journal of Human Performance in Extreme Environments*, 7, 5–14.

Hancock, P.A. & Warm, J.S. (1989). A dynamic model of stress and sustained attention. *Human Factors*, 31, 519–37.

Helton, W.S., Matthews, G. & Warm, J.S. (2009). Stress state mediation between environmental variables and performance: The case of noise and vigilance. *Acta Psychologica*, 130, 204–13.

Hockey, G.R.J. (1997). Compensatory control in the regulation of human performance under stress and high workload: A cognitive–energetical framework. *Biological Psychology*, 45, 73–93.

Hockey, G.R.J. & Hamilton, P. (1983). The cognitive patterning of stress states, in G.R.J. Hockey (ed.), *Stress and Human Performance*. Chichester, UK: Wiley.

Hockey, G.R.J., Wastell, D.G. & Sauer, J. (1998). Effects of sleep deprivation and user interface on complex performance: A multilevel analysis of compensatory control. *Human Factors*, 40, 233–53.

Hoddes, E., Zarcone, V., Smythe, H., Phillips, R. & Dement, W.C. (1973). Quantification of sleepiness: A new approach. *Psychophysiology*, 10, 431–6.

Holding, D.H. (1983). Fatigue, in G.R.J. Hockey (ed.), *Stress and Fatigue in Human Performance*. Chichester, UK: Wiley.

Knauth, P. (2007). Extended work periods. *Industrial Health*, 45, 125–36.

Knauth, P. & Hornberger, S. (2003). Preventive and compensatory measures for shift workers. *Occupational Medicine*, 53, 109–16.

Lal, S.K.L. & Craig, A. (2001). A critical review of the psychophysiology of driver fatigue. *Biological Psychology*, 55, 173–94.

Maslach, C., Schaufeli, W.B. & Leiter, M.P. (2001). Job burnout. *Annual Review of Psychology*, 52, 397–422.

Matthews, G. (2001). Levels of transaction: A cognitive science framework for operator stress, in P.A. Hancock & P.A. Desmond (eds), *Stress, Workload and Fatigue*. Mahwah, NJ: Lawrence Erlbaum, 5–33.

Matthews, G. (2011). Personality and individual differences in cognitive fatigue, in P.L. Ackerman (ed.), *Cognitive Fatigue: The Current Status and Future for Research and Applications*, Washington, DC: American Psychological Association, 209–27.

Matthews, G. & Desmond, P.A. (1998). Personality and multiple dimensions of task-induced fatigue: A study of simulated driving. *Personality and Individual Differences*, 25, 443–58.

Matthews, G., Campbell, S.E., Falconer, S., Joyner, L., Huggins, J., Gilliland, K., Grier, R. & Warm, J.S. (2002). Fundamental dimensions of subjective state in performance settings: Task engagement, distress and worry. *Emotion*, 2, 315–40.

Matthews, G., Davies, D.R., Westerman, S.J. & Stammers, R.B. (2000). *Human Performance: Cognition, Stress and Individual Differences*. London: Psychology Press.

Matthews, G., Gruszka, A. & Szymura, G. (2010). Conclusion: The state of the art in research on individual differences in executive control and cognition, in A. Gruszka, G. Matthews & B. Szymura (eds), *Handbook of Individual Differences in Cognition: Attention, Memory and Executive Control*. New York: Springer, 437–62.

Matthews, G., Panganiban, A.R. & Gilliland, K. (in press). Measurement of cognitive functioning, in L. Tetrick, R. Sinclair & M. Wang (eds), *Research Methods in Occupational Health Psychology: State of the Art in Measurement, Design, and Data Analysis*. London: Routledge.

Matthews, G., Saxby, D.J., Funke, G.J., Emo, A.K. & Desmond, P.A. (2011). Driving in states of fatigue or stress, in D. Fisher, M. Rizzo, J. Caird & J. Lee (eds), *Handbook of Driving Simulation for Engineering, Medicine and Psychology*. Boca Raton, FL: CRC Press/Taylor and Francis, 29-1 –29-10.

Matthews, G., Warm, J.S., Reinerman, L.E., Langheim, L.K. & Saxby, D.J. (2010a). Task engagement, attention and executive control, in A. Gruszka, G. Matthews & B. Szymura (eds), *Handbook of Individual Differences in Cognition: Attention, Memory and Executive Control*. New York: Springer, 205–30.

Matthews, G., Warm, J.S., Reinerman, L.E., Langheim, L, Washburn, D.A. & Tripp, L. (2010b). Task engagement, cerebral blood flow velocity, and diagnostic monitoring for sustained attention. *Journal of Experimental Psychology: Applied*, 16, 187–203.

Matthews, G., Warm, J.S., Reinerman-Jones, L.E., Langheim, L.K., Guznov, S., Shaw, T.H. & Finomore, V.S. (in press). The functional fidelity of individual differences research: The case for context-matching. *Theoretical Issues in Ergonomics Science*.

Michielsen, H.J., De Vries, J., Van Heck, G.L., Van de Vijver, F.J.R. & Sijtsma, K. (2004). Examination of the dimensionality of fatigue: The construction of the Fatigue Assessment Scale (FAS). *European Journal of Psychological Assessment*, 20, 39–48.

Miller, N.L., Matsangas, P. & Shattuck, L.G. (2008). Fatigue and its effect on performance in military environments, in P.A. Hancock & J.L. Szalma (eds), *Performance Under Stress*. Burlington, VT: Ashgate, 231–50.

Minors, D. & Waterhouse, J. (1981). *Circadian Rhythms and the Human*. Bristol: John Wright.

Minzenberg, M.J. & Carter, C.S. (2008). Modafinil: A review of neurochemical actions and effects on cognition. *Neuropsychopharmacology*, 33, 1477–502.

Neubauer, C., Langheim, L.K., Matthews, G. & Saxby, D.J. (in press). Personality and automation use in simulated driving. *Proceedings of the Human Factors and Ergonomics Society*, 55.

Oakhill, J. (1986). Effects of time of day on the integration of information in text. *British Journal of Psychology*, 77, 481–88.

O'Donnell, R.D. & Eggemeier, F.T. (1986). Workload assessment methodology, in K.R. Boff, L. Kaufman & J.P. Thomas (eds), *Handbook of Human Performance*. Vol. 2. *Cognitive Processes and Performance*. Chichester, UK: Wiley, 1–49.

Rosa, R. (1995). Extended workshifts and excessive fatigue. *Journal of Sleep Research*, 4(Suppl. 2), 51–6.

Rosekind, M.R., Gander, P.H., Gregory, K.B., Smith, R.M., Miller, D.L., Oyung, R. & Johnson, J.M. (1995). Managing fatigue in operational settings: I. Physiological considerations and countermeasures. *Behavioral Medicine*, 21, 157–65.

Saper, C.B., Scammell, T.E. & Lu, J. (2005). Hypothalamic regulation of sleep and circadian rhythms. *Nature*, 437, 1257–63.

Schmorrow, D.D. & Reeves, L.M. (2007). 21st century human-system computing: Augmented cognition for improved human performance. *Aviation, Space, and Environmental Medicine*, 78(5, Sect II, Suppl), B7–B11.

Smith, A.P. (2011). From the brain to the workplace: Studies of cognitive fatigue in the laboratory and aboard ship, in P.L. Ackerman (ed.), *Cognitive Fatigue: Multidisciplinary Perspectives on Current Research and Future Applications*. Washington, DC: American Psychological Association, 291–305

Sternberg, R.J. (2000). The concept of intelligence, in R.J. Sternberg (ed.), *Handbook of Intelligence*. New York: Cambridge University Press, 3–15.

Temple, J.G., Warm, J.S., Dember, W.N., Jones, K.S., LaGrange, C.M. & Matthews, G. (2000). The effects of signal salience and caffeine on performance, workload and stress in an abbreviated vigilance task. *Human Factors*, 42, 183–94.

Thayer, R.E. (1989). *The Biopsychology of Mood and Arousal*. Oxford: Oxford University Press.

Thorndike, E.L. (1914). *Mental Work and Fatigue and Individual Differences and their Causes*. New York: Teachers College.

Tucker, A.M., Whitney, P., Belenky, G., Hinson, J.M. & Van Dongen, H.P.A. (2010). Effects of sleep deprivation on dissociated components of executive functioning. *Sleep: Journal of Sleep and Sleep Disorders Research*, 33, 47–57.

Van Dongen, H.P.A., Maislin, G., Mullington, J.M. & Dinges, D.F. (2003). The cumulative cost of additional wakefulness: Dose-response effects on neurobehavioral functions and sleep physiology from chronic sleep restriction and total sleep deprivation. *Sleep: Journal of Sleep and Sleep Disorders Research*, 26, 117–26.

Warm, J.S., Parasuraman, R. & Matthews, G. (2008). Vigilance requires hard mental work and is stressful. *Human Factors*, 50, 433–41.

Weisberg, D.S., Keil, F.C., Goodstein, J., Rawson, E. & Gray, J.R. (2008). The seductive allure of neuroscience explanations. *Journal of Cognitive Neuroscience*, 20, 470–77.

Wever, R.A. (1979). *The Circadian System of Man: Results of Experiments Under Temporal Isolation.* New York: Springer.

Williamson, A., Lombardi, D.A., Folkard, S., Stutts, J., Courtney, T.K. & Connor, J.L. (2011). The link between fatigue and safety. *Accident Analysis and Prevention*, 43, 498–515.

2

An Historical View of Operator Fatigue

James C. Miller

Fatigue makes cowards of us all. (Vince Lombardi)

This chapter examines the history of research concerning the effects of operator fatigue. The look-back is, of course, not comprehensive, but relies on existing sources. For additional historical information the reader is directed to the following resources:

1. Historical review of mental, or cognitive, fatigue (Ackerman, 2011).
2. Comprehensive treatment of sleep physiology (Kleitman, 1963).
3. Historical critique of vigilance (McGrath, Harabedian & Buckner, 1959).

The chapter will briefly trace notions of fatigue from antiquity and pre-industrial times through the early experimental studies that began in the late nineteenth century, up to modern developments.

EARLY CONCEPTIONS OF FATIGUE AND SLEEP

Religious and Mythological Perspectives

The concepts of work, stress, rest, and recovery are present in our earliest religious writings. Even Genesis states that on the seventh day, God "rested from all his work," though we also learn that God is tireless and only chooses to rest. Exodus indicates that the Sabbath was to be set aside, even today, as a period during which people rest instead of labor. Ancoli-Israel (2001) noted that: "Although we think we have discovered many new features about sleep disorders, much of what we know today was suggested thousands of years ago and documented in the Bible and the Talmud."

The Qur'an treats creation of the world in a slightly different manner: "We have created the heavens and the earth, and everything between them in six days, and no fatigue touched us." The Qur'an defines a daily program for believers, which includes rest and recovery through prayer. Muslims go to work after the end of Morning Prayer feeling "optimistic and high spirited." Early rising allows work to be started and completed early. Prayers are repeated five times every day, but the next prayer comes after the fatigue induced by seven hours of work and "serves as a relaxation technique according to the teachings of Islam." It was also within the Islamic culture of fifteenth-century Arabia that the use of coffee – still a popular countermeasure – to offset fatigue was first reported (Weinberg & Bealer, 2001).

Greek mythology recounts that before the beginning of the universe, there was nothing in existence until Chaos, which personifies the abyss, a void, and a formless confusion. Out of this void came, among other things, Nyx (night); the mother of many, including Hypnos (sleep; the Romans' god, Somnos) and Thanatos (death), Moros (doom), and the Fates or Moerae. The gods of sleep and dreams lived near the entrance of the Underworld and near the river Lethe (forgetfulness). Thus, in the Greek mind, the idea of sleep was probably associated primarily with night, darkness, death, doom, fate, and forgetfulness. This is perhaps the origin of the incorrect assumption that sleep is a passive, vegetative state.

In slightly later times, Marcus Cornelius Fronto wrote about the effort Lucius expended in restoring order in the Roman army:

> This great decay in military discipline Lucius took in hand as the case demanded, setting up his own energy in the service as a pattern. […] he keeps the first watch easily, for the last he is awake long beforehand and waiting; work is more to his taste than leisure, and his leisure he misuses for work. The sleep he took was earned by toil, not wooed with silence.

The final statement in this quote seems to differentiate between sleep that provides recovery from great mental exertion and long hours of wakefulness, and sleep between periods of idleness (cf., Desmond & Hancock, 2001). Brink (2000) has linked archaic descriptions of sleep to more modern research. She noted that as far back as Virgil and Homer, sleep was described as being divided into two periods during the night, separated by a period of wakefulness sometimes called "the watch." Interestingly, there is a military tradition of nighttime watchstanding that divides the night into the first watch and the second watch. The timing of these watches probably corresponded loosely with the first and second sleep. The changing of the guard probably occurred during "the watch," between sleep periods, when the present guard and the guard's relief both tended to be awake.

The Pre-industrial Age

Ekirch (2001, 2006) has reviewed sleep habits in pre-industrial England. Until the close of the early modern era, Western Europeans typically experienced two major intervals of sleep bridged by up to an hour or more of quiet wakefulness. Today, we think of an unbroken eight-hour period of sleep as standard, but Ekirch (2001) describes it as an invention of the late seventeenth century. Sleep patterns started to change around this time with the spread of gas lighting and industrialization. Ekirch also points out that the subject of sleep frequently absorbed people's thoughts. Indeed, the Aristotelian belief that the need for sleep originated in the abdomen by means of a process called "concoction" was a part of medical thinking in the late Middle Ages Digestion of food caused fumes to ascend to the head and congeal in the brain, blocking the passages of the senses, resulting in sleep. Sleep was seen as essential to physical vitality, lively spirits, and increased longevity.

Literary sources confirm the popular restorative view of sleep, although it was considered to be an entirely passive, vegetative state. In Shakespeare's *Macbeth*, Macduff announces the murder of Duncan: "Malcolm. Banquo. Awake! Shake off this downy sleep, death's counterfeit, And look on death itself." Macbeth tells of "Sleep that knits up the ravell'd sleeve of care, The death of each day's life, sore labour's bath" (Shakespeare, *Macbeth*, Act II, Scene II). In *Measure for Measure* the Duke comforts Claudio, who is to die: "The best of rest is sleep" (Shakespeare, *Measure for Measure*, Act III, Scene I).

Similar views prevailed elsewhere in Europe. In the early seventeenth century Cervantes opined that, "Blessings on him who first invented sleep. It covers a man all over, thoughts and all, like a cloak. It is meat for the hungry, drink for the thirsty, heat for the cold, and cold for the hot. It makes the shepherd equal to the monarch and the fool to the wise" (*Don Quixote*, Part ii, Chapter lxviii). Bonaparte's maxims from the late eighteenth and early nineteenth centuries included the observations that "The first qualification of a soldier is fortitude under fatigue and privation" (Cairnes & Chandler, 1995)

Thus by the beginning of the twentieth century, fatigue remained understood in relation to myths about sleep that persist today. These myths include: "Sleep is a passive, vegetative state; I can sleep when I die," and "Sleepiness and fatigue always occur, and nothing can be done about them." Such incorrect beliefs have helped generate a pandemic of daily sleep deprivation in the United States today (Committee on Sleep Medicine and Research, 2006).

TOWARD A SCIENCE OF OPERATOR FATIGUE

The Industrial Age

One of the earliest published expressions of scientific interest in the effects of fatigue on the human operator came from Jastrzębowski in 1857:

> The first of the Perfections is called Efficiency, since it makes us fit to carry out or effect the most difficult tasks without strain, fit to accomplish without fatigue even the most arduous tasks or useful work the aim of which is to make people and things useful, or to make people and things serve the common good. (Jastrzębowski, 2000: paragraph IV, 3.23)

In laboratory research, Mosso (1915) reported a study of muscular fatigue in 1884 in which the primary measure was performance decrement, measured by an ergograph. He also noted changes in the subjective feelings of his subjects, mostly in the form of increased irritability. In 1917, Starch & Ash noted that under fatiguing conditions, "when an error occurs it is followed immediately by other errors and more and more frequently as the period of work continues" (Starch & Ash, 1917, 401, quoted in Ackerman, 2011). In 1921, Muscio championed the idea of defining fatigue as a measurable, time-dependent change from before to after a period of performance. Soon afterwards Whiting and English used subjective fatigue as an explanatory concept for work decrement. According to them, fatigue was supposed to create a "negative emotional appetite" for continued performance and "does not directly cause a work decrement but raises the threshold at which certain work motives are effective" (1925: 49). During that same period, Weiskotten and Ferguson (1930) applied subjective fatigue techniques in their study of sleep loss, while Bills (1931, 1937) pioneered the systematic study of mental fatigue in terms of lapses in attention – the famous Bills blocks.

Meanwhile, in the domain of applied science, Edward Thorndike (e.g. 1900) published a series of articles on fatigue in schoolchildren, demonstrating performance decrements associated with prolonged work. His 1900 article includes two resonant points. He noted that a feeling of incompetency is not the same as the fact of incompetency, prefiguring much debate over the validity of subjective measures. He also described (1900: 481) "an emotional repugnance to the idea of doing mental work." Later writers (Brown, 1994; Hockey, Chapter 3, this volume) have emphasized the pivotal role of loss of motivation in mental fatigue.

The major impetus to applied research at this time was in occupational psychology. According to Derickson (1994: 484), Goldmark used "scientific evidence [to argue] against long hours, night work, speed-ups, and other adverse conditions" and "explained how 'long and late hours of labor must physiologically result in lessened output.'" By tying fatigue to the quest for efficiency – a veritable national mania in the 1910s – reformers gave the issue added gravity as a public concern and hoped to make corrective measures more palatable to business leaders (Derickson, 1994).

During World War I, British munitions workers had to work long hours producing military equipment and supplies, sometimes 12 hours a day, leading to a proliferation of falling productivity, accidents, spoiled work, absenteeism, and other manifestations of fatigue. The Health of Munition Workers Committee (HMWC) appointed by the British government found that, "although employees' rate of output fell after eight hours on the job, their total output after ten or more hours' work still exceeded that accomplished on the shorter shift […] A January 1916 memorandum urged that adult males work no more than sixty-seven hours per week and that females of all ages and boys under sixteen years work no more than sixty hours."

In the United States, the Council of National Defense (CND) set up a Committee on Industrial Fatigue (CIF) in 1917 with an "unabashedly productivist mission" (Derickson, 1994). The CIF chairman, Thomas Darlington, was also the welfare secretary of the American Iron and Steel Institute, and his primary aim was to defend his industry's much-criticized 12-hour day, obstructing any serious consideration of the role of long hours in inducing fatigue. Darling was opposed by Frederic S. Lee, secretary of the CIF, who conducted fatigue research at Columbia University. Lee quickly produced a manuscript titled "How Industrial Fatigue May Be Reduced," which was published by the U.S. Health Service in 1918. The manuscript concluded that although more investigations were needed, evidence favored a reasonably short work day. Because the shortened work day was a political hot potato, the CIF concentrated on the usefulness of rest breaks, which were uncommon then in industry.

Dericksen (1994) also cited studies that supported Lee's perspective. A 1917 study at Ford Motor Company by Lee and Schereschewsky of the effectiveness of two ten-minute rests during work shifts led them to report that, despite the twenty-minute reduction in work time, daily output had increased. Subsequently, scientific interest in productivity and fatigue waned, but some further evidence emerged. In 1924, Elton Mayo, then at the Wharton School, investigated productive efficiency in textile workers. He concluded (1924: 273) that monotonous, pessimistic thoughts and preoccupations exacerbated the physical causes of fatigue. He considered that both physiological strain and harmful mental "reveries" might be relieved by several short rest breaks during the work shift

Between the two world wars, the concern in the United States about problems associated with operator fatigue was centered primarily around profit. Fatigue was seen as a phenomenon which reduced work output and little concern was given to the effect of fatigue on the workers themselves (Ryan, 1947). Studies performed during this period dealt primarily with the length of the work day or week, scheduling, lighting and ventilation or, more generally, workplace design.

By contrast, the staff of the Fatigue Laboratory of the Harvard Business School viewed fatigue as an overarching concept that affected and was affected by many of the specific biological systems and functions that were investigated in detail at the Laboratory. The Laboratory was founded in 1927 in response to the concept shared by Professors L.J. Henderson and Elton Mayo that fatigue was important not just for the physical and mental health of the individual but also for group psychology and social problems that might affect work. Additionally, Wallace Donham, a long-time friend of Henderson, had determined that "a well-organized and clearly defined center for the study of normal human physiology" was important (Horvath & Horvath, 1973).

The significance of "fatigue" in the Laboratory's name is clear upon considering the nature of its research – the study of industrial hazards. Physical and mental fatigue may not be commonly thought of as industrial "hazards," per se, but they are often a basic cause of accidents. This may be due to increased susceptibility to industrial diseases, and dissatisfaction with a job, sufficient to cause a worker to resign.

In writing about the contributions of the Harvard Fatigue Laboratory, Horvath and Horvath (1973) noted that because everyone has experienced fatigue in one form or another, even the layman thinks he understands it. Thus, empirical studies of fatigue and its relationship with physiology were especially important to counter the myths and traditions that may otherwise inform practice.

With the end of World War II and changing priorities at Harvard, the Harvard Fatigue Laboratory ceased to exist in 1947. However, its functions were carried on in a multitude of civilian and military laboratories, thanks to the staff diaspora after the war (Horvath & Horvath, 1973).

The Impetus from World War II

Around the start of World War II, studies in fatigue reflected new directions. Investigations emphasized fatigue effects in flying and driving with the human operator as the focus of attention. The new direction in fatigue investigation was toward improved safety and preventing loss of life (Bartley & Chute, 1947; Cameron, 1973a). In the United States, the National Research Council Committee on Aviation Psychology was established in 1939. Experiments conducted by Alexander C. Williams, Jr. together with Jenkins at the University of Maryland appear to have been the first to use an airborne polygraph to measure psychophysiological "tension," and its relationship with performance in flight training (Roscoe, 1997).

Among the earliest experimental studies of the human factors in equipment design were those made during World War II at the Applied Psychology Unit (APU) of Cambridge University under the leadership of Sir Frederick Bartlett (Director from 1944 to 1951). He was succeeded as Director of the APU by Norman Mackworth (1951–1958) and Donald Broadbent (1958–1974). Initially, Bartlett's staff included Mackworth, who defined the field of human vigilance as it exists today, and K.J.W. Craik, Margaret Vince, and W.E. Hick. These latter researchers performed pioneering studies of the effects of system design variables on manual control performance.

In the fatigue context, the Cambridge Cockpit studies of Drew (1940), Bartlett (1942) and Davis (1948), which relied primarily on measures of performance but included subjective fatigue estimations from pilots, were especially important. The device that became known as the Cambridge Cockpit was constructed by Kenneth Craik, using the front end of a Spitfire fighter aircraft (Craik, 1940). The idea for an experimental cockpit seems to have come initially from Frederick Bartlett FRS, the first Professor of Experimental Psychology at Cambridge, who had an interest in the psychology of working conditions (Rolfe, n.d.). By early 1940 the apparatus had been built and by December a report was presented by G.C Drew to the Flying Personnel Research Committee (FPRC) describing the results of "An Experimental Study of Mental Fatigue." This research proved to be so influential (Oldfield, 1959) that it is worth quoting the summary of the main findings of the first study:

> The tired operator tends to make more movements than the fresh one, and he does not grade them appropriately to the stimuli provided by changes in instrument readings. Furthermore he does not make the movements at the right time. He is unable to react any longer in a smooth and economical way to a complex pattern of signals, such as those provided by the blind-flying instruments. Instead his attention becomes fixed on one of the separate instruments, and he attempts to make adjustments to changes in this, leaving the others until their indications have departed far beyond the range he would normally tolerate. Although he is unaware of it, his standards of normality relax considerably, so that now a greater external change is needed to initiate a corrective response. Signals in the periphery of his field of attention tend increasingly to be neglected. Thus he will fail to respond to a falling indication of the petrol gauge by turning on to the reserve supply. He becomes increasingly distractible. And, increasingly, feelings of bodily discomfort begin to obtrude upon his consciousness. If he becomes aware of his diminished accuracy of performance, he is apt to attribute this, not to his own shortcomings, but to imperfections that have developed in the apparatus. (Drew 1940)

The Cambridge Cockpit Studies showed that Muscio's idea of defining fatigue as a measurable, time-dependent change from before to after a period of performance was applicable to simulated flying performance (Bartlett, 1942; Davis, 1948; Drew, 1940). An extensive range of further studies investigated the influence of noise, sleep deprivation, the administration of alcohol, amphetamine, and vitamins, as well as prolonged working periods. Davis (1948) published a detailed description of the research in 1948 for the British government, entitled "Pilot Error: Some Laboratory Experiments."

Another avenue of investigation initiated at the APU concerned the ability of the human operator to sustain attention to a task, in response to the difficulties of wartime radar operators. The reader is referred to Reynolds and Tansey (2001) for a transcript of a 2001 seminar on the history of the APU featuring accounts from surviving researchers from its early days. Mackworth's classic studies of signal detection in tasks simulating those of radar operators provided a paradigm for much subsequent vigilance research. His Clock Test was used to establish one of the fundamental findings about human behavior: the vigilance decrement; signal detection accuracy decreases after 30 minutes on task (Mackworth, 1948). Mackworth's demonstration of the beneficial effects of amphetamine on sustained attention suggested the theory of vigilance decrement as a product of low arousal (Davies & Parasuraman, 1982). The vigilance decrement suggests a continuous decline in perceptual sensitivity, but a rather different kind of performance impairment was shown by Broadbent and others at the APU in studies of reaction time tasks. People performing monotonous tasks tend increasingly to produce unusually slow response times, which have been termed "blocks" or "gaps" and may reflect a lapse in attention or even a "microsleep." Reaction time tasks also provided the basis for studies of interactions of stressors, and evidence that, in some instances, fatigue might be countered by agents that elevated arousal (Broadbent, 1971).

The wartime origins of the APU were also reflected in an interest in the impact of sleep on vigilance, a pressing practical issue for the military. Work conducted primarily by Robert Wilkinson and Peter Colquhoun investigated the role of stress and arousal factors, as well as individual differences in resistance to fatigue. These themes converged in Colquhoun's (e.g., 1996) studies of circadian rhythms. He and Michael Blake showed that circadian fluctuations in performance coincided with the circadian rhythm in body temperature (Colquhoun, Blake & Edwards, 1968). Individual differences in the cycles were associated with extraversion–introversion. This research, together with that of Kleitman (1963) established the circadian rhythm in body temperature as a function that is essential for understanding and modeling human alertness and performance.

A final contribution from this era comes from Bartley and Chute's (1947) conceptual analysis of the nature of fatigue, discriminating at least three distinct classes of fatigue. They regarded these as distinct variables since the presence of one type of fatigue was not conclusive evidence of the presence of any other. The first of these fatigue classes is the physiological effect of strenuous physical activity, including measurable changes in the blood chemistry and accumulation of metabolites in muscles. Many laboratory studies have attempted to associate fatigue with the accumulation of metabolites, such as lactic acid, in the blood (McFarland, 1971). This approach stems from work relating physical fatigue to the exhaustion of biochemical energy reserves.

The second class of fatigue is called "work decrement" or "performance decrement." This refers to the "measurable change of output in a productive activity" (e.g., Mosso, 1915; Bills, 1931). The third class of fatigue is the subjective feeling of experiencing discomfort, or "malaise," which accompanies prolonged activity of a sedentary nature. The tripartite approach of Bartley and Chute to the measurement of operator fatigue effects has been much used subsequently (e.g., Mackie & Miller, 1978).

Discrimination of different classes of fatigue response is one way in which research in the area has diversified into separate (though interrelated) research topics. Studies of fatigue also became increasingly divided into work on the major sources of fatigue, including lack of sleep, circadian rhythms and cognitive fatigue induced by task demands. Next, this chapter will briefly describe some key historical developments within these sub-fields.

SLEEP AND CIRCADIAN RHYTHMS

Psychophysiology of Sleep

The dawning of the age of polysomnography occurred in 1924 with the report that the first human electroencephalogram (EEG) had been recorded (Berger, 1929). After confirmation of this finding (Adrian & Matthews, 1934), electroencephalography gained widespread recognition and practical use in sleep research (Kleitman, 1963). Investigations of the 1950s indicated that sleep is not a unitary condition, but is punctuated by periods that paradoxically resemble wakefulness in terms of the signals recorded from the cerebral cortex, though sleep-like behavior continues to be exhibited (Aserinsky & Kleitman, 1953). Sleep is generated actively by complex circuits in the brain, and is composed mainly of two states, today generally categorized as slow-wave sleep (SWS) and rapid-eye-movement

(REM) sleep. By the end of the 1960s, a standardized method for polysomnography had been designed and subsequently was adopted worldwide (Rechtschaffen & Kales, 1968). Subsequently, we have learned that sleep is a fragile state, disrupted easily by rapid travel across time zones, shift work, poor sleep hygiene, alcohol, caffeine and energy drinks, as well as over-the-counter medications.

Performance Across the 24-hour Cycle

Modern interest in circadian rhythms dates to the late nineteenth century. A brief review may be found in Kleitman, 1933. Initially, studies focused on motor performance, suggesting maximal muscular efficiency in the early afternoon, with a minimum early in the morning. Subsequent studies addressed mental performance, culminating in Kleitman's influential studies (e.g., 1933, 1963) that showed comparable rhythms in core body temperature and a variety of performance measures, taken from tasks including reaction time, card-sorting, mirror-tracing, multiplication and code transcription. A link with sleep was suggested by Patrick and Gilberts' observation in 1896 that during prolonged sleep deprivation, the need for sleep seems to wax and wane within a period of approximately 24 hours (Dijk & von Schantz, 2005). As has been noted, these findings were elaborated by the work at the Cambridge APU.

In addition to dealing with the circadian rhythm in operator performance, one must also consider the 12-hour (circasemidian) rhythm that modulates the circadian rhythm. Numerous data sets have shown a daily two-peak error pattern in industrial and transportation environments (e.g., Browne, 1949; Hildebrandt, Rohmert & Rutenfranz, 1974; Lavie, Wollma & Pollack, 1986). The pattern was also obvious in many of the charts shown in the review by Rutenfranz and Colquhoun (1979), though they did not suggest a circasemidian rhythm as a mediator for the pattern. Other investigators have reported a circasemidian rhythm in body temperature (Colquhoun et al., 1968; Martineaud et al., 2000), melatonin (Maggioni et al., 1999) and slow-wave sleep (Hayashi et al., 2002). Broughton (1998) proposed that the two-peak pattern may be explained by an interaction of (1) the expression of the GABA-ergic circadian arousal system that is based in the suprachiasmatic nuclei (SCN) and responsive to light and to the alertness-enhancing drug, modafinil, with (2) Process-S of the now classic 2-process model of arousal and alertness. The circasemidian rhythm had been observed to occur in a number of laboratory tasks involving working memory (Hursh et al., 2004). Miller (2006) has generated a method for studying the circasemidian rhythm in the laboratory.

Fatigue Modeling

Improved understanding of sleep and circadian rhythms has allowed the development of quantitative models of fatigue that may be used to guide efforts at countermeasures. A pivotal advance was provided by Borbély's two-process model (1982) that discriminated a homeostatic process governing need for sleep and a clock-like circadian process that may be modeled as a skewed sine wave with a 24-hour period. Other processes that may be added to the model include an ultradian process controlling the alternation of periods of rapid eye movement (REM) and non-REM sleep, and a sleep inertia process controlling the transition from sleep to wakefulness (Borbély et al., 1989). Other models have focused on modeling multiple circadian oscillators (e.g., Kronauer et al., 1982), building on the

1976 findings of Aschoff and Wever from studies of temporal isolation that discriminated separate oscillators controlling wake–sleep behavior and the body temperature rhythm. By 2004, Mallis, Mejdal, T. Nguyen & Dinges (2004) were able to review no fewer than seven biomathematical models of this type.

These models have been elaborated to predict sleepiness and alertness in occupational settings (e.g., Kostreva, McNelis & Clemens, 2002) and in daily life (Åkerstedt, Folkard & Portin, 2004). Modeling also supports algorithms for optimizing shift systems (Knauth et al., 1982). Folkard (2006; Folkard & Lombardi, 2004, 2006) specified both a Fatigue Index and a Risk Index designed to estimate the risk of human error on different work schedules based on trends in the relative risk of accidents and injuries.[1] Specialized modeling efforts have also been directed toward specific domains, such as the System for Aircrew Fatigue Evaluation (Belyavin & Spencer, 2004), which is applicable mainly to the civil aviation environment. The Sleep, Alertness, Fatigue and Task Effectiveness (SAFTE) model applies similar principles to the military setting (Hursh et al., 2004).

Sustained Operator Performance

In addition to original myths about fatigue and sleep, two more modern myths concerning operator performance have arisen in the industrial age. These are, "Humans are good at 'standing watch,'" and "The performance of a human operator in a system cannot be measured objectively." Information gathered over the last half-century has put the lie to these myths.

Although temporal decrements in performance are best known from vigilance research, there is also a substantial history of research on more cognitively demanding tasks, such as the prolonged writing and arithmetic tasks studied by Thorndike (e.g., 1914). Ackerman's (2011) historical review of fatigue includes a list of task characteristics having been considered during investigations of cognitive fatigue and is fully applicable to investigations of operator fatigue. These output measures include time on-task, demands on intellectual functioning, continuous vs. intermittent work, tasks not subject to large direct learning increments, exactingness of attending and responding (attention to detail), tolerance for errors, costs of distractions, general stressors, time pressure, high stakes, failure vs. success rates, intrinsically interesting or enjoyable tasks, and knowledge of results. Ackerman included the following in his list of dependent performance variables used historically to investigate cognitive fatigue: reaction time (specifically, the "Bills block"); changes to numbers of items attempted within a fixed period; patterns of errors over time; qualities of responses; and change in riskiness of responding. Ackerman's list of variables that represent inputs to operator fatigue state included lack of sleep, time of day, recency of last meal, and drugs.

Vigilance

Mackworth's discovery (1948) of the vigilance decrement developed into a specific subfield of experimental psychology. Some of the premier applied work in this field was

---

1 Folkard's technical report, a spreadsheet for risks of injuries and accidents during shiftwork, and user's guide are available at http://www.hse.gov.uk/research/rrhtm/rr446.htm

performed and reviewed by Mackie and his colleagues (Mackie, 1977; McGrath et al., 1959; Miller & Mackie, 1980). Much of that work focused on sonar operators in the U.S. Navy. Subsequently, Mackie applied theories of vigilance to investigations of commercial driver fatigue.

Davies and Parasuraman (1982) reviewed the numerous factors that influence vigilance decrement. These include task demand factors such as memory load and event rate, variables that influence motivation such as performance feedback, and adverse environmental conditions (Hancock, 1984). The recent review by Joel Warm and his colleagues (Warm, Parasuraman & Matthews, 2008) identifies several developments of these studies of the vigilance decrement. See et al., (1995) reported a meta-analysis of factors influencing vigilance decrement that showed the likelihood and magnitude of performance decrement increased systematically with task demands. A critical factor is whether the discrimination required is successive (absolute judgment) or simultaneous (comparative judgment). Subjective reports also show that the workload of vigilance is high and sensitive to factors that increase processing demands. Warm, Parasuraman & Matthews (2008) concluded that vigilance requires hard mental work and is stressful (see also Hancock & Warm, 1989). In addition, neuroimaging studies using transcranial Doppler sonography provide strong, independent evidence for resource changes linked to performance decrement in vigilance tasks. Physiological and subjective reports also confirm that vigilance tasks reduce task engagement and increase distress and that these changes rise with increased task difficulty. We may then challenge the myth that "Humans are good at 'standing watch'": sometimes true, most often not.

Contemporary studies of vigilance are most often based on measurement of "miss" and "false positive" errors in line with the theory of signal detection (Green & Swets, 1966), but two alternative approaches should be noted. Wilkinson and Houghton (1982) developed a simple, portable 10-minute test of "arousal" and "continuous, concentrated attention" called the unprepared simple reaction time test (USRT), and based upon a cassette tape. Dinges then introduced (Dinges & Powell, 1985) a solid-state version of the USRT called the Psychomotor Vigilance Task (PVT), which is now used widely and successfully in sleep loss studies (Lim & Dinges, 2008). However, the USRT/PVT has a signal probability of 1.0, much unlike many classic vigilance tests that have signal probabilities of less than 0.05 (Miller & Mackie, 1980. While the USRT/PVT is sensitive to fatigue (especially sleepiness), it is difficult to classify it as a "vigilance" task because it is such a short task and its signals are not embedded in a background of high-frequency, non-meaningful events. Thus, though the USRT/PVT addresses some aspects of sustained attention, and captures errors of omission in the form of lapses (i.e., the Bills Block), it fails to address the visual search and decision-making components of vigilance performance.

One must also note the seminal contribution of Stern (Stern, Walrath & Goldstein, 1984). Stern's revelation of the information content present in the conformation of the endogenous blink was at least as meaningful as Mackworth's characterization of the vigilance decrement and the description of the circasemidian rhythm in body temperature by Colquhoun, Blake & Edwards (1968). Stern et al. indicated that:

> The endogenous eyeblink is an ocular response with a characteristic rate and waveform, and is coordinated centrally with cognitive events. […] its production is coordinated with stimulus occurrence […], with motor response production, and with other oculomotor events. Parameters of the endogeous blink are sensitive indicants of cognitive activity. The allocation of attentional resources, the subject's level of activation, and the effects of accumulated time on-task are all evidenced by variations in endogenous blinking. (1984: 31)

INDUSTRIAL APPLICATIONS

As is discussed above, early applied studies of fatigue focused on deleterious effects of extended work shifts (Derickson, 1994). Fatigue associated with shift work remains an acute problem in occupational settings, especially given the shift toward a "24/7" society. Strategies for managing shift work have been strongly informed by growing understanding of sleep loss and circadian rhythms. Two collections of information about occupational risk and safety in shift work were provided by Haider (1986) and then Colquhoun (1996). Costa (2001) discussed the "24-hour society" and asked, "[W]hat kind of 24-hour society do we need? At what costs? Are they acceptable/sustainable?" "What are the advantages and disadvantages for the individual, the companies, and the society? What is the cost/benefit ratio in terms of physical health; psychological well-being, family and social life?" The research on irregular working hours and health has highlighted the adverse consequences of organizing work activities with regard to human physiology. Costa (2001) rejects passive acceptance of irregular work and its consequent maladjustments at both individual and social level. He advocates coping by adopting effective preventive and compensative strategies that aim to build a more sustainable society, maximizing benefits and keeping costs at an acceptable level. Of course, the fatiguing effects of shift work on operator performance has been an area of intense investigation for many decades. There are numerous general references concerning the operator fatigue associated with shift work (e.g., Copsey & Corlett, 1986; Dunham, 1977; Miller, Fisher & Cardenas, 2005; Monk, Folkard & Wedderburn, 1996). The bibliography of the field by Schroeder and Goulden (1983) lists references from 1950 onwards.

The problem of fatigue suggests a need for systematic efforts at fatigue management and for tests for the individual's "fitness for duty" (Gilliland & Schlegel, 1995). Tests are needed to detect occurrences of unfitness for work, for whatever reason (especially fatigue), on a daily basis at the worksite, particularly for jobs that affect public safety (Miller, 1996). Devices or procedures must be low-cost, convenient reliably and legally and socially acceptable. As Miller (1996) discussed, approaches to test for fitness for duty include performance, neurological, and biochemical measurement. Biochemical and neurological measures assess lifestyle factors, including drug use. However, such measures, such as urine tests for drugs, may require too much time for analysis to provide a true fitness-for-duty screening method. Performance measures are optimal for assessment of immediate, job-relevant issues.

Occupational fatigue research has also moved on from its origins in factory work, and during the World War II period, in aviation. Beginning in the 1960s, the methods used to study fatigue in aviation were applied to highway operations. Subsequently, these methods migrated into research concerning fatigue in various forms of transportation, and in the new working environments created by advances in technology, within which familiar forms of fatigue are apt to re-emerge. The increasing need for workers to monitor automated technologies, for example, brings us back to vigilance decrements (Warm, Parasuraman & Matthews et al., 2008). Next, the chapter briefly outlines some of the main applied areas of research that have developed in the period since the 1960s.

The Nuclear Power Plant Control Room

Interest in control room design and the effects of fatigue on operators in control rooms that operate 24 hours per day escalated after the near-meltdown of the nuclear reactor at

Three Mile Island in 1979. This incident began with maintenance work being performed at 0400 am, the nadir of the circadian variation in human performance, and was exacerbated by operators' misinterpretation of a poorly designed valve status indicator. The misinterpretation may well have been caused in part by the fatigue associated with night shift work. Though it was not mentioned specifically in reports, it was rested personnel who diagnosed the malfunction correctly and prevented a tragedy. The TMI incident highlighted the effects of shift work (rotation and length) and overtime on operator performance, and the need to evaluate these effects in plant operators (DiSalvo & Pittman, 1982). Subsequent studies reviewed the safety-related aspects of the use of 12-hour shift work schedules at nuclear power plants (Smiley & Moray, 1989) and examined empirically the effects of simulated 8-hour and 12-hour work shifts (Baker, 1995).

Motor Vehicles

The U.S. Transportation Research Board provides access to reports concerning operator fatigue in transportation systems.[2] An excellent review was provided to the U.S. Department of Transportation (DoT), relevant to all modes of transportation (McCallum et al., 2003). Concerning the development of methods for measuring driver performance with respect to fatigue, reviews were provided by A. Crawford (1961) of England's Transport and Road Research Laboratory, Colin Cameron (1973b) of the Australian Road Research Board, and Robert R. Mackie and James C. Miller of Human Factors Research, Inc. (Mackie & Miller, 1978). The latter review was a prelude to an intensive DoT-funded study of commercial driver fatigue in which the standard deviation of lane position was shown to be a sensitive measure of driver fatigue, and the electroencephalogram (EEG) was recorded during all highway operations. A huge number of highway research studies and reviews are in print. One critical review of special interest is by Barr, Howarth, Popkin & Carroll (2005) in which they "identify, categorize, and assess emerging behavioral and physiological [driver] drowsiness measures and technologies that will be available in the near future."

Rail

Locomotive drivers too are vulnerable to the effects of night and shift work. In a study of the German railroad, Rohmert, Hildebrandt, and Rutenfranz (1974) noted that discontinuity was the outstanding characteristic of their work schedule, due to daily shift changes that might subject the driver to alternations in route, starting time, working time, and even particular engine driven. Härmä, Sallinen, Ranta, Mutanen, and Müller (2002) confirmed the prevalence of severe sleepiness during the night shift. Sherry (2000) reviewed fatigue countermeasures programs instituted by various rail agencies and companies in the United States and Canada. Sherry concluded that, in the several years, railroads have made considerable progress in adopting a napping policy and educating management and employees in the effects and management of fatigue. There have been concerns also about the work–rest schedules and sleep patterns of U.S. railroad dispatchers (Gertler & Viale, 2007).

---

2 Transportation Research Board, available at www.trb.org

Maritime

The traditional four-hour shift (watch) length is used in many, but not all, military and commercial maritime surface operations (Miller, 2008). The main problems with the traditional 4-and-8 maritime watch plan are reduced total sleep time and fragmented sleep. With only eight hours off between watches, the watchstander is never able to get the eight hours of continuous, uninterrupted sleep found by clinicians and researchers to be the average sleep need. Thus, many maritime watchstanders operate at a continual, unacceptably low, predicted-average level of cognitive performance effectiveness (usually about 80 percent to 90 percent per the SAFTE model, below) while the vessel is under way.

At the APU, Colquhoun (1985) reviewed the problems associated with hours of work and watchkeeping duties on ocean-going vessels sailing between distant ports. Research findings from submarine studies in the United States in 1949–1950 by Nathaniel Kleitman and others were considered, as were a number of previous studies from the APU. The program proposed by Colquhoun led to a series of "Work at Sea" publications (e.g., Colquhoun, Rutenfranz et al., 1988). The United States Coast Guard (USCG) has also played an important role in undertaking systematic research that seeks to establish a technical basis for maritime operational practice and regulating work–rest scheduling and work hour limitations (e.g., Miller, Dyche et al., 2003). Other arenas for maritime research include surface naval ships (Nguyen, 2002), submarines (N.L Miller, Nguyen, Sanchez & J.C. Miller, 2003), and fishing vessels (Gander, van den Berg & Signal, 2008).

Aviation

So many reports have been published about fatigue effects in aviation that they cannot be summarized here. One may capture the results of World War II-era research by reading the review chapter by Ross McFarland, "Operational aspects of fatigue" (McFarland, 1953: 326–68). McFarland concluded that aging equipment was a major fatiguing factor in airline operations, accompanied by the length and nature of the operation, type of aircraft, number of crew members, terrain, night-flying, weather and traffic density. At that time, most aircrews were limited to 85 flight hours per month (union) or 100 hours, and that did not appear to be a problem. He noted that mental fatigue was a far greater problem in pilots than physical fatigue. Two important components of mental fatigue were skill deterioration and anxiety reactions. After other interesting conclusions, McFarland noted that aircrew fatigue issues would only be solved by the "coordinated efforts of the engineering, medical and administrative departments of the operating airlines."

J.A. Caldwell and J.L. Caldwell (2004) reviewed modern research on the contributions of the two main determinants of fatigue, one being the circadian rhythm and the other being the sleep homeostat. Then, they explained the nature and usefulness of a variety of fatigue countermeasures, and end with recommendations for the design of fatigue risk management programs. One of the notable scientific advances of the last decade that has helped aircrews deal with fatigue caused by long-duration flights is the acceptance, in some airlines and some military cockpits, of cockpit napping (e.g., Rosekind et al., 1994).

CONCLUSIONS

One may summarize the state of historical knowledge concerning the effects of sleepiness, motivation and task characteristics on operator performance from a simplistic, human factors viewpoint as follows (see below, Figure 2.1). System operations define the necessity for the performance of specific tasks by the human operator. The demand (workload) placed upon the operator by a given task is a function of many factors, including difficulty, frequency, levels of operator training and currency, distractions, etc. The task demand affects operator performance and feeds back to the operator in terms of the operator's complex perception of mental effort (also labeled workload in some contexts). Operator arousal, circadian and circasemidian rhythm status and the actions of the sleep "homeostat" affect sleep and task performance. The latter physiological functions, along with real and perceived task demands (and other factors) affect motivation, which also affects task performance.

Task characteristics are usually defined in two domains: the cognitive–fatigue structure outlined by Ackerman, above, and through task analysis applied to the system or a system component. Performance measurement tools range from sophisticated analyses of response time through assessments of performance variability to computer modeling of functional neuroanatomy. Physiological measurement tools allow the investigator to assess cerebrocortical arousal, the interplay between the sympathetic and parasympathetic branches of the autonomic nervous system, and conduction velocities in the peripheral nerves. Operator affect is assessed with a wide range of validated subjective rating scales that deal with fatigue, mood, sleepiness, perceived exertion, workload, etc.

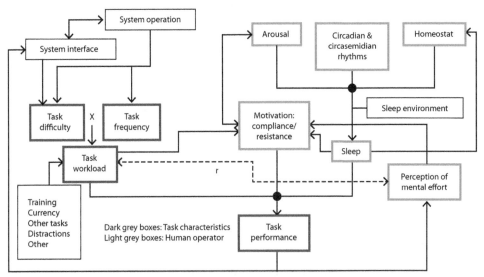

Figure 2.1   **Schematic diagram of the effects of sleepiness, motivation and task characteristics on operator performance**

The advances made in understanding the function of the operator as a component in a human–machine system and understanding the effects of fatigue on the performance operator, and the sophistication of the relevant measurement tools, have occurred primarily since World War II. As in all scientific disciplines, we stand upon the shoulders of those who came before us, grateful for their insights, intuition and perseverance.

REFERENCES

Ackerman, P.L. (2011). 100 years without resting, in P.L. Ackerman (ed.), *Cognitive Fatigue: Multidisciplinary Perspectives on Current Research and Future Applications*. Washington, DC: American Psychological Association, 11–44.

Adrian, E.D. & Matthews, B.H.C. (1934). Berger rhythm: Potential changes from the occipital lobes of man. *Brain*, 57, 355–85.

Åkerstedt, T., Folkard, S., & Portin, C. (2004). Predictions from the three-process model of alertness. *Aviation, Space, and Environmental Medicine*, 75(3 Suppl), A75–83.

Åkerstedt, T.A., Knutsson, A., Westerholm, P., Theorell, T., Alfredsson, L. & Kecklund, G. (2004). Mental fatigue, work and sleep. *Journal of Psychosomatic Research*, 57, 427–33.

Ancoli-Israel, S. (2001). "Sleep is not tangible," or what the Hebrew tradition has to say about sleep. *Psychosomatic Medicine*, 63, 778–87.

Aschoff, J. & Wever, R. (1976). Human circadian rhythms: A multioscillatory system. *Federation Proceedings*, 35, 236–332.

Aserinsky, E. & Kleitman, N. (1953). Regularly occurring periods of eye motility, and concomitant phenomena, during sleep. *Science*, 118, 273–4.

Baker, T. (1995). Alertness, Performance and Off-Duty Sleep on 8-Hour and 12-Hour Night Shifts in a Simulated Continuous Operations Control Room Setting (NUREGCR6046). Boston, MA: Institute for Circadian Physiology. Available at: http://www.ntis.gov/search/product.aspx?ABBR=NUREGCR6046

Barr, L., Howarth, H., Popkin, S. & Carroll, R.J. (2005). A review and evaluation of emerging driver fatigue detection measures and technologies. Proceedings of the 2005 International Conference on Fatigue Management in Transportation Operations. Ottawa: Transport Canada. Available at: http://www.tc.gc.ca/eng/innovation/tdc-publication-tp14620-b02-133.htm.

Bartlett, F. (1942). Fatigue in the Air Pilot (Tech. Report 448). Cambridge, UK: Flying Personnel Research Committee, Air Ministry.

Bartley, S. & Chute, E. (1947). *Fatigue and Impairment in Man*. New York: McGraw-Hill.

Belyavin, A.J. & Spencer, M.B. (2004). Modeling performance and alertness: The QinetiQ approach. *Aviation, Space, and Environmental Medicine*, 75, 93–106.

Berger, H. (1929). Über das Elektrenkephalogramm des Menchen. *Archives für Psychiatrie*, 87, 527–70.

Bills, A.G. (1931). Blocking: A new principle in mental fatigue. *American Journal of Psychology*, 43, 230–45.

Bills, A.G. (1937). Fatigue in mental work. *Physiological Review*, 17, 436–53.

Borbély, A.A. (1982). A two process model of sleep regulation. *Human Neurobiology*, 1, 195–204.

Borbély, A.A., Achermann, P., Trachsel, L. & Tobler, I. (1989). Sleep initiation and initial sleep intensity: Interactions of homeostatic and circadian mechanisms. *Journal of Biological Rhythms*, 4, 37–48.

Brink, S. (2000). Sleepless society. *US News and World Report*, 129, 63–72.

Broadbent, D.E. (1971). *Decision and Stress*. London: Academic Press.

Broughton, R.J. (1998). SCN controlled circadian arousal and the afternoon "nap zone." *Sleep Research*, 1, 166–78.

Brown, I.D. (1994). Driver fatigue. *Human Factors*, 36, 298–314.

Browne, R.C. (1949). The day and night performance of teleprinter switchboard operators. *Occupational Psychology*, 23, 1–6.

Cairnes, W.E. & Chandler, D.G. (eds) (1995 [1901]). Napoleon's 58th Maxim, in *The Military Maxims of Napoleon*. Cambridge, MA: Da Capo Press.

Caldwell, J.A. & Caldwell, J.L. (2004). *Fatigue in Aviation: A Guide to Staying Awake at the Stick.* Farnham, UK: Ashgate Publishing.

Cameron, C. (1973a). A theory of fatigue. *Ergonomics*, 16, 633–48.

Cameron, C. (1973b). Fatigue and driving: A theoretical analysis. *Australian Road Research*, 5, 36–44.

Colquhoun, W.P. (1985). Hours of work at sea – Watchkeeping schedules, circadian rhythms and efficiency. *Ergonomics*, 28, 637–53.

Colquhoun, W.P. (1996). *Shiftwork: Problems and Solutions.* New York: Peter Lang.

Colquhoun, W.P., Blake, M.J. & Edwards, R.S. (1968). Experimental studies of shift-work: A comparison of "rotating" and "stabilized" 4-hour shift systems. *Ergonomics*, 11, 437–53.

Colquhoun, W.P., Rutenfranz, J., Goethe, H., Neidhart, B., Condon, R., Plett, R. & Knauth, P. (1988). Work at sea: A study of sleep, and of circadian rhythms in physiological and psychological functions, in watchkeepers on merchant vessels. *International Archives of Occupational and Environmental Health*, 60, 321–29.

Committee on Sleep Medicine and Research (2006). *Sleep Disorders and Sleep Deprivation: An Unmet Public Health Problem.* Washington, DC: National Academies Press.

Copsey, S. & Corlett, E. (1986). Shiftwork review: Research of the European Foundation, 1981–1984. European Communities, December.

Costa, G. (2001). The 24-hour society between myth and reality. *Journal of Human Ergology*, 30, 15–20.

Craik, K.J.W. (1940). The fatigue apparatus (Cambridge cockpit) (Report 119). London: British Air Ministry, Flying Personnel Research Committee.

Crawford, A. (1961). Fatigue and driving. *Ergonomics*, 4, 143–54.

Davies, D.R. & Parasuraman, R. (1982). *The Psychology of Vigilance.* London: Academic Press.

Davis, D.R. (1948). *Pilot Error: Some Laboratory Experiments.* London: HMSO.

Derickson, A. (1994). Physiological science and scientific management in the progressive era: Frederic S. Lee and the Committee on Industrial Fatigue. *The Business History Review*, 68, 483–514.

Desmond, P.A. & Hancock, P.A. (2001). Active and passive fatigue states, in P.A. Hancock & P.A. Desmond (eds), *Stress, Workload, and Fatigue.* Mahwah, NJ: Lawrence Erlbaum.

Dijk, D.J. & von Schantz, M. (2005). Timing and consolidation of human sleep, wakefulness, and performance by a symphony of oscillators. *Journal of Biological Rhythms*, 20, 279.

Dinges, D.F. & Powell, J.W. (1985). Microcomputer analyses of performance on a portable simple visual RT task during sustained operations. *Behavior Research Methods, Instruments & Computers*, 17, 652–55.

DiSalvo, R. & Pittman, G. (1982). A Feasibility Study of Using Licensee Event Reports for a Statistical Assessment of the Effect of Overtime and Shift Work on Operator Error (NUREG0872). Nuclear Regulatory Commission, Washington, DC: Office of Nuclear Regulatory Research. Available at: http://www.ntis.gov/search/product.aspx?ABBR=NUREG0872

Drew, G.C. (1940). An experimental study of mental fatigue. Paper 227. Cambridge, UK: Air Ministry, Flying Personnel Research Committee.

Dunham, R.B. (1977). Shift work: A review and theoretical analysis. *The Academy of Management Review*, 2, 624–34.

Ekirch, A.R. (2001). Sleep we have lost: Pre-industrial slumber in the British Isles. *The American Historical Review*, 106, 343–86.

Ekirch, A.R. (2006). *At Day's Close: Night in Times Past*. New York: W.W. Norton.

Folkard, S. (2006). Fatigue/Risk index for shiftworkers – health and safety in the workplace (No. RR446). U.K. Health and Safety Executive. Available at: http://www.hse.gov.uk/research/rrhtm/rr446.htm

Folkard, S. & Lombardi, D. (2004). Toward a "Risk Index" to assess work schedules. *Chronobiology International*, 21, 1063–72.

Folkard, S. & Lombardi, D. (2006). Modeling the impact of the components of long work hours on injuries and "accidents." *American Journal of Industrial Medicine*, 49, 953–63.

Gander, P.H., van den Berg, M. & Signal, L. (2008). Sleep and sleepiness of fishermen on rotating schedules. *Chronobiology International*, 25, 389–98.

Gertler, G. & Viale, A. (2007). Work schedules and sleep patterns of railroad dispatchers. (DOT/FRA/ORD-07/11). Washington, DC: Federal Railroad Administration.

Gilliland, K. & Schlegel, R.E. (1995). Readiness-to-perform testing and the worker. *Ergonomics in Design*, 3, 14–19.

Green, D.M. & Swets, J.A. (1966). *Signal Detection Theory and Psychophysics*. Oxford: Wiley.

Haider, M. (1986). Night and shiftwork: Long-term effects and their prevention, in *Studies in Industrial and Organizational Psychology*. Vol. 3. New York: Peter Lang.

Hancock, P.A. (1984). Environmental stressors, in J.S. Warm (ed.), *Sustained attention in human performance*, New York: Wiley, 103–42.

Hancock, P.A. & Warm, J.S. (1989). A dynamic model of stress and sustained attention. *Human Factors*, 31, 519–37.

Härmä, M., Sallinen, M., Ranta, R., Mutanen, P., & Müller, K. (2002). The effect of an irregular shift system on sleepiness at work in train drivers and railway traffic controllers. *Journal of Sleep Research*, 11(2), 141–51.

Hayashi, M., Morikawa, T. & Hori, T. (2002). Circasemidian 12 hour cycle of slow wave sleep under constant darkness. *Clinical Neurophysiology*, 113, 1505–16.

Hildebrandt, G., Rohmert, W. & Rutenfranz, J. (1974). 12 and 24 h rhythms in error frequency of locomotive drivers and the influence of tiredness. *International Journal of Chronobiology*, 2, 175–80.

Horvath, S.M. & Horvath, E.C. (1973). *Harvard Fatigue Laboratory: Its History and Contributions*. Englewood Cliffs, NJ: Prentice Hall.

Hursh, S.R., Redmond, D.P., Johnson, M.L., Thorne, D.R., Belenky, G., Balkin, T.J., Storm, W.F. et al. (2004). Fatigue models for applied research in warfighting. *Aviation, Space, and Environmental Medicine*, 75, 44–60.

Jastrzębowski, W. (2000 [1857]). Outline of Ergonomics, or The Science of Work Based Upon Truths Drawn from the Science of Nature. Warsaw: Central Institute for Labour Protection.

Kleitman, N. (1933). Studies on the physiology of sleep. VIII. Diurnal variation in performance. *American Journal of Physiology*, 104, 449–56.

Kleitman, N. (1963). *Sleep and wakefulness*. 2nd ed. Chicago, IL: University of Chicago Press.

Knauth, P., Schwarzenau, P., Brockmann, W. & Rutenfranz, J. (1982). Computerized construction of shift plans for continuous production which meet physiological, social, and legal requirements. *Journal of Human Ergology*, 11, 441–6.

Kostreva, M., McNelis, E. & Clemens, E. (2002). Using a circadian rhythms model to evaluate shift schedules. *Ergonomics*, 45, 739–63.

Kronauer, R.E., Czeisler, C.A., Pilato, S.F., Moore-Ede, M.C. & Weitzman, E.D. (1982). Mathematical model of the human circadian system with two interacting oscillators. *The American Journal of Physiology*, 242, 3–17.

Lavie, P., Wollma, M. & Pollack, I. (1986). Frequency of sleep-related traffic accidents and hour of the day. *Sleep Research*, 15, 275.

Lim, J. & Dinges, D.F. (2008). Sleep deprivation and vigilant attention. *Annals of The New York Academy of Sciences*, 1129, 305–22.

Mackie, R.R. (ed.) (1977). *Vigilance: Theory, Operational Performance, and Physiological Correlates*. New York: Springer.

Mackie, R.R. & Miller, J.C. (1978). Effects of Hours of Service, Regularity of Schedules and Cargo Loading on Truck and Bus Driver Fatigue. (No. HFR-TR-1765-F, NTIS PB-290–957). Washington, DC: National Highway Traffic Safety Administration.

Mackworth, N.H. (1948). The breakdown of vigilance during prolonged visual search. *Quarterly Journal of Experimental Psychology*, 1, 6–21.

Maggioni, C., Cornelissen, G., Antinozzi, R., Ferrario, M., Grafe, A. & Halberg, F. (1999). A half-yearly aspect of circulating melatonin in pregnancies complicated by intrauterine growth retardation. *Neuroendicrinology Letters*, 20, 55–68.

Mallis, M., Mejdal, S., Nguyen, T. & Dinges, D.F. (2004). Summary of the key features of seven biomathematical models of human fatigue and performance. *Aviation, Space, and Environmental Medicine*, 75, 4–14.

Martineaud, J.P., Cisse, F. & Samb, A. (2000). Circadian variability of temperature in fasting subjects. *Scripta Medica*, 73, 15–24.

Mayo, E. (1924). Revery and industrial fatigue. *Journal of Personnel Research*, 3, 273–81.

McCallum, M., Sanquist, T., Mitler, M. & Krueger, G. (2003). Commercial Transportation Operator Fatigue Management Reference. (No. DTRS56-01-T-003). Washington, DC: U.S. Department of Transportation.

McFarland, R.A. (1953). *Human Factors in Air Transportation: Occupational Health and Safety*. New York: McGraw-Hill, Chapter 7.

McFarland, R.A. (1971). Fatigue in industry. *Ergonomics*, 14, 1–10.

McGrath, J.J., Harabedian, A. & Buckner, D.N. (1959). Review and critique of the literature on vigilance performance (No. AD0237691). Los Angeles, CA: Human Factors Research Inc. Available at: http://oai.dtic.mil/oai/oai?verb=getRecord&metadataPrefix=html&identifier=AD0237691

Miller, J.C. (1996). Fit for duty? *Ergonomics in Design*, 4, 11–17.

Miller, J.C. (2006). In Search of Circasemidian Rhythms (No. 2006-0074, ADA458153). Brooks City-Base TX: Air Force Research Laboratory. Available at: http://stinet.dtic.mil/oai/oai?&verb=getRecord&metadataPrefix=html&identifier=ADA458153

Miller, J.C. (2008). Fundamentals of shiftwork scheduling. *Ergonomics in Design*, 16, 13–17.

Miller, J.C. & Mackie, R.R. (1980). Vigilance Research and Nuclear Security: Critical Review and Potential Applications to Security Guard Performance. (No. 2722; National Bureau of Standards contract NBS-GCR-80-201 for the Defense Nuclear Agency. Goleta CA: Human Factors Research Inc. Available from the Electronics and Electrical Engineering Laboratory, Office of Law Enforcement Standards, National Institute of Standards and Technology. Available at: www.nist.gov/eeel/

Miller, J.C., Dyche, J., Cardenas, R. & Carr, W. (2003). Effects of Three Watchstanding Schedules on Submariner Physiology, Performance and Mood (No. 1226, ADA422572). Groton, CT: Naval Submarime Medical Research Laboratory. Available at: http://www.ntis.gov/search/index.aspx

Miller, J.C., Fisher, D. & Cardenas, C.M. (2005). *Air Force Shift Worker Fatigue Survey* (No. 2005-0128, ADA438140). Brooks AFB, TX: Air Force Research Laboratory. Available at: http://stinet.dtic.mil/oai/oai?&verb=getRecord&metadataPrefix=html&identifier=ADA438140

Miller, N.L., Nguyen, J.L., Sanchez, S. & Miller, J.C. (2003). Sleep Patterns and Fatigue Among U.S. Navy Sailors: Working the Night Shift During Combat Operations Aboard the USS STENNIS

During Operation Enduring Freedom. Presented at the Aerospace Medical Association. Available at: http://faculty.nps.edu/nlmiller/millerpu.htm

Monk, T.H., Folkard, S. & Wedderburn, A.I. (1996). Maintaining safety and high performance on shiftwork. *Applied Ergonomics*, 27, 17–23.

Mosso, A. (1915). *Fatigue*. London: Allen & Unwin.

Muscio, B. (1921). Is a fatigue test possible? *British Journal of Psychology*, 12, 31–46.

Nguyen, J.L. (2002). The Effects of Reversing Sleep–Wake Cycles on Sleep and Fatigue on the Crew of USS John C. Stennis (No. ADA407035). Monterey CA: Naval Postgraduate School. Available at: http://stinet.dtic.mil/oai/oai?&verb=getRecord&metadataPrefix=html&identifier=ADA407035

Oldfield, R.C. (1959). The analysis of human skill, in P. Halmos & A. Iliffe (eds), *Readings in General Psychology*. London: Routlege & Kegan Paul.

Rechtschaffen, A. & Kales, A. (1968). A manual of standardized terminology, techniques and scoring system for sleep stages of human subjects (No. 204). Washington D.C.: National Institutes of Health.

Reynolds, L.A. & Tansey, E.M. (eds) (2001). MRC CBU. Cambridge: History of the Unit. Welcome Trust. Available online at: http://www.mrc-cbu.cam.ac.uk/history/history/wellcome.html

Rohmert, W., Hildebrandt, G. & Rutenfranz, J. (1974). Night and Shift Work of Locomotive Engineers. Second Report: Investigations on the Organization of Daily Service Schedules (N7521019). Scientific Translation Service, Santa Barbara, CA. Available at: http://www.ntis.gov/search/product.aspx?ABBR=N7521019

Rolfe, J.M. (n.d.). Two Cambridge inventors. *Royal Aeronautical Society Flight Simulation Group – The Cambridge Cockpit*. Available at: http://www.raes-fsg.org.uk/18/The_Cambridge_Cockpit/ (accessed March 28, 2011).

Roscoe, S.N. (1997). *The Adolescence of Engineering Psychology*. Santa Monica, CA: Human Factors and Ergonomics Society.

Rosekind, M.R., Gander, P.H., Miller, D.L., Gregory, K.B., Smith, R.M., Weldon, K.J. & Co, E.L. (1994). Fatigue in operational settings: Examples from the aviation environment. *Human Factors*, 36, 327–38.

Rutenfranz, J. & Colquhoun, P. (1979). Circadian rhythms in human performance. *Scandanavian Journal of Work Environment and Health*, 5, 167–77.

Ryan, T.A. (1947). *Work and Effort*. New York: Ronald Press.

Schroeder, D. & Goulden, D. (1983). *A Bibliography of Shift Work Research, 1950–1982* (ADA135644). Washington, D.C.: Office of Aviation Medicine, Federal Aviation Administration.

See, J.E., Howe, S.R., Warm, J.S. & Dember, W.N. (1995). Meta-analysis of the sensitivity decrement in vigilance. *Psychological Bulletin*, 117, 230–49.

Shakespeare, W. (ca. 1604). *Measure for Measure*.

Shakespeare, W. (ca. 1607). *Macbeth*.

Sherry, P. (2000). *Fatigue Countermeasures in the Railroad Industry: Past and Current Developments*. Washington, DC: Association of American Railroads.

Smiley, A. & Moray, N. (1989). *Review of 12-hour shifts at nuclear generating stations*. Final report (DE93604017). Ottawa (Ontario): Atomic Energy Control Board. Available at: http://www.ntis.gov/search/product.aspx?ABBR=DE93604017

Starch, D. & Ash, I.E. (1917). The mental work curve. *Psychological Review*, 29, 391–402.

Stern, J.A., Walrath, L.C. & Goldstein, R. (1984). The endogenous eyeblink. *Psychophysiology*, 21, 22–33.

Thorndike, E.L. (1900). Mental fatigue. *Psychological Review*, 7, 466–82.

Thorndike, E.L. (1914). *Mental Work and Fatigue and Individual Differences and their Causes*. New York: Teachers College.

Warm, J.S., Parasuraman, R. & Matthews, G. (2008). Vigilance requires hard mental work and is stressful. *Human Factors*, 50, 433–41.

Weinberg, A. & Bealer, B.K. (2001). *The World of Caffeine: The Science and Culture of the World's Most Popular Drug*. London: Routledge.

Weiskotten, J. & Ferguson, J.E. (1930). A further study of the effects of loss of sleep. *Journal of Experimental Psychology*, 13, 247–66.

Whiting, H.F. & English, H.B. (1925). Fatigue tests and incentives. *Journal of Experimental Psychology*, 8, 33–49.

Wilkinson, R.T. & Houghton, D. (1982). Field test of arousal: A portable reaction timer with data storage. *Human Factors*, 24, 487–93.

3
Challenges in Fatigue and Performance Research

Bob Hockey

INTRODUCTION

I need to start this chapter with an assertion; after more than 100 years of research on fatigue we do not really know very much about it. The problem is poorly understood, even within the scientific community, and there is still no mature theory of its origins and function. All we have is a longstanding general assumption that fatigue is somehow associated with a "depletion of energy" – an unwanted by-product of (physical or mental) work. For the most part this view has been held uncritically, with little attempt at systematic theory-building or generation of testable hypotheses. At the same time, several seemingly different types of fatigue (mental, physical, sleepiness) live largely separate lives, following separate research agendas, "explained" by separate theories, and usually published in separate journals. Yet these different manifestations of fatigue are often confused in the literature, or treated as if they were the same thing. For the rest of the chapter I will try to identify the main areas where our understanding is most limited, and where research effort is most needed to address these problems.

Historical Context of Fatigue Research

It is only during the past 150 years or so that fatigue has come to have an almost universally negative connotation. Before that time tiredness – the core experience of fatigue – was perceived as a natural consequence of any activity, and not something to be concerned about. This is still true today in many contexts; we often refer to a *pleasant* tiredness, after, say, a long walk, a game of tennis, or even following one of those rare productive days of chapter writing! In fact, the meaning of tiredness appears to have changed since the pre-modern period; while the use of words such as "tired" and "fatigued" to describe feelings dates from the fourteenth century, social historians have suggested that their widespread use in everyday language dates only from the regularization of work during the early 1800s. The farmhand or laborer of the seventeenth century would certainly have got tired from working all day, but there is no evidence that he or she perceived this as an aversive state; as with the feeling of sleepiness toward the end of the day, it was a natural part of the flow of life. Fatigue as a recognized medical problem did not exist before 1850, whereas by 1900 it was established as *the* disease of the modern age, exemplified by George Beard's description of neurasthenia as nervous exhaustion (Rabinbach, 1990: 153). For most of the post-industrial period up to the present day fatigue has continued to be perceived as a major problem for society. How did this change occur?

Our current understanding of fatigue is in terms of a depletion of energy resources from overwork. This idea derives from the enthusiastic adoption of the energy conservation metaphor as the basis for understanding human work limitations in the mid-1800s, driven by the impact on society (and work, in particular) of the rapid spread of energy transforming machines. Rabinbach (1990) refers to this new way of conceptualizing human activity as the "human motor." It was originally applied to physical work, but very quickly extended to account for fatigue from mental activities. The energy metaphor had an immediate and major impact on both the scientific literature and everyday language. The terms "fatigue" and "energy" became associated with human activity from around 1860, particularly within psychology and clinical medicine. Before the advent of this conception of human activity, failures to complete work were typically attributed to a problem of volition – an unwillingness to start a task or to continue it. The principle of energy conservation meant that fatigue could now be explained as a result of running down the supply of whatever fuel the body (or mind) used. The energy-depletion explanation of fatigue, in effect, medicalized what had previously been a natural result of interaction with environmental events; fatigue had become the symptom of the strain of coping with the more complex and demanding world. In passing, it may be observed that modern models of motivation and self-regulation (e.g., Karoly, 1993) typically include elements related to *both will* (effort, planning, control) and *energy* (fatigue, arousal, stress).

A review of the theoretical work on fatigue over the last 100 years shows that the emphasis on fatigue as expended energy has not been as useful an idea as first thought. I would go further, however, and argue that it has been detrimental to the development of a genuine theory of fatigue. An alternative interpretation (Hockey, in press)[1] proposes that fatigue is better considered as part of the adaptive management of goal-oriented human activity. Its primary role is to interrupt task activity in order to force a decision upon the system – whether to switch away from the task in order to attend to other relevant goals or to continue with the present activity. On this view fatigue is not a necessary consequence of work, as has typically been assumed, but of behavior that is externally driven (activities we would rather not be engaged in); it has been known since the early days of research on the problem (e.g., Thorndike, 1912) that fatigue is not commonly experienced when people carry out work that is personally meaningful, or when they engage in self-initiated activities.

Challenges for Research on Fatigue: A Four-part Agenda

What are the main challenges for fatigue and performance research? In one sense there may be only one, though the challenge is a major one: to develop a convincing theory. This should answer questions such as: What are the origins of fatigue? What is its function? Does it have an identifiable representation in terms of brain activity? For the purposes of this chapter I make the assumptions: that fatigue is a real process, identifiable, for

---

1   The motivational control theory of fatigue will be developed in full in a new monograph by the author, to be published by Cambridge University Press in 2012.

example, as the subjective experience of acute tiredness in task contexts; that its impact on behavior is distinctive, and separable from those of other states (such as anxiety or boredom); and that it has a central representation. I believe that there are four broad areas where work needs to be carried out in order to make progress toward a scientific understanding of fatigue:

1. The work–fatigue hypothesis – the need to identify the factors that cause fatigue and evaluate the assumed link between fatigue and work.
2. The relation between fatigue and energy – the need to understand the role of brain energy transformations in the fatigue process.
3. Motivational control aspects of fatigue – the need to consider the alternative role of fatigue as an adaptive process helping to regulate goal-directed activity.
4. Varieties of fatigue – the need to address the similarities, differences and connections between different types of fatigue: mental, physical, sleepiness.

THE WORK–FATIGUE HYPOTHESIS

As we have seen, the traditional view of fatigue is that it is caused by work. The work–fatigue hypothesis has been assumed more or less without question in mainstream theory and practice. However, the boundary conditions for this relationship have not been made clear. A major challenge for research is to identify the factors that give rise to fatigue. Under what conditions does work lead to fatigue? And when does it not? What else leads to fatigue? Does it depend on the level of work demands, or their duration, or their interest and meaning for the performer? What can be done to prevent fatigue developing during periods of work?

In general, it appears that the work–fatigue hypothesis may hold only for carrying out activities that are genuine *tasks* – in the sense of burdens or duties imposed on the individual from outside, and which are carried out to meet some obligation. Of course, this is true for most paid work, and it is here (or in the simulated laboratory version of it) where the work–fatigue hypothesis as been applied most successfully. However, not all work is like this, and many intellectually or physically demanding activities are not thought of by their performers as tasks: those that we choose to do, or that, while imposed by work conditions, are also highly consistent with personal goals; or leisure and play. It goes without saying that such activities are not necessarily unstructured, simple, or undertaken in a casual manner. They may in fact be very demanding and utterly absorbing, often carried out for hours without breaks. Consider games such as chess or bridge; or modern computer games; reading academic papers and writing book chapters; running, cycling and tennis. As we have noted, fatigue is not a typical feature of such work contexts. In general, as Bartley and Chute (1947) concluded, people do not feel tired when they are interested in or enthusiastic about what they are doing. I would add that this is true however demanding it may seem to an outsider.

Control Over Work and Goals

In modern terms, we would say that the work–fatigue relationship is moderated by the degree of control the performer has over the work. Here, control means that individuals feel that they have discretion over their work activities: to choose what to do, to be able to carry out tasks in different ways and at different times, and so on. The work psychology literature has long recognized the role that human interpretation and coping play in managing the impact of work demands, not only on fatigue but more generally on strain, well-being and health. The most influential and widely applied framework for this is Karasek's demands–control model (Karasek & Theorell, 1990), which emphasizes the moderating influence of personal control on work strain. Mainly on the basis of questionnaire studies of naturally occurring work, high levels of controllability have generally been shown to offset the negative effects of work (e.g., Ganster, 1989; Wall, Jackson, Mullarkey & Parker, 1996). The same pattern has also been found by a recent laboratory study of simulated office work (Hockey & Earle, 2006); a number of fatigue effects observed under high workload (induced by time pressure) were reduced or abolished when participants were allowed to determine their own task schedules (compared to yoked controls, who were required to follow someone else's schedule).

It is therefore clear that work does not necessarily lead to fatigue, and that its effects are reduced by high control. In fact, the control that characterizes many self-generated goals may result not just in the absence of fatigue but in its opposite – strongly positive states of energy and elation. This kind of state has been identified in Frankenhaeuser's (1986) taxonomy of work stress patterns as "effort without fatigue," the analysis by Hockey, Payne and Rick (1996) of individual demand management modes as "enabling work," and in Csikszentmihalyi's (1990) account of "flow." Most of us have experienced the fatigue-free exhilaration of long creative periods spent writing, playing music, painting or other such activity. Yet, the maintenance of such states is hard to guarantee. In Csikszentmihalyi's framework the experience of flow occurs as a conjunction of high challenge and high skill; too little of one or the other can cause the state to tip over into boredom or anxiety. While Csikszentmihalyi does not consider fatigue as such, it is closely related to boredom in the work context, and may develop even under apparently ideal work conditions. Very few activities are homogeneous in their attentional needs, so may produce unpredictable shifts between flow and the need for directed attention; many will also have experienced the sudden deflating effects of "overdoing it" – of getting so engrossed in (ostensibly enjoyable) work that we fail to recognize the developing need for rest or change of goal. This experience suggests that what has been play has now become a task, requiring effort-driven attention. This can easily happen in any complex activity, when sub-goals that are activated have a more task-like property. In writing this chapter I occasionally have to move away from the (mainly pleasurable) act of generating freely flowing text from ideas, toward a more functional, task-like mode where I need to dig out references, check half-remembered contents of saved pdf files, and the like. If I do this for too long before getting back to the flow mode I can begin to feel frustrated or tired.

Very little is known about the microstructure of attention management in work. Clearly, with a view to understanding the causes of fatigue, we need to find out how the experience of fatigue (or of flow) is related to individual skills in managing goals, and in monitoring and evaluating changing mental states. This problem is recognized in Kaplan's (2001) approach to the management of attention, emphasizing the need for skill and self-knowledge in balancing the reliance on effortful and effortless modes of attention, and timely, effective restoration of the capacity for active attention.

FATIGUE AND BRAIN ENERGY

A second problem for future research is to clarify the role of brain energy in fatigue. It has long been assumed that fatigue is the indirect result of doing work, somehow caused by a depletion of energy stores. It is time to re-evaluate this assumption. In its original form energy depletion was meant as a *metaphor*, though it soon assumed a literal reality; for example, Dodge (1913) identified mental work unambiguously with neural metabolism. It seems reasonable that physical fatigue (from exercise or manual work) might be associated with a shortfall of energy, because of the high metabolic demands of skeletal muscles (though, as discussed in the next section, this does not appear to be true in any simple way). But what about mental fatigue? Does the brain really use up more energy when demanding mental tasks are being carried out? And is that lost energy sufficient to account for the various phenomena of fatigue? The brain has long been recognized as a hungry consumer of the energy produced by the body: although making up only 2 percent of the average adult body weight, it receives 15 per cent of the cardiac output, accounts for 20 percent of the total oxygen consumption, and 25 percent of glucose utilization (Clark & Sokoloff, 1999). Yet, our understanding of the relation between brain energy and mental activity remains unclear. During the nineteenth century the adoption of the energy metaphor for mental work led to the assumption that energy was required for both brain work and muscular work, and that the two kinds of activity had mutually debilitating effects. For example, neurasthenia (the first formal description of a fatigue "disease") was regarded as a chronic loss of mental energy resulting from a failure to cope with the demands of the modern world (Rabinbach, 1990: 153), while psychologists such as Binet and Henri (1898) identified the rapidly growing demands for study and intellectual work in schoolchildren as the cause of endemic physical weariness and lethargy.

However, scientific attempts to confirm the impact of mental activity on brain energy proved largely unsuccessful. Van den Berg (1986) observed that Hans Berger, who developed the EEG through his interest in measuring the energy demands of mental work, concluded in a later review that the level of metabolism in the resting brain was so high that changes with mental tasks would be very difficult to detect. This was confirmed by Kety and Schmidt (1948), whose method for estimating glucose oxidation, based on cerebral blood flow – the first reliable procedure for use with unanaesthetized humans – is still used today.[2] Much of the research that followed found little evidence of any appreciable change in overall brain energy under a very wide range of conditions, including the definitive study of Sokoloff et al. (1955), using mental arithmetic as the task. This view has been the dominant one for the last half century (Van den Berg, 1986) and has at least made it clear that the relationship between fatigue and energy is, at best, a subtle one (and certainly nothing as drastic as the "fatigue is caused by energy depletion" of common belief).

Brain Energy Metabolism

While it is inappropriate here to give a detailed account of brain energy metabolism, a brief synopsis is necessary. The source of energy for all body cells is adenosine triphosphate

---

2   Kety and Schmidt's method requires participants to inhale a highly diffusible, inert gas (nitrous oxide), which is taken up by the brain. Rate of glucose uptake by the brain is calculated from the difference between the amount of gas brought to the brain by the arterial blood and the amount carried away in the cerebral venous blood.

(ATP), which is essential for neural activation, but is constantly being broken down and resynthesized. In the body, ATP can be produced from a range of nutritional sources (glucose, fats or protein). However, in the brain it is derived almost exclusively from glucose; this is because, only glucose is able to cross the blood–brain barrier. Since the brain has very limited glucose reserves (in the form of about 1g of glycogen, stored in glial cells), compared to the liver and skeletal muscle, most of it must come from the circulation, a process known as *oxidative phosphorylation*. A second route for ATP production, *glycolysis*, makes direct use of stored glycogen. While this accounts for only a small amount of total ATP it may have a strategic role in emergencies or sudden heavy demands because it is independent of blood flow and therefore much faster (Brown, 2004).

The way in which the brain and body manage their supply of energy is still not completely understood. It was thought until quite recently that the energy demands of both were regulated in parallel by the hypothalamus, though this view has had to be revised in the light of evidence showing that extreme energetic demands such as those of heavy exercise, fasting and starvation had minimal impact upon the integrity of brain function. To account for these disparate facts Peters et al. (2004) developed the "selfish brain" model, arguing that brain not only controlled the regulation of all metabolic activity in the body but also had privileged access to it on an "energy on demand" system. In addition to the accepted regulatory control of fat and glucose by the hypothalamus, Peters and his colleagues proposed a control system for ATP, with set points for optimal levels of ATP and tightly controlled feedback processes that responded to shortages by triggering an increase of glucose uptake across the blood–brain barrier. Such a system would ensure a near-constant supply of energy for neural activation, while causing temporary shortfalls in peripheral processes (and a resultant increase in nutritional intake).

Our understanding of the relationship between mental activity and brain energy has become more detailed, if still not entirely clear, with the advent of more sophisticated techniques (such as PET and fMRI) for measuring cerebral blood flow. These have allowed assessment of regional changes, rather than simply overall levels of metabolic activity. It now seems that, whereas mental activity has little impact on brain energy as a whole, it has been consistently shown to cause a shift toward brain areas that are currently relevant for task activity. However, the effects are quite small; a review by Raichle and Mintun (2006) indicated blood flow increases of no more than around 5–10 percent of resting levels. Furthermore, they argued, since glucose utilization in the regional response to task events involves a much lower rate of oxygen consumption than occurs for baseline activity, this corresponds to an increase of at most 1 percent in actual energy consumption by ongoing mental processes. Despite this, there has been a recent resurgence of the idea that brain glucose depletion may be the basis of fatigue, or at least mental work. For example, a number of studies have shown that executive-hungry activities reduce the level of peripheral blood glucose (e.g., Fairclough & Houston, 2004; Gailliot & Baumeister, 2007; Kennedy & Scholey, 2000). Since the brain stores almost no glucose, such results mean that there is a compensatory increase of blood flow to the brain to provide additional glucose for these demanding tasks. Under normal nutritional conditions, based on the selfish brain model (Peters et al, 2004), this would be an automatic process, based on the detected shortfall in ATP, though short-term deficits may occur. In such instances, it is possible that glycogen reserves, although small, may play a role in maintaining neural activation, and Gailliot (2008) has suggested that this may be the basis for effective executive processing.

Is Fatigue Energy Depletion?

Clearly, there is no simple relationship between energy and fatigue. Let us summarize what we know: brains use a lot of energy, but most of it is used to maintain the infrastructure, even at rest; mental activity may or may not increase the overall energy demands of the brain, but its impact is, at most, small; specific task demands attract energy costs to brain areas heavily involved in their execution, with changes of the order of 5–10 percent of baseline levels; tasks that make greater demands on executive processing consume more energy. Little in this list appears to tell us much of direct relevance to the *psychology* of energy and fatigue. First, there is no sign of what might be considered a "depletion" of energy. In fact, the selfish brain model argues that (under the normal range of nutritional conditions) it is not possible for the brain to be deprived of even a small fraction of its required level. (A more interesting possibility, however, is that a relatively high level of activity triggered by the ATP control loop may serve a signal function for the fatigue alerting mechanism; this is discussed in the next section). Second, mental fatigue is typically characterized by a period of degradation followed by a recovery to full function. In the brain energy literature, I could not find any observations of an energy refractory phase, or the time course of a presumed depletion and restoration process. Finally, there is nothing at all in this literature relating to one of the central tenets of a motivational control perspective – that the nature of the activity plays a major moderating role in the patterning of fatigue. Are the reported increases in brain glucose for high demand tasks still observed when tasks are self-chosen or highly controllable? Are losses of ATP across active brain areas smaller for preferred activities than for imposed tasks?

FATIGUE AND MOTIVATIONAL CONTROL

The third issue I see as a challenge for research on fatigue concerns its role in the effective control of motivation. Rather than simply being an unwanted by-product of overwork and energy depletion, fatigue may be considered as having an adaptive goal-directing function, helping in the management of motivational choices (Hockey, 2010). On this view, feelings of fatigue have a signal value, alerting the executive controller to the need to resolve conflicts between current goals and other possible courses of action. This conception of fatigue was recognized in early reviews of the problem. For example, Thorndike (1900) interpreted fatigue as a problem of doing the right thing, rather than of doing too much. Bartley and Chute (1947) considered fatigue the outcome of an ongoing conflict between competing behavioral tendencies: between doing and not doing; between doing one thing and doing another. Even for physical work, unless extreme, fatigue was not regarded as an inability to do work, but a lack of desire: "[an] attempt to retreat or escape from a situation" (Bartley & Chute 1947: 53). These early views acknowledged the volitional feature of the fatigue feeling, characterized by the reluctance, rather than the inability, to carry on with an activity. Bartley and Chute suggested that the experience of fatigue served as a warning to the performer of the need to escape from the situation and depended on their attitude to a task – whether or not it was desired.

Compensatory Control and Performance Protection

While fatigue from extended work may cause decrements in performance on externally imposed tasks, this is not usually the case for activities that engage the individual's concerns for the quality of their performance. As Bartlett (1943) and Davis (1946) demonstrated, effects of continued work on complex skills tend to take the form of a selective redistribution of attention rather than general decrement. Performance breakdown is hard to detect in real-life jobs where performance outcomes have major consequences for safety, or laboratory tasks that challenge self-esteem, or encourage the commitment of additional effort (Hockey, 1997; Kahneman, 1971). This compensatory control is often successful in protecting performance, though at the expense of a range of costs or latent decrements (Hockey, 1997): shortcuts in information-processing strategies, decrements in secondary task elements, increased physiological activation, and fatigue aftereffects. As is apparent with most interesting ideas in fatigue research, this has also been known for nearly 100 years (e.g., Dodge, 1917; Thorndike, 1912). Thorndike argued that threats to performance may be overcome by further effort, making performance an unsuitable index of decrement, but that fatigue could be inferred from measures of the reduced efficiency of work. Building on the use of this kind of approach in models of self-regulation (Carver & Scheier, 1982; Karoly, 1993) and performance management under stress (Broadbent, 1971; Kahneman, 1973; Teichner, 1968), Hockey's Compensatory Control Model (Hockey, 1993; 1997; 2005) emphasizes the need to examine not only primary task performance but also these indirect costs of regulatory control.

Effort, Fatigue and Goal Management

The Compensatory Control Model (CCM) was developed to account for the typically observed stability of performance under stress and high workload, but has recently been extended to accommodate the specific effects of fatigue (Hockey, 2010). The interpretation of fatigue as a factor in motivational control is based on its hypothesized role as a marker of cognitive conflict or discomfort. In the model this operates by sensing the strain of sustaining effective task output, and influencing a change of goal direction by promoting increased bias against current activities. The elements of the model are shown in Figure 3.1. Goal management is assumed to operate through a two-level negative feedback control system, in which performance protection is based on an executive decision to increase on-task effort and maintain performance whenever a state of strain (discrepancy between target performance and actual output) is detected. Such a strategy is costly in terms of its demands on regulatory activity (the latent decrements referred to above), so is adopted only when goals are regarded as important. Figure 3.1 makes explicit the assumption that executive selection of a specific task goal inevitably excludes many others, including several strong contenders for control of action. It shows the current goal (G) activated and maintained by executive bias, along with non-selected goals (g1, g2, etc.) competing for motivational control. Two executive functions are included: effort control and goal selection. An effort monitor detects strain in the routine control of performance (increasing difficulty in maintaining reliable output at the current effort setting), and triggers a decision-making process in the upper loop with a choice of two outputs: either to increase the effort budget (1), protecting performance from threatened impairment, but with increased costs; or to maintain the present level of effort (2), allowing performance to fail, but minimizing costs.

In fact, although we refer to this as an effort control loop, a sensed need for greater effort may be thought of as reflecting the same affective state as one of increased discomfort or fatigue. This state is also hypothesized to underlie the brief interruptions (gaps, blocks, lapses) in the flow of performance observed in continuous response tasks (Bills, 1931; Broadbent, 1963; Williams, Lubin & Goodnow, 1959). Such "phasic fatigue" effects have been shown to build up over several responses (Bertelson & Joffe, 1963), and typically followed by compensatory faster reactions, with the whole process increasing in frequency over time. Overall impairments are often prevented by this compensatory regulation of responses, and are routinely observed only under extreme circumstances, as with sleep deprivation (Van Dongen, Belenky & Krueger, 2011).

In Figure 3.1, the manifestation of fatigue as performance decrement is identified with the use of the low effort control option (2), and can be implemented in terms of goal selection in one of two ways; either (3) the current goal may be maintained, but at a lower level of performance, or (4) the goal may be displaced by one of the competing goals, based on concurrent values and costs of alternative actions. In either case, the feeling of fatigue is expected to dissipate, as the strain state is abandoned. However, the selection of the high-effort route (1) must still be understood as part of the fatigue process, since fatigue is identified with the requirement to sustain task output under strain. In that case, as Thorndike (1912) argued, we need to look for evidence of fatigue in the "collateral damage" – the costs of compensatory activity. Of course, while CCM offers an interesting and testable alternative to the conception of fatigue as a general loss of energy or resources, it is somewhat speculative. The challenge for future work is to understand the factors that influence goal selection options on a moment-by-moment basis. What determines whether a current externally imposed task will stay in place with extended time on task, or be executed at a lower level, or displaced by something else? The effect of time is clearly fundamental; why should maintenance of a goal become more difficult the longer it has been in place? It is likely that goals lose activation with use and need to be refreshed to

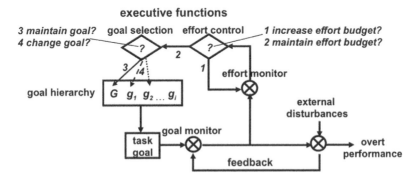

Figure 3.1   Hockey's (1997) compensatory control model of task management under stress and demand adapted to illustrate the mechanism for fatigue effects on goal maintenance: see text for explanation

maintain dominance over competing goals. Why does a high level of controllability help? One reason may simply be that strain from routine disturbances can be minimized by flexible changes in task scheduling or more effective timing of effort, to coincide with peak executive function (Hockey & Earle, 2006; Hockey, Wastell & Sauer, 1998).

A related challenge is to uncover possible neuroanatomical processes that may underpin goal management and fatigue. While space does not permit even a brief discussion of this burgeoning field, a few relevant pointers should be mentioned. The central role of the prefrontal cortex (PFC) in executive function and regulatory control has been well established (Miller & Cohen, 2001), while the anterior cingulate cortex (ACC) is now known to be a major centre for monitoring of conflicts between alternative actions and their execution (Botvinick, Cohen & Carter, 2004), with a feedback loop linking monitoring and control functions. But what determines whether goals are maintained or changed? And what drives effort? Boksem and Tops (2008) suggested that the patterning of fatigue effects depends on the evaluation of changes in the costs-benefits trade-off of alternative actions, shown to occur in ACC (Walton, Kennerley et al., 2006), through dopaminergic projections between midbrain and precortical sites. Effort-based activities may be pursued only if rewards are sufficiently high, via activation of goal maintenance activity in PFC (Floresco & Magyar, 2006). As mentioned in the previous section, there may also be a specific role for energy transactions in the signalling of fatigue, associated with ATP levels dropping below set point limits and triggering an increase in glucose oxidation (Peters et al., 2004). This is clearly a major growth area. While little of the current research effort into the neuroanatomy of attention and executive functions has been directly focused on mechanisms of fatigue, it nevertheless indicates a real promise for understanding the neural basis for fatigue phenomena, particularly in relation to motivational control and goal, management.

VARIETIES OF FATIGUE: ONE PROCESS OR SEVERAL?

A final challenge for research is to determine whether fatigue is a unitary process, applicable to all kinds of activity, or a portmanteau term that refers to several rather different psychophysiological states. Are mental fatigue and physical fatigue two different things or are they manifestations of the same state? And what about fatigue from sleep disruption? The problem has a very long history and like many other aspects of fatigue, remains unresolved and not systematically explored.

Mental Fatigue

Even considering just mental fatigue (associated with sustained cognitive task performance and heavy workload), it is not clear whether a single fatigue state applies to all conditions. For example, is the response to heavy demands over short periods (workload decrements) the same as that to lighter demands over long periods (vigilance-type decrements)? This has sometimes been characterized as the difference between overload and underload, though it may be now thought to be a superficial distinction; recent analyses (e.g., Grier et al., 2003; Smit, Eling & Coenen, 2004) strongly suggest that the vigilance decrement is caused by the effort demands of maintaining active attention, rather than the lack of engagement usually assumed. Then there is emotional fatigue, resulting from situations requiring strong and sustained responses involving emotions; such a requirement may be considered to make demands on a different set of resources from those used by cognitive tasks. There may be important distinctions between these different paradigms that can throw light on the details of the adaptive mechanisms for responding to environmental demands; certainly, a case may be made for a program of work that examines possible interactions between factors that challenge two or more hypothesized processes. At the present time, while there are clearly many differences of origin and expression between these different sorts of mental activity, there do not seem to be any strong reasons for separating them in terms of their resulting fatigue states. In all cases, the limiting factor may be argued to be the cognitive strain that comes from the need to manage goals actively (at an executive level) for sustained periods.

Physical Fatigue

Whether or not there is but a single kind of mental fatigue, many would argue that there is a need to distinguish mental fatigue from physical fatigue, not least on the grounds that they appear to have quite different origins – in mental or muscular work – though Bartley and Chute (1947) found no reason to consider them essentially different. Based largely on procedures in which participants exercise to exhaustion, models of exercise endurance have traditionally assumed that fatigue is a direct result of a metabolic limitation in the exercising muscles – when the rate of ATP depletion by the muscles exceeds that being produced. In an extensive series of articles, Noakes and his colleagues (e.g., Noakes, 2008) have criticized the traditional hypothesis on a number of grounds; for example, that near-maximal levels of cardiac output are never reached during prolonged exercise, and that knowledge of the length of exercise session optimizes endurance performance. They propose an alternative view, based on extensive data from pacing and management of energy during natural endurance running. Their "central regulator" hypothesis argues that muscle activity is managed by a brain strategy for energy management, built up over many years of exercise and feedback from exercise. The central governor is calibrated to plan effort and pacing/speed profiles, and to prevent catastrophic breakdown. The subjective experience of fatigue is argued to serve an adaptive function, acting as a warning to slow down or stop, as we have also argued for mental work. It is assumed to derive from the discrepancy between the current rate of energy use and the projected long-term time–energy profile, possibly as indexed by reduced neural recruitment of muscle fibers. In other words, the pacing of exercise is a self-limiting cognitive process, stopped by the occurrence of the fatigue feeling.

The model has been criticized by some as not allowing any possibility of catastrophic breakdown, since such problems are known to occur occasionally, for example when athletes are very highly motivated (Esteve-Lanao et al., 2008). However, there is no reason why a central regulator model could not be extended to allow the possibility of the individual overriding fatigue, as commonly happens during highly motivated mental work. (It is likely that the homeostatic set points are similarly conservative in the two cases). Regulatory breakdowns may occur if the pacing model is not well developed through experience, or if cognitive capacity is impaired. In any case, the central governor model provides a promising link between physical and mental task management.

It also fits well with the many demonstrations that the limiting condition for physical endurance is a cognitive one – the willingness to overcome resistance to further effort or pain (Holding 1983). However, more sophisticated cognitive methods may be required to test the hypothesis that both physical and mental fatigue depend on the same mechanism for assessing effort demands, through involvement of the executive control system – for example, this would predict an increasing loss of pacing control when athletes are required to respond to greater cognitive demands.

Sleep Fatigue

There is, perhaps, an even stronger case to be made for distinguishing all kinds of work-based fatigue from that associated with sleep disturbances, since they are caused not by doing work but by the prevention of a fundamental sleep need. However, many sleep researchers treat sleep deprivation as the core model of mental fatigue and assume a direct link between sleep control mechanisms and mental fatigue (e.g., Dawson & McCulloch, 2005). It is now accepted that sleep behavior is controlled by two (sometimes three) processes: a sleep homeostat that produces an increasing drive for sleep with time spent awake, and a circadian regulator that aims for wakefulness peaks during the middle of the day. The additive combination of these two processes provides the basis for models of sleep and wakefulness that have been highly effective in predicting levels of alertness and fatigue (and also performance) in jobs affected by shift work or demands for extended working time. Many models (e.g., Åkerstedt & Folkard, 1995; Achermann & Borbély, 1994) now include a transient third component, sleep inertia, reflecting the time required to attain full alertness after waking. Oddly, however, very few models include work hours as an input variable (Mallis et al., 2004), and none make any formal reference to the nature of the work or task activity. On this kind of model, a very busy day would be expected to have no greater impact on alertness and fatigue than a relaxing day with no external demands. Yet, time on-task effects are known to interact with sleep fatigue factors, so that, for example, performance decrements over time on-task are greater when sleep drive is high and the need for wakefulness low (Van Dongen, Belenky & Krueger, 2010). Such effects demonstrate that both subjective fatigue and its effects on performance are a function of both task and sleep variables.

So why are work variables not included in models of sleep and alertness? One reason may be that they are more difficult to measure reliably, but this is a weak argument. A more pertinent question is how much difference this extra predictor would make to the effectiveness of the alertness index, since standard models routinely account for between 60 percent and 95 percent of the variance (Åkerstedt & Folkard, 1995; Fletcher & Dawson, 2001). However, the models aim to predict normative or group levels of alertness, whereas, from a practical point of view, it is often necessary to identify performance vulnerability at an individual level. It seems likely that predictions would benefit considerably from including work variables, partly since the between person variability of response is probably greater than with sleep and circadian factors. A rather separate question is whether mental fatigue associated with sustained task performance and fatigue from disruption of sleep and circadian rhythm are connected in any direct way. There is currently no clear evidence on this, but Van Dongen et al. (2010) have suggested that they involve the same neurobiological pathways, with local neuronal assemblies switched off by continual use (both through the day and during sustained tasks). Another relevant suggestion (Bennington & Heller,

1995) is that the normal restorative function of sleep is to restore brain glycogen stores, so that sleep deprivation may impair performance by preventing maintenance of the backup energy supply. While this hypothesis has still not been confirmed, it is consistent with the finding (Porkka-Heiskanen et al., 1997) that sleep deprivation leads to a drop in ATP, acting as a signal for sleep. Once again, this promises to be a major growth point in understanding the relationship between different kinds of fatigue.

CONCLUSIONS

In this chapter, I have examined four topics that may be considered among the greatest challenges for future research on fatigue and the management of performance demands. In all of them, there are major issues that not only confront accepted dogma but also offer real opportunities for enhanced understanding.

Theme (1) concerned the work–fatigue hypothesis. It is now clear that fatigue cannot be considered to be caused by work per se, but by having to do imposed work – *tasks* – that, if not quite resented, are at least not those that one would choose to do. And the fatigue that is caused by this kind of work is more a feeling of discomfort or resistance to further effort than a state of incompetence or inability to perform. However, much more needs to be done to pin down the conditions under which fatigue develops and dissipates. As is discussed in theme (2), it is also unlikely that fatigue can be regarded as a state of low energy, at least in any simple form. Although it has always been characterized in terms of energy depletion, there is no strong evidence to support such a view. Under mental task conditions relative blood flow reflects which brain structures are required for processing, but the overall level of available brain energy appears to be largely independent of what mental activity is taking place. There are, however, indications that some aspects of brain energy regulation may have a signal function, perhaps promoting the sensation of tiredness. What is needed to clarify this is research that specifically examines brain energy metabolism in relation to variables that maximize or minimize the fatigue state; in particular, tasks that vary in terms of external versus personal goals. A more likely framework for understanding fatigue, as is discussed in theme (3), is that it serves a motivational control function, directing attention to the unfavorable cost–benefit trade-off associated with activities low in the goal hierarchy, and making a switch of goal possible. To date, there is very little data of direct relevance to tests of this interpretation, though several lines suggest themselves: for example, determining the factors that best predict changes of goal under fatigue conditions, and the relation of these to underlying personal goal hierarchies. The final theme, (4), examined the differences and similarities between the various putative forms of fatigue and concluded that there is no fundamental basis for separating mental and physical fatigue. Sleep-based fatigue may have a different origin, and possibly a different function, though, even here, there appear to be a number of points of convergence. Again, there is little direct research on the problem. It would be informative, for example, to examine the interaction between different forms of fatigue, and effects of pre-loading on one set of fatiguing conditions before the introduction of a second.

In any case, what I have tried to indicate is that there is a lot for a future program of research on fatigue and performance to do. I started by saying that despite the long history of research on the subject, we still do not know very much about it. I reiterate this claim here, and argue that a more focused approach, targeting issues such as these, is clearly needed – and long overdue.

REFERENCES

Achermann, P. & Borbély, A.A. (1994). Simulation of daytime vigilance by the additive interaction of a homeostatic and a circadian process. *Biological Cybernetics*, 71, 115–21.

Åkerstedt, T. & Folkard, S. (1995). Validation of the S and C components of the three-process model of alertness regulation. *Sleep*, 18, 1–6.

Bartlett, F.C. (1943). Fatigue following highly skilled work. *Proceedings of the Royal Society of London* (B), 131, 247–57.

Bartley, S.H. & Chute, E. (1947). *Fatigue and Impairment in Man*. New York: McGraw-Hill.

Bennington, J.H. & Heller, H.C. (1995). Restoration of brain energy metabolism as the function of sleep. *Progress in Neurobiology*, 45, 347–60.

Bertelson, P. & Joffe, R. (1963). Blockings in prolonged serial responding. *Ergonomics*, 6, 102–15.

Bills, A.G. (1931). Mental work. *Psychological Bulletin*, 28, 505–32.

Binet, A. & Henri, V. (1898). *La Fatigue intellectuelle*. Paris: Schleicher.

Boksem, M.A.S. & Tops, M. (2008). Mental fatigue: Costs and benefits. *Brain Research Reviews*, 59, 125–39.

Botvinick, M.M., Cohen, J.D. & Carter, C.S. (2004). Conflict monitoring and anterior cingulate cortex: An update. *Trends in Cognitive Sciences*, 8, 539–46.

Broadbent, D.E. (1963). Differences and interactions between stresses. *Quarterly Journal of Experimental Psychology*, 15, 205–11.

Broadbent, D.E. (1971). *Decision and Stress*. London: Academic Press.

Brown, A.M. (2004). Brain glycogen re-awakened. *Journal of Neurochemistry*, 89, 537–52.

Carver, C.S. & Scheier, M.F. (1982). Control theory: A useful conceptual framework of personality – social, clinical and health psychology. *Psychological Bulletin*, 92, 111–35.

Clarke, D.D. & Sokoloff, L. (1994). Circulation and energy metabolism of the brain, in G. Siegel, B. Agranoff, R.W. Albers, & P. Molinoff (eds), *Basic Neurochemistry: Molecular, Cellular, and Medical Aspects*. 5th ed. New York: Raven Press, 645–80.

Csikszentmihalyi, M. (1990). *Flow*. New York: Harper.

Davis, D.R. (1946). This disorganization of behaviour in fatigue. *Journal of Neurology, Neurosurgery and Psychiatry*, 9, 23–29.

Dawson, D. & McCulloch, K. (2005). Managing fatigue: It's about sleep. *Sleep Medicine Reviews*, 9, 365–80.

Dodge, R. (1913). Mental work: A study in psychodynamics. *Psychological Review*, 20, 1–42.

Dodge, R. (1917). The laws of relative fatigue. *Psychological Review*, 14, 89–113.

Esteve-Lanao, J., Lucia, A., deKoning, J.J. & Foster, C. (2008). How do humans control physiological strain during strenuous endurance exercise? *PLoS ONE 3: e2943. doi:10.1371/journal.pone.0002943*.

Fairclough, S.H. & Houston, K. (2004). A metabolic measure of mental effort. *Biological Psychology*, 66, 177–90.

Fletcher, A. & Dawson, D. (2001). A quantitative model of work-related fatigue: Empirical evaluations. *Ergonomics*, 44, 475–88.

Floresco, S.B. & Magyar, O. (2006). Mesocortical dopamine modulation of executive functions: Beyond working memory. *Psychopharmacology*, 188, 567–85.

Frankenhaeuser, M. (1986). A psychobiological framework for research on human stress and coping, in M.H. Appley & R. Trumbull (eds), *Dynamics of Stress: Physiological, Psychological and Social Perspectives*. New York: Plenum.

Gailliot, M.T. (2008). Unlocking the energy dynamics of executive functioning linking executive functioning to brain glycogen. *Perspectives on Psychological Science*, 3, 245–63.

Gailliot, M.T. & Baumeister, R.F. (2007). The physiology of willpower: Linking blood glucose to self-control. *Personality and Social Psychology Review*, 11, 303–27.

Ganster, D.C. (1989). Worker control and well-being: A review of research in the workplace, in S.L. Sauter, J.J. Hurrell & C.C. Cooper (eds), *Job Control and Worker Health*. New York: Wiley, 3–24.

Grier, R.A., Warm, J.S., Dember, W.N., Matthews, G., Galinsky, T.L., Szalma, J.L. & Parasuraman, R. (2003). The vigilance decrement reflects limitations in effortful attention, not mindlessness. *Human Factors*, 45, 349–59.

Hockey, G.R.J. (1993). Cognitive–energetical control mechanisms in the management of work demands and psychological health, in A.D. Baddeley & L. Weiskrantz (eds), *Attention, Selection, Awareness and Control: A Tribute to Donald Broadbent*. Oxford: Oxford University Press.

Hockey, G.R.J. (1997). Compensatory control in the regulation of human performance under stress and high workload: a cognitive–energetical framework. *Biological Psychology*, 45, 73–93.

Hockey, G.R.J. (2005). Operator functional state: the prediction of breakdown in human performance, in J. Duncan, P. McLeod, & L. Phillips (eds), *Speed, Control and Age: In Honour of Patrick Rabbitt*. Oxford: Oxford University Press.

Hockey, G.R.J. (2010). A motivational control theory of cognitive fatigue, in P.L. Ackerman (ed.), *Cognitive Fatigue: Multidisciplinary Perspectives on Current Research and Future Applications*. Washington, DC: American Psychological Association.

Hockey, G.R.J. & Earle, F. (2006). Control over the scheduling of simulated office work reduces the impact of workload on mental fatigue and task performance. *Journal of Experimental Psychology: Applied*, 12, 50–65.

Hockey, G.R.J., Payne, R.L. & Rick, J.T. (1996). Intra-individual patterns of hormonal and affective adaptation to work demands: An n = 2 study of junior doctors. *Biological Psychology*, 42, 393–411.

Hockey, G.R.J., Wastell, D.G. & Sauer, J. (1998). Effects of sleep deprivation and user-interface on complex performance: A multilevel analysis of compensatory control. *Human Factors*, 40, 233–53.

Holding, D.H. (1983). Fatigue, in G.R.J. Hockey (ed.), *Stress and Fatigue in Human Performance*. Chichester: Wiley.

Kahneman, D. (1971). Remarks on attentional control, in A.F. Sanders (ed.), *Attention and Performance*, III. Amsterdam: North Holland.

Kahneman, D. (1973). *Attention and Effort*. Englewood Cliffs, NJ: Prentice-Hall.

Kaplan, S. (2001). Meditation, restoration, and the management of mental fatigue. *Environment and Behavior*, 33, 480–506.

Karasek, R.A. & Theorell, T. (1990). *Healthy Work*. New York: Basic Books.

Karoly, P. (1993). Mechanisms of self-regulation: A systems view. *Annual Review of Psychology*, 44, 23–52.

Kennedy, D.O. & Scholey, A.B. (2000). Glucose administration, heart rate and cognitive performance: Effects of increasing mental effort. *Psychopharmacology*, 149, 63–71.

Kety, K.S. & Schmidt, C.F. (1945). The determination of cerebral blood flow in man by the use of nitrous oxide in low concentrations. *American Journal of Physiology*, 143, 53–66.

Mallis, M.M., Mejdal, S., Nguyen, T.T. & Dinges, D.F. (2004). Summary of the key features of seven biomathematical models of human fatigue and performance. *Aviation, Space and Environmental Medicine*, 75, A4–A14.

Miller, E.K. & Cohen, J.D. (2001). An integrative theory of prefrontal cortex function. *Annual Review of Neuroscience*, 24, 167–202

Noakes, T.D. (2008). Testing for maximum oxygen consumption has produced a brainless model of human exercise performance. *British Journal of Sports Medicine*, 42, 551–5.

Peters, A., Schweiger, U., Pellerine, L., Hubolda, C., Oltmannsb, K.M., Conrad, M., Schultes, B., Bornd, J. & Fehm, H.L. (2004). The selfish brain: Competition for energy resources. *Neuroscience and Biobehavioral Reviews*, 28, 143–80.

Porkka-Heiskanen, T., Strecker, R.E., Thakkar, M., Bjørkum, A.A., Greene, R.W. & McCarley, R.W. (1997). Adenosine: A mediator of the sleep-inducing effects of prolonged wakefulness. *Science*, 276, 1265–68.

Rabinbach, A. (1990). *The Human Motor*. Berkeley, CA: University of California Press.

Raichle, M.E. & Mintun, M.A. (2006). Brain work and brain imaging. *Annual Reviews of Neuroscience*, 29, 449–76.

Smit, A.S., Eling, P.A.T.M. & Coenen, A.M.L. (2004). Mental effort causes vigilance decrease due to resource depletion. *Acta Psychologica*, 115, 35–42.

Sokoloff, L., Mangold, R., Wechsler, R., Kennedy, C. & Kety, S.S. (1955). The effect of mental arithmetic on cerebral circulation and metabolism. *Journal of Clinical Investigation*, 34, 1101–8.

Teichner, W.H. (1968). Interaction of behavioral and physiological stress reactions. *Psychological Review*, 75, 51–80.

Thorndike, E.L. (1900). Mental fatigue. *Psychological Review*, 7, 466–82.

Thorndike, E.L. (1912). The curve of work. *Psychological Review*, 19, 165–94.

Thorndike, E.L. (1914). *Educational Psychology, Vol. 3: Work and Fatigue, Individual Differences and their Causes*. New York: Teachers College Press.

Van den Berg, C.J. (1986). On the relation between energy transformations in the brain and mental activities, in G.R.J. Hockey, A.W.K. Gaillard, & M.G.H. Coles (eds), *Energetics and Human Information Processing*. Dordrecht, NL: Martinus Nijhoff.

Van Dongen, H.P.A., Belenky, G. & Krueger, P.M. (2010). Investigating the temporal dynamics and underlying mechanisms of cognitive fatigue, in P.L. Ackerman (ed.), *Cognitive Fatigue: Multidisciplinary Perspectives on Current Research and Future Applications*. Washington, DC: American Psychological Association.

Wall, T.D., Jackson, P.R., Mullarkey, S. & Parker, S.K. (1996). The demands–control model of job strain: a more specific test. *Journal of Occupational and Organizational Psychology*, 69, 153–67.

Walton, M.E., Kennerley, S.W., Bannerman, D.M., Phillips, P.E.M. & Rushworth, M.S.F. (2006). Weighing up the benefits of work: Behavioral and neural analyses of effort-related decision making. *Neural Networks*, 19, 1302–14.

Williams, H.L., Lubin, A. & Goodnow, J.L. (1959). Impaired performance with acute sleep loss. *Psychological Monographs*, 73, 1–26.

PART II

The Nature of Fatigue

4

Conceptualizing and Defining Fatigue

P.A. Hancock, Paula A. Desmond, and Gerald Matthews

INTRODUCTION

In one of the classic and, indeed, seminal texts in all of psychology, William James in his *Principles of Psychology* (1890) sought to define the concept of attention. In so doing, he appealed to the commonalty of social experience, observing that:

> Everyone knows what attention is. It is the taking possession by the mind, in clear and vivid form, of one out of what seem several simultaneously possible objects or trains of thought. Focalization, concentration, of consciousness are of its essence. It implies withdrawal from some things in order to deal effectively with others. (403–4)

When we seek to define the concept of fatigue, we are faced with the same problem, for fatigue, like attention, is another of the energetic states of the organism that everyone has experienced.[1] By this standard, a definition of fatigue ought to be relatively simple to construct. However, like attention, fatigue and its definition hide a wealth of hidden problems. If we are not willing to engage these complexities we can generate a definition of the order of James' pronouncement and then have done with the issue. Using this logic, fatigue could be defined as follows:

> Everyone knows what fatigue is. It is the taking possession of the mind by a sense of lassitude. It is a reduction in the capacity and desire to react. It is characterized by tiredness and an aversion to the continuation of goal-directed work. It is accompanied by a strong desire for rest through the cessation of ongoing activity.

Consistent with James, this level of definition is largely descriptive of the subjective symptomatology of the phenomenon. It appeals to everyone's summed, collective experience and provides alternative semantic descriptors of the fatigue experience. To a degree, this level of descriptive definition is sufficient and useful. It serves to inaugurate

---

1   It could indeed be that there is only one energetic state and that attention and fatigue are simply facets of the same fundamental characteristic. This is a potentiality we explore later.

discussion and acts as a basis for subsequent topic development. However, it is clearly limited in its scope and, thus, its eventual utility. While some form of acceptable definition is needed at this surface level, we have to dive beyond the surface of these shallow waters to plumb the depths of what still remains, now into the twenty-first century, a most complex and still poorly understood facet of human response capacity.

MUSCIO'S PARADOX

It would be wrong, even by omission, to suggest that this issue of fatigue definition is not one that has plagued researchers for more than a century and a half (see Bartley & Chute, 1947; Bitterman, 1944; Broadbent, 1979; Ryan, 1944). Since the earliest formal empirical explorations of the fatigue state and, indeed, during the centuries of informal literature before the rise of experimental science, people have been very aware of what it is to be fatigued (Bartley, 1976).[2] However, for the sake of the present work, we can start our own inquiry here with the foundational paper of Muscio (1921). For it was Muscio who faced this challenge head on, although the answer he provided, while sound in and of itself, has never been accepted. This is because Muscio found such flaws in the basis of the definition and indeed the very word "fatigue" itself, that he suggested we simply abandon the term altogether (see Landauer & Cross, 1971). However, by the very presence of the text you are currently holding and those of other recent allied efforts (see Chapter 6 in this volume by Ackerman, Calderwood & Conklin), it is evident that Muscio's recommendation has never been enacted.[3]

It is, however, important to follow Muscio's reasoning carefully. In essence, what he expressed was a conundrum that bedevils all of science, and not merely the energetic aspects of the psychological dimensions of behavior alone. However, it is in this realm that we are considering here that the argument is very much in evidence. Muscio opined that in order to measure fatigue (i.e., to create a fatigue test), we must have a clear and unequivocal definition of what fatigue actually is. Lacking this definition, we cannot be sure what the nominal tests of fatigue are actually measuring. However, none of the definitions Muscio could find at that time, and arguably up to the present, permit sufficient quantitative definition so that a test can be created and validated. This remains Muscio's paradox with respect to fatigue. Here, we have to ask whether we are in any better scientific shape to answer this riddle today than we were almost a century ago. Obviously, advances in neuroscience provide us with a more likely avenue to resolve this concern, but whether our answer is yet sufficient remains in abeyance. Rather than cycle through the well-trodden debates or rehearse the arguments that have more recently been aired more articulately by others (see Ackerman, Calderwood & Conklin, Chapter 6 in this volume), we take a differing perspective and base our approach to the question of fatigue on the process of evolution.

---

2 See http://www.websters-online-dictionary.org/fa/fatigue.html
3 One of the central problems is that by giving the phenomenon a name, "fatigue," a certain power and impetus is created by the very naming process. As the philosopher Hume was very much aware, giving something a name provides it with an identity all of its own. However, beyond this, fatigue is such a ubiquitous experience that even if we do not use this particular title, we have to use some label to refer to the state or collection of states in which our capacity to continue to exert voluntary effort is inhibited, curtailed or extinguished. Muscio's aspiration was defeated by these forces, which continue very much to exert their influence today.

EVOLUTION AND THE NATURE OF FATIGUE

To begin to effectively address the issues around the conceptualization and definition of fatigue, it is our contention that we have to approach the subject from the perspective of evolution. It is critically important to see the issue in terms of the evolutionary imperatives faced by the exposed organism and in this specific case by human beings. Since there remains a persistent and, one might say, enshrined division between physical and cognitive fatigue (see Ryan, 1944), our discussion will begin from these two separate perspectives. However, to pre-empt our own conclusion, we look in our summary to provide a more unified account that does not rely on what we see as an artificial division. Physical fatigue begins with an examination of the muscle and the way in which it stores, recruits, recreates and uses energy. Repeated contraction serves to quickly use up resident stores of energy and mobilizes the process by which replacement fuel can be secured. In an isolated culture, muscle tissue will quickly exhaust its capacity to respond and the profile of this diminution of capacity is well documented. In the living organism, the picture is somewhat modified by the capacity for replacement, but continuing activity requires the search for, and securing, of external sources of energy to enter the metabolic processes to sustain ongoing capacity. It might be thought that muscles work up to the limit of their exhaustion of this metabolic energy; however, this is not the case. For this is where the artificial division between physical and cognitive fatigue begins to falter. It is not the muscle that serves as the most common rate-limiting element in physical fatigue, but, rather, the brain. Long before muscular exhaustion is reached, the subjective propensity to suspend physical effort is engaged and the individual will often "give up," while still claiming to be physically exhausted. This threshold, however, is itself a protective mechanism engaged by evolution in order to protect the organism from damaging physical injury and to leave energy in reserve if unexpectedly faced with danger. Although this fatigue "barrier" can be ignored, it courts the threat of serious, long-term muscular damage to do so and thus reduced fitness to respond. Ignoring such protective warnings can only be justified in extremis. The paradox here is that physical fatigue is largely and most often cognitive in nature.

Cognitive fatigue is often thought to be a very different order of phenomenon, but we would argue that this is not the case. As the muscle is the organ of physical action, so the brain is the organ of cognitive action and each utilizes highly similar response strategies, especially in terms of energy consumption. Brain and muscle both require fuel for their ongoing activity and the fuel in each case derives, at least to some degree, from glucose.[4] The body stores this energy in a form that allows immediate access, but such stores, as we have seen in the case of muscular response, are rapidly depleted. For either physical or cognitive activity to persist there exists an immediate need to begin to access an ongoing supply. To the degree that a steady state of physical action or cognition can be established, the pipeline of necessary energy must be kept sufficiently open and effective. However, high levels of transient, acute demand or, indeed, demand above the level of a sustainable steady state, will begin to erode levels of both physical and cognitive response, edging dangerously toward failure. The questions posed by evolution concern where such levels of acute demand occur. This brings us back to our evolutionary imperatives and the very evolution of fatigue itself.

---

4   For this stage of the discussion we have not included the ongoing need for oxygen in order to complete the metabolism of this glucose-based energy.

FATIGUE IN EARLY HUMAN SOCIETIES

Early human societies were formed of hunter-gatherer groups. In large part because of the necessities of birthing and nursing, hunting and gathering functions were divided between the sexes. As a sufficient generalization for our present concerns, males hunted and females gathered.[5] Thus, the imperatives of task demand imposed upon the respective sexes provided somewhat differing physical and cognitive challenges. Males proved to be physically larger and thus able to exert greater muscular effort over short intervals of time. Females are more resilient to extended demands of deprivation and are preserved by nature because of the relative value of eggs over sperm.[6] The genesis of human fatigue must be couched in terms of these early constraints, although, of course, the evolutionary genesis of fatigue itself well precedes the development of the human species. Physical effort and its associated fatigue is a straight energy equation (Burton, 1934). Human beings (like other homeotherms) maintain the mobility advantage of a constant internal body temperature only at a tremendous cost in terms of finding and consuming calories (energy). It is essential to understand fatigue in terms of this critical energy economy. When an individual goes out to hunt (regardless of what they are hunting), they are expending an enormous amount of energy in the hope and expectation of a return on their investment. This is one reason why hunting is usually for meat in terms of animal targets, the calorific value of meat being so relatively high and thus, the rate of return even for a single "kill" is most often energetically profitable. Gathering is more sedentary by nature, but it has to be, since the return on calorific investment is relatively low. Actions of various members of the animal kingdom reflect these relative rates of return in terms of their carnivore, omnivore, and herbivorous natures. However, can we think our way around some of these intrinsic trade-offs? The answer for humans is certainly yes and, indeed, this is what the human brain has been designed to do (see Hancock, 2010).[7]

Brains are very expensive. They are expensive to create in terms of the calories used by the mother during gestation. They are expensive in terms of the energy it takes to invest in the child prior to its full development, and they are expensive to operate even after they have reached full maturation. These observations imply that the advantage of brains must be extensive given the energy investment made in their creation, maturation, and operation. Indeed, the brain takes approximately one-third of the resting rate of calorific expenditure of the body even when apparently doing "nothing." This level of demand obviously increases with ascending levels of mental work. This investment must be repaid somehow, since evolution is very intolerant of failure. The function of the brain in metabolic terms is to provide greater return on investment. That is, increased brain functionality repays itself by generating superior strategies to secure food sources more efficiently and more regularly. In terms of fatigue, investment in cognitive energy obviates the need for wasted physical energy. In short, it pays the organism to think its way around

---

5   This is not to say that subsequent fiction (e.g., Jean M. Auel's *Clan of the Cave Bear*) has not sought to rewrite even this aspect of history.
6   These differentiates can still be seen at the extremes of physical effort in terms of world records for self-propelled locomotion (e.g., running, swimming, etc.). For distances up to many tens of miles, males outperform females. However, at extended distances well beyond these distances and times it is eventually the female that dominates. In the end, the female will survive under the most adverse conditions, while the male often perishes. The account of the Donner party of American pioneers of 1846–1847 is a most interesting example, although there are issues of social role and expectation that overlay survival in such cases.
7   In a true sense the brain has not been designed, since this implies the teleological nature of a "designer." In more evolutionary terms, this is the niche that the human species has managed to exploit.

the energy balance equation. Indeed, as we have seen, one primary function of cognitive protection is to ensure the organism does not damage itself by going over the edge of failure due to continued energy expenditure in the latter phases of a physical fatigue state. In this way, we can see that cognitive fatigue is a later addition to the spectrum of fatigue in evolutionary terms.

CONCEPTUALIZATIONS OF FATIGUE

Now from this basis of understanding we can work toward an initial definition for fatigue based on an evolutionary and physiological perspective. Thus, fatigue is *a lack of sufficient steady state energy to power physical and/or cognitive work*. Now, of course, we have to hedge this definition around with the necessary caveats. The first caveat is that our primary encounter with fatigue is a subjective apperception of self-state. This perception is itself triggered by the insufficiency of steady-state energy. As a subjective experience, it is itself susceptible to all the nuances and subtleties of individual differences. Thus, people will not react uniformly to the same external, objective series of demands. The second caveat is that to a degree, the individual can ignore fatigue. That is, continued, goal-directed effort can be exerted in certain circumstances, especially those that threaten the individual's ongoing survival. In short, in extremis, the subjective threshold of fatigue can be fractured (or at least manipulated). As was noted earlier, this might serve to induce physical injury, but the body in particular and evolution in general tend to favor injury over death. The threshold is also amenable to change contingent on the appraisal of the nature of the task to be performed. In reality, this is a subset of the survival issue, but it is important, since varying levels of task involvement will result in different reports of fatigue onset. Thus, we would anticipate that task appraisal will be an important factor in the onset and sustenance of the fatigue experience. It is also clear that since fatigue represents only one of the many energetic facets of performance, there will be necessary connections between fatigue and other energetic dimensions of human behavior. In the present case, not unnaturally, we wish to relate fatigue to the notion of stress (see Hancock & Desmond, 2001). In general, we believe that fatigue can be considered a chronic form of stress. Thus, models and theories that describe stress should also apply directly to an understanding of the conceptualization of fatigue. Again, not unexpectedly, we look to the formulation of Hancock and Warm (1989) to provide commonalty and insight here. This can be had through a comparison with some of the basic foundations of this latter model.

FATIGUE AND STRESS

Hancock and Warm described stress as residing in three possible locations as seen in Figure 4.1. Here, we can use this formulation to describe fatigue. In terms of input, fatigue can be described by the driving conditions of the situation in which individuals find themselves. The prime example of input is time on task, which has and continues to be one of the major drivers of fatigue (see Ackerman, Calderwood & Conklin, Chapter 6, this volume). Of course, the individual must be intrinsically or extrinsically invested in the task. That is, the person must want to continue to perform with increasing time on task. This want or desire must either come from within the individual or be imposed by an external agency, such as occurs in slavery, prison hard labor, military discipline,

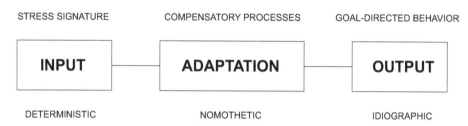

STRESS SIGNATURE   COMPENSATORY PROCESSES   GOAL-DIRECTED BEHAVIOR

| INPUT | — | ADAPTATION | — | OUTPUT |

DETERMINISTIC   NOMOTHETIC   IDIOGRAPHIC

Figure 4.1   The three loci of stress (after Hancock & Warm, 1989)

etc. Absent this want or desire and the task-associated fatigue is totally dissipated. The analogous case with the dissolution of task-related mental workload has been made by Hancock and Caird (1993). This latter connection indeed shows how mental workload is itself simply another face of the collective energetic response of the individual. Fatigue can also be expressed in the second locus shown in Figure 4.1, the adaptive process. Here we see the function of appraisal and the subjective apperception of the individual. For a task that is boring, repetitive and adverse, the onset of fatigue through time on task is very quick. In contrast, for a task appraised to be pleasant and hedonomic (see Hancock, Pepe & Murphy, 2005) or with a task with low cyclic frequency, the onset of fatigue is much postponed. However, as St. (Sir) Thomas More pointed out in his essay "Utopia," even the most appealing and nominally pleasant of activities eventually palls with repetition over time. The latter issue is one that has concerned theologians worried about the pleasures of heaven which have to be played out presumably over eternity. Thus, there is a limit even to what we might initially perceive as pleasure before it too becomes fatiguing and eventually aversive!

The final locus shown in Figure 4.1 is output. Here we return once again to the vast literature on fatigue that has looked extensively at the diminishment of productivity over time.[8] Without rest and respite, human productivity on an unvarying task declines over time. It is not a simple monotonic decrease, since human beings are cyclic in their capacities. Over periods of days, humans express circadian variation and there are also intrinsic ultradian and infradian rhythms embedded in the more dominant circadian cycle. These characteristics lead to apparent paradoxes in which individuals will decrease in capacity for many hours and then subsequently improve as upturns in cyclic capacities are expressed (Dawson & Reid, 1997). However, despite these transient increases, the general trend across time will be toward a reduction in response unless some period of respite is provided. This pattern is a confirmation of the energy expenditure assertion made earlier. The human operator is resilient, but recovery time is critical so that energy resources can be replenished. The actual process of failure is expressed in the tenets of the extended-U description, which is illustrated in Figure 4.2. Here, we can see the steady-state situation is expressed by the horizontal plateau at the top of the extended-U. The onset of fatigue, the fatigue "threshold" referred to earlier, is the boundary of the zone of psychological maximal adaptability. Here are the limits to the protection provided by cognition to the

---

8   It is part of the ambivalence of fatigue studies whether the study itself has been motivated by a desire to prevent fatigue so as to improve the health, safety, and quality of working life of the individual; or, in contrast, whether fatigue studies have been motivated by a desire to understand the phenomena so as to make individuals capable of working harder for longer in order to sub-serve profit. These moral issues lie at the very heart of all endeavors in applied human experimentation (and see Hancock, 2009).

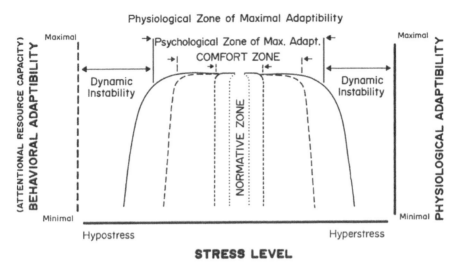

Figure 4.2   The extended-U model of stress and response capacity (after
                Hancock & Warm, 1989)

underlying physiological support system. As is evident, human individuals can fracture this threshold, but they court long-term damage as they approach the comparable edge of the zone of maximal physiological adaptability. It is in this fashion that fatigue is simply expressed as one particular form of chronic stress. Most often, this form of stress derives from sustained physical or cognitive work. Of course, in a more general theory of stress, there are other sources of chronic stress in the environment, of which continual work is simply one expression (see Hancock & Warm, 1989).

COGNITIVE FATIGUE AND ATTENTION: GLUCOSE AS RESOURCES

Having looked at the commonalty of stress and fatigue, we may also now be able to begin to draw some further links with other energetic expressions of human response capacity. In this endeavor, a link to the notion of attention and attentional resource theory may prove especially helpful. We have suggested that fatigue (in both its putative physical and cognitive forms) is a subjective state indicative of an incipient failure of the steady-state supply of necessary energy. If this is so, we ought to be able to make a link to the allied construct of attention. After all, in the Hancock and Warm (1989) theory, cited above, the process of cognitive failure under the driving influence of stress was posited as due to the diminution and, indeed, loss of attentional resources. Some decades ago, the notion of resources in attention theory, as posited by Kahneman (1973) was looked upon as a metaphorical construct (Hockey, Gaillard & Coles 1986). Indeed, it was argued by some that resources were an unnecessary construct (Navon, 1984). More modern findings from neuroscience have begun to suggest that the metaphor may be much more real than was conceived of at that time.

Here, we wish to suggest – and, to our knowledge, this suggestion is made for the first time – that attentional resources may reflect levels of extracellular glucose within the active central nervous system.[9] Thus, the depletion of resources with specific forms of task demand represent a diminution of available systemic energy at specific modular sites involved in response to specific cognitive demands. Fatigue, as a general overarching subjective apperception, may involve the summated assessment of overall available energy, or may be triggered as a threshold response to specific forms of depletion, contingent upon the repetitive task that is being undertaken. These are not mutually exclusive propositions but may be different respective "triggers" for the fatigue state. Cognitive fatigue, then, is a chronic state of attentional resource depletion very much related to the biochemical substrate energy that supports ongoing activity. This is a postulation we derive from a number of recent observations which we do not elaborate upon here (Bequet, Gomez-Merino, Berthelot & Guezennec, 2002; Dwyer, 2002; Lieberman, Falco & Slade, 2002; McNay, McCarty & Gold, 2001; McNay, Fries & Gold, 2000; Nybo, Moller, Pedersen, Neilsen & Secher, 1993; Robinson & Rapoport, 1986). The full exposition of such a proposition will need to be made in a separate work.

A PERSISTING PROBLEM IN DEFINITION

Before we leave the problems of conceptualization and definition of fatigue, there is one final, philosophical barrier that we have to address. We should note immediately that this problem is not necessarily a tractable one but it is nevertheless one we have to be cognizant of as we look to move the fatigue question forward. The question centers around the subjective nature of experience and the attempts that are made to render this experience objective (see Yoshitake, 1971). Of course, this is not a new problem and is one that has especially bedeviled psychological inquiry and behavioral research over the past century. However, it is especially relevant to the issue of fatigue. In our own work here and, indeed, in much of the extant literature, it is clear that fatigue is primarily thought of as a subjective state of the individual. Indeed, we began the present chapter with a general, descriptive definition that relied on this dimension of explanation. The challenge, as indeed is expressed in all of psychophysics and much of the discussion of the mind–body problem is how to render this subjective experience open to mutual, independent, and objective exploration. In an echo of the behaviorist concern, much of the fatigue literature has focused on the nature of the antecedent conditions. This inevitably includes the task characteristics themselves (see Ackerman, Calderwood & Conklin, Chapter 6, this volume) and especially the time on-task which may be regarded as almost the quintessential objective measure. It is the focus in what we have called the "input" factors (see Figure 4.1), that cast the emphasis on the "scientific" aspect of fatigue.

This selfsame aspiration is now being expressed in enthusiastic statements about how advances in neuroscience, and especially neuroimaging, will open an equally "scientific" window on the "adaptive" response processes as expressed in the active brain of the

---

9   We are very aware that the protestation of a single aspect of the overall biochemical balance of the brain is most probably a very simplistic representation of the situation. However, as with Watson's protestation about theoretical claims, the advocacy for a bold statement is that it is probably the only form of statement that gets noticed and temporization of explanation remains largely ignored.

exposed individual. This is a hopeful statement and one we have tacitly endorsed in our suggestion that various aspects of neurobiological understanding of energy metabolism will open up new vistas of explanation and understanding of this difficult fatigue issue (see Gailliot, 2008). Yet, there remains no fundamental guarantee that such reductionist sophistication and exploration can render a full account of what is fundamentally an individual, subjective experience (see Nagel, 1974). If, as we might eventually conclude, fatigue is an emergent property of conscious experience, underwritten by the interaction of multidimensional factors of both an environmental and a neurophysiological nature, then its very essence may necessarily elude the form of investigative strategy that we are pursuing so heavily. That this statement may well also be true of all other energetic aspects of behavior is a caution we must acknowledge in our search for essential answers.

SUMMARY AND CONCLUSIONS

Here, we have looked to approach the question of the definition and conceptualization of fatigue largely from an evolutionary perspective. The primary driver of the evolutionary world is energy. Human beings are homeotherms. That is, they have inherited the capacity to carry around with them a constant body temperature. This provides great freedom compared with some other living creatures, which have to depend upon the vagaries of the environment to warm them up to a sufficient level for activity. However, carrying around this Promethean "fire of life" imposes an enormous burden in terms of the search for and the acquisition of foodstuffs to support this endless internal blaze. Understanding fatigue must be couched in these terms. Physical fatigue derives from the unsuccessful search for food energy. Physical fatigue signals the fact that the individual has expended more energy than might be considered wise in the search for sustenance- more than might be considered wise because a fatigued animal is now more prey than predator. Humans experience cognitive fatigue because they (among others in the animal kingdom) have developed very costly brains so that they can substitute cognitive search strategies for the eventually more costly physical search strategies. In terms of the energy equation, brains eventually pay off. In humans, it has been argued, brain growth has been so accelerated that the species is happy to consider itself almost divorced from the rest of the animal kingdom (Hancock, 2009). This overlaid complexity of brain size proliferation has obfuscated understanding of basic processes such as fatigue. Cognitive fatigue is the result of energy imbalance in the brain in exactly the same way that physical fatigue eventually derives from energy deprivation in the muscle. The processes are exactly coincident (although, as we have noted earlier, the brain tends to protect the muscle as a matter of survival safety). The apparent division between physical and cognitive fatigue is spurious. Further, we have identified a candidate metabolic source of energy as the rate-limiting factor in fatigue and have suggested that this is strongly linked to the psychological construct of attentional resources.

That there are inevitable links to other allied constructs like mental (cognitive) workload and situation awareness are an obvious outcome of this overall energetic approach to understanding human behavioral response (see Freeman, 1948; Cameron, 1973; Hancock & Warm, 1989). We do not see this as the "answer" to the fatigue question but rather the beginning of a much larger synthesis of many allied energetic facets of response. That these are inevitably based upon some underlying biochemical limitation has to be true as performance extremes are encountered at the end stages of fatigue. How they are related to the more subtle changes in the mid-ranges of stress as represented by the plateau of the extended-U framework has yet to be articulated, although significant strides have been taken in this direction in recent decades. That fatigue is a crucial issue in a society ever more seduced by a 24/7/365-oriented world is undeniable. Proposed solutions have to be founded upon an understanding of our origins and the forces that gave rise to fatigue in the first place. Our aspiration is that the present work has set us upon a course toward such viable solutions.

REFERENCES

Bartley, S. (1976). What do we call fatigue? In E. Simonson & P. Weiser (eds), *Psychological Aspects and Physiological Correlates of Work and Fatigue*. Springfield, IL: Charles C. Thomas, 409–14.

Bartley, S.H. & Chute, E. (1947). *Fatigue and Impairment in Man*. New York: McGraw Hill.

Bequet, F., Gomez-Merino, D., Berthelot, M. & Guezennec, C.Y. (2002). Evidence that brain glucose availability influences exercise-enhanced extracellular 5-HT level in the hippocampus: A microdialysis study in exercising rats. *Acta Physiologica Scandinavia*, 176, 65–69.

Bitterman, E. (1944). Fatigue defined as reduced efficiency. *American Journal of Psychology*, 57, 569–73.

Broadbent, D.E. (1979). Is a fatigue test now possible? *Ergonomics*, 22, 1277–90.

Burton, A.C. (1934). The application of the theory of heat flow to the study of energy metabolism. *Journal of Nutrition*, 7, 497–533.

Cameron, C. (1973). A theory of fatigue. *Ergonomics*, 16, 633–48.

Dawson, D. & Reid, K. (1997). Fatigue, alcohol and performance impairment. *Nature*, 388: 235.

Dwyer, D. (2002). Glucose metabolism in the brain. *Proceedings of the National Academy of Sciences*, 105, 1044–49.

Freeman, G.L. (1948). *The Energetics of Human Behavior*. Ithaca, NY: Cornell University Press.

Gailliot, M.T. (2008). Unlocking the energy dynamics of executive functioning linking executive functioning to brain glycogen. *Perspectives on Psychological Science*, 3, 245–63.

Hancock, P.A. (2009). *Mind, Machine, and Morality*. Farnham, UK: Ashgate.

Hancock, P.A. (2010). The battle for time in the brain. In J.A. Parker, P.A. Harris & C. Steineck (eds), *Time, Limits and Constraints: The Study of Time XIII*. Leiden, NL: Brill.

Hancock, P.A. & Caird, J.K. (1993). Experimental evaluation of a model of mental workload. *Human Factors*, 35, 413–29.

Hancock, P.A. & Desmond, P.A. (eds). (2001). *Stress, Workload and Fatigue*. Mahwah, NJ: Lawrence Erlbaum.

Hancock, P.A. & Warm, J.S. (1989). A dynamic model of stress and sustained attention. *Human Factors*, 31, 519–37.

Hancock, P.A., Pepe, A. & Murphy, L.L. (2005). Hedonomics: The power of positive and pleasurable ergonomics. *Ergonomics in Design*, 13, 8–14.

Hockey, G.R.J., Gaillard, A.W.K., & Coles, M.G.H. (eds) (1986). *Energetics and Human Information Processing*. Dordrecht: Martinus Nijhoff.

James, W. (1890). *Principles of Psychology*. New York: Holt.

Kahneman, D. (1973). *Attention and Effort*. Englewood Cliffs, NJ: Prentice Hall.

Landauer, A.A. & Cross, M.J. (1971). A forgotten Australian: Muscio's contribution to industrial psychology. *Australian Journal of Psychology*, 23, 235–40.

Lieberman, H.R., Falco, C.M. & Slade, S.S. (2002). Carbohydrate administration during a day of sustained aerobic activity improves vigilance, as assessed by a novel ambulatory monitoring device and mood. *American Journal of Clinical Nutrition*, 76, 120–27.

McNay, E.C., Fries, T.M. & Gold, P.E. (2000). Decreases in rat extracellular hippocampal glucose concentration associated with cognitive demand during a spatial task. *Proceedings of the National Academy of Sciences*, 97, 2881–5.

McNay, E.C., McCarty, R.C. & Gold, P.E. (2001). Fluctuations in brain glucose concentration during behavioral testing: Dissociations between brain areas and between brain and blood. *Neurobiology of Learning and Memory*, A75, 325–37.

Muscio, B. (1921). Is a fatigue test possible? *British Journal of Psychology*, 12, 31–46.

Nagel, T. (1974). What is it like to be a bat? *Philosophical Review*, 33, 435–50.

Navon, D. (1984). Resources – A theoretical soup stone? *Psychological Review*, 91, 216–34.

Nybo, L., Moller, K., Pedersen, B.K., Neilsen, B. & Secher, N.H. (1993). Association between fatigue and failure to preserve cerebral energy turnover during prolonged exercise. *Acta Physiologica Scandinavia*, 179, 67–74.

Robinson, P.J. & Rapoport, S.I. (1986). Glucose transport and metabolism in the brain. *American Journal of Physiology*, 250 (1, Pt. 2), R127–36.

Ryan, T.A. (1944). Varieties of fatigue. *American Journal of Psychology*, 57, 565–9.

Yoshitake, H. (1971). Three characteristic patterns of subjective fatigue symptoms. *Ergonomics*, 21, 231–3.

5
Individual Differences in Stress, Fatigue and Performance

James L. Szalma

INTRODUCTION

Defining and differentiating affective/motivational constructs such as stress and fatigue has historically been quite difficult (e.g., Burnham, 1908), and in the case of fatigue it has been argued that a valid measure is not even possible (Broadbent, 1979; Muscio, 1921). This has led some to emphasize more general and inclusive concepts such as operator functional state, although the problem of developing measures that capture relevant dimensions still exists (Hockey, Gaillard & Burov, 2003). Current approaches to stress and fatigue view them as distinct aspects of state within a common theoretical framework (Gaillard, 2001; Hancock & Verwey, 1997; Matthews, 2002).

The dominant theories emphasize self-regulatory control mechanisms (Carver & Scheier, 1998; Hockey, 1986; 1997; 2003), the transactional nature of fatigue and stress (Matthews, 2001a, 2001b, 2002) and the characteristics of the environment that induce these states (Hancock & Verwey, 1997; Hancock & Warm, 1989). However, individual differences in vulnerability to task-induced stress and fatigue have received less theoretical or empirical attention compared to environmental factors, although there have been several studies in specific domains such as driving (Matthews, 2001a, 2002, Desmond & Matthews, 2009) and vigilance (Berch & Kanter, 1984). In this chapter an individual differences approach to stress and fatigue will be described, and incorporation of individual differences into the cybernetic (Hockey, 1997) and dynamic adaptability (Hancock & Verwey, 1997; Hancock & Warm, 1989) approaches to stress and fatigue will be articulated.

GENERAL THEORETICAL PERSPECTIVES ON FATIGUE AND STRESS

Although psychological theories seem to continuously proliferate, there are three general approaches to understanding the performance, subjective, and physiological consequences of stress. These are transactional theory (Lazarus, 1999; Matthews, 2001b), the dynamic adaptability model (Hancock & Verwey, 1997; Hancock & Warm, 1989), and the compensatory control theory (Hockey, 1986, 1997, 2003). Although these theories emphasize different aspects of human response to environmental change, they share a common, energetic, resource-based perspective that emphasizes the importance of considering both the specific properties of the environment and how people respond and adapt to the demands imposed on them. These theories are therefore compatible and a synthesis of these models in conjunction with incorporation of an individual differences approach to stress and fatigue may be a fruitful avenue for improving our collective understanding of performance under stress. First, however, the main components of these theories are briefly described.

Transactional Perspective

One of the best-known cognitive theories of stress is the transactional theory advocated by Lazarus (1999). However, it was conceived as broadly applicable to acute and chronic *life* stress. Matthews (2001a, 2001b, 2002) has proposed a transactional perspective that emphasizes the importance of cognitive mechanisms of appraisal for understanding the *performance* effects of stress and fatigue. With respect to fatigue this has been done most extensively in the context of driving tasks in both field and simulation studies, in which both the person (traits) and the environment are considered determinants of the behavioral response (including performance) and cognitive state (i.e., stress and fatigue) of the driver (Matthews 2001b, 2008; Neubauer, Matthews & Saxby, Chapter 23, this volume).

Dynamic Adaptability Theory

The dynamic adaptability model of Hancock and Warm (1989; see also Hancock, Desmond & Matthews, Chapter 4, this volume) asserts that humans can successfully adapt and maintain behavioral and physiological stability across a wide range of magnitudes of stress exposure (see Figure 5.1 and Hancock, Desmond & Matthews, Chapter 4, this volume). Failures of adaptation occur progressively, with loss of comfort and impairment of cognitive states occurring at lower magnitudes of stress than performance impairment. This nested structure creates zones of dynamic instability in which one form of adaptation fails while, simultaneously, another is maintained. Note that this theory views stress effects as symmetric around the comfort zone (see Figure 5.1). That is, demands can be excessively high (hyperstress) or low (hypostress).

One of the most widely cited aspects of the dynamic adaptability theory is its explicit recognition that the most proximal source of stress is the task the person is performing. Hancock and Warm (1989) identified two fundamental task dimensions that serve as inputs to an individual's adaptive state (Figure 5.2). Information structure, a spatial dimension, refers to the organization of task elements (e.g., displays/controls, uncertainty, amount of information to be processed). The other dimension, information rate, refers to the temporal structure of the task (e.g., time required to perform an action, the duration and rate of information presentation).

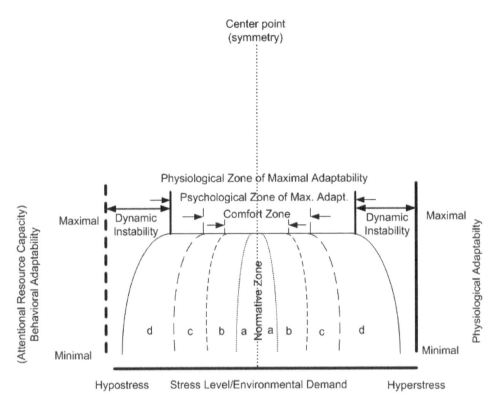

Figure 5.1   The Dynamic Adaptability Model (adapted from Hancock & Warm, 1989)

Note: The adaptation is manifested in the plateau at the apex of the extended inverted-U, which describes the zones of stable response to environmental demands (stressors). Note that the vertical line labeled "Center point" indicates that adaptation is symmetrical with respect to under- and overload, with the balance point residing within the normative zone. See text for explanations of regions a–d.

Task-Induced Fatigue

Hancock and Verwey (1997) asserted that many environmental factors can induce fatigue, including tasks, and that these factors can be described in terms of the aforementioned information rate and information structure. For instance, vigilance tasks are often characterized as having low event rates and restricted spatial structures. Prolonged work doing repetitive actions can also induce fatigue. Desmond (2001) identified these as different forms of fatigue: *passive*, in which fatigue results from a requirement to continuously maintain a level of alertness; and *active*, in which fatigue results from continuous performance of a perceptual–motor task or task component. In both cases fatigue states result from continuous demands, often over prolonged durations, and they differ primarily in the kinds of behavior that induce them. Research has supported the distinction between these types of fatigue, but has also established that a common theoretical framework may account for both forms (Matthews, 2002;

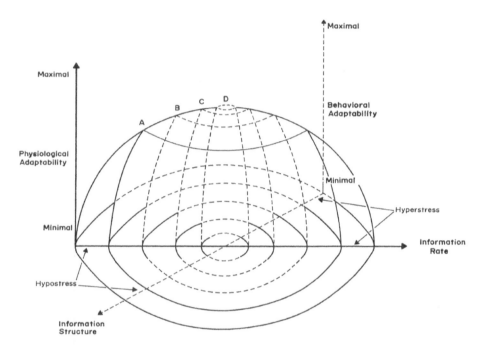

Figure 5.2   The Adaptability Model (after Hancock & Warm, 1989)

Note: The stress/environmental base axis has been decomposed into two separate component dimensions of task demands, information rate and information structure.

Matthews, Desmond & Hitchcock, Chapter 9, this volume; Matthews, Hancock & Desmond, Chapter 10, this volume).

Compensatory Control Model

Hockey (1986) identified four broad patterns of the effects of stress on performance:

1. Performance is often maintained when the performer is exposed to stress.
2. The effects of stress have different patterns (e.g., cognitive patterning; Hockey & Hamilton, 1983) for different stressors and task demands.
3. The relationship between stress and activation depends on the level of engagement in the task.
4. Stress effects depend "on the appraisal of stress, rather than purely on physical environmental conditions" (Hockey, 1986: 287).

The compensatory control model (CCM; Hockey, 1997, 2003, Chapter 3, this volume) was developed to account for these effects (Figure 5.3; for a more detailed treatment, see Chapter 3, this volume). From the CCM perspective, performance is maintained via a negative feedback self-regulatory control mechanism in which environmental demands are evaluated via a low-level, automatic appraisal mechanism (e.g., Sander, Grandjean

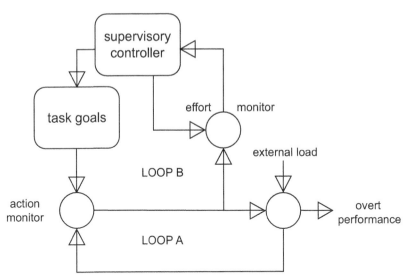

Figure 5.3   Compensatory Control Model of Self-Regulation (after Hockey, 1997)

& Scherer, 2005), and when a discrepancy is detected an effort monitor engages a higher-level regulatory loop that provides additional mental resources to accommodate the demand.[1]

The lower loop shown in Figure 5.3 (loop A) controls behavior in a relatively automatic fashion and is engaged in contexts in which little effort or mental capacity is required. When the demands are increased beyond the capacity of the lower loop, and this increase is detected by the effort monitor, the individual may engage the upper loop (loop B) to allocate the necessary resources to maintain task performance. Increases beyond the capacity due to extreme demand or extended periods of work of this upper loop will eventually result in performance failures because the repetitive use of the regulatory process fatigues or strains the compensatory mechanism. Whether additional effort is allocated depends on the decision of an executive function (the supervisory controller), which selects one of two alternatives: increasing effort or adjusting task goals.

A crucial aspect of the appraisal process is the perceived controllability of the stress, as it determines the choice regarding effort allocation. Hockey (1986) argued that appraisals of threat to performance (and the subsequent effects on cognitive processing efficiency) are less likely to occur when an individual believes that he/she can effectively cope with the source of the stress. It is in these appraisal processes that individual differences are likely to exert particularly strong effects. Indeed, Hockey (2003) identified the incorporation of individual differences into the compensatory control model as one of the next challenges in developing the theory.

---

1   Carver and Scheier (1998) have proposed a self-regulation theory of motivation and emotion, but in their model the regulatory loops are organized hierarchically according to goal type. Further, their version of control theory is not energetic in that they do not address the issue of mental resource allocation and compensatory effort to deal with environmental demands. However, both these theories may be considered self-regulatory, cybernetic perspectives.

Theoretical Integration

The three theories are compatible with one another, and in some respects they are complementary. An individual responds to changes in task and environmental demand via an active (automatic and conscious) appraisal mechanism in which the task elements are evaluated and coping decisions are made based upon these appraisals. The CCM explicitly acknowledges the importance of (automatic) appraisals in terms of the action and effort monitors as well as the supervisory executive (Hockey, 1986, 1997, 2003, Chapter 3, this volume). Fatigue and stress may be conceptualized as the result of multilevel transactions between person and task in which resources are reduced at a rate faster than replenishment can occur, and the individual strains the compensatory regulation mechanism to maintain performance. Presumably this process could underlie not only stress but also both kinds of fatigue, with active/passive types being distinguished by the type of behavior or components of task being examined (e.g., steering vs. monitoring).

From the perspective of the dynamic adaptability theory the physiological and cognitive costs of compensatory effort manifest in different patterns of performance–stress response. For instance, in some ranges of demand the individual may exhibit performance insensitivity (Hancock, 1996), because performance is maintained at the cost of higher workload and stress resulting from the allocation of more effort (region c in Figure 5.1). However, as time passes or task demand increases, a tipping point is reached where performance failure also occurs. Presumably active fatigue would be represented on the right half of the dynamic adaptability model, and passive on the left side of the model. The energetic perspective can therefore accommodate the three different bio-behavioral states (Gaillard, 2001) of stress, workload, and fatigue in a common theoretical framework.

Note that the failure in the self-regulatory mechanism discussed by Hockey (1986, 1997) occurs in the zones of dynamic instability described by Hancock and Warm (1989), and both the engagement and efficiency of these mechanisms are dynamically (i.e., continuously over time) controlled by both automatic and conscious appraisal mechanisms. The kinds of appraisals and core relational themes, as well as the coping strategies selected, will determine the level of effort and, ultimately, the efficiency of the control loop. Together these will determine adaptive state and thresholds/slopes of the zones of instability. Indeed, the information task parameters will impact on the person primarily through appraisals at multiple levels of the organism (Matthews, 2001b).

Modes of Control and Zones of Adaptation

Hockey (2003) described the modes of state control and their performance, effort, and subjective state correlates. These are reproduced in Table 5.1. These can be modified and linked to the different zones of adaptability described by Hancock and Warm (1989), as shown in Figure 5.4.

Engaged control, which corresponds to active engagement and optimal performance, would be expected to be within the normative zone. However, there can be more than one output of engaged control: Performance may remain stable (engaged mode A) or improve (engaged mode B). In the normative zone engaged control results in performance with minimal effort (region a in Figure 5.4), and this may manifest either as a performance–workload association (performance improves and workload decreases) or a high but stable level of performance and no change in workload as a function of demand.

Table 5.1   Modes of Adaptive Response in the Compensatory Control Model (after Hockey, 2003)

Control Mode	Environmental Context	Performance (task goal)	Affective State	Stress Hormones
Engaged	High demands high control	Optimal (high)	Anxiety 0 Effort + Fatigue −	Adrenaline + Cortisol −
Strain	High demands low control	Adequate (high)	Anxiety + Effort + Fatigue +	Adrenaline + Cortisol +
Disengaged	High demands low control	Impaired (reduced)	Anxiety + Effort − Fatigue 0	Cortisol +

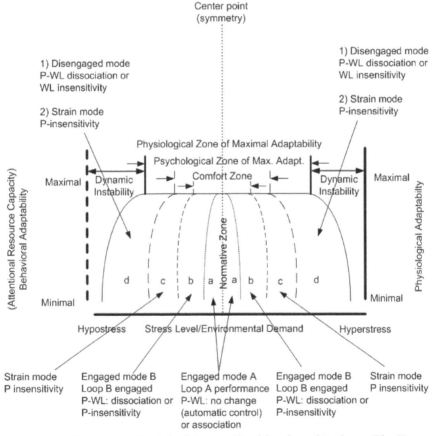

Figure 5.4   Composite Model of Stress, Workload, and Fatigue: The Dynamic Adaptability Model, incorporating Hockey's compensatory control modes and performance–workload associations, dissociations, and insensitivities

Note: P=Performance; WL= workload; see text for explanations of regions a–d.

Within the comfort zone but outside the normative zone (region b) the system remains in engaged mode (as defined by Hockey, effort without distress), but some effort is required. There are two possible outcomes: a performance–workload dissociation in which performance improves but workload also increases, or a performance insensitivity in which workload increases but performance remains stable. Strain mode occurs at levels of demand outside the comfort zone (region c), resulting in a performance workload insensitivity in which performance is maintained only at the cost of substantial increases in workload and stress.

Region d represents levels of demand at which compensatory effort no longer supports performance. Hockey (2003) referred to these circumstances as a "disengaged mode" (see Table 5.1), but in fact the control mode within this region depends on choice of effort. If the operator abandons the task or reduces task goals then a disengaged mode occurs. Under these conditions one of three relationships of performance with cognitive state will be observed (see Figure 5.4). Either performance workload dissociation will occur, in which performance and workload and stress both decline (i.e., the participant "gives up;" Hancock, 1996), or a workload insensitivity will occur in which performance continues to decline but the workload and stress do not change (either because a maximum level has been reached or because the person becomes less sensitive to his/her state). However, if the operator persists in effort even in the face of declining performance (e.g., he/she cannot or will not modify task goals) then the individual will remain in a strain mode that manifests as a performance insensitivity in which performance has reached a minimum level and workload and stress increase.

INDIVIDUAL DIFFERENCES IN STRESS AND FATIGUE

The notion that person characteristics can influence stress and fatigue states is not new. For instance, Stern noted that convenient methods for measuring fatigue are offset by the fact that "the personal *significance* of the amount of fatigue they reveal is indeterminate [and] possibilities for recovery are no less diverse than those of fatigue" (Stern, 1938: 505, emphasis in original). He noted, for instance, that a spontaneous activity chosen by the person following performance of a compulsory task can have a strong restorative effect. Stern's conceptualization anticipated the importance of personality effects on appraisal (Matthews, 2001b, 2008) and on the role of choice and autonomy on energetic motivational states (Ryan & Deci, 2000).

More recently Desmond and Matthews (2009) showed how person characteristics can be incorporated into a transactional theory of stress and fatigue in driving. Fatigue induced by driving (and, one may presume, by other tasks as well) results from the dynamic transaction between the task and other environmental parameters and a person's vulnerability to stress and fatigue. Coping response seems to be particularly important in determining the level of and vulnerability to task-induced stress and fatigue. Desmond and Matthews (2009) reported evidence that task-focused coping has been linked to greater task engagement, and emotion-focused coping has been linked to greater distress and worry. Hence, traits that influence both general coping patterns and coping strategies specific to driving will likely influence driver fatigue.

From their data Desmond and Matthews (2009) concluded that the trait of "fatigue proneness" is a reliable predictor of fatigue state change after driving, and that it is also associated with the use of emotion-focused and avoidant coping strategies. Emotion-focused coping, which correlates with distress reactions, may relate to fatigue via two mechanisms (Desmond & Matthews, 2009: 1) the mental workload imposed by prolonged self-criticisms may itself induce fatigue; and 2) emotion-focused coping is associated with self-focused attention (a facet of worry; Matthews et al., 2002), which may heighten the driver's awareness of somatic symptoms and discomfort.

With respect to the distinction between stress and fatigue, Desmond and Matthews (2009) showed that stress and fatigue symptoms can co-exist and are related to common trait variables. They identified three possible explanations for this co-variation: 1) task demands may invoke both stress and fatigue simultaneously; 2) dynamic interactions between stress and fatigue may occur, such that subjective fatigue may generate a compensatory effort response, which drains capacity and thereby induces more fatigue; 3) different types of fatigue states may exist (i.e., active vs. passive). With respect to the different forms of fatigue, it seems reasonable to assume that both types require compensatory effort if performance is to be maintained, and that these mechanisms (e.g., appraisal, self-regulation of effort) are general and common not only to both forms of fatigue but also to the multiple dimensions of stress and workload. Hence active and passive fatigue may not be distinguishable from one another in terms of the cognitive/affective mechanisms, but may be differentiated by the type of response required for performance (prolonged active psychomotor vs. prolonged attentional demand). In terms of the dynamic adaptability model, active fatigue may be more likely to occur in the regions to the right of the comfort zone (hyperstress region), while passive fatigue may occur in the regions of dynamic instability to the left of the comfort zone (hyperstress region).

Hancock and Verwey (1997) noted that the location of an individual's comfort zone (and, by implication, other zones) will depend in part on the individual's personal history, by which they meant both genetic and learned differences across individuals. It is likely that in addition to these trait effects, the location of the zones will be influenced by variation in state within and between individuals. Traits may impose constraints on the range of information structure and rate within which a particular individual is in his/her comfort zone, but the specific point within that range will vary dynamically over time as a function of physiological and subjective state, as well as the specific organization of the immediate environment. Hence, the "starting point"[2] will influence the level of adaptation as internal (person) and external characteristics (environment, task) fluctuate over time. In addition to influencing the location of the zones, the width of the normative zone, of the zones of dynamic instability, and of the slopes of the functions that specify the boundaries between zones, may vary as a function of the person characteristics (Szalma, 2008).

INCORPORATION OF INDIVIDUAL DIFFERENCES INTO THEORIES OF FATIGUE AND STRESS

Individual differences in fatigue are likely related to the same general mechanisms (e.g., appraisal, emotion-focused coping, compensatory effort) underlying stress and coping. If this is so, then these differences may also be conceptualized within a common framework with stress and workload effects. This does not mean that the same traits will impact all

---

2   The notion of a "starting point" is something of a convenient fiction because a person's level of adaptation continuously fluctuates. The "starting point" is therefore the position of the comfort zone at any given time.

three aspects of state, but that the relationships operate through the same transactional appraisal-based self-regulatory mechanisms. However, personality may also influence the amount of available resources and strategic decisions regarding allocation of effort. For instance, based on this approach, one might predict that the trait of extraversion would be associated with greater resistance to active fatigue because that trait is associated with more resource capacity, higher energetic arousal, and task coping tendencies. However, extraverts may be more vulnerable to passive fatigue due to a relative lack of stimulation and the need for compensatory effort to deal with the aversive (i.e., under-stimulating) environmental conditions.

The Example of Extraversion

Szalma (2008) showed how individual differences can potentially influence dynamic adaptation by changing the threshold and rate of failure of a given form of adaptation (e.g., comfort, performance, etc.). Here this approach is extended to include passive and active fatigue, as well as how appraisals and self-regulatory control of effort may vary as a function of personality. Extraversion was chosen as an exemplar because this trait

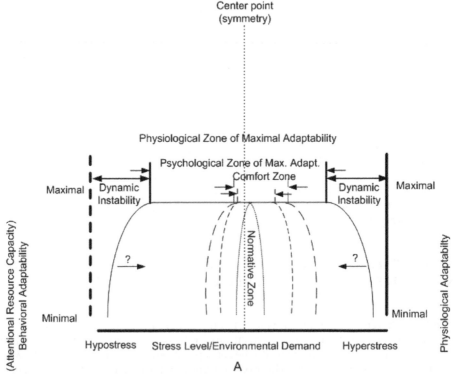

Figure 5.5a   The Adaptability Model, incorporating hypothesized adaptive function of individuals high in extraversion

Note: The vertical line labeled "Center point" illustrates the asymmetry in the model introduced by consideration of personality control modes and performance–workload associations, dissociations and insensitivities.

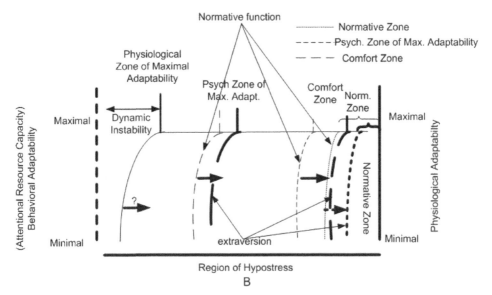

Figure 5.5b   Representation of the Adaptability Model shown in Figure 5.5a for extraverts, which focuses on the hypostress region (adapted from Szalma, 2008)

Note: The thin curves represent "normative" (i.e., "average") patterns of adaptation and the thicker curves represent hypothesized adaptive patterns for individuals high in extraversion. High extraversion would be expected to shift the thresholds of adaptive instability (the "shoulders" of the functions) such that for these individuals adaptive failure occurs at less extreme levels of hypostress or under stimulation. However, the degree of shift is not equivalent for each level of adaptation. Thus, one might expect the normative and comfort zones to narrow to a greater degree than the zone of psychological (i.e., performance) adaptation.

has been investigated extensively in human performance research (Matthews, Deary & Whiteman, 2009), and because the cognitive patterning of the trait has been well articulated (Matthews, 1992).

Extraversion is characterized by a cognitive patterning that supports engagement with the environment, particularly the social environment. Hence, higher extraversion is associated with greater resource and working memory capacity, superior divided attention and performance under time pressure, and higher levels of positive affect and greater use of task-focused coping strategies. Thus, extraverts have cognitive characteristics that provide them with an adaptive advantage in stimulating environments. However, these same characteristics prove to be a disadvantage in relatively unstimulating environments such as those requiring prolonged monitoring.

The dynamic adaptability model can accommodate these patterns by considering the trait of the individual as a factor that influences the parameters of the adaptive function (i.e., range of adaptation, location of the thresholds of dynamic instability, and the slope of the decline in effective adaptation). The normative and comfort zones should, on average, be to the right of the point of symmetry for those high in extraversion (see Figure 5.5a) and to the left of center for those low on this trait (Figure 5.6a). If one assumes that the parameters and efficiency of self-regulation and specific content of appraisals vary as a function of the trait, but that the mechanisms are general and universal, one might expect that the rate of failure of a mechanism will be of the same general form but vary in slope as a function of the cognitive patterning of extraversion.

If an individual high in extraversion is confronted with an extremely unstimulating task (e.g., vigilance), that individual's processing efficiency will be impaired and manifest as a fairly steep slope and lower threshold of adaptive instability. In other words, relative to those low in extraversion, for those higher on the trait smaller increases in demand would be necessary for a decrease of one "unit" of adaptation (e.g., perceived workload/stress, physiological response; see Figure 5.5b). In contrast, more introverted individuals would be expected to show lower thresholds and steeper slopes in the hyperstress region (Figure 5.6b). Note that in Figures 5.5a and b, and 5.6a and b there are question marks above the arrows at the threshold of physiological adaptation. These convey the uncertainty regarding individual differences at the extremes of demand. As personality traits are believed to be a multi-level set of adaptive processes that include physiological response (Matthews, 2008), personality may influence the location of this final threshold of adaptation and the rate of decline. On the other hand, this extreme region of dynamic instability represents complete system failure (e.g. unconsciousness; see Harris, Hancock & Harris, 2005). It may be that such conditions suppress individual differences. For

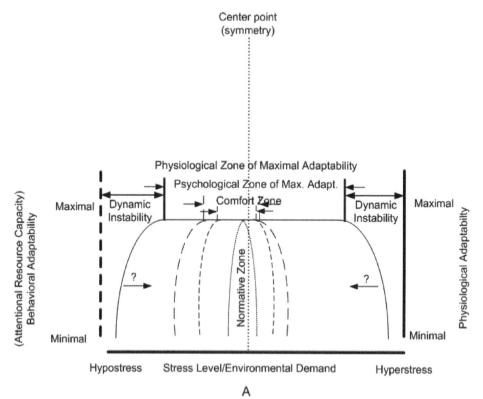

Figure 5.6a   The Adaptability Model, incorporating hypothesized adaptive function of individuals low in extraversion (introverts)

Note: The vertical line labeled "Center point" illustrates the asymmetry in the model introduced by consideration of traits.

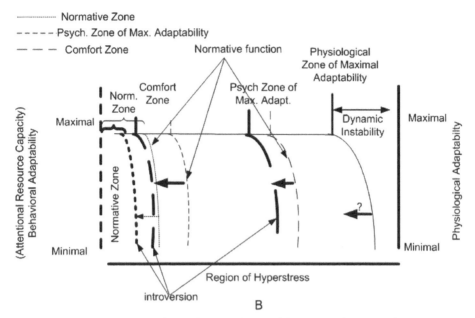

Figure 5.6b   Representation of the Adaptability Model shown in Figure 5.6a for introverts, which focuses on the hyperstress region (adapted from Szalma, 2008)

Note: The thin curves represent "normative" (i.e., "average") patterns of adaptation and the thicker curves represent hypothesized adaptive patterns for individuals low in extraversion. Introverts would be expected to shift the thresholds of adaptive instability (the "shoulders" of the functions) lower, such that for these individuals adaptive failure occurs at lower levels of environmental demand in the hyperstress region. However, the degree of shift is not equivalent for each level of adaptation. Thus, one might expect the normative and comfort zones to narrow to a greater degree than the zone of psychological (i.e., performance) adaptation.

instance, there are ranges of temperatures and levels of noise above which humans cannot function regardless of their level of extraversion.

Hence, the efficiency of self-regulation as a function of extraversion would be expected to vary according to how well cognitive characteristics support performance given a particular level of task demand. Assuming that more effort is allocated, individuals higher in extraversion would have to exert more effort in the hypostress region but less in the hyperstress region. More introverted individuals would be expected to show the reverse pattern. For instance, extraverted individuals may be more vulnerable to greater stress and passive fatigue in vigilance tasks of low event rate and/or a single display. However, the correlation may reverse when the event rate or the number of displays is increased.

Although extraversion may be related to capacity differences (either in working memory capacity or the amount of energetic resources available for task performance) there are also likely to be individual differences in strategy choice by the supervisory executive (Mulder, Leonova & Hockey, 2003). If allowed a choice, more introverted individuals would be expected to be more likely to disengage from tasks that impose substantial stimulation (e.g., multiple tasks performed under time pressure), so that extraversion would be negatively correlated with the adoption of strain or disengaged modes of control in the hyperstress region and positively related to the engaged mode in the hyperstress region. However, given further increases in demand (region d in Figure 5.4) the strain or disengaged modes may be adopted by individuals regardless of their level of extraversion.

Future Research Issues

If stress and fatigue effects occur as a joint function of person and task characteristics, it may also be true that there are trait–trait interactions in a manner analogous to the now famous stressor interactions (e.g.; sleep deprivation and noise; Broadbent, 1963). Although most personality research has considered trait effects separately, there is some evidence for interaction. For instance, Szalma et al. (2005) found that neuroticism predicted higher levels of stress after performance of a difficult firearms task, but only for those individuals who were also high in extraversion. Similarly, post-task stress was related to lower levels of intellect (a facet of openness), but only for those individuals who were also low in conscientiousness (see Szalma, 2008, for a summary of these effects). Hence, traits such as extraversion and conscientiousness may serve as protective factors that attenuate the association of traits such as neuroticism or (low) openness to both performance and cognitive state. However, human performance research has mostly neglected investigation of such interactions, perhaps because of the dearth of human performance research for traits other than extraversion and neuroticism (Matthews, Deary & Whiteman, 2009).

Another important issue that has not received adequate attention is the influence of general vs. specific traits. General traits (e.g., the big five) have been shown to predict stress (e.g., Matthews, Joyner, Gilliland, Campbell, Falconer & Huggins, 1999; Matthews, Deary & Whiteman, 2009) and fatigue (e.g., Matthews & Desmond, 1998; Thiffault & Bergeron, 2003), but there are cases in which specific traits are the best predictors of outcome measures. For instance, there is evidence from Matthews and his colleagues that 1) situation-specific traits are the strongest predictors of fatigue effects; and 2) these specific traits are related to but distinct from broader traits such as fatigue proneness and Neuroticism (Matthews & Desmond 2002; Desmond & Matthews, 2009). An issue for future research will be to clarify the relationships and distinctions across broad versus narrow traits, as well as how these traits are moderated by different levels of information rate and structure. A particularly difficult problem for evaluating the role of task types and characteristics is in developing valid metrics for representing information rate and structure that is generalizable across task domains. Quantification of information processing has been identified as a major challenge for behavioral science, particularly for performance of complex tasks (For an overview, see McBride & Schmorrow, 2005).

CONCLUSIONS

Thresholds of adaptive response may vary as a function of the transaction between personality and task structure. The rate of decline in effective adaptation would also be expected to be a property of a specific transaction. However, given that the impact on performance varies as a function of the type of task and environmental demand, the information rate and structure parameters will vary from task to task, so the threshold and slopes of failure may therefore vary as a function of task for a given person. Perhaps the best way to approach such complexity is to establish a metric for the transactional unit of analysis, although it is not clear how to do this in a way that captures multiple levels and dimensions of transactions.

The different theories converge on common, or at least highly compatible, mechanisms. In moving forward, there are at least three ways to enhance our theoretical understanding of stress and fatigue, and therefore also point the way toward effective mitigation through training and environmental design. First, synthesize the major theories of stress into a common framework. Second, incorporate person characteristics into the composite model. This paper has attempted to initiate these first two endeavors. Finally, conduct programmatic empirical research to identify the specific patterns of person–task transactions. A fourth but much more difficult goal would be the development of reliable and valid metrics to quantify the structures in the task and environment and to quantify the level of adaptation of an individual at a given level of demand at a specific point in time.

REFERENCES

Berch, D.B. & Kanter, D.R. (1984). Individual differences, in J.S. Warm (ed.), *Sustained Attention in Human Performance*. Chichester, UK: Wiley, 143–78.

Broadbent, D.E. (1963). Differences and interactions between stresses. *Quarterly Journal of Experimental Psychology*, 15, 205–11.

Broadbent, D.E. (1979). Is a fatigue test now possible? *Ergonomics*, 22, 1277–90.

Burnham, W.H. (1908). The problem of fatigue. *The American Journal of Psychology*, 19, 385–99.

Carver, C.S. & Scheier, M.F. (1998). *On the Self-regulation of Behavior*. New York: Cambridge University Press.

Carver, C.S. & Scheier, M.F. (2009). Individual differences in stress and fatigue in two field studies of driving. *Transportation Research*, Part F, 12, 265–76.

Gaillard, A.W.K. (2001). Stress, workload, and fatigue as three biobehavioral states: A general overview, in P.A. Hancock & P.A. Desmond (eds), *Stress, Workload, and Fatigue*. Mahwah, NJ: Erlbaum, 623–39.

Hancock, P.A. (1996). Effects of control order, augmented feedback, input device and practice on tracking performance and perceived workload. *Ergonomics*, 39, 1146–62.

Hancock, P.A. & Desmond, P.A. (eds) 2001. *Stress, Workload, and Fatigue*. Mahwah, NJ: Erlbaum.

Hancock, P.A. & Verwey, W.B. (1997). Fatigue, workload, and adaptive driver systems. *Accident Analysis & Prevention*, 4, 495–506.

Hancock, P.A. & Warm, J.S. (1989). A dynamic model of stress and sustained attention. *Human Factors*, 31, 519–37.

Harris, W.C., Hancock, P.A. & Harris, S.C. (2005). Information processing changes following extended stress. *Military Psychology*, 17, 15–128.

Hockey, G.R.J. (1986). A state control theory of adaptation and individual differences in stress management, in G.R.J. Hockey, A.W.K. Gaillard & M.G.H. Coles (eds), *Energetic Aspects of Human Information Processing*. Dordrecht: Matinus Nijhoff, 285–98.

Hockey, G.R.J. (1997). Compensatory control in the regulation of human performance under stress and high workload: A cognitive–energetical framework. *Biological Psychology*, 45, 73–93.

Hockey, G.R.J. (2003). Operator functional state as a framework for the assessment of performance, in G.R.J. Hockey, A.W.K. Gaillard and O. Burov (eds), *Operator Functional State: The Assessment and Prediction of Human Performance Degradation in Complex Tasks*. Amsterdam: IOS Press, 8–23.

Hockey, R. & Hamilton, P. (1983). The cognitive patterning of stress states, in G.R.J. Hockey (ed.), *Stress and Fatigue in Human Performance*. Chichester: Wiley, 331–62.

Hockey, G.R.J., Gaillard, A.W.K. & Burov, O. (eds) (2003). *Operator Functional State: The Assessment and Prediction of Human Performance Degradation in Complex Tasks*. Amsterdam: IOS Press.

Lazarus, R.S. (1999). *Stress and Emotion: A New Synthesis*. New York: Springer.

Matthews, G. (1992). Extraversion, in A.P. Smith & D.M. Jones (eds), *Handbook of Human Performance*. Vol. 3: *State and Trait*. London: Academic Press, 95–126.

Matthews, G. (2001a). A transactional model of driver stress, in P.A. Hancock & P.A. Desmond (eds), *Stress, Workload, and Fatigue*. Mahwah, NJ: Erlbaum, 133–63.

Matthews, G. (2001b). Levels of transaction: a cognitive science framework for operator stress, in P.A. Hancock and P.A. Desmond (eds), *Stress, Workload, and Fatigue*. Mahwah, NJ: Erlbaum, 5–33.

Matthews, G. (2002). Towards a transactional ergonomics for driver stress and fatigue. *Theoretical Issues in Ergonomics Science*, 3, 195–211.

Matthews, G. (2008). Personality and information processing: A cognitive–adaptive theory, in G.J. Boyle, G. Matthews & D.H. Saklofske (eds), *Handbook of Personality Theory and Assessment*. Vol. 1: *Personality Theories and Models*. Thousand Oaks, CA: Sage, 56–79.

Matthews, G. & Desmond, P.A. (1998). Personality and multiple dimensions of task-induced fatigue: a study of simulated driving. *Personality & Individual Differences*, 25, 443–58.

Matthews, G. & Desmond, P.A. (2002). Task-induced fatigue in simulated driving performance. *The Quarterly Journal of Experimental Psychology*, 55, 659–89.

Matthews, G., Campbell, S.E., Falconer, S., Joyner, L.A., Huggins, J., Gilliland, K., Grier, R. & Warm, J.S. (2002). Fundamental dimensions of subjective state in performance settings: Task engagement, distress, and worry. *Emotion*, 2, 315–40.

Matthews, G., Deary, I.J. & Whiteman, M.C. (2009). *Personality Traits*. 3rd ed. Cambridge: Cambridge University Press.

McBride, D.K. & Schmorrow, D. (eds) (2005). *Quantifying Human Information Processing*. Lanham, MD: Lexington Books.

Mulder, L.J.M., Leonova, A.B. & Hockey, G.R.J. (2003). Mechanisms of psychophysiological adaptation, in G.R.J. Hockey, A.W.K. Gaillard & O. Burov (eds), *Operator Functional State: The Assessment and Prediction of Human Performance Degradation in Complex Tasks*. Amsterdam: IOS Press, 345–55.

Muscio, B. (1921). Is a fatigue test possible? *British Journal of Psychology*, 12, 31–46.

Ryan, R.R. & Deci, E.L. (2000). Self-determination theory and the facilitation of intrinsic motivation, social development, and well-being. *American Psychologist*, 55, 66–78.

Sander, D., Grandjean, D. & Scherer, K.R. (2005). A systems approach to appraisal mechanisms in emotion. *Neural Networks*, 18, 317–52.

Stern, W. (1938). *General Psychology: From the Personalistic Standpoint*. New York: The Macmillan Company.

Szalma, J.L. (2008). Individual differences in stress reaction, in P.A. Hancock and J.L. Szalma (eds), *Performance Under Stress*. Farnham: Ashgate, 323–57.

Szalma, J.L., Oron-Gilad, T. & Hancock, P.A. (2005). Individual differences in workload, stress, and coping in police officers engaged in shooting tasks, in P. Carayon, M. Robertson, E. Kleiner & P.L.T. Hoonakker (eds), *Human Factors in Organizational Design and Management*, VIII. Santa Monica, CA: IEA Press, 587–92.

Thiffault, P. & Bergeron, J. (2003). Fatigue and individual differences in monotonous simulated driving. *Personality & Individual Differences*, 34, 159–76.

6

Task Characteristics and Fatigue

Phillip L. Ackerman, Charles Calderwood and Erin Marie Conklin

INTRODUCTION

Fatigue can be decisively defined for physical materials because the resulting failure from fatigue is typically preceded by dislocations and deformations that can be observed with the proper tools. For humans, physical fatigue can also be indexed with physiological measurements, as first identified by Mosso (1906). In contrast, psychological fatigue, whether identified as "mental" or "cognitive" fatigue, can be revealed in a variety of ways that are often uncorrelated or only minimally correlated with one another. The most obvious domain where one would want to identify cognitive fatigue is for criteria of task performance. Direct methods of measuring cognitive fatigue often involve examining patterns of performance over time on-task (e.g. Noll, 1932; Thorndike, 1926). The problem with such measures is that they are also affected by other factors, such as learning or inhibition, that are often associated with increasing time on-task, making the delineation of fatigue from other factors problematic. Another method for assessing cognitive fatigue is to ask the individuals who are performing the task to report subjective feelings of fatigue. The extant research base of measures for assessing fatigue is substantial (e.g. Chalder et al., 1993; Hockey, Maule, Clough & Bdzola, 2000; Michielsen et al., 2004), and the consensus of this research is that there are several dimensions of subjective experience that are related to fatigue. However, none of these dimensions are unequivocally identifiable as fatigue, independent of other factors. For the current chapter, we will include both performance effects and subjective ratings of fatigue as indicators of cognitive fatigue.

One distinction should be made between "fatigue" and "boredom." Several investigators have attempted to separate these two concepts (e.g. Pattyn et al., 2008). Generally speaking, fatigue is typically associated with high levels of cognitive demands, and boredom associated with under-arousal and low levels of cognitive demands over time. Myers, for example, suggested that "true fatigue [requires] rest from work, mere boredom [requires] only a change of work, for their respective alleviation" (1937: 298). Bartley and Chute state that "a bored individual attributes his state with environmental events, whereas a fatigued individual lays the blame for his condition on himself" (1947: 55). Tasks that involve low levels of arousal and low levels of effort will fall outside the treatment of fatigue here.

GENERAL EFFECTS VS. INDIVIDUAL DIFFERENCES

Past psychological research on fatigue has resulted in a bewildering array of findings, from increased fatigue to null effects to decreased fatigue, even with ostensibly identical tasks and participant populations. One reason for these conflicting results is that there are wide individual differences in reactions to task situations. Although mean subjective feelings of fatigue are generally robust in a particular task situation, performance effects are notoriously unreliable. One possible explanation for these differences, offered by Davis (1946), is that some individuals will fail to feel fatigued in a specific task situation, behaving consistently and reporting no changes to subjective fatigue. Other individuals will feel fatigued, reporting increased feelings of fatigue and showing reduced performance efficiency. However, some of the individuals who feel fatigued will respond by recruiting additional attentional effort to the task, thus increasing performance efficiency. The expectation is that if the task proceeds for a long enough period of time, even these latter individuals will exhaust their reserve attentional resources, or find the task sufficiently aversive that they will eventually reduce effort, leading to diminished performance. However, in any group of people recruited for a study of fatigue, there may be more individuals who reduce effort with increasing fatigue or there may be more individuals who increase effort with increasing fatigue. Depending on the proportion of individuals who adopt these different strategies, overall levels of performance across the entire sample may increase, decrease, or show no change over time, even as subjective fatigue increases in a robust fashion.

One prominent approach to account for differential effort output in relation to task performance is Hockey's compensatory control model (1997, 2011, Chapter 3, this volume). Focusing on task performance in demanding conditions, Hockey has argued that individuals can control the distribution of cognitive resources to a task through their mobilization of mental effort. Situational influences on short-term states and individual differences in long-term traits are determinants of target goals for task performance, which are continuously compared against perceived performance for the purpose of detecting discrepancies between goals and outcomes. Both personality and motivational variables are proposed to play a role in the working effort budget devoted to a particular task, the relevance of task goals to a given individual, and variability in enacted responses to sustained, high-demand tasks. Perhaps most relevant to mental fatigue engendered by task performance, these sources of influence combine to determine whether individuals will respond to a perceived inability to meet task demands (characteristic of subjective fatigue) by either increasing effort or setting lower goal standards for performance. While a linear positive relationship between subjective and performance indicators of fatigue would be expected for individuals adopting the latter strategy, those applying greater effort when faced with subjective fatigue should delay the impact of mental fatigue on objective performance. However, while the application of greater effort in the face of fatigue likely has short-term gains, long-term costs may ensue in the form of a greater fatigue when reserve effort is depleted (Hockey, 1997). In addition, there may be negative effects on well-being or negative health outcomes stemming from chronic exposure to fatiguing conditions.

We began with a broad definition of cognitive fatigue that includes both performance decrements and subjective feelings of fatigue or related constructs. Although fatigue may be revealed in simultaneous increases in subjective ratings of fatigue and decreases in performance levels, the extant literature indicates that findings of increases in subjective feelings of fatigue with time on-task are far more robust than findings of decrements in performance. Such dissociations may result from an insensitivity of the task to decreased effort levels or differences in the way individuals respond to feelings of fatigue. Whether the interests of the task designer or supervisor are primarily related to performance decrements or subjective reactivity to the task milieu depends on the particular application. However, long-term effects of prolonged subjective fatigue have been implicated in health issues (e.g. Leone et al., 2008), and are likely to play a role in burnout and job turnover (Demerouti et al., 2001).

TASK CHARACTERISTICS ASSOCIATED WITH COGNITIVE FATIGUE

Necessary Condition

There appears to be one task characteristic that is at least generally necessary for participants to report feeling fatigued or to show performance decrements associated with fatigue. That characteristic is whether, for the individual operator, performing the task is not intrinsically interesting. When an individual is engaged in a task that he/she finds enjoyable, subjective feelings of fatigue are much diminished or absent, even over long periods of time on-task. Thus, one can read for pleasure, play video games, or play Sodoku for hours on end without fatigue, but sometimes when similar activities are *assigned* by a teacher or a parent to a student or child for whom the task holds no intrinsic interest, fatigue often follows. Dodge (1913) first noted that although there does not appear to be more mental energy required to engage in play than in work, the essence of "work" is that it can be accomplished only "against resistance" – that is, it is not intrinsically enjoyable. A similar discussion was offered by Broadbent (1953).

Many investigators have explored the task characteristics which might have an impact on a person's intrinsic motivation to engage in various tasks (see Deci & Ryan, 1987 for a review). In general, tasks that include an evaluative aspect, a deadline, or a reward are typically associated with less intrinsic interest than tasks which do not included these components. Similarly, tasks which require controlled cognitive activity and low levels of creativity are also related to low intrinsic interest in the task. In contrast, intrinsic interest is likely to be high when a person has a choice in which task(s) to perform and receives positive feedback regarding task performance.

An Overarching Framework of Contributing Factors

Based on a review of the literature on cognitive/mental fatigue, Ackerman (2011) identified three major factors associated with tasks that result in cognitive fatigue. These factors, individually and in combination, can be further divided into five broad categories of task characteristics that appear to be related to expressions of fatigue. The three factors of task characteristics are those that affect: (a) cognitive effort available; (b) cognitive effort allocated to the task; and (c) the presence of off-task distractions. These factors and their constituent categories will be described in turn below. At the outset, however, it must be noted that each of these factors requires a ceteris paribus qualification. That is, if everything else is equal, a task that has the identified characteristic will be more likely to result in fatigue, compared to an identical task without the particular characteristic.

Factors Associated with Available Cognitive Effort

The general consensus from models of cognitive fatigue (e.g. see Grandjean, 1968; Schmidke, 1976) and more general theories of attentional effort (e.g. Kahneman, 1973) is that attentional effort represents personal resources that are available in some quantity which can be depleted through use. Recovery of attentional resources occurs during rest, sleep, and when the individual is not otherwise engaged in cognitive effort. Given that we are primarily interested in situations in which an individual is engaged in a single task, as opposed to a multitasking situation, the extent to which these resources come from an undifferentiated source (Kahneman, 1973) or different sources (Wickens, 1984) is not a central concern for our current purposes.

Cumulative Intellectual Demands

The speed at which resources are depleted during task engagement and, thus, the amount of attentional effort for the task, is determined mainly by four task characteristics, namely:

1. Task demands on intellectual functioning;
2. Continuous effort demands;
3. Total time on-task; and
4. The task is not subject to large learning increments.

Ceteris paribus, tasks that make greater demands on intellectual functioning (such as mental multiplication of four-digit numbers, e.g. Arai, 1912, serial mental addition tasks, e.g. Baranski, 2007; Baranski et al., 2002) result in faster onset of fatigue, compared to tasks that make lesser demands on intellectual functioning (such as simple memorization tasks, e.g. Nolte et al., 2008, or proofreading, e.g. Takeda, Sugai & Yagi, 2001). Similarly, tasks that require the operator to work continuously (such as tasks with continuous tracking demands, e.g. Matthews & Desmond, 1998; Thackray & Touchstone, 1991) result in faster onset of fatigue, compared to discrete tasks that have an inherent opportunity for the operator to take mental breaks between stimulus presentations (e.g. Åhsberg, 2000). The essence of these characteristics is that when more attention is demanded by the task, or when there are no opportunities for breaks, available attentional resources will be depleted at a faster rate. As time on-task progresses without breaks, attentional resources are further depleted, resulting in greater levels of fatigue. At some point, the individual may reach a phenomenological perception that it is simply too difficult to continue the task (e.g. see Davis, 1946), and would be expected to disengage from the task (i.e., leaving the field of conflict, e.g. Lewin, 1935). However, research in the field of ergonomics and the study of work behaviors (e.g. Tucker, 2003) suggests that at least some individuals continue to work at a task even when there are performance decrements associated with cognitive fatigue, "perhaps continuing to work until subjective feelings of fatigue become intolerable" (Tucker, 2003: 128–9). A failure to attend to signals of subjective cognitive fatigue may represent an even more important problem, at least as far as work performance is concerned (e.g. in terms of industrial or transportation accidents).

For tasks that involve consistent information processing (e.g. Shiffrin & Schneider, 1977), rapid learning may take place with increasing time on-task. Under these circumstances, the effort demanded by the task may decrease as the individual transitions from using controlled information processing to developing and using automatic information processing. Because the attentional effort demands associated with automatic processing are substantially less than using controlled information processing, the expectation with such tasks is that fatigue will be attenuated if substantial learning occurs (e.g. see Fisk & Schneider, 1981). From this perspective, tasks that are *not* subject to large learning increments are most likely to show greater increments in fatigue with increasing time on-task. Together, one can consider increases in fatigue to be a function of integrating the amount of effort demanded by the task over time. Greater demands on attentional effort and more time on-task will both lead to higher levels of cognitive fatigue.

Penalties for Attentional Blinks

Where discrete tasks are expected to allow for short breaks from attentional effort, other characteristics of the task demands also appear to result in faster development of cognitive fatigue. When brief attentional interruptions result in marked decrements in task performance, such tasks are likely to be associated with the concurrent development of fatigue. Such tasks typically: (a) impose penalties for errors; (b) involve a high cost of short distractions of attention; or (c) require a high degree of attention to detail. While these task characteristics are not themselves associated with demands of higher levels of intellectual operations, they nonetheless prevent the operator from removing attention from the task, resulting in a greater drain of available attentional resources.

Factors Associated with Cognitive Effort Allocated to the Task

Typical laboratory-based cognitive information processing tasks are "resource limited" (see Norman & Bobrow, 1975). This means that performance levels are sensitive to the amount of attentional effort devoted to the task, through the range of effort levels available to the individual performing the task. The general working assumption for such scenarios is that the individual will devote most or all of his/her effort to performing the task, unless or until the task becomes unacceptably aversive. In the real world (and most likely in the laboratory, too), individuals vary widely in the proportion of their available attentional resources they actually devote to a particular task. When greater proportions of available resources are allocated to a task, ceteris paribus, the typical finding is an increase in cognitive fatigue. Task characteristics that relate to the proportion of effort allocated to a task fall into two categories: (a) motivational factors; and (b) factors that affect both motivation and arousal.

The most prominent motivational factor is whether the task is intrinsically interesting or enjoyable to the individual performing the task. As has been mentioned earlier, a lack of intrinsic interest in the task is a major necessary condition for the development of fatigue. When a task is intrinsically interesting, cognitive fatigue may not develop at all, or may not develop until there are physical fatigue factors (e.g. muscle fatigue) that must be overcome in order to persist in task performance. Numerous researchers have attempted to delineate the task characteristics that lead to intrinsic interest (e.g. Cooper, 1973; Dickey, 2005; Malone, 1981), but to date there are few, if any, task characteristics that are univocally associated with intrinsic interest. Instead, intrinsic interest appears to be transactional between some general characteristics of a task, the individual's motivational states or traits, the individual's interest in aspects of the task, and the extent to which the individual attempts to make the task engaging. For example, some video games are intrinsically interesting to a significant portion of the population of a particular age group, but may generate little or no interest in other age groups. Other tasks may lack any inherent intrinsic interest among a large portion of the population, but some individuals may *make* the task interesting by challenging themselves with goals for performance or learning. Even some of the most tedious tasks (e.g. proofreading, filing, product inspection, operating a copier) can be made interesting or engaging by some individual performers. To the degree that the task is made interesting or engaging, fatigue may have delayed onset or may not develop at all.

Notwithstanding an individual's desire to make a task interesting or engaging, if the task does not provide feedback or knowledge of the results, it is more likely to be associated with a decrement in the individual's motivation for performance over time (Sansone, 1986). When this occurs, fatigue is much more likely to develop, compared to tasks where feedback and/or knowledge of results are provided relatively frequently.

Task Characteristics that Affect both Motivation and Arousal

Two other task characteristics that affect motivation, which are also associated with more general arousal effects, are:

1. High-stakes situations; and
2. High rates of failure.

When the task milieu involves substantial rewards for success or substantial penalties for failure (referred to as a steep "utility of performance function," see Kanfer, 1987), the individual is likely to increase the proportion of available effort allocated to the task, compared to tasks where there are minimal rewards/punishments for different levels of performance. What is most notable about this type of task is that subjective fatigue is the variable that is more likely to show substantial increments with increasing time on-task, compared to performance decrements, because the individual may override an increasing desire to quit with increased resolve to maintain effort (e.g. see Dodge, 1913).

In contrast, when the individual faces high levels of failure in the task, he/she is likely to experience reduced motivation, and thus decrease effort as time on-task increases. Under these conditions, the most likely source of fatigue effects will be in terms of performance decrements. Subjective fatigue may either increase, if the individual attributes his/her failures in performance to internal influences, or may show no change, if the individual attributes his/her failures to external influences, such as an uncontrollable task (e.g. see Davis, 1946). However, when there are high stakes for task performance and/or the individual is faced with a high rate of task failures, the individual may also experience off-task distractions (such as frustration or anxiety), which result in decreased effort allocated to the task. In such cases, fatigue may increase at a faster rate than when attention is not drawn away from the task.

Factors Associated with Off-task Distractions

Some of the motivational factors mentioned above may have associated off-task distractions. For example, an individual may worry about errors made in the past or ruminate about ramifications of task failure. It is possible for characteristics of the task to have direct effects on the individual's arousal, leading to indirect effects on off-task distractions, without a direct effect on the individual's task motivation. The presence of general stressors during task performance (e.g. noise, temperature, vibration, poor illumination, poor ventilation) has often been implicated in a general tendency of performers to encounter mental/cognitive fatigue (see Laird, 1933; Persinger, Tiller & Koren, 1999; Thorndike, 1916). Although the presence of stressors may increase the individual's arousal level, which may in some cases lead to more effort available for the task, the individual who encounters these stressors is likely to also find that his/her attention is drawn away from the task to concerns about the stressful situation. Under such circumstances, subjective fatigue is likely to increase, even as task performance remains unchanged, decreases, or increases with additional time on-task. Whether the presence of such stressors is a part of the task itself, or an unrelated environmental variable, will depend on the overall task situation. If the stressor is separable from the task, then it is possible to map out the independent and interactive effects on fatigue from the task and the stressor.

In contrast, time pressure is typically an intrinsic characteristic of some tasks. When the individual must perform under time pressure, both arousal and off-task distractions may occur (especially when coupled with high failure or error rates). The presence of high levels of time pressure are typically expected to be associated with higher levels of fatigue, compared to low levels of time pressure (Teuchmann, Totterdell & Parker, 1999).

OTHER TASK CHARACTERISTICS

Other characteristics of tasks that have been associated with higher levels of fatigue do not fit into our larger framework, but have been replicated enough to merit mention. Three factors that appear to lead to higher levels of fatigue are:

1. Tasks that have high demands on visual attention (in contrast to auditory attention);
2. Tasks with homogeneous stimulus content; and
3. Tasks that have verbal content (in contrast to math content).

Higher levels of fatigue for visual attention tasks are consistent with the overarching view that visual attention tasks are generally more effortful and demanding than auditory tasks. This effect likely occurs because active control of attention is required for the visual tasks, while a substantial amount of auditory processing can occur with minimal levels of attentional effort (except when there are high levels of noise or the individual is engaged in divided attention across multiple auditory channels). Tasks with homogeneous content have been associated with greater susceptibility to "inhibition" (see Kray & Lindenberger, 2000), which is associated with an increase in difficulty for the individual to persist in task engagement over a period of time. If the information processing becomes more effortful with increasing time on-task, one can expect that there will be a greater depletion of attentional effort available, independent of the stimulus-specific influences on information processing. Finally, it is unclear what the source of greater fatigue is for tasks with verbal content, compared to math content. It may be that studies of these phenomena have not equated the intellectual difficulty of the tasks, or that there is greater consistency of information processing for math problem-solving than for tasks like reading comprehension or solving analogies. At this point, the underlying causes of such differences remain yet to be determined.

We have reviewed task characteristics that have been implicated in increased levels of subjective fatigue and performance decrements associated with fatigue. Three major categories of effects of particular task characteristics have been discussed, namely those that pertain to the amount of attentional effort available to the individual, those that affect the proportion of available attentional effort actually devoted to the task, and those that are associated with off-task distractions during task performance. In practice, these characteristics are often present in various combinations within a single task. In fact, some tasks may involve many or most of these factors. On the one hand, professional certification tests often require individuals to complete extensive testing over the course of multiple days. For example, admission to the Bar (a certification test completed by lawyers to be allowed to practice law in the USA), requires examinees to complete two to three days of consecutive testing. An international test for Chartered Financial Analyst certification requires three separate eight-hour tests over ten-hour testing periods. College entrance examinations (such as the SAT in the USA) are often less extensive (e.g. four or five hours of time on-task), but have some of the same characteristics. These tests have most of the characteristics described above (substantial time on-task, high demands on intellectual functioning, high demands for attention to detail, low tolerance for errors, time pressure, high stakes, lack of knowledge of results/feedback, and so on). In the case of the SAT test, the high-stakes nature of the test appears to lead to little or no general decrements in performance over a five-hour test session, even as average subjective fatigue steadily increases with increasing time on-task (Ackerman & Kanfer, 2009).

On the other hand, few day-to-day occupational tasks have as many of the task characteristics that are implicated in fatigue effects as these one-time certification tests. Many attentionally demanding real-world tasks are themselves discrete, rather than continuous. Other tasks allow for breaks.

There are also government regulations around the world that take fatigue issues into account. According to the U.S. Department of Labor (2009), only 9 states out of 50 have statutory regulations regarding breaks during the working day for employees in the private sector. Although there is some variance, the modal rule is for a 10-minute rest period for every 4 hours of continuous work. Interestingly, in the European Community, adults are allowed a break if they work six hours continuously (European Commission Working Time Directive, 2003). Many jobs, and perhaps most jobs, do not require continuous exertion of high levels of mental effort (e.g. see Brown, 1994). For some jobs, an informal mental break or slowdown during the working day may have little or no consequences for the individual worker's performance. For other jobs, sufficient recovery may not take place in the 10-minute rest period.

Similarly, operators may adjust the ebb and flow of their work in a way that provides inherent opportunities for periodic recovery from fatigue. Nonetheless, subjective reports of fatigue among workers suggest that fatigue is frequently experienced in the workplace. For example, Åhsberg noted that:

> concerning work-related fatigue it has recently been shown that of 14,400 employees 41.8% reported to feel "physically exhausted" every week, and 33.2% employees reported to feel "tired and listless" every week. Another Swedish example is from 522 workers, employed in production and distribution of electricity, where 11% reported themselves to be physically tired and 5% mentally tired after a working day. (1998: 1)

Of course, these data do not control for other factors, such as chronic or acute sleep deprivation, that are external to the job tasks themselves.

CONCLUSIONS

Numerous characteristics of tasks that have been identified as contributing to the development and expression of cognitive fatigue have been reviewed in this chapter. Ultimately, whether fatigue occurs or not for a given individual in a particular task situation is dependent on these task characteristics, the interests and motivation of the individual performing the task, whether the task is intrinsically interesting to that individual, or whether the task *can be made* intrinsically interesting to the individual. Although some individuals are drawn to particular job tasks by their interests, others are selected for the job by organizational human resources procedures (e.g. selection tests or other filtering procedures, such as realistic job previews). Ideally, there is a good match between the task demands and the individual's intrinsic interests, in which case cognitive fatigue will be less likely to occur. This would be considered a situation in which there is a good *fit* between the job requirements and the person performing the job. When there is an insufficient supply of individuals who provide a good fit to the job task requirements, the result is that the a priori condition for fatigue (i.e., that the task is not interesting or enjoyable to the individual performing it) will be present. Fatigue effects may be ameliorated to a degree by job redesign or changes to the work environment (e.g. eliminating or reducing external stressors), but will probably not be eliminated.

REFERENCES

Ackerman, P.L. (2011). 100 Years without Resting, in P.L. Ackerman (ed.) *Cognitive Fatigue: The Current Status and Future for Research and Application*. Washington, D.C.: American Psychological Association, 11–37.

Ackerman, P.L. & Kanfer, R. (2009). Test length and cognitive fatigue: An empirical examination of effects on performance and test-taker reactions. *Journal of Experimental Psychology: Applied*, 15, 163–81.

Åhsberg, E. (1998). *Perceived fatigue related to work*. Report, University of Stockholm National Institute for Working Life, 19.

Åhsberg, E. (2000). Dimensions of fatigue in different working populations. *Scandinavian Journal of Psychology*, 41, 231–41.

Arai, T. (1912). Mental fatigue. *Contributions to Education*, New York: Teachers College, Columbia University, 54, 1–115.

Baranski, J. (2007). Fatigue, sleep loss, and confidence in judgment. *Journal of Experimental Psychology: Applied*, 13(4), 182–96.

Baranski, J., Gil, V., McLellan, T., Moroz, D., Buguet, A. & Radomski, M. (2002). Effects of Modafinil on cognitive performance during 40 hours of sleep deprivation in a warm environment. *Military Psychology*, 14, 23–48.

Bartley, S. & Chute, E. (1947). *Fatigue and Impairment in Man*. New York: McGraw Hill.

Broadbent, D.E. (1953). Neglect of the surroundings in relation to fatigue decrement in output, in W.F. Floyd & A.T. Welford (eds) *Symposium on Fatigue*. London: H. K. Lewis & Co., 173–8.

Brown, I. (1994). Driver fatigue. *Human Factors*, 36, 298–314.

Chalder, T., Berelowitz, G., Pawlikowska, T., Watts, L., Wessely, S., Wright, D. & Wallace, E. (1993). Development of a fatigue scale. *Journal of Psychosomatic Research*, 37, 147–53.

Cooper, R. (1973). Task characteristics and intrinsic motivation. *Human Relations*, 26, 387–413.

Davis, D. (1946). This disorganization of behaviour in fatigue. *Journal of Neurology, Neurosurgery, and Psychiatry*, 9, 23–29.

Deci, E.L. & Ryan, R.M. (1987). The support of autonomy and the control of behavior. *Journal of Personality and Social Psychology*, 53, 1024–37.

Demerouti, E., Bakker, A., Nachreiner, F. & Schaufeli, W. (2001). The job demands – resources model of burnout. *Journal of Applied Psychology*, 86, 499–512.

Dickey, M. (2005). Engaging by design: How engagement strategies in popular computer and video games can inform instructional design. *Educational Technology Research and Development*, 53, 67–83.

Dodge, R. (1913). Mental work: A study in psychodynamics. *Psychological Review*, 20, 1–42.

European Commission Working Time Directive (2003). Available at: http://eur-lex.europa.eu/LexUriServ/LexUriServ.do?uri=CELEX:32003L0088:EN:NOT

Fisk, A. & Schneider, W. (1981). Control and automatic processing during tasks requiring sustained attention: A new approach to vigilance. *Human Factors*, 23, 737–50.

Grandjean, E. (1968). Fatigue: Its physiological and psychological significance. *Ergonomics*, 11, 427–36.

Hockey, G.R.J. (1997). Compensatory control in the regulation of human performance under stress and high workload: A cognitive–energetical framework. *Biological Psychology*, 45, 73–93.

Hockey, G.R.J. (2011). A motivational control theory of cognitive fatigue, in P.L. Ackerman (ed.), *Cognitive Fatigue: Multidisciplinary Perspectives on Current Research and Future Applications*. Washington, D.C.: American Psychological Association, 167–83.

Hockey, G., Maule, A., Clough, P. & Bdzola, L. (2000). Effects of negative mood state on risk in everyday decision making. *Cognition and Emotion*, 14, 823–56.

Kahneman, D. (1973). *Attention and Effort*. Englewood Cliffs, NJ: Prentice Hall.

Kanfer, R. (1987). Task-specific motivation: An integrative approach to issues of measurement, mechanisms, processes, and determinants. *Journal of Social and Clinical Psychology*, 5, 237–64.

Kray, J. & Lindenberger, U. (2000). Adult age differences in task switching. *Psychology and Aging*, 15, 126–47.

Laird, D. (1933). The influence of noise on production and fatigue, as related to pitch, sensation level, and steadiness of noise. *Journal of Applied Psychology*, 17, 320–30.

Leone, S., Huibers, M., Knottnerus, J. & Kant, I. (2008). A comparison of the course of burnout and prolonged fatigue: A 4-year prospective cohort study. *Journal of Psychosomatic Research*, 65, 31–38.

Lewin, K. (1935). *A Dynamic Theory of Personality: Selected Papers*. Translated by D.K. Adams & K.E. Zener. New York: McGraw Hill.

Malone, T. (1981). Toward a theory of intrinsically motivating instruction. *Cognitive Science: A Multidisciplinary Journal*, 5, 333–69.

Matthews, G. & Desmond, P. (1998). Personality and multiple dimensions of task-induced fatigue: A study of simulated driving. *Personality and Individual Differences*, 25, 443–58.

Michielsen, H., De Vries, J., Van Heck, G., Van de Vijver, F. & Sijtsma, K. (2004). Examination of the dimensionality of fatigue: The construction of the Fatigue Assessment Scale (FAS). *European Journal of Psychological Assessment*, 20, 39–48.

Mosso, A. (1906). *Fatigue*. Translated by Margaret Drummond & W.B. Drummond. New York: G.P. Putnam's Sons.

Myers, C. (1937). Conceptions of mental fatigue. *American Journal of Psychology*, 50, 296–306.

Noll, V. (1932). A study of fatigue in three-hour college ability tests. *Journal of Applied Psychology*, 16, 175–83.

Nolte, R., Wright, R., Turner, C. & Contrada, R. (2008). Reported fatigue, difficulty, and cardiovascular response to a memory challenge. *International Journal of Psychophysiology*, 69, 1–8.

Norman, D. & Bobrow, D. (1975). On data-limited and resource-limited processes. *Cognitive Psychology*, 7, 44–64.

Pattyn, N., Neyt, X., Henderickx, D. & Soetens, E. (2008). Psychophysiological investigation of vigilance decrement: Boredom or cognitive fatigue? *Physiology and Behavior*, 93, 369–78.

Persinger, M., Tiller, S. & Koren, S. (1999). Background sound pressure fluctuations (5 dB) from overhead ventilation systems increase subjective fatigue of university students during three-hour lectures. *Perceptual and Motor Skills*, 88, 451–6.

Sansone, C. (1986). A question of competence: The effects of competence and task feedback on intrinsic interest. *Journal of Personality and Social Psychology*, 51, 918–31.

Schmidtke, H. (1976). Disturbance of processing of information, in E. Simonson & P.C. Weiser (eds), *Psychological Aspects and Physiological Correlates of Work and Fatigue*. Springfield, IL: Charles C. Thomas, 336–405.

Shiffrin, R. & Schneider, W. (1977). Controlled and automatic human information processing: II. Perceptual learning, automatic attending, and a general theory. *Psychological Review*, 84, 127–90.

Takeda, Y., Sugai, M. & Yagi, A. (2001). Eye fixation related potentials in a proofreading task. *International Journal of Psychophysiology*, 40, 181–6.

Teuchmann, K., Totterdell, P. & Parker, S. (1999). Rushed, unhappy, and drained: An experience sampling study of relations between time pressure, perceived control, mood, and emotional exhaustion in a group of accountants. *Journal of Occupational Health Psychology*, 4, 37–54.

Thackray, R. & Touchstone, R. (1991). Effects of monitoring under high and low task load on detection of flashing and colored radar targets. *Ergonomics*, 34, 1065–81.

Thorndike, E.L. (1926). *Educational Psychology*. Vol. 3: *Mental Work and Fatigue and Individual Differences and their Causes*. New York: Teachers College, Columbia University.

Thorndike, E., McCall, W. & Chapman, J. (1916). Ventilation in relation to mental work. Teachers College, Columbia University Contributions to Education, 78. New York: Teachers College.

Tucker, P. (2003). The impact of rest breaks upon accident risk, fatigue, and performance: A review. *Work & Stress*, 17, 123–37.

U.S. Department of Labor (2009). Minimum paid rest requirements under state law for adult employees in private sector – January 1, 2010. Available at: http://www.dol.gov/whd/state/rest.htm

Wickens, C. (1984). Processing resources in attention, in R. Parasuraman & D.R. Davies (eds), *Varieties of Attention*. New York: Academic Press, 63–102.

7
Fatigue in Sports Psychology

Timothy David Noakes

Now if you are going to win any battle you have to do one thing. You have to make the mind run the body. Never let the body tell the mind what to do. The body will always give up. It is always tired morning, noon and night. But the body is never tired if the mind is not tired. When you were younger the mind could make you dance all night, and the body was never tired ... You've always got to make the mind take over and keep going. (George S. Patton, US Army General, 1912 Olympian)

INTRODUCTION

Why does fatigue occur during exercise? And what exactly is the origin and purpose of that fatigue? Does an understanding of exercise-induced fatigue enhance our comprehension of the fatigue that is the single most common reason that patients seek the help of medical practitioners? The past decade has seen major advances in our understanding of this complex phenomenon. This new knowledge improves our understanding of the reasons why fatigue develops during exercise and suggests novel insights into the likely origins of the abnormal fatiguability that develops in those afflicted by, amongst other conditions, the chronic fatigue syndrome(s).

NOVEL INSIGHTS INTO THE MECHANISMS CAUSING FATIGUE DURING EXERCISE

Currently there are two popular models to explain the fatigue that develops during exercise – the peripheral and central fatigue models (Noakes, Crewe & Tucker, 2009). Both are catastrophic models (Noakes, St Clair Gibson & Lambert, 2004; Noakes, St Clair Gibson & Lambert, 2005; St Clair Gibson & Noakes, 2004), in which fatigue leading to the termination of exercise develops only after there has been a failure of homeostasis in one or other bodily part. These models conflict with the historical understanding that human physiology functions with one overriding purpose: "All the vital mechanisms, varied as they are, have only one object, that of preserving constant the conditions of life in the internal environment" (Bernard, 1957). Importantly neither model allows the human brain the freedom to anticipate what will occur in the future and hence to take preventive action to insure that the biological catastrophe does not occur. These explanations also conflict with another fundamental concept which holds that the human brain exists specifically to protect us from harm (Herbert, 2007).

More recently we have proposed a different model, in which the brain regulates exercise performance "in anticipation" specifically to insure protection of the "milieu intérieur." One of the key components of this model is its proposal that the brain uses fatigue as a disagreeable emotion to insure that exercise always terminates before there is a catastrophic failure of biological function. We have called this the Central Governor Model (CGM) Noakes, 1997; Noakes, 1998; Noakes, 2000; Noakes, Peltonen & Rusko, 2001; Noakes, St Clair Gibson & Lambert, 2004; Noakes, Calbet, et al., 2004; Noakes, St Clair Gibson & Lambert, 2005; Noakes, 2007a; Noakes, 2011a; Noakes, 2011b). Here I explain how we believe the CGM works and how this new model posits an entirely novel explanation for the reason why fatigue must develop during exercise. If we are correct, the conclusion must be that all states of abnormal fatigue, including those currently classified as the chronic fatigue syndrome(s), must occur because of specific disorders within the brain.

THE CENTRAL GOVERNOR MODEL OF HUMAN EXERCISE PERFORMANCE

Humans do not suffer a catastrophic biological failure each time we exercise. Instead it now appears that our species evolved as long-distance persistence hunters (Bramble & Lieberman, 2004; Lieberman & Bramble, 2007; Liebenberg, 2006; Liebenberg, 2008) with a capacity to regulate our body temperatures during exercise in the heat greater than all other creatures on this planet. Millions of years of evolutionary pressure fashioned our species with a profoundly robust metabolic capacity, best exemplified by our capacity to exercise for prolonged periods (Noakes, 2006) especially in dry heat. At one time in the last 1–2 million years there may have been as many as four different ancestral hominid species active simultaneously on the African continent. Yet only one of those four species – that which led to *Homo sapiens* – survived. *Homo sapiens* survived because we are of remarkably robust stock, fashioned by the most demanding evolutionary past.

So it is perhaps not surprising that there is no published evidence showing that any biological system fails in healthy humans even when exercise is performed to a "maximum" capacity. For example, it is not possible willfully to drive humans to the point at which they develop thermoregulatory failure leading to heat stroke. Rather, the rising sensations of fatigue cause the termination of exercise before a catastrophe can occur. To develop an experimental model of heat stroke, mammals other than humans must be studied. This occurs because the human body contains numerous redundant control mechanisms (Lambert, St Clair Gibson & Noakes, 2005), the goal of which is to modify exercise performance under all conditions specifically to insure that the exercise terminates while the body is still in homeostasis.

An important contribution of our research over the past decade has been to establish that fatigue is the key regulator which insures that exercise, at least in healthy humans, always terminates before a biological catastrophe can develop. Thus we have concluded that fatigue is a brain-directed emotion (St Clair Gibson, Lambert & Noakes, 2001) not a physical event shown, for example, as a reduced ability of the exercising muscles to produce force, the more usual explanation. This does not mean that a physical form of "fatigue" does not occur; indeed, it is clear that skeletal muscles do tire during exercise (Amann, Eldridge et al., 2006; Amann, Romer et al., 2007; Amann & Dempsey, 2008). But our interpretation is that such skeletal muscle (peripheral) "fatigue" is not the key regulator of exercise performance since if it so chose, the brain could override that fatigue by increasing skeletal muscle recruitment, thereby reducing the extent of skeletal muscle reserve that is always present at fatigue (see later), thereby maintaining the exercise performance.

We have shown that the sensations of fatigue develop as a function of the exercise duration (Noakes, 2004) and so are not related directly to the biological changes that are present in the body at that time. These rising feelings of discomfort ultimately become unbearable, causing the termination of exercise before the point of biological failure is reached in healthy subjects.

This concept that fatigue is purely an emotion, rather than a physical event, has important implications for our understanding of the psychology of sporting performance. For it means that the sensation of fatigue must be under some form of subconscious control and could therefore be modified if the mechanism of its production were better understood. That some athletes under unique circumstances are indeed able to modify these sensations and so produce inexplicable performances is clear (see "A case study – The presence of biological reserve in Olympic competition" in the following pages).

WHAT BIOLOGICAL VARIABLES ARE SENSED DURING EXERCISE IN ORDER TO PRODUCE THE SENSATIONS OF FATIGUE SO THAT A CATASTROPHIC FAILURE CAN BE PREVENTED?

The traditional exercise model holds that exercise is limited by a failure of oxygenation of the exercising skeletal muscles (Noakes, 2008a; Noakes, 2008b). The hypothetical failure of oxygen delivery produces a state of anaerobiosis in the exercising muscles that in turn stimulates the production of lactic acid that "poisons" the exercising muscles, causing fatigue. But this seems to be improbable as there is very little evidence that exercising skeletal muscles develop significant hypoxia (reduced partial pressure of oxygen) and certainly none that an absence of oxygen (anaerobiosis) occurs during exercise (Noakes & St Clair Gibson, 2004).

Instead exercise in extreme hypoxia at real or simulated high altitudes (Amann, Eldridge et al., 2006; Amann, Romer et al., 2007; Calbet, Boushel et al., 2003a; Calbet, Boushel et al., 2003b; Calbet, De Paz et al., 2003; Green et al., 1989; Kayser et al., 1994; Noakes, 2007b; Noakes, 2009) and in severely hot conditions (Gonzalez-Alonso et al., 1999; Nybo & Nielsen, 2001) always terminates without any evidence for homeostatic failure. Exercise of high intensity also always terminates without any evidence that profoundly reduced pH levels are reached or that all the available sources of energy production are maximally utilized (Calbet, De Paz et al., 2003). Thus one or more biological variables must be sensed in the body so that the exercise behavior is modified to insure that exercise terminates even though those variables are within a safe homeostatic range.

More recently the concept has arisen that changes in cerebral oxygenation may regulate exercise performance and influence the decision when to terminate exercise (Rasmussen, Nybo et al., 2010; Rasmussen, Nielsen et al., 2010; Nybo & Rasmussen, 2007a; Rupp &

Perrey, 2008). Thus a falling arterial partial pressure of oxygen (PaO_2) must be prevented if cerebral function is to be sustained and the brain protected. The logical way in which a dangerous fall in PaO_2 can be prevented is to control the function of the organ that consumes the largest volume of oxygen during exercise – the active skeletal muscles (Noakes, Peltonen & Rusko, 2001). Logically this would be regulated by a brain controller sensitive to, amongst other variables, an index of arterial oxygenation and which acts by reducing skeletal muscle recruitment (activation) whenever the PaO_2 threatens to fall below the safe, homeostatically regulated range. That such a controller acts to modify exercise behavior in this way during exercise in hypoxia was established already in 1994 (Kayser et al., 1994) but its significance has been appreciated only more recently (Kayser, 1996; Kayser, 2003; Noakes, 2007b; Noakes, 2009).

Similarly, during exercise in the heat, the effort terminates before a dangerous core body temperature is reached, that is, before there is a catastrophic failure of thermal regulation (Tucker, Rauch et al., 2004; Tucker, Marle et al., 2006). Thus, when they are able freely to modify their exercise behavior – that is, when they are able to pace themselves during exercise – humans reduce their exercise intensity in response to a rising environmental heat load, long before the core body temperature reaches a dangerous level. This is an excellent example of how the brain does not wait for the catastrophe to have already occurred. Rather, the brain anticipates the future and takes corrective action to avoid the catastrophe.

Numerous studies have established that humans voluntarily terminate prolonged exercise in the heat at a fixed work rate at similar core temperatures, regardless of the pre-exercise body temperature (Gonzalez-Alonso et al., 1999; Walters et al., 2000), the ambient temperature (Gonzalez-Alonso et al., 1999; Galloway & Maughan, 1997), state of heat acclimatization (Nielsen et al., 1993) or rate of rise of core temperature (Gonzalez-Alonso et al., 1999). Some have therefore concluded that fatigue always occurs when a critical core body temperature of approximately 40°C is reached during exercise in the heat (Nybo & Nielsen, 2001; Nielsen et al., 1993). The presumption is that this elevated temperature causes a catastrophic failure of brain function. This has been termed "central fatigue." But the model seems improbable since the (selfish) brain is designed to protect its functioning at the expense of the rest of the body. To allow itself to become "fatigued" before the rest of the body, in particular the exercising muscles, would seem to be a serious design flaw.

All the studies leading to this conclusion were performed in the laboratory at a fixed exercising work rate, at either a constant running or cycling speed. As a result the tested subject's brain was unable to modify the work rate, that is, to modify the exercise intensity in response to changing environmental and internal biological conditions. Furthermore, other studies have shown that when exercising in competition, many humans are able to continue exercising at much higher core body temperatures without any evidence for fatigue (Byrne et al., 2006; Maron, Wagner & Horvath, 1977; Ely, Ely, Cheuvront et al., 2009). Thus the 40°C value measured in the laboratory is not a maximum value; rather, it is clear that the motivation of competition can override this response. Thus what is termed "central fatigue" may simply be the consequence of a loss of motivation to continue the exercise.

But when subjects are allowed to pace themselves during exercise, they respond as if the body anticipates what will happen in the future so that it acts proactively to insure that dangerously high body temperatures are not reached, causing harm to the brain and body (Tucker, Rauch et al., 2004; Tucker, Marle et al., 2006; Marino, Lambert & Noakes, 2004;

Marino, 2004). This concept of "teleoanticipation" was first proposed by Ulmer (1996), in work that was largely ignored until we and others rediscovered it.

HOW MIGHT THE BRAIN REGULATE THE EXERCISE PERFORMANCE TO INSURE THAT A CATASTROPHIC FAILURE DOES NOT OCCUR?

In 1924 British Nobel Laureate, A.V. Hill, proposed that "maximal" exercise performance was limited by a falling cardiac output that resulted from a reduced blood flow to the coronary arteries supplying oxygen to the heart. He further proposed the existence of a "governor" in the heart or the brain that reduced the contractile function of the heart immediately the blood flow to the heart was insufficient, causing myocardial ischemia. When I rediscovered that theory in 1998 (Noakes, 1998), I realized that a more probable model would be one in which the brain did not "talk" to the heart but rather to the muscles. For in reality the heart is purely the slave to the muscles. Thus the most effective way to regulate the work of the heart is to control the work of the muscles. This can be done instantly by reducing the feed forward recruitment of the muscles by the brain and spinal cord (central nervous system). It was this logic that led to the development of the CGM.

Thus the CGM proposes that the brain regulates the mass of skeletal muscle that is recruited during exercise specifically to insure that bodily homeostasis is never threatened. Accordingly, the key prediction of this model is that the brain acts "in anticipation" to insure that exercise is always at an intensity and of a (safe) duration that does not threaten a catastrophic failure of homeostasis in any bodily system. The method of control is through moment to moment changes in the extent of skeletal muscle recruitment by the brain, since it is the activity of the skeletal muscles that places the body at risk during exercise by setting (1) the rate at which the body fuels, including oxygen, are used; (2) the level of work of the heart (since the heart is just the slave, principally to the demands of the muscles); and (3) the rapidity of bodily heat production. Thus this model predicts that while the heart may indeed be the slave to the working muscles (Noakes, Calbet et al., 2004), at least it has the brain on its side to insure that the muscles never make excessive demands on the pumping capacity of the heart.

THE IMPORTANCE OF PACING IN HUMAN BIOLOGY

The most obvious features of human endurance performance, characterized by swimming, running or cycling, are that athletes begin races of different distances at different paces (Tucker, Lambert & Noakes, 2006) and they consistently increase their pace during the final 5–10 percent of races lasting more than about 20 minutes (Noakes, St Clair Gibson & Lambert, 2004; Noakes, 2007a), the so-called "end-spurt phenomenon" (Catalano 1974). Neither feature can be explained by the traditional Hill model, which argues that the pacing strategy is determined exclusively by the action of "poisonous metabolites," which accumulate in the exercising muscles, impairing their function. For that model must predict that athletes begin all running races, regardless of distance, at the same very high speed which then falls progressively as the "poisonous" metabolites begin to accumulate, causing a slowing of the pace. But this most obviously does not occur.

Similarly, according to the Hill model, fatigue should be greatest near the end of a race as the concentrations of "poisonous" intramuscular metabolites reach their highest levels immediately prior to the termination of exercise. Thus that model is quite unable to explain why athletes are able to speed up immediately before the end of competitive events lasting more than a few minutes, reaching their highest speeds during the final few moments of those events.

But these observations are more easily explained by the CGM. Thus the different paces at the start of races of different durations can be explained by different levels of recruitment (activation) of the skeletal muscles in the exercising limbs; much less in a race of 1,000 km, much more in a race of 100m. That the extent of EMG activity, an indirect measure of the extent of skeletal muscle activation, rises as a linear function (Bigland-Ritchie & Woods, 1974; Henriksson & Bonde-Petersen, 1974) or curvilinear function (Gamet, Duchene, Garapon-Bar et al., 1993; Hug et al., 2004; Laplaud, Hug & Grelot, 2006) of increasing power output (exercise intensity), is an established finding; thus this possibility is not purely hypothetical. Furthermore, magnetic resonance imaging also shows that muscle activation increases as a function of the power output (Akima, Kinugasa & Kuno, 2005). Similarly, the increase in power output that can be achieved during the endspurt (Catalano, 1974) immediately prior to the end of an exercise bout is associated with an increase in EMG activity, indicating increased skeletal muscle activation (Tucker, Rauch et al., 2004; Tucker, Kayser et al., 2007; Ansley, Schabort et al., 2004; Nummela, Heath et al., 2008; Nummela, Paavolainen et al., 2006).

Thus the CGM proposes that the initial pace during exercise is based on prior experience of the average pace that can be sustained for the expected duration of the exercise bout (Swart et al., 2009b; Tucker, 2009; Tucker & Noakes, 2009). As a result, the appropriate number of motor units is recruited by the brain (central command) in the exercising limbs at the start of exercise; more for higher intensity exercise of short duration, less for more prolonged exercise of a lower intensity. Afferent feedback is sent continually from multiple systems in the body to the brain as the exercise progresses and this produces a continuous modification of the pace as is typically observed in all forms of competitive exercise (Tucker, Lambert & Noakes, 2006; Tucker, Bester et al., 2006). As a result calculations are made and efferent commands are relayed back to the active muscles in order to complete the required task within the metabolic and biomechanical limits of the body (Hampson et al., 2001). This control can only be observed during "self-paced" exercise, in which the subject is free to vary his or her work rate, and not when the exercise intensity is fixed and imposed by the researcher, the more usual form of laboratory exercise.

Once the importance of studying self-paced exercise instead of exercise at a constant, externally imposed workload was grasped, it has become rather easier to show that exercise regulation "in anticipation" is a normal, indeed defining, feature of human physiology. This anticipatory response cannot be explained by a system that is regulated solely in the periphery (Noakes & Marino, 2007) and which is activated only after a catastrophic failure of homeostasis has occurred in either the exercising muscles or in the brain.

THE PRESENCE OF SKELETAL MUSCLE RESERVE AT EXHAUSTION IN ALL FORMS OF EXERCISE

It is a common observation that victorious athletes do not stop running in an exhausted state the instant they cross the finishing line of races of any distance from 100m to 42km. Instead, the marathon victor proudly runs a victory lap with her national flag draped around her torso, acknowledging the cheers of the admiring crowd. Then she springs without apparent fatigue onto the victory dais to accept her gold medal. There is little evidence for the stress to which her body was exposed just minutes or hours earlier.

Probably the most emphatic such display occurred at the finish of the 1964 Olympic Games Marathon when the victor, Ethiopian Abebe Bikila, performed a series of calisthenic exercises as he awaited the arrival some minutes later of the rest of the vanquished marathon field. A model which requires that the exercise pace is regulated by the accumulation of "poisonous" metabolites that reach their maximum concentrations the instant before the finish line is crossed cannot explain these easily observed phenomena.

Instead, the CGM predicts that all forms of exercise are regulated to insure that there is always a biological reserve, that is, that all available motor units in the exercising limbs are never fully activated to the point of their complete exhaustion. Indeed, a number of studies show that even during a so-called maximal exercise test, only about 60 percent of the available motor units are activated in the exercising limbs at the point of exercise termination (Albertus, 2008; Sloniger et al., 1997). The presence of this biological reserve indicates that the athlete could have gone faster or further (by recruiting additional, previously unrecruited motor units in the exercising limbs) if the athlete's brain had wished to do so. Instead the brain must have had a very good reason to terminate the exercise even in the presence of such untapped, reserve capacity.

There is an interesting corollary to this finding. It is that if an athlete finishes a close second in a sporting event but does not die, then that athlete could have run faster (since he finished with unused reserve).

Thus, finishing second may be even more of a personal choice than is finishing first. One focus of sports psychology in the future should be to understand what factors determine the choice to come second. Or first.

CONCLUSIONS

A model of exercise in which an intelligent brain controller regulates the exercise performance "in anticipation" specifically to insure that a catastrophic failure of biological function is avoided, is the only explanation for the three related phenomena commonly observed in sporting competitions – the adoption of different paces from the outset of exercise, which are determined by the known duration of the exercise bout; the end-spurt, and the presence of skeletal muscle reserve at all times in all forms of exercise. This is the opposite of the outcome predicted by the catastrophic fatigue models that have dominated the teaching of the exercise sciences for the past 100 years or more.

The Conscious Perception of Fatigue

The traditional catastrophic models of exercise performance predict that fatigue occurs because of a failure of the functioning of the skeletal muscles; this failure continues progressively and irreversibly until the system fails catastrophically at the point of exhaustion. Once this state of absolute fatigue has been reached, recovery can occur only after a period during which homeostasis is restored (St Clair Gibson & Noakes, 2004). The analogy of this catastrophe model is the jockey trying to cajole a dead horse to rise up and gallop to the finish line. Regardless of how much the jockey might wish it to happen and how forcefully he applies his whip to the rump of the deceased animal, the brain-dead horse is unable to respond. Similarly according to this model, as the muscles fail progressively during exercise, the brain, like the jockey's vigorously flailing whip, cannot persuade the exhausted muscles to produce more work.

Thus the traditional fatigue models predict that the brain has no capacity to prevent the inexorable development of a progressive failure of muscle function.

The traditional catastrophic models also predict that fatigue is a sensation developed by the brain in direct proportion to the extent of the homeostatic failure that is developing or has developed in the active muscles. But this makes little sense. For if exhaustion is the irreversible consequence of a failure of peripheral muscle function that cannot be overridden by the brain, then it makes no sense for the brain to "know" exactly what is the extent of the skeletal muscles' distress. For the brain has no need of that knowledge for it cannot alter what is the inevitable outcome – the jockey on the dead horse analogy. *Thus the very presence of the sensations of fatigue prove that exercise is a regulated phenomenon and one in which those very sensations must play an integral regulatory role.*

The CGM posits that the conscious sensation of fatigue is protective since it influences behaviour in order to insure that no exerciser overrides the subconscious reductions in skeletal muscle recruitment and power output and so places the body at risk by continuing to generate too much power (and heat) for too long (Noakes, 2004; Tucker, Rauch et al., 2004; Tucker, Marle et al., 2006; Albertus et al., 2005; Morrison, Sleivert & Cheung, 2004; Tucker, 2009). Or, in the words of Lehmann et al. (Hultman & Nielsen 1971), exercise may terminate "when the sum of all negative factors, such as fatigue and muscle pain, are felt more strongly than the positive factors of motivation and will power."

It is traditionally argued that the perception of effort measured as the Rating of Perceived Exertion (RPE) is derived from the integration of a number of afferent signals (St Clair Gibson, Lambert & Noakes, 2001; Hampson et al., 2001) including such cardiopulmonary factors as the respiratory rate, minute ventilation, heart rate and oxygen uptake, as well as peripheral factors such as blood lactate concentrations, mechanical strain, and skin and core temperature. But we have recently shown that this interpretation is not the complete truth.

Rather, an initial analysis of the data of Baldwin et al. (2003), who studied subjects exercising at a constant work rate until they fatigued, when starting with different muscle glycogen concentrations, showed that the RPE values reported by those subjects were not different between trials at either the start or the finish (Noakes, 2004) although values were higher in both trials at the end of exercise. The RPE rose as a linear function of exercise duration and the rate of rise was faster in the shorter duration (low glycogen) trial. But when analyzed relative to the percentage of exercise duration, the RPE increased at the same rate in both trials, indicating the presence of a previously unrecognized biological phenomenon. Numerous other studies have now confirmed this observation even during high-intensity exercise of short duration, including that used for the measurement of the maximum oxygen consumption (Eston, Lamb et al., 2005; Eston, Faulkner, Mason & Parfitt, 2006; Eston, Shepherd et al., 2007; Eston, Faulkner, St Clair Gibson, Noakes & Parfitt, 2007; Faulkner, Parfitt & Eston, 2007).

For example Crewe, Tucker & Noakes (2008) examined changes in RPE and thermoregulatory variables during exercise in which the work rate was fixed for the duration of the exercise bout and was not free to vary. Subjects were required to exercise to exhaustion at a range of exercise intensities in either hot (H) or cool (C) conditions. It was found that the RPE increased linearly in all exercise trials, irrespective of the exercise intensity and the environmental temperature. The linear increase occurred from the onset of the trials, suggesting that the subconscious brain is able to forecast the duration of exercise and then set the rate of increase in RPE so that volitional fatigue occurs before, in this case, catastrophic hyperthermia can develop. The rate of increase in RPE then determines the time to volitional fatigue. The study confirmed that, when expressed as a percentage of the completed exercise duration, the RPE increased at the same rate in all conditions, regardless of temperature or exercise intensity.

These findings suggest that, at the onset of exercise, the subconscious brain calculates the exercise duration that can be maintained without disrupting whole body homeostasis. The brain then generates a conscious sensation of exertion in a forecasted (feed-forward) manner as a function of the percentage of the total exercise time that has been completed (or which still remains). The result is that the maximal tolerable RPE occurs at the same moment that the full exercise duration (100 percent of completed bout) is achieved (Swart et al., 2009b; Tucker, 2009). Thus, it is suggested that the rate at which the RPE increases will determine the total duration of exercise (Noakes, 2004).

The crucial point is that, presently, only the CGM can explain these findings. It is also the sole model that explains all the currently reported findings of how the pace is set during exercise and why and when the decision to terminate exercise is taken.

A case study – The presence of biological reserve in Olympic competition
One of the greatest-ever sporting performances was achieved by Michael Phelps at the 2008 Olympic Games where he won eight gold medals for swimming. A key moment in his success was the final of the 4x100m freestyle final, in which the French team, anchored by Alain Bernard, the 100m freestyle world record holder, began the final 100m lap, 0.4 seconds ahead of the United States team anchored by Jason Lezak. In the course of their respective careers Lezak was consistently at least 0.3 seconds slower than Bernard. Thus the moment Lezak entered the water, the reasonable assumption was that the French team would win the gold medal by at least 0.7 seconds and that Michael Phelps would have to be satisfied with just seven medals in the 2008 Olympics.

But the result was quite different, as Lezak swam the fastest 100m in the history of the sport, finishing in 46.06 seconds compared to Bernard's 46.73 seconds. As a result, the Untied States won the gold medal by a fingertip and Michael Phelps was assured of his eighth gold medal.

Four days later, the 100m male freestyle final was held. Since Lezak had swum 0.57 seconds faster than any of the other 31 competitors in that 4x100m relay final, it was perhaps reasonable to assume that he would also win the 100m final. Yet in the end the historic performance difference between Bernard and Lezak was restored; Bernard won the gold medal swimming 0.46 second faster than Lezak, who claimed the third place. This performance difference between these two swimmers remained unchanged in 2009.

The point of the anecdote is to show that under certain unique circumstances, some individuals can produce performances that are so totally unexpected as to appear inexplicable. Such performances are best understood by the presence of a biological reserve that is only accessible under unique circumstances when the brain can be convinced that for a short time it can delve even further into that reserve without risking a biological failure.

A key component of this effect must be the suppression of the brain-generated sensations of fatigue, which are the ultimate regulator of the exercise performance. Somehow the best athletes manage to sustain higher levels of performance even as they experience the same intense feelings of discomfort as do all their defeated competitors.

Some historical and current studies which confirm that exercise is regulated in anticipation by central motor command that responds also to the presence of feedback from a variety of different organs

Figure 7.1 shows some of the key evidence that supports the predictions of the CGM. Starting from the top and rotating clockwise, the figure presents the evidence that:

1. The ultimate regulation of the exercise performance resides in the central nervous system, the function of which is potentially modifiable by interventions that act exclusively on central (brain) mechanisms.
2. The exercise pace is set "in anticipation."
3. The protection of homeostasis requires that there is always a skeletal muscle reserve during all forms of exercise.
4. Afferent sensory feedback can modify the exercise performance.
5. The exercise intensity may increase near the end of exercise – the end-spurt – even in the face of significant "fatigue."

An already large and growing number of studies support the interpretation that exercise is regulated by a complex intelligent system, provisionally termed the Central Governor Model. According to this model, the brain sets the exercise performance by determining the number of motor units that are recruited in the exercising limbs specifically to insure that the exercise terminates before there is a catastrophic biological failure. The figure identifies studies which show that:

1. Factors that act only on the central nervous system can influence exercise performance;
2. Exercise is regulated "in anticipation" (teleoanticipation) even during the VO_2max test;
3. There is always biological reserve at the point of exhaustion during all forms of exercise;
4. Afferent sensory feedback modifies exercise performance; and
5. There is usually an end-spurt in most forms of competitive exercise.

Thus, Figure 7.1 includes at least 30 studies which show that exercise performance can be modified by interventions such as music (Barwood et al., 2009; Lim et al., 2009), the use of placebos (Trojian & Beedie, 2008; Beedie et al., 2006; Foad, Beedie & Coleman, 2008), self-belief (Micklewright et al., 2010), prior experience (Mauger, Jones & Williams, 2009), time deception (Morton, 2009), knowledge of the endpoint (Wittekind, Micklewright & Beneke, 2009), the presence of other competitors (Wilmore, 1968), monetary reward

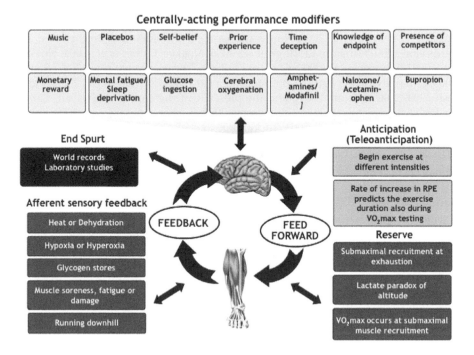

Figure 7.1   The Central Governor Model

(Cabanac, 1986), mental fatigue (Marcora, Staiano & Manning, 2009), sleep deprivation (Martin, 1981), glucose ingestion (Chambers, Bridge & Jones, 2009; Rollo et al., 2009), hand cooling (Kwon et al., 2010), cerebral oxygenation (Seifert et al., 2009; Rasmussen, Nybo et al., 2010; Rasmussen, Nielsen et al., 2010; Rupp & Perrey, 2008; Nybo & Rasmussen, 2007b; Billaut et al., 2010), centrally acting drugs or chemicals such as the amphetamines (Swart et al., 2009a), modafinil (Jacobs & Bell, 2004), naloxone (Sgherza et al., 2002), acetaminophen (Mauger, Jones & Williams, 2010), the cytokines IL-6 (Robson-Ansley et al., 2004) and IL-1b (Carmichael et al., 2006), bupropion (Roelands et al., 2009), all of which can reasonably be assumed to act exclusively or predominantly on the central nervous system. More recently it has been shown that even superstition (Damisch, Stoberock & Mussweller, 2010) can influence performance in skilled sporting activities. In contrast, the "brainless" A.V. Hill model (Noakes, 2008b) cannot explain how interventions that acts exclusively on the brain can influence athletic performance.

Similarly there are many other studies which support the predictions of the CGM specifically the presence of anticipatory feedforward control of skeletal muscle recruitment during exercise (Tucker, 2009; Tucker & Noakes, 2009; Ansley, Robson et al., 2004; Ansley, Schabort et al., 2004; Castle et al., 2006; Joseph et al., 2008; Marino, Lambert & Noakes, 2004; Marino, 2004); the presence of skeletal muscle recruitment reserve during exhaustive exercise (Amann, Eldridge et al., 2006; Swart, Lamberts, Lambert et al., 2009a; Marcora & Staiano, 2010; Albertus, 2008; Ross, Goodall, Stevens et al., 2010) especially at high altitude

(Kayser et al., 1994; Noakes, 2009); afferent sensory feedback that influences exercise performance (Tucker, Rauch et al., 2004; Tucker, Marle et al., 2006; Marino, Lambert & Noakes, 2004; Altareki et al., 2009; Flouris & Cheung, 2009; Morante & Brotherhood, 2008; Edwards et al., 2007; Amann, Eldridge et al., 2006; Noakes, Calbet et al., 2004; Noakes & Marino, 2007; Johnson et al., 2009; Clark et al., 2007; Tucker, Kayser et al., 2007; Rauch et al., 2005; Lima-Silva et al., 2010; Eston, Faulkner et al., 2007; Racinais et al., 2008; Marcora & Bosio, 2007; Baron et al., 2009); and the presence of the end spurt (Tucker, Lambert & Noakes, 2006; Noakes, Lambert & Human, 2009; Tucker, Kayser et al., 2007; Amann, Eldridge et al., 2006; Kay et al., 2001; Tucker, Rauch et al., 2004).

Relevance to Human and Sports Psychology

The adoption of the Hill model of human exercise physiology after 1923 produced a reductionist approach to the understanding of human exercise performance. In this model fatigue was explained by changes in the skeletal muscles as a result of either a failure of oxygen delivery (in high-intensity exercise) or of energy production from the stored carbohydrate fuel, glycogen (during very prolonged exercise). But this explanation has to be simplistic, since any deficit in energy production in the exercising muscles must lead ultimately to a depletion of skeletal muscle ATP concentrations, leading to skeletal muscle rigor (Noakes & St Clair Gibson, 2004). But this does not happen. Thus all the physiological processes in the body must be regulated by a central integrator specifically to prevent any such catastrophic failure. The brain is the sole organ able to insure that this occurs. But this will not be understood if we continue to adopt a reductionist approach to exercise physiology, especially if that reductionism excludes any role for the brain, as is the current approach (Levine, 2008; Joyner & Coyle, 2008).

Indeed, the crucial weakness of the reductionist (A.V. Hill) peripheral model of exercise fatigue is that the model allows no role for the brain and hence for a host of factors such as motivation, "mental toughness," prior experience and pharmacological manipulations which so obviously have major effects on human performance. Continuing to ignore the obvious would not seem to be particularly wise.

Instead, the CGM shows the way in which any of a number of variables can influence the manner in which the body functions during exercise. Perhaps for the first time it also allows the conclusion that the condition of fatigue, whether it occurs in sport or in any other human behaviour, is generated in the brain, not in the peripheral organs, and has biological value. Thus, to understand fatigue wherever it occurs requires that the function of the brain in generating that sensation must be considered. The great American track high hurdler, Edwin Moses, once wrote that: "Everything affects the way I run." Figure 7.1 explains why this is so.

The extension of this statement is that "everything" affects the way I experience fatigue, be it in work or in play. This brain-directed model of fatigue should allow a more thorough and ultimately more rewarding investigation of the phenomenon of fatigue, wherever it occurs. Clearly this has important implications for sports psychologists interested in maximizing the athletic performances of their athletic clients.

Interestingly Hill understood the role of the brain in the regulation of skilled movements, for he wrote in 1933: "In the case of bodily movement the nervous system is the steersman, who has to compound all the messages – the nerve waves – he receives to form one general impression on which to act [...] The continual reaction between muscles, nerves, end-organs and central nervous system is the physiological basis of muscular skill, and on its smooth and efficient working depend many of the things that mankind finds worthy of accomplishment" Hill, 1965: 117).

The future of sporting performance, it seems to me, lies with sport psychology and a more complete understanding of how the "steersman" uses conscious and subconscious controls to regulate exercise performance specifically to prevent the catastrophic failure that Hill believed incorrectly was the final outcome of a maximal sporting effort.

REFERENCES

Akima, H., Kinugasa, R. & Kuno, S. (2005). Recruitment of the thigh muscles during sprint cycling by muscle functional magnetic resonance imaging. *International Journal of Sports Medicine*, 26(4), 245–52.

Albertus, Y. (2008). Critical analysis of techniques for normalising electromyographic data. PhD thesis, University of Cape Town, 1–219.

Albertus, Y., Tucker, R., St Clair Gibson, A., Lambert, E.V., Hampson, D.B. & Noakes, T.D. (2005). Effect of distance feedback on pacing strategy and perceived exertion during cycling. *Medicine and Science in Sports and Exercise*, 37(3), 461–8.

Altareki, N., Drust, B., Atkinson, G., Cable, T. & Gregson, W. (2009). Effects of environmental heat stress (35 degrees C) with simulated air movement on the thermoregulatory responses during a 4-km cycling time trial. *International Journal of Sports Medicine*, 30, 9–15.

Amann, M. & Dempsey, J.A. (2008). Locomotor muscle fatigue modifies central motor drive in healthy humans and imposes a limitation to exercise performance. *The Journal of Physiology*, 586(1), 161–73.

Amann, M., Eldridge, M.W., Lovering, A.T., Stickland, M.K., Pegelow, D.F. & Dempsey, J.A. (2006). Arterial oxygenation influences central motor output and exercise performance via effects on peripheral locomotor muscle fatigue in humans. *The Journal of Physiology*, 575(Pt 3), 937–52.

Amann, M., Romer, L.M., Subudhi, A.W., Pegelow, D.F. & Dempsey, J.A. (2007). Severity of arterial hypoxaemia affects the relative contributions of peripheral muscle fatigue to exercise performance in healthy humans. *The Journal of Physiology*, 581(Pt 1), 389–403.

Ansley, L., Robson, P.J., St Clair Gibson, A. & Noakes, T.D. (2004). Anticipatory pacing strategies during supramaximal exercise lasting longer than 30 s. *Medicine and Science in Sports and Exercise*, 36(2), 309–14.

Ansley, L., Schabort, E., St Clair Gibson, A., Lambert, M.I. & Noakes, T.D. (2004). Regulation of pacing strategies during successive 4-km time trials. *Medicine and Science in Sports and Exercise*, 36(10), 1819–25.

Baldwin, J., Snow, R.J., Gibala, M.J., Garnham, A., Howarth, K. & Febbraio, M.A. (2003). Glycogen availability does not affect the TCA cycle or TAN pools during prolonged, fatiguing exercise. *Journal of Applied Physiology*, 94(6), 2181–87.

Baron, B., Deruelle, F., Moullan, F., Dalleau, G., Verkindt, C. & Noakes, T.D. (2009). The eccentric muscle loading influences the pacing strategies during repeated downhill sprint intervals. *European Journal of Applied Physiology*, 105(5), 749–57.

Barwood, M.J., Weston, N.J.V., Thelwell, R. & Page, J. (2009). A motivational music and video intervention improves high-intensity exercise performance. *Journal of Sports Science and Medicine*, 8, 422–42.

Beedie, C.J., Stuart, E.M., Coleman, D.A. & Foad, A.J. (2006). Placebo effects of caffeine on cycling performance. *Medicine and Science in Sports and Exercise*, 38(12), 2159–64.

Bernard, C. (1957). *An Introduction to the Study of Experimental Medicine*. New York: Dover Publications.

Bigland-Ritchie, B. & Woods, J.J. (1974). Integrated EMG and oxygen uptake during dynamic contractions of human muscles. *Journal of Applied Physiology*, 36(4), 475–79.

Billaut, F., Davis, J.M., Smith, K.J., Marino, F.E. & Noakes, T.D. (2010). Cerebral oxygenation decreases but does not impair performance during self-paced, strenuous exercise. *Acta Physiologica* (Oxford), 198(4), 477–86.

Bramble, D.M. & Lieberman, D.E. (2004). Endurance running and the evolution of Homo. *Nature*, 432(7015), 345–52.

Byrne, C., Lee, J.K., Chew, S.A., Lim, C.L. & Tan, E.Y. (2006). Continuous thermoregulatory responses to mass-participation distance running in heat. *Medicine and Science in Sports and Exercise*, 38(5), 803–10.

Cabanac, M. (1986). Money versus pain: Experimental study of a conflict in humans. *Journal of the Experimental Analysis of Behavior*, 46(1), 37–44.

Calbet, J.A., Boushel, R., Radegran, G., Sondergaard, H., Wagner, P.D. & Saltin, B. (2003a). Determinants of maximal oxygen uptake in severe acute hypoxia. *American Journal of Physiology*, 284(2), R291–R303.

Calbet, J.A., Boushel, R., Radegran, G., Sondergaard, H., Wagner, P.D. & Saltin, B. (2003b). Why is VO_2 max after altitude acclimatization still reduced despite normalization of arterial O_2 content? *American Journal of Physiologyl*, 284(2), R304–R316.

Calbet, J.A., De Paz, J.A., Garatachea, N., Cabeza, d., V. & Chavarren, J. (2003). Anaerobic energy provision does not limit Wingate exercise performance in endurance-trained cyclists. *Journal of Applied Physiology*, 94(2), 668–76.

Carmichael, M., Davis, J.M., Murphy, A., Brown, A.S., Carson, J.A., Mayer, E.P. et al. (2006). Role of brain IL-1beta on fatigue after exercise-induced muscle damage. *American Journal of Physiology*, 291, R1344–R1348.

Castle, P.C., Macdonald, A.L., Philp, A., Webborn, A., Watt, P.W. & Maxwell, N.S. (2006). Precooling leg muscle improves intermittent sprint exercise performance in hot, humid conditions. *Journal of Applied Physiology*, 100(4), 1377–84.

Catalano, J.F. (1974). End-spurt following simple repetitive muscular movement. *Perceptual and Motor Skills*, 39(2), 763–66.

Chambers, E.S., Bridge, M.W. & Jones, D.A. (2009). Carbohydrate sensing in the human mouth: Effects on exercise performance and brain activity. *The Journal of Physiology*, 587(Pt 8), 1779–94.

Clark, S.A., Bourdon, P.C., Schmidt, W., Singh, B., Cable, G., Onus, K.J. et al. (2007). The effect of acute simulated moderate altitude on power, performance and pacing strategies in well-trained cyclists. *European Journal of Applied Physiology*, 102(1), 45–55.

Crewe, H., Tucker, R. & Noakes, T.D. (2008). The rate of increase in rating of perceived exertion predicts the duration of exercise to fatigue at a fixed power output in different environmental conditions. *European Journal of Applied Physiology*, 103(5), 569–77.

Damisch, L., Stoberock, B. & Mussweiler, T. (2010). Keep your fingers crossed!: How superstition improves performance. *Psychological Science*, 21(7), 1014–20.

Edwards, A.M., Mann, M.E., Marfell-Jones, M.J., Rankin, D.M., Noakes, T.D. & Shillington, D.P. (2007). The influence of moderate dehydration on soccer performance: Physiological responses to 45-min of outdoors match-play and the immediate subsequent performance of sport-specific and mental concentration tests. *British Journal of Sports Medicine*, 41(6), 385–91.

Ely, B.R., Ely, M.R., Cheuvront, S.N., Kenefick, R.W., DeGroot, D.W. & Montain, S.J. (2009). Evidence against a 40 degrees C core temperature threshold for fatigue in humans. *Journal of Applied Physiology*, 107(5), 1519–25.

Eston, R.G., Faulkner, J.A., Mason, E.A. & Parfitt, G. (2006). The validity of predicting maximal oxygen uptake from perceptually regulated graded exercise tests of different durations. *European Journal of Applied Physiology*, 97(5), 535–41.

Eston, R., Faulkner, J., St Clair Gibson, A., Noakes, T. & Parfitt, G. (2007). The effect of antecedent fatiguing activity on the relationship between perceived exertion and physiological activity during a constant load exercise task. *Psychophysiology*, 44(5), 779–86.

Eston, R.G., Lamb, K.L., Parfitt, G. & King, N. (2005). The validity of predicting maximal oxygen uptake from a perceptually-regulated graded exercise test. *European Journal of Applied Physiology*, 94(3), 221–27.

Eston, R.G., Lambrick, D., Sheppard, K. & Parfitt, G. (2007). Prediction of maximal oxygen uptake in sedentary males from a perceptually regulated, sub-maximal graded exercise test. *Journal of Sports Science*, 26(2), 131–9.

Faulkner, J., Parfitt, G. & Eston, R. (2007). Prediction of maximal oxygen uptake from the ratings of perceived exertion and heart rate during a perceptually-regulated sub-maximal exercise test in active and sedentary participants. *European Journal of Applied Physiology*, 101(3), 397–407.

Flouris, A.D. & Cheung, S.S. (2009). Human conscious response to thermal input is adjusted to changes in mean body temperature. *British Journal of Sports Medicine*, 43(3), 199–203.

Foad, A.J., Beedie, C.J. & Coleman, D.A. (2008). Pharmacological and psychological effects of caffeine ingestion in 40-km cycling performance. *Medicine and Science in Sports and Exercise*, 40(1), 158–65.

Galloway, S.D. & Maughan, R.J. (1997). Effects of ambient temperature on the capacity to perform prolonged cycle exercise in man. *Medicine and Science in Sports and Exercise*, 29(9), 1240–49.

Gamet, D., Duchene, J., Garapon-Bar, C. & Goubel, F. (1993). Surface electromyogram power spectrum in human quadriceps muscle during incremental exercise. *Journal of Applied Physiology*, 74(6), 2704–10.

Gonzalez-Alonso, J., Teller, C., Andersen, S.L., Jensen, F.B., Hyldig, T. & Nielsen, B. (1999). Influence of body temperature on the development of fatigue during prolonged exercise in the heat. *Journal of Applied Physiology*, 86(3), 1032–39.

Green, H.J., Sutton, J.R., Young, P.M., Cymerman, A. & Houston, C.S. (1989). Operation Everest II: Muscle energetics during maximal exhaustive exercise. *Journal of Applied Physiology*, 66, 1032–9.

Hampson, D.B., St Clair Gibson, A., Lambert, M.I. & Noakes, T.D. (2001). The influence of sensory cues on the perception of exertion during exercise and central regulation of exercise performance. *Sports Medicine*, 31(13), 935–52.

Henriksson, J. & Bonde-Petersen, F. (1974). Integrated electromyography of quadriceps femoris muscle at different exercise intensities. *Journal of Applied Physiology*, 36(2), 218–20.

Herbert, J. (2007). *The Minder Brain*. Hackensack, NJ: World Scientific Press.

Hill, A.V. (1965). *Trails and Trials in Physiology*. London: Edward Arnold.

Hill, A.V., Long, C.H.N. & Lupton, H. (1924). Muscular exercise, lactic acid and the supply and utilisation of oxygen. Parts VII–VIII. Proceedings of the Royal Society B: Biological Sciences, 97, 155–76.

Hug, F., Decherchi, P., Marqueste, T. & Jammes, Y. (2004). EMG versus oxygen uptake during cycling exercise in trained and untrained subjects. *Journal of Electromyography and Kinesiology*, 14(2), 187–95.

Hultman, E. & Nielsen, B. (1971). Liver glycogen in man: Effects of different diets andmuscular exercise, in B. Pernow & B. Saltin (eds), *Muscle Metabolism During Exercise*. New York: Plenum Press, 69–85.

Jacobs, I. & Bell, D.G. (2004). Effects of acute modafinil ingestion on exercise time to exhaustion. *Medicine and Sciencein Sports and Exercise*, 36(6), 1078–82.

Johnson, B.D., Joseph, T., Wright, G., Battista, R.A., Dodge, C., Balweg, A. et al. (2009). Rapidity of responding to a hypoxic challenge during exercise. *European Journal of Applied Physiology*, 106(4), 493–99.

Joseph, T., Johnson, B., Battista, R.A., Wright, G., Dodge, C., Porcari, J.P. et al. (2008). Perception of fatigue during simulated competition. *Medicine and Sciencein Sports and Exercise*, 40, 381–6.

Joyner, M.J. & Coyle, E.F. (2008). Endurance exercise performance: the physiology of champions. *The Journal of Physiology*, 586(1), 35–44.

Kay, D., Marino, F.E., Cannon, J., St Clair Gibson, A., Lambert, M.I. & Noakes, T.D. (2001). Evidence for neuromuscular fatigue during high-intensity cycling in warm, humid conditions. *European Journal of Applied Physiology*, 84(1–2), 115–21.

Kayser, B. (1996). Lactate during exercise at high altitude. *European Journal of Applied Physiology and Occupational Physiology*, 74(3), 195–205.

Kayser, B. (2003). Exercise starts and ends in the brain. *European Journal of Applied Physiology*, 90(3–4), 411–19.

Kayser, B., Narici, M., Binzoni, T., Grassi, B. & Cerretelli, P. (1994). Fatigue and exhaustion in chronic hypobaric hypoxia: Influence of exercising muscle mass. *Journal of Applied Physiology*, 76(2), 634–40.

Kwon, Y.S., Robergs, R.A., Kravitz, L.R., Gurney, B.A., Mermier, C.M. & Schneider, S.M. (2010). Palm cooling delays fatigue during high intensity bench press exercise. *Medicine and Science in Sports and Exercise,* 42(8), 1557–65.

Lambert, E.V., St Clair Gibson, A. & Noakes, T.D. (2005). Complex systems model of fatigue: Integrative homoeostatic control of peripheral physiological systems during exercise in humans. *British Journal of Sports Medicine*, 39(1), 52–62.

Laplaud, D., Hug, F. & Grelot, L. (2006). Reproducibility of eight lower limb muscles activity level in the course of an incremental pedaling exercise. *Journal of Electromyography and Kinesiology*, 16(2), 158–66.

Levine, B.D. (2008). VO2max: What do we know, and what do we still need to know? *The Journal of Physiology*, 586(1), 25–34.

Liebenberg, L. (2006). Persistence hunting by modern hunter-gatherers. *Current Anthropology*, 47(6), 1017–25.

Liebenberg, L. (2008). The relevance of persistence hunting to human evolution. *Journal of Human Evolution*, 55(6), 1156–9.

Lieberman, D.E. & Bramble, D.M. (2007). The evolution of marathon running: Capabilities in humans. *Sports Medicine*, 37(4–5), 288–90.

Lim, H.B., Atkinson, G., Karageorghis, C.I. & Eubank, M.R. (2009). Effects of differentiated music on cycling time trial. *International Journal of Sports Medicine*, 30(6), 435–42.

Lima-Silva, A., Pires, F.O., Bertuzzi, R.C.M., Lira, F.S., Casarini, D. & Kiss, M.A. (2010). Low carbohydrate diet affects the oxygen uptake on kinetics and rating of perceived exertion in high intensity exercise. *Psychophysiology, July 6* [epub ahead of print].

Marcora, S.M. & Bosio, A. (2007). Effect of exercise-induced muscle damage on endurance running performance in humans. *Scandanavian Journal of Medicine and Science in Sports*, 17(6), 662–71.

Marcora, S.M. & Staiano, W. (2010). The limit to exercise tolerance in humans: Mind over muscle? *European Journal of Applied Physiology*, 109(4), 763–70.

Marcora, S.M., Staiano, W. & Manning, V. (2009). Mental fatigue impairs physical performance in humans. *Journal of Applied Physiology*, 106(3), 857–64.

Marino, F.E. (2004). Anticipatory regulation and avoidance of catastrophe during exercise-induced hyperthermia. *Comparative Biochemistry and Physiology*, Part B: *Biochemistry and Molecular Biology*, 139(4), 561–9.

Marino, F.E., Lambert, M.I. & Noakes, T.D. (2004). Superior performance of African runners in warm humid but not in cool environmental conditions. *Journal of Applied Physiology*, 96(1), 124–30.

Maron, M.B., Wagner, J.A. & Horvath, S.M. (1977). Thermoregulatory responses during competitive marathon running. *Journal of Applied Physiology*, 42(6), 909–14.

Martin, B.J. (1981). Effect of sleep deprivation on tolerance of prolonged exercise. *European Journal of Applied Physiology and Occupational Physiology*, 47(4), 345–54.

Mauger, A.R., Jones, A.M. & Williams, C.A. (2009). Influence of feedback and prior experience on pacing during a 4-km cycle time trial. *Medicine and Science in Sports and Exercise*, 41(2), 451–8.

Mauger, A.R., Jones, A.M. & Williams, C.A. (2010). Influence of acetaminophen on performance during time trial cycling. *Journal of Applied Physiology*, 108(1), 98–104.

Micklewright, D., Papadopoulou, E., Swart, J. & Noakes, T. (2010). Previous experience influences pacing during 20 km time trial cycling. *British Journal of Sports Medicine*, 44(13), 952–60.

Morante, S.M. & Brotherhood, J.R. (2008). Autonomic and behavioural thermoregulation in tennis. *Britsh Journal of Sports Medicine*, 42(8), 679–85.

Morrison, S., Sleivert, G.G. & Cheung, S.S. (2004). Passive hyperthermia reduces voluntary activation and isometric force production. *European Journal of Applied Physiology*, 91(5–6), 729–36.

Morton, R.H. (2009). Deception by manipulating the clock calibration influences cycle ergometer endurance time in males. *Journal of Science and Medicine in Sport*, 12(2), 332–7.

Nielsen, B., Hales, J.R., Strange, S., Christensen, N.J., Warberg, J. & Saltin, B. (1993). Human circulatory and thermoregulatory adaptations with heat acclimation and exercise in a hot, dry environment. *The Journal of Physiology*, 460, 467–85.

Noakes, T.D. (1997). Challenging beliefs: Ex Africa semper aliquid novi: 1996 J.B. Wolffe Memorial Lecture. *Medicine and Science Sports Exercise*, 29(5), 571–90.

Noakes, T.D. (1998). Maximal oxygen uptake: "classical" versus "contemporary" viewpoints: A rebuttal. *Medicine and Science Sports Exercise*, 30(9), 1381–98.

Noakes, T.D. (2000). Physiological models to understand exercise fatigue and the adaptations that predict or enhance athletic performance. *Scandanavian Journal of Medicine and Science in Sports*, 10(3), 123–45.

Noakes, T.D. (2004). Linear relationship between the perception of effort and the duration of constant load exercise that remains. *Journal of Applied Physiology*, 96(4), 1571–2.

Noakes, T.D. (2006). The limits of endurance exercise. *Basic Research in Cardiology*, 101, 408–17.

Noakes, T.D. (2007a). The central governor model of exercise regulation applied to the marathon. *Sports Medicine*, 37(4–5), 374–7.

Noakes, T.D. (2007b). The limits of human endurance: What is the greatest endurance performance of all time? Which factors regulate performance at extreme altitude? In R.C. Roach et al. (eds), *Hypoxia and the Circulation*. New York: Springer, 259–80.

Noakes, T.D. (2008a). How did A.V. Hill understand the VO2max and the "plateau phenomenon?" Still no clarity? *British Journal of Sports Medicine*, 42(7), 574–80.

Noakes, T.D. (2008b). Testing for maximum oxygen consumption has produced a brainless model of human exercise performance. *British Journal of Sports Medicine*, 42(7), 551–5.

Noakes, T.D. (2009). Evidence that reduced skeletal muscle recruitment explains the lactate paradox during exercise at high altitude. *Journal of Applied Physiology*, 106(2), 737–8.

Noakes, T.D. (2011a). The central governor model and fatigue during exercise, in F.E. Marino (ed.), *The Regulation of Fatigue in Exercise*. Hauppauge, NY: Nova Publishers.

Noakes, T.D. (2011b). The VO_2 max and the central governor: A different understanding, in F.E. Marino (ed.), *The Regulation of Fatigue in Exercise*. Hauppauge, NY: Nova Publishers.

Noakes, T.D. & Marino, F.E. (2007). Arterial oxygenation, central motor output and exercise performance in humans. *The Journal of Physiology*, 585(Pt 3), 919–21.

Noakes, T.D. & St Clair Gibson, A. (2004). Logical limitations to the "catastrophe" models of fatigue during exercise in humans. *British Journal of Sports Medicine*, 38(5), 648–49.

Noakes, T.D., Calbet, J.A., Boushel, R., Sondergaard, H., Radegran, G., Wagner, P.D. et al. (2004). Central regulation of skeletal muscle recruitment explains the reduced maximal cardiac output during exercise in hypoxia. *American Journal of Physiology – Regulatory Integrative and Comparative Physiology*, 287(4), R996–R999.

Noakes, T.D., Crewe, H. & Tucker, R. (2009). The brain and fatigue, in R.J. Maughan (ed.), *The Olympic Textbook of Science in Sport*. Oxford: Wiley-Blackwell, 340–61.

Noakes, T.D., Lambert, M. & Human, R. (2009). Which lap is the slowest? An analysis of 32 world record performances. *British Journal of Sports Medicine*, 43, 760–64.

Noakes, T.D., Peltonen, J.E. & Rusko, H.K. (2001). Evidence that a central governor regulates exercise performance during acute hypoxia and hyperoxia. *The Journal of Experimental Biology*, 204(Pt 18), 3225–34.

Noakes, T.D., St Clair Gibson, A. & Lambert, E.V. (2004). From catastrophe to complexity: A novel model of integrative central neural regulation of effort and fatigue during exercise in humans. *British Journal of Sports Medicine*, 38(4), 511–14.

Noakes, T.D., St Clair Gibson, A. & Lambert, E.V. (2005). From catastrophe to complexity: A novel model of integrative central neural regulation of effort and fatigue during exercise in humans: Summary and conclusions. *British Journal of Sports Medicine*, 39(2), 120–24.

Nummela, A.T., Heath, K.A., Paavolainen, L.M., Lambert, M.I., St Clair, G.A., Rusko, H.K. et al. (2008). Fatigue during a 5-km running time trial. *International Journal of Sports Medicine*, 29, 738–45.

Nummela, A.T., Paavolainen, L.M., Sharwood, K.A., Lambert, M.I., Noakes, T.D. & Rusko, H.K. (2006). Neuromuscular factors determining 5 km running performance and running economy in well-trained athletes. *European Journal of Applied Physiology*, 97(1), 1–8.

Nybo, L. & Nielsen, B. (2001). Hyperthermia and central fatigue during prolonged exercise in humans. *Journal of Applied Physiology*, 91(3), 1055–60.

Nybo, L. & Rasmussen, P. (2007a). Inadequate cerebral oxygen delivery and central fatigue during strenuous exercise. *Exercise and Sport Sciences Reviews*, 35(3), 110–18.

Nybo, L. & Rasmussen, P. (2007b). Inadequate cerebral oxygen delivery and central fatigue during strenuous exercise. *Exercise and Sports Sciences Reviews*, 35(3), 110–18.

Racinais, S., Bringard, A., Puchaux, K., Noakes, T.D. & Perrey, S. (2008). Modulation in voluntary neural drive in relation to muscle soreness. *European Journal of Applied Physiology*, 102(4), 439–46.

Rasmussen, P., Nielsen, J., Overgaard, M., Krogh-Madsen, R., Gjedde, A., Secher, N.H. et al. (2010). Reduced muscle activation during exercise related to brain oxygenation and metabolism in humans. *The Journal of Physiology*, 588(11), 1985–95.

Rasmussen, P., Nybo, L., Volianitis, S., Moller, K., Secher, N.H. & Gjedde, A. (2010). Cerebral oxygenation is reduced during hyperthermic exercise in humans. *Acta Physiol* (Oxford), 199(1), 63–70.

Rauch, H.G., St Clair Gibson, A., Lambert, E.V. & Noakes, T.D. (2005). A signalling role for muscle glycogen in the regulation of pace during prolonged exercise. *British Journal of Sports Medicine*, 39, 34–8.

Robson-Ansley, P.J. de M.L., Collins, M. & Noakes, T.D. (2004). Acute interleukin-6 administration impairs athletic performance in healthy, trained male runners. *Canadian Journal of Applied Physiology*, 29(4), 411–18.

Roelands, B., Hasegawa, H., Watson, P., Piacentini, M.F., Buyse, L., De, S.G. et al. (2009). Performance and thermoregulatory effects of chronic bupropion administration in the heat. *European Journal of Applied Physiology*, 105(3), 493–8.

Rollo, I., Cole, M., Miller, R. & Williams, C. (2009). The influence of mouth-rinsing a carbohydrate solution on 1 hour running performance. *Medicine and Science in Sports and Exercise*, 42(4), 798–804.

Ross, E.Z., Goodall, S., Stevens, A. & Harris, I. (2010). Time course of neuromuscular changes during running in well-trained subjects. *Medicine and Science in Sports and Exercise*, 42(6), 1184–90.

Rupp, T. & Perrey, S. (2008). Prefrontal cortex oxygenation and neuromuscular responses to exhaustive exercise. *European Journal of Applied Physiology*, 102(2), 153–63.

Seifert, T., Rasmussen, P., Secher, N.H. & Nielsen, H.B. (2009). Cerebral oxygenation decreases during exercise in humans with beta-adrenergic blockade. *Acta Physiol* (Oxford), 196(3), 295–302.

Sgherza, A.L., Axen, K., Fain, R., Hoffman, R.S., Dunbar, C.C. & Haas, F. (2002). Effect of naloxone on perceived exertion and exercise capacity during maximal cycle ergometry. *Journal of Applied Physiology*, 93(6), 2023–28.

Sloniger, M.A., Cureton, K.J., Prior, B.M. & Evans, E.M. (1997). Lower extremity muscle activation during horizontal and uphill running. *Journal of Applied Physiology*, 83(6), 2073–79.

St Clair Gibson, A. & Noakes, T.D. (2004). Evidence for complex system integration and dynamic neural regulation of skeletal muscle recruitment during exercise in humans. *British Journal of Sports Medicine*, 38(6), 797–806.

St Clair Gibson, A., Lambert, M.L. & Noakes, T.D. (2001). Neural control of force output during maximal and submaximal exercise. *Sports Medicine*, 31(9), 637–50.

Swart, J., Lamberts, R.P., Lambert, M.I., St Clair Gibson, A., Lambert, E.V., Skowno, J. et al. (2009a). Exercising with reserve: evidence that the central nervous system regulates prolonged exercise performance. *British Journal of Sports Medicine*, 43(10), 782–88.

Swart, J., Lamberts, R.P., Lambert, M.I., Lambert, E.V., Woolrich, R.W., Johnston, S. et al. (2009b). Exercising with reserve: exercise regulation by perceived exertion in relation to duration of exercise and knowledge of endpoint. *British Journal of Sports Medicine*, 43(10), 775–81.

Trojian, T.H. & Beedie, C.J. (2008). Placebo effect and athletes. *Current Sports Medicine Reports*, 7(4), 214–17.

Tucker, R. (2009). The anticipatory regulation of performance: the physiological basis for pacing strategies and the development of a perception-based model for exercise performance. *British Journal of Sports Medicine*, 43(6), 392–400.

Tucker, R. & Noakes, T.D. (2009). The physiological regulation of pacing strategy during exercise: a critical review. *British Journal of Sports Medicine*, 43(6), doi:10.1136/bjsm.2009.057562.

Tucker, R., Bester, A., Lambert, E.V., Noakes, T.D., Vaughan, C.L. & St Clair Gibson, A. (2006). Non-random fluctuations in power output during self-paced exercise. *British Journal of Sports Medicine*, 40(11), 912–17.

Tucker, R., Kayser, B., Rae, E., Raunch, L., Bosch, A. & Noakes, T. (2007). Hyperoxia improves 20 km cycling time trial performance by increasing muscle activation levels while perceived exertion stays the same. *European Journal of Applied Physiology*, 101(6), 771–81.

Tucker, R., Lambert, M.I. & Noakes, T.D. (2006). An analysis of pacing strategies during men's world record performances in track athletics. *International Journal of Sports Physiology and Performance*, 1, 233–45.

Tucker, R., Marle, T., Lambert, E.V. & Noakes, T.D. (2006). The rate of heat storage mediates an anticipatory reduction in exercise intensity during cycling at a fixed rating of perceived exertion. *The Journal of Physiology*, 574(Pt 3), 905–15.

Tucker, R., Rauch, L., Harley, Y.X. & Noakes, T.D. (2004). Impaired exercise performance in the heat is associated with an anticipatory reduction in skeletal muscle recruitment. *Pflügers Archives (European Journal of Physiology)*, 448(4), 422–30.

Ulmer, H.V. (1996). Concept of an extracellular regulation of muscular metabolic rate during heavy exercise in humans by psychophysiological feedback. *Experientia*, 52(5), 416–20.

Walters, T.J., Ryan, K.L., Tate, L.M. & Mason, P.A. (2000). Exercise in the heat is limited by a critical internal temperature. *Journal of Applied Physiology*, 89(2), 799–806.

Wilmore, J. (1968). Influence of motivation on physical work capacity and performance. *Journal of Applied Physiology*, 24, 459–63.

Wittekind, A.L., Micklewright, D. & Beneke, R. (2009). Teleoanticipation in all-out short duration cycling. *British Journal of Sports Medicine*, 45, 114–19.

PART III

Assessment of Fatigue

8

Approaches to the Measurement of Fatigue

Christopher Christodoulou

INTRODUCTION

Fatigue is a familiar and widely experienced phenomenon (Ranjith, 2005; Ricci, Chee, Lorandeau, & Berger, 2007) Fatigue affects almost every aspect of modern life, from occupational and athletic to military and medical activities. It is highly prevalent in the general community (Kroenke & Price, 1993), and in the workforce (Ricci, et al., 2007). It is associated with increased risk of injuries and accidents and has been estimated to cost employers in the United States over 100 billion dollars annually (Folkard & Lombardi, 2006; Ricci, et al., 2007).

Fatigue has proven quite difficult to define and measure (Bartley & Chute, 1947; Wessely, 2005) and there is no consensus on how it should be assessed (Christodoulou, 2005). No widely acceptable biological marker for fatigue has emerged, though research in this area continues (Kohl et al., 2009). There are generally two types of approaches to defining and measuring fatigue. Fatigue can be seen as either a "subjective feeling" or a "performance decrement" (Wessely, Hotopf, & Sharpe, 1998). The different approaches to measuring fatigue appear to arise in part from the interests of those doing the measuring. Clinicians, for example, are generally interested in the fatigue experienced by their patients that led them to seek care, and this may be why clinical measurement of fatigue often tends to rely on patients' self-reported feelings of fatigue (Mock, et al., 2000; Multiple Sclerosis Council for Clinical Practice Guidelines, 1998). This contrasts with the perspective sometimes taken by business and the military, which is more pragmatic and directly interested in loss of productivity and/or increased accidents, and therefore more likely to value behavioral measures of performance decrement (Caldwell & Caldwell, 2005; Folkard & Lombardi, 2006) Nonetheless, the ease with which fatigue can be assessed by self-reports has probably contributed to their widespread use in healthy as well as clinical populations. This chapter will review and compare these two approaches to the measurement of fatigue. In addition, it will also examine measures of the related issue of sleepiness.

PATHOLOGICAL AND NON-PATHOLOGICAL FATIGUE

While fatigue can be identified in otherwise healthy people (Kroenke & Price, 1993; Ricci, et al., 2007) it can reach pathological levels in a wide variety of medical conditions, including multiple sclerosis (MS) (Hadjimichael, Vollmer, & Oleen-Burkey, 2008) cancer (Cella, Lai, Chang, Peterman, & Slavin, 2002) depression (Johnson, 2005), traumatic brain injury (Belmont, Agar, & Azouvi, 2009) stroke (Annoni, Staub, Bogousslavsky, & Brioschi, 2008; Winward, Sackley, Metha, & Rothwell, 2009) and Parkinson's disease (Friedman, et al., 2007). Pathological fatigue is distinguished from non-pathological fatigue by features that include greater intensity, longer duration, and more disabling effects on functional activities. Non-pathological fatigue is usually a short-lived phenomenon following a period of exertion or sleep deprivation, which is reduced if not eliminated by rest or sleep. In contrast, pathological fatigue can remain after rest as a severe chronic condition that disrupts an individual's ability to carry out important daily social and occupational activities and obligations. In this chapter, the focus will primarily be upon non-pathological fatigue in otherwise healthy individuals.

QUESTIONNAIRES

A large number of questionnaires have been used to measure fatigue in the general population. They have been applied in a variety of circumstances, including the assessment of health (Chalder, et al., 1993; Lai, et al., 2011), employee attributes (Ahsberg, 2000; Beurskens, et al., 2000; Maslach & Jackson, 1981; Michielsen, De Vries, Van Heck, Van de Vijver, & Sijtsma, 2004), and athletic performance (Borg, 1970, 1982a). Fatigue is sometimes assessed within broader quality of life instruments such as the Vitality subscale of the Medical Outcomes Study Short-Form (SF-36) (Ware; Ware & Sherbourne, 1992) or in the World Health Organization Quality of Life WHOQOL-100 (Power, Harper, & Bullinger, 1999) and WHOQOL-BREF ("Development of the World Health Organization WHOQOL-BREF quality of life assessment. The WHOQOL Group," 1998). Fatigue can also be assessed within more general measures of mood, as in the Profile of Mood States (POMS) (McNair, Lorr, & Droppleman, 1992). More frequently, however, fatigue is assessed in scales specific to the domain, such as Chalder's Fatigue Scale (FS) (Chalder, et al., 1993), the Fatigue Assessment Scale (FAS) (Michielsen, De Vries, & Van Heck, 2003), the Fatigue Severity Scale (FSS) (Krupp, LaRocca, Muir-Nash, & Steinberg, 1989), the Patient-Reported Outcome Measurement Information System Fatigue Item Bank (PROMIS-FIB)[1] (Lai, et al., 2011), and many others (Ahsberg, 2000; Beurskens, et al., 2000; Borg, 1970, 1982b; Krupp, et al., 1989; Smets, Garssen, Bonke, & De Haes, 1995; Watt, et al., 2000). Some instruments measure overall level of fatigue with a single item (Borg, 1970, 1982a; Krupp, et al., 1989) while others are longer and multidimensional in nature (Smets, et al., 1995).

ASPECTS OF FATIGUE

Most scales include items that assess the feelings of fatigue and the perceived impact of fatigue on the lives of respondents. Examples of the core feelings of fatigue include

---

1   See http://www.nihpromis.org

feeling tired, exhausted, sluggish, and drained. Such items are emphasized in scales such as the fatigue subscale of the Profile of Mood States (POMS) (McNair, et al., 1992) and the PROMIS-FIB (Lai, et al., 2011). Some scales measure other types of feelings, including those of a more motivational nature, such as feeling indifferent, unmotivated, or apathetic. Motivational items of this type are included in scales such as the Multidimensional Fatigue Inventory (MFI) (Smets, et al., 1995; Watt, et al., 2000) the Checklist Individual Strength (CIS) (Beurskens, et al., 2000; Vercoulen, et al., 1994) and the Swedish Occupational Fatigue Inventory (SOFI) (Ahsberg, 2000; Ahsberg, Gamberale, & Kjellberg, 1997).

In addition to directly asking about feelings of fatigue, most scales also ask respondents to report on the impact of fatigue on their daily activities. The PROMIS-FIB includes a large number of items assessing interference with a variety of activities including, domestic, occupational, recreational, and social (Lai, et al., 2011). This instrument includes items sensitive to the impact of fatigue over a range of intensity, from low levels that may only disrupt strenuous athletic activities to high levels of fatigue that interfere with even basic activities of daily living such as bathing. Questionnaires often distinguish between the interference of fatigue with mental/cognitive activities (e.g., impaired concentration) versus physical/muscular activities (e.g., reduction in sustained activity) (Christodoulou, 2005). The MFI (Smets, et al., 1995) and Chalder's Fatigue Scale (FS) (Chalder, et al., 1993) are among those that make this distinction.

There are disparate views as to whether different types of fatigue symptoms (e.g., physical, mental, motivational) represent separate dimensions of fatigue, or whether fatigue is best viewed along a single overall dimension of severity (Lai, et al., 2011; Michielsen, et al., 2004; Smets, et al., 1995). There is some evidence that work requirements of different occupations elevate some aspects of fatigue more than others. The SOFI, for example, contains two subscales representing physical fatigue (physical exertion, physical discomfort) and two representing mental fatigue (lack of motivation and sleepiness) (Ahsberg, 2000; Ahsberg, et al., 1997). Firefighters on the SOFI were found to display higher scores on physical fatigue subscales, while nightshift operators at nuclear power plants had higher scores on mental fatigue subscales (Ahsberg, et al., 1997). In addition, strenuous physical activity has been found to increase the SOFI physical fatigue subscale scores (Ahsberg & Gamberale, 1998), while demanding mental activity has been found to increase SOFI mental fatigue subscale scores (Ahsberg, Gamberale, & Gustafsson, 2000).

There is considerable evidence, however, that a single overall dimension of fatigue can explain much of the variance in a variety of fatigue scales. For example, an analysis of the SOFI has shown that the majority of the variance in its subscales can be accounted for by a single underlying latent dimension of overall fatigue, called lack of energy (Ahsberg, 2000). An analysis of the PROMIS-FIB found that this scale too was sufficiently unidimensional (Lai, et al., 2011). De Vries and colleagues assessed four fatigue questionnaires in a sample of employed individuals and found that all four were unidimensional (De Vries, Michielsen, & Van Heck, 2003), including two scales presumed to be multidimensional, the FS (with subscales of Mental and Physical Fatigue) (Chalder, et al., 1993) and the CIS (with subscales of Subjective Experience of Fatigue, Concentration, Motivation, and Physical activity) (Vercoulen, et al., 1994). The De Vries group also found that when all scales were combined, they again resulted in one fatigue factor, which explained more than 60 percent of the variance. With regard to the CIS (Beurskens, et al., 2000), a single overall score has been found to predict occupational accidents (Swaen, Van Amelsvoort, Bultmann, & Kant, 2003), sickness absences from work (Janssen, Kant, Swaen, Janssen, &

Schroer, 2003), and to distinguish between non-fatigued and fatigued (pregnant, lower back herniation, and "mentally" fatigued) individuals in the workforce (Beurskens, et al., 2000).

Variations and Innovations in Questionnaires

One recent innovation in the measurement of fatigue involves the use of computerized adaptive testing (CAT), as applied to the PROMIS-FIB (Lai, et al., 2011). Item response theory (IRT) methods (Hays, Morales, & Reise, 2000) were used to calibrate each item from the large pool of questions, thus enabling the CAT instrument to select the best items for a particular subject, sequentially selecting items until a given degree of accuracy is achieved, usually within a total of five to ten items (Fries, Cella, Rose, Krishnan, & Bruce, 2009). The result is a short, precise measure tailored to the individual. CAT techniques can increase reliability and statistical power, and improve measurement precision over a wider range of fatigue intensity (Fries, et al., 2009; Lai, et al., 2011). The CAT approach has been used for some time in educational testing (Green, Bock, Humphreys, Linn, & Reckase, 1984), but has only recently been applied to the measurement of health-related outcomes (Fries, Bruce, & Cella, 2005; Fries, et al., 2009).

A notable variation in questionnaire development has been taken by Borg (1970, 1982b), whose work has been quite influential among physiologists and those interested in the measurement of physical work capacity (Robertson, Goss, & Metz, 1998). Borg developed his first single item scale to measure current ratings of perceived exertion (RPE). Using a series of verbal descriptors on a scale from 6 ("no exertion at all") to 20 ("maximal exertion") he designed the scale to increase linearly with heart rate from 60 to 200 beats per minute. It has in fact been found to correlate quite highly with heart rate during exercise, from 0.80 to 0.90 (1982b). His new scale was designed to provide a ratio level of measurement using verbal descriptors linked to a scale from 0–10. This Category Ratio Scale (CR-10) has also been shown to correlate strongly with physiological measures of exertion (1982b). The CR-10 has also been shown to correlate well with subscales of the SOFI during physical work (Ahsberg & Gamberale, 1998) and mental work (Ahsberg, Gamberale & Gustafsson, 2000).

Another variation in the use of questionnaires involves Ecological Momentary Assessment (EMA) (Broderick, 2008; Shiffman, Stone, & Hufford, 2008; Stone & Shiffman, 1994). EMA measures a person's current state in their natural environment using repeated measurement. Subjects often fill out EMA on portable electronic devices such as personal digital assistants (PDAs) or smartphones. EMA data can also be collected using a paper and pencil format, though the latter approach can sometimes result in lower levels of compliance (Broderick, 2008). The use of EMA makes it easy to assess momentary changes in fatigue over time. Because EMA is used to measure current momentary fatigue, it also has the advantage of avoiding any possibility of recall bias.

Time

The role of recall, and the possibility of recall bias, has often been underappreciated in self-report measures of fatigue. Some questionnaires do not even specify the time period over which subjects should base their responses (Chalder, et al., 1993). It is becoming evident, however, that recall can influence fatigue ratings. Recalled fatigue ratings can be influenced by cognitive heuristics that can distort self-report (Broderick, 2008). For example, recalled ratings tend to be higher than those for current momentary assessments using EMA (Broderick, Schwartz, Schneider, & Stone, 2009; Broderick, et al., 2008). However, initial studies have found that recalled fatigue ratings out to 28 days correlate fairly well with momentary assessments (Broderick, et al., 2008).

Apart from recall bias, it is important to measure the duration of fatigue as well as its temporal pattern because these parameters can provide valuable information. Fatigue of longer duration is indicative of greater pathology, as is fatigue that does not lessen following rest (Wessely, et al., 1998). Another important temporal aspect to consider is the diurnal variation of fatigue, an issue which few questionnaires address (Lai, et al., accepted for publication; Schwartz, Jandorf, & Krupp, 1993).

PERFORMANCE-BASED MEASURES OF FATIGUE

Performance-based measures of fatigue represent the other major approach to assessment. These behavioral and cognitive measures generally operationalize fatigue as a decrement in task performance. This can involve the measurement of decline over time during the performance of a task, or it can involve lowered performance under circumstances that are thought to engender fatigue, such as long periods of work without rest. Almost any behavior or cognitive task can be used. The measures can sometimes be real life or simulations or real life, or they can be laboratory tasks.

Muscle Fatigue

One of the most common methods of assessing behavioral fatigue is by the examination of muscle fatigue (Sangnier & Tourny-Chollet, 2008), which is defined as an "exercise-induced reduction in the maximal capacity to generate force or power output" (Vollestad, 1997). While this may sound straightforward, it is not without complications. It can be difficult to obtain maximal contractions from volunteers due to motivational and other factors (Vollestad, 1997). With submaximal contractions performance can remain steady for a long period of time. Electrical stimulation of motor neurons of the muscle is sometimes applied to generate and estimate true maximal capacity.

Actigraphy

An actigraph is a portable motion-sensing device that provides a method of measuring behavioral activity. When worn by persons over the course of a day, such devices can be used to measure whether individuals with fatigue display reduced levels of activity. Studies have found that persons with fatigue-related disorders such as CFS and MS show lower levels of activity as compared to healthy individuals (Vercoulen, et al., 1997). Some studies have been unable to find a correlation between actigraph measured levels of activity and self-reported fatigue in healthy persons (Dimsdale, Ancoli-Israel, Elsmore, & Gruen, 2003; Vercoulen, et al., 1997), though one found such a relation in persons with chronic fatigue syndrome (Vercoulen, et al., 1997).

Examples of Real Life Performance Decrement

Variations in performance during a variety of real-life situations have been related to fatigue. For example, a study of medical interns found that those who worked traditional shifts of 24 hours or more made substantially more serious medical errors than those whose shifts lasting only approximately 16 hours; in fact, serious diagnostic errors were more than five times more likely in the group working the longer shift (Landrigan, et al., 2004). Another study of medical interns found an increase in self-reported (and later documented) motor vehicle accidents following extended work shifts (Barger, et al., 2005; Lockley, et al., 2007). Overtime work hours have also been associated with declines in performance in other occupations as well, as is evidenced by a study of operations personnel in nuclear power plants, which found an increase in operator errors during overtime (Baker, Olson, & Morisseau, 1994). Other work extends the measurement of fatigue to the world of sports. A study of professional soccer players, involving the use of video analysis, found a decline in performance during the second half of matches as compared to the first. The decline was found for physical performance related to running and technical abilities including passing (Rampinini, Impellizzeri, Castagna, Coutts, & Wisloff, 2009).

Examples of Performance Decrement in Simulated and Laboratory Environments

A variety of simulated environments have also been studied. For example, driving simulators have been used to identify declines in driving ability over periods of prolonged wakefulness (Baulk et al., 2008). More complex environments have also been examined, as in an investigation of F-117A fighter pilots which found that flying skills on a fully functioning flight simulator were degraded over a 37-hour period of continuous wakefulness (Caldwell, et al., 2003). Another study simulated work in an airport security x-ray screening system, with participants examining images of luggage every two hours over an extended period of time. Results indicated that accuracy and false alarm rates increased during night work (Basner, et al., 2008). Others have found that college students made more errors on a memory recognition task following a simulated work day of demanding activities (Schellekens et al., 2000). The students in the study also displayed a shift toward lower effort and riskier responding following the fatigue-inducing activities. Another study found that healthy participants displayed declines in planning ability and an increase in perseverative errors after performing a cognitively demanding task, as compared to non-fatigued individuals (van der Linden, Frese, & Meijman, 2003) (van der Linden, Frese & Meijman, 2003).

COMPARING SELF-REPORT AND PERFORMANCE-BASED FATIGUE MEASURES

The important advantage of performance-based measures of fatigue is that they provide data that is objectively verifiable by outside observers. Self-report scales provide no such external verification, but they are unique in assessing the individual's experience and feelings of fatigue. It has long been known that the two approaches to measuring fatigue do not always agree (Thorndike, 1900). For example, some have found a correspondence between declining cognitive performance and self-reported fatigue in healthy individuals (Schellekens, Sijtsma, Vegter, & Meijman, 2000), others have not (Bryant, Chiaravalloti, & DeLuca, 2004; Krupp & Elkins, 2000)

One possible reason for the incongruity between the two approaches is that increased effort may help individuals maintain a given level of performance over time (Christodoulou, 2005; DeLuca, 2005) Under such conditions the person may report feeling increasingly fatigued, but it would not be evident in their behavioral performance. There has been increasing focus on the role of effort in fatigue (Belmont, et al., 2009; Burgess & Jones, 1997; Fahlen, et al., 2006; Rosen, King, Wilkinson, & Nixon, 1990; Segerstrom & Nes, 2007; Thickbroom, et al., 2006; Wada, et al., 2008; Wright, Stewart, & Barnett, 2008), including an examination of the biological basis of effort (Chaudhuri & Behan, 2004; DeLuca, 2005; Kohl, Wylie, Genova, Hillary, & Deluca, 2009; White, Lee, Light, & Light, 2009) Consistent with an effort-based hypothesis, persons with fatigue-related disorders (multiple sclerosis, chronic fatigue syndrome, traumatic brain injury) have been shown to display greater increases over time in cerebral activation during effortful task performance as compared to healthy controls (Cook, O'Connor, Lange, & Steffener, 2007; DeLuca, Genova, Hillary, & Wylie, 2008; Kohl, et al., 2009) In addition, one study found that momentary self-reported fatigue was related to brain activation on fMRI during the performance of a fatiguing task (Cook, et al., 2007).

Self-report measures have several practical advantages over direct measures of performance, due to their widespread availability and ease of use on the part of experimenters and participants. Performance-based measures tend to be fairly specialized and can require expensive equipment and training. In addition, the complexity of some activities, especially real world behaviors, can sometimes make it difficult to decipher the underlying nature of the decline. Performance-based measures, because they are often designed to induce fatigue, can also require a substantial commitment on the part of participants. This may represent an unacceptable burden for some potential subjects.

Among the disadvantages of self-report measures is that they can be influenced by various factors such as recall bias and mood (American Psychiatric Association, 2000; Broderick, et al., 2008). Recall bias can be reduced by the use of shorter recall periods such as end-of-day reports (Broderick, et al., 2009) or eliminated by the use of aggregated momentary assessments (Broderick, 2008). Negative mood can be problematic because it can exacerbate feelings of fatigue and fatigue itself is a symptom of a major depressive episode (American Psychiatric Association, 2000). Performance-based measures can also be influenced by mood. For example, depression is known to impact cognitive tasks, such as those requiring effortful processing (Zakzanis, Leach, & Kaplan, 1998). The administration of mood and depression scales can be used to quantify and at least partially control for their possible influence on fatigue measures (Nyenhuis, et al., 1998; Watson, Clark, & Tellegen, 1988).

SLEEPINESS VERSUS FATIGUE

One area in need of further research is the clarification of the relation of fatigue to sleepiness (Pigeon, Sateia, & Ferguson, 2003; Shen, Barbera, & Shapiro, 2006). In part this is because there is no consensus on the definition of sleepiness (Shen, et al., 2006), just as is the case for fatigue (Christodoulou, 2005). The terms are often used interchangeably, and it is often sleep researchers who are interested in distinguishing between the two concepts (Shen, et al., 2006). Sleepiness is often defined operationally by the method by which it is assessed (Shen, et al., 2006).

One of the most common methods of measuring sleepiness is the Multiple Sleep Latency Test (MSLT), which specifically measures sleep propensity, or the tendency to fall asleep (Littner, et al., 2005). During the MSLT, subjects are provided with four to five napping opportunities in a bed in a quiet darkened room at two-hour intervals over the course of the day. Sleep onset for the MSLT is determined by the detection of rapid eye movement using electrooculogram (EOG) recording (Littner, et al., 2005). The MSLT examines propensity to sleep under relatively ideal circumstances. This contrasts with the ability to remain awake under such circumstances, which is tested by another widely used task called the Maintenance of Wakefulness Test (MWT) (Littner, et al., 2005). During the MWT the subject is given the instruction to sit up in bed and remain awake for a specified period of time (20–40 minutes) in a quiet, darkened room. The MWT is more appropriate for the question of whether a person is likely to fall asleep under conditions where they need to remain awake (e.g., while at work), though it only measures this ability in one set of relatively ideal sleeping conditions. The two tests are only modestly correlated with one another (Sangal, Thomas, & Mitler, 1992). It has not been established that either test is useful in predicting accident risk and safety under real world circumstances, and sleep clinicians are advised not to rely solely on mean sleep latency scores from either test to make a prognosis of such risk (Littner, et al., 2005; Wise, 2006).

The psychomotor vigilance test (PVT) is another common method of measuring sleepiness (Balkin, et al., 2004). The PVT is a simple task on which a light blinks on at a random schedule every few seconds over a period of five to ten minutes. The subject is told to press a button as soon as this occurs. The PVT has been found to be among the more sensitive measures of sleep loss (Balkin, et al., 2004).

Another approach to assessing sleepiness is through the use of eye blink measures (Schleicher, Galley, Briest, & Galley, 2008; Stern, Boyer, & Schroeder, 1994). Aspects of blinking (e.g., duration) have been linked to sleepiness while driving (Schleicher, et al., 2008). Correlations have also been found between blink measures and self-reported fatigue in fighter pilots (Morris & Miller, 1996). Blink rates also increase with time on task in tasks requiring vigilance (Stern, et al., 1994).

In addition to the objective measures of sleepiness, there are a number of questionnaires that have been used for this purpose, including the Stanford Sleep Scale (SSS) and the Epworth Sleepiness Scale (ESS) (Åkerstedt & Gillberg, 1990; Hoddes, Zarcone, Smythe, Phillips, & Dement, 1973; Johns, 1991). The Stanford Sleepiness Scale (SSS) is one of the oldest and most widely used self-report measures of sleepiness (Hoddes, et al., 1973). It assesses current state of sleepiness by asking the respondent to choose from seven ranked statements ranging from 1 "feel active and vital; alert; wide awake" to 7 "almost in reverie; sleep onset soon; lost struggle to remain awake" (Hoddes, et al., 1973). While the focus of the SSS is upon current state, the ESS is design to assess an individual's general tendency to fall asleep under eight sets of circumstances (e.g. watching television, sitting and talking with someone) (Johns, 1991).

One attempt to distinguish sleepiness from fatigue is the FACES questionnaire, designed to distinguish the "faces" of fatigue and sleepiness (Shapiro, et al., 2002). The FACES instrument consists of five subscales: Fatigue, Anergy, Consciousness, Energized, and Sleepiness, which emerged from a factor analysis of a large number of fatigue-related adjectives in a group of subjects with severe insomnia. It should be noted, however, that the Fatigue factor accounted for the large majority of the variance in the analysis.

REFERENCE

Ahsberg, E. (2000). Dimensions of fatigue in different working populations. *Scandinavian Journal of Psychology, 41*(3), 231–41.

Ahsberg, E., & Gamberale, F. (1998). Perceived fatigue during physical work: An experimental evaluation of a fatigue inventory. *International Journal of Industrial Ergonomics, 21*(2), 117–31.

Ahsberg, E., Gamberale, F., & Gustafsson, K. (2000). Perceived fatigue after mental work: an experimental evaluation of a fatigue inventory. *Ergonomics, 43*(2), 252–68.

Ahsberg, E., Gamberale, F., & Kjellberg, A. (1997). Perceived quality of fatigue during different occupational tasks – Development of a questionnaire. *International Journal of Industrial Ergonomics, 20*(2), 121–35.

Åkerstedt, T., & Gillberg, M. (1990). Subjective and objective sleepiness in the active individual. *International Journal of Neuroscience, 52*(1–2), 29–37.

American Psychiatric Association. (2000). *Diagnostic and Statistical Manual of Mental Disorders, Fourth Edition – Text Revision (DSMIV-TR)*. Washington, DC: American Psychiatric Association.

Annoni, J. M., Staub, F., Bogousslavsky, J., & Brioschi, A. (2008). Frequency, characterisation and therapies of fatigue after stroke. *Neurological Sciences, 29 Suppl 2*, S244–246.

Baker, K., Olson, J., & Morisseau, D. (1994). Work practices, fatigue, and nuclear power plant safety performance. *Human Factors, 36*(2), 244–257.

Balkin, T.J., Bliese, P.D., Belenky, G., Sing, H., Thorne, D.R., Thomas, M., Redmond, D.P., Russo, M., & Wesensten, N.J. (2004). Comparative utility of instruments for monitoring sleepiness-related performance decrements in the operational environment. *Journal of Sleep Research, 13*(3), 219–27.

Barger, L.K., Cade, B.E., Ayas, N.T., Cronin, J.W., Rosner, B., Speizer, F.E., & Czeisler, C.A. (2005). Extended work shifts and the risk of motor vehicle crashes among interns. *New England Journal of Medicine, 352*(2), 125–34.

Bartley, S.H., & Chute, E. (1947). *Fatigue and Impairment in Man*. New York: McGraw-Hill.

Basner, M., Rubinstein, J., Fomberstein, K.M., Coble, M.C., Ecker, A., Avinash, D., & Dinges, D.F. (2008). Effects of night work, sleep loss and time on task on simulated threat detection performance. *Sleep, 31*(9), 1251–9.

Belmont, A., Agar, N., & Azouvi, P. (2009). Subjective fatigue, mental effort, and attention deficits after severe traumatic brain injury. *Neurorehabilitation and Neural Repair, 9*, 939–44.

Beurskens, A.J., Bultmann, U., Kant, I., Vercoulen, J.H., Bleijenberg, G., & Swaen, G.M. (2000). Fatigue among working people: Validity of a questionnaire measure. *Occupational and Environmental Medicine, 57*(5), 353–7.

Borg, G. (1970). Perceived exertion as an indicator of somatic stress. *Scandinavian Journal of Rehabilitation Medicine, 2*(2), 92–8.

Borg, G. (1982a). A category scale with ratio properties for intermodal and interindividual comparisons. In H.G. Geissler & P. Petzgold (eds), *Psychophysical Judgment and the Process of Perception*. New York: North-Holland Publishing Co.

Borg, G. (1982b). Psychophysical bases of perceived exertion. *Medicine and Science in Sports and Exercise, 14*(5), 377–81.

Broderick, J.E. (2008). Electronic diaries: Appraisal and current status. *Pharmaceutical Medicine, 22*(2), 69–74.

Broderick, J.E., Schwartz, J.E., Schneider, S., & Stone, A.A. (2009). Can end-of-day reports replace momentary assessment of pain and fatigue? *Journal of Pain, 10*(3), 274–81.

Broderick, J.E., Schwartz, J.E., Vikingstad, G., Pribbernow, M., Grossman, S., & Stone, A.A. (2008). The accuracy of pain and fatigue items across different reporting periods. *Pain, 139*(1), 146–57.

Bryant, D., Chiaravalloti, N.D., & DeLuca, J. (2004). Objective measurement of cognitive fatigue in multiple sclerosis. *Rehabilitation Psychology, 49*(2), 114–22.

Burgess, P.R. & Jones, L.F. (1997). Perceptions of effort and heaviness during fatigue and during the size-weight illusion. *Somatosensory and Motor Research, 14*(3), 189–202.

Caldwell, J.A. & Caldwell, J.L. (2005). Fatigue in military aviation: An overview of US military-approved pharmacological countermeasures. *Aviation Space and Environmental Medicine, 76*(7), C39–C51.

Caldwell, J.A., Caldwell, J.L., Brown, D.L., Smyth, N.K., Smith, J.K., Mylar, J.T., Mandichak, M.L., & Schroeder, C. (2003). The effects of 37 hours of continuous wakefulness on the physiological arousal, cognitive performance, self-reported mood, and simulator flight performance of F-117A pilots. U.S. Air Force Research Laboratory.

Cella, D., Lai, J.S., Chang, C.H., Peterman, A. & Slavin, M. (2002). Fatigue in cancer patients compared with fatigue in the general United States population. *Cancer, 94*(2), 528–38.

Chalder, T., Berelowitz, G., Pawlikowska, T., Watts, L., Wessely, S., Wright, D., & Wallace, E.P. (1993). Development of a fatigue scale. *Journal of Psychosomatic Research, 37*(2), 147–53.

Chaudhuri, A. & Behan, P.O. (2004). Fatigue in neurological disorders. *Lancet, 363*(9413), 978–88.

Christodoulou, C. (2005). The assessment and measurement of fatigue. In J. DeLuca (ed.), *Fatigue as a Window to the Brain*. New York: MIT Press, 19–35.

Cook, D.B., O'Connor, P.J., Lange, G. & Steffener, J. (2007). Functional neuroimaging correlates of mental fatigue induced by cognition among chronic fatigue syndrome patients and controls. *Neuroimage, 36*(1), 108–22.

De Vries, J., Michielsen, H.J., & Van Heck, G.L. (2003). Assessment of fatigue among working people: A comparison of six questionnaires. *Occupational and Environmental Medicine, 60 Suppl 1*, i10–15.

DeLuca, J. (2005). Fatigue, cognition, and mental effort. In J. DeLuca (ed.), *Fatigue as a Window to the Brain*. New York: MIT Press, 37–57.

DeLuca, J., Genova, H.M., Hillary, F.G. & Wylie, G. (2008). Neural correlates of cognitive fatigue in multiple sclerosis using functional MRI. *Journal of the Neurological Sciences, 270*(1–2), 28–39.

Development of the World Health Organization WHOQOL-BREF quality of life assessment. The WHOQOL Group. (1998). *Psychological Medicine, 28*(3), 551–8.

Dimsdale, J.E., Ancoli-Israel, S., Elsmore, T.F. & Gruen, W. (2003). Taking fatigue seriously: I. Variations in fatigue sampled repeatedly in healthy controls. *Journal of Medical Engineering and Technology, 27*(5), 218–22.

Fahlen, G., Knutsson, A., Peter, R., Åkerstedt, T., Nordin, M., Alfredsson, L., & Westerholm, P. (2006). Effort-reward imbalance, sleep disturbances and fatigue. *International Archives of Occupational and Environmental Health, 79*(5), 371–8.

Folkard, S. & Lombardi, D. A. (2006). Modeling the impact of the components of long work hours on injuries and "accidents". *American Journal of Industrial Medicine, 49*(11), 953–63.

Friedman, J.H., Brown, R.G., Comella, C., Garber, C.E., Krupp, L.B., Lou, J.S., Marsh, L., Nail, L., Shulman, L. & Taylor, C.B. (2007). Fatigue in Parkinson's disease: A review. *Movement Disorders, 22*(3), 297–308.

Fries, J.F., Bruce, B., & Cella, D. (2005). The promise of PROMIS: Using item response theory to improve assessment of patient-reported outcomes. *Clinical and Experimental Rheumatology, 23* (5 Suppl 39), S53–57.

Fries, J.F., Cella, D., Rose, M., Krishnan, E. & Bruce, B. (2009). Progress in assessing physical function in arthritis: PROMIS short forms and computerized adaptive testing. *Journal of Rheumatology, 36*(9), 2061–6.

Green, B.F., Bock, R.D., Humphreys, L.G., Linn, R.L. & Reckase, M.D. (1984). Technical guidelines for assessing computerized adaptive tests. *Journal of Educational Measurement, 21*(4), 347–60.

Hadjimichael, O., Vollmer, T. & Oleen-Burkey, M. (2008). Fatigue characteristics in multiple sclerosis: The North American Research Committee on Multiple Sclerosis (NARCOMS) survey. *Health and Quality of Life Outcomes, 6*, 100.

Hays, R.D., Morales, L.S., & Reise, S.P. (2000). Item response theory and health outcomes measurement in the 21st century. *Medical Care, 38*(9 Suppl), II28–42.

Hoddes, E., Zarcone, V., Smythe, H., Phillips, R., & Dement, W. C. (1973). Quantification of sleepiness: A new approach. *Psychophysiology, 10*(4), 431–6.

Janssen, N., Kant, I.J., Swaen, G.M., Janssen, P.P., & Schroer, C.A. (2003). Fatigue as a predictor of sickness absence: Results from the Maastricht cohort study on fatigue at work. *Occupational and Environmental Medicine, 60* Suppl 1, i71–76.

Johns, M.W. (1991). A new method for measuring daytime sleepiness: The Epworth sleepiness scale. *Sleep, 14*(6), 540–45.

Johnson, S.K. (2005). Depression and fatigue. In J. DeLuca (ed.), *Fatigue as a Window to the Brain.* Cambridge, MA: MIT Press, 157–72.

Kohl, A.D., Wylie, G.R., Genova, H.M., Hillary, F.G., & Deluca, J. (2009). The neural correlates of cognitive fatigue in traumatic brain injury using functional MRI. *Brain Injury, 23*(5), 420–32.

Kroenke, K. & Price, R.K. (1993). Symptoms in the community: Prevalence, classification, and psychiatric comorbidity. *Archives of Internal Medicine, 153*(21), 2474–80.

Krupp, L.B. & Elkins, L.E. (2000). Fatigue and declines in cognitive functioning in multiple sclerosis. *Neurology, 55*(7), 934–9.

Krupp, L.B., LaRocca, N.G., Muir-Nash, J. & Steinberg, A.D. (1989). The fatigue severity scale: Application to patients with multiple sclerosis and systemic lupus erythematosus. *Archives of Neurology, 46*(10), 1121–3.

Lai, J.S., Cella, D., Choi, S., Junghaenel, D.U., Christodoulou, C., Gershon, R. & Stone, A. (2011). How item banks and their application can influence measurement practice in rehabilitation medicine: A PROMIS fatigue item bank example. *Archives of Physical Medicine and Rehabilitation, 92*(10 Suppl), S20–27.

Landrigan, C.P., Rothschild, J.M., Cronin, J.W., Kaushal, R., Burdick, E., Katz, J.T., Lilly, C.M., Stone, P.H., Lockley, S.W., Bates, D.W. & Czeisler, C.A. (2004). Effect of reducing interns' work hours on serious medical errors in intensive care units. *New England Journal of Medicine, 351*(18), 1838–48.

Littner, M.R., Kushida, C., Wise, M., Davila, D.G., Morgenthaler, T., Lee-Chiong, T., Hirshkowitz, M., Daniel, L.L., Bailey, D., Berry, R.B., Kapen, S. & Kramer, M. (2005). Practice parameters for clinical use of the multiple sleep latency test and the maintenance of wakefulness test. *Sleep, 28*(1), 113–21.

Lockley, S.W., Barger, L.K., Ayas, N.T., Rothschild, J.M., Czeisler, C.A. & Landrigan, C.P. (2007). Effects of health care provider work hours and sleep deprivation on safety and performance. *Joint Commission Journal on Quality and Patient Safety, 33*(11 Suppl), 7–18.

Maslach, C. & Jackson, S.E. (1981). The measurement of experienced burnout. *Journal of Occupational Behaviour, 2*(2), 99–113.

McNair, D.M., Lorr, M. & Droppleman, L.F. (1992). *Profile of Mood States.* San Diego, CA: Educational and Industrial Testing Service.

Michielsen, H.J., De Vries, J. & Van Heck, G.L. (2003). Psychometric qualities of a brief self-rated fatigue measure: The Fatigue Assessment Scale. *Journal of Psychosomatic Research, 54*(4), 345–52.

Michielsen, H.J., De Vries, J., Van Heck, G.L., Van de Vijver, F.J.R. & Sijtsma, K. (2004). Examination of the dimensionality of fatigue – The Construction of the Fatigue Assessment Scale (FAS). *European Journal of Psychological Assessment, 20*(1), 39–48.

Mock, V., Atkinson, A., Barsevick, A., Cella, D., Cimprich, B., Cleeland, C., Donnelly, J., Eisenberger, M.A., Escalante, C., Hinds, P., Jacobsen, P.B., Kaldor, P., Knight, S.J., Peterman, A., Piper, B.F., Rugo, H., Sabbatini, P. & Stahl, C. (2000). NCCN practice guidelines for cancer-related fatigue. *Oncology (Williston Park), 14*(11A), 151–61.

Morris, T.L., & Miller, J.C. (1996). Electrooculographic and performance indices of fatigue during simulated flight. *Biological Psychology, 42*(3), 343–60.

Multiple Sclerosis Council for Clinical Practice Guidelines. (1998). *Fatigue and Multiple Sclerosis: Evidence-based Management Strategies for Fatigue in Multiple Sclerosis.* Washington, DC: Paralyzed Veterans Association of America.

Nyenhuis, D.L., Luchetta, T., Yamamoto, C., Terrien, A., Bernardin, L., Rao, S.M. & Garron, D.C. (1998). The development, standardization, and initial validation of the Chicago multiscale depression inventory. *Journal of Personality Assessment, 70*(2), 386–401.

Pigeon, W.R., Sateia, M.J. & Ferguson, R.J. (2003). Distinguishing between excessive daytime sleepiness and fatigue: Toward improved detection and treatment. *Journal of Psychosomatic Research, 54*(1), 61–9.

Power, M., Harper, A. & Bullinger, M. (1999). The World Health Organization WHOQOL-100: tests of the universality of Quality of Life in 15 different cultural groups worldwide. *Health Psychology, 18*(5), 495–505.

Rampinini, E., Impellizzeri, F.M., Castagna, C., Coutts, A.J. & Wisloff, U. (2009). Technical performance during soccer matches of the Italian Serie A league: Effect of fatigue and competitive level. *Journal of Science and Medicine in Sport, 12*(1), 227–33.

Ranjith, G. (2005). Epidemiology of chronic fatigue syndrome. *Occupational Medicine (Oxford, England), 55*(1), 13–19.

Ricci, J.A., Chee, E., Lorandeau, A.L. & Berger, J. (2007). Fatigue in the U.S. workforce: Prevalence and implications for lost productive work time. *Journal of Occupational and Environmental Medicine, 49*(1), 1–10.

Robertson, R.J., Goss, F.L. & Metz, K.F. (1998). Perception of physical exertion during dynamic exercise: A tribute to Professor Gunnar A.V. Borg. *Perceptual and Motor Skills, 86*(1), 183–91.

Rosen, S.D., King, J.C., Wilkinson, J.B. & Nixon, P.G. (1990). Is chronic fatigue syndrome synonymous with effort syndrome? *Journal of the Royal Society of Medicine, 83*(12), 761–4.

Sangal, R.B., Thomas, L. & Mitler, M.M. (1992). Maintenance of wakefulness test and multiple sleep latency test: Measurement of different abilities in patients with sleep disorders. *Chest, 101*(4), 898–902.

Sangnier, S. & Tourny-Chollet, C. (2008). Study of the fatigue curve in quadriceps and hamstrings of soccer players during isokinetic endurance testing. *Journal of Strength and Conditoning Research, 22*(5), 1458–67.

Schellekens, J.M., Sijtsma, G.J., Vegter, E. & Meijman, T.F. (2000). Immediate and delayed after-effects of long lasting mentally demanding work. *Biological Psychology, 53*(1), 37–56.

Schleicher, R., Galley, N., Briest, S. & Galley, L. (2008). Blinks and saccades as indicators of fatigue in sleepiness warnings: Looking tired? *Ergonomics, 51*(7), 982–1010.

Schwartz, J.E., Jandorf, L. & Krupp, L.B. (1993). The measurement of fatigue: A new instrument. *Journal of Psychosomatic Research, 37*(7), 753–62.

Segerstrom, S.C. & Nes, L.S. (2007). Heart rate variability reflects self-regulatory strength, effort, and fatigue. *Psychological Science, 18*(3), 275–81.

Shapiro, C.M., Flanigan, M., Fleming, J.A., Morehouse, R., Moscovitch, A., Plamondon, J., Reinish, L. & Devins, G.M. (2002). Development of an adjective checklist to measure five FACES of fatigue and sleepiness. Data from a national survey of insomniacs. *Journal of Psychosomatic Research, 52*(6), 467–73.

Shen, J., Barbera, J. & Shapiro, C.M. (2006). Distinguishing sleepiness and fatigue: Focus on definition and measurement. *Sleep Medicine Reviews, 10*(1), 63–76.

Shiffman, S., Stone, A.A. & Hufford, M.R. (2008). Ecological momentary assessment. *Annual Review of Clinical Psychology, 4*, 1–32.

Smets, E.M., Garssen, B., Bonke, B. & De Haes, J.C. (1995). The Multidimensional Fatigue Inventory (MFI) psychometric qualities of an instrument to assess fatigue. *Journal of Psychosomatic Research, 39*(3), 315–25.

Stern, J.A., Boyer, D. & Schroeder, D. (1994). Blink rate: A possible measure of fatigue. *Human Factors, 36*(2), 285–97.

Stone, A.A. & Shiffman, S. (1994). Ecological momentary assessment (EMA) in behavioral medicine. *Annals of Behavioral Medicine, 16*(3), 199–202.

Swaen, G.M., Van Amelsvoort, L.G., Bultmann, U. & Kant, I.J. (2003). Fatigue as a risk factor for being injured in an occupational accident: Results from the Maastricht Cohort Study. *Occup Environ Med, 60 Suppl 1*, i88–92.

Thickbroom, G.W., Sacco, P., Kermode, A.G., Archer, S.A., Byrnes, M.L., Guilfoyle, A. & Mastaglia, F.L. (2006). Central motor drive and perception of effort during fatigue in multiple sclerosis. *Journal of Neurology, 253*(8), 1048–53.

Thorndike, E. (1900). Mental fatigue I. *Psychological Review, 7*, 466–82.

van der Linden, D., Frese, M., & Meijman, T.F. (2003). Mental fatigue and the control of cognitive processes: Effects on perseveration and planning. *Acta Psychologica, 113*(1), 45–65.

Vercoulen, J.H., Bazelmans, E., Swanink, C.M., Fennis, J.F., Galama, J.M., Jongen, P.J., Hommes, O., Van der Meer, J.W. & Bleijenberg, G. (1997). Physical activity in chronic fatigue syndrome: Assessment and its role in fatigue. *Journal of Psychiatric Research, 31*(6), 661–73.

Vercoulen, J.H., Swanink, C.M., Fennis, J.F., Galama, J.M., van der Meer, J.W. & Bleijenberg, G. (1994). Dimensional assessment of chronic fatigue syndrome. *Journal of Psychosomatic Research, 38*(5), 383–92.

Vollestad, N.K. (1997). Measurement of human muscle fatigue. *Journal of Neuroscience Methods, 74*(2), 219–27.

Wada, K., Sakata, Y., Theriault, G., Aratake, Y., Shimizu, M., Tsutsumi, A., Tanaka, K. & Aizawa, Y. (2008). Effort-reward imbalance and social support are associated with chronic fatigue among medical residents in Japan. *International Archives of Occupational and Environmental Health, 81*(3), 331–6.

Ware, J.E. SF-36 Health Survey Update. Available at: http://www.sf-36.org/tools/sf36.shtml#VERS2 [accessed 08/15/2010].

Ware, J.E. & Sherbourne, C.D. (1992). The MOS 36–item short-form health survey (SF-36). I. Conceptual framework and item selection. *Medical Care, 30*(6), 473–83.

Watson, D., Clark, L.A. & Tellegen, A. (1988). Development and validation of brief measures of positive and negative affect: The PANAS scales. *Journal of Personality and Social Psychology, 54*(6), 1063–70.

Watt, T., Groenvold, M., Bjorner, J.B., Noerholm, V., Rasmussen, N.A. & Bech, P. (2000). Fatigue in the Danish general population: Influence of sociodemographic factors and disease. *Journal of Epidemiology and Community Health, 54*(11), 827–33.

Wessely, S. (2005). Forward. In J. DeLuca (ed.), *Fatigue as a Window to the Brain*. New York: MIT Press, xi–xvii.

Wessely, S., Hotopf, M. & Sharpe, D. (1998). *Chronic Fatigue and its Syndromes*. New York: Oxford University Press.

White, A.T., Lee, J.N., Light, A.R. & Light, K.C. (2009). Brain activation in multiple sclerosis: A BOLD fMRI study of the effects of fatiguing hand exercise. *Multiple Sclerosis, 15*(5), 580–86.

Winward, C., Sackley, C., Metha, Z. & Rothwell, P.M. (2009). A population-based study of the prevalence of fatigue after transient ischemic attack and minor stroke. *Stroke, 40*(3), 757–61.

Wise, M. S. (2006). Objective measures of sleepiness and wakefulness: Application to the real world? *Journal of Clinical Neurophysiology, 23*(1), 39–49.

Wright, R.A., Stewart, C.C. & Barnett, B.R. (2008). Mental fatigue influence on effort-related cardiovascular response: Extension across the regulatory (inhibitory)/non-regulatory performance dimension. *International Journal of Psychophysiology, 69*(2), 127–33.

Zakzanis, K.K., Leach, L. & Kaplan, E. (1998). On the nature and pattern of neurocognitive function in major depressive disorder. *Neuropsychiatry Neuropsychology and Behavioral Neurology, 11*(3), 111–19.

9
Dimensional Models of Fatigue

Gerald Matthews, Paula A. Desmond and Edward M. Hitchcock

There was nothing the matter with them except that they were dead tired. It was not the dead-tiredness that comes through brief and excessive effort, from which recovery is a matter of hours; but it was the dead-tiredness that comes through the slow and prolonged strength drainage of months of toil. (Jack London, *Call of the Wild*)

INTRODUCTION

"Fatigue" is an umbrella term that refers to symptoms including tiredness, sleepiness, loss of motivation, and lethargy. Definitions of fatigue differ, but it may be seen most simply as a constellation of mental and physical symptoms (Gawron, French & Funke, 2001). People also seem to differ considerably in their vulnerability to fatigue states. Some individuals appear to tire easily whereas others are typically full of energy. As our introductory quotation illustrates, fatigue may be a transient state or an enduring condition. At the extreme, chronic fatigue is recognized as a clinical issue (Jason, Brown, Evans & Brown, Chapter 18, this volume).

This chapter addresses dimensional models that may guide research on fatigue, including the choice of instruments for fatigue assessment. The construction and validation of sound psychometric models is more complex than may at first appear. Fatigue is multifaceted, in that it is expressed in changes in behavior, in physiological functioning, and in subjective experience. There are multiple sources of fatigue, including sleep loss, circadian rhythms, prolonged work and health issues, which may need to be conceptualized separately. Fatigue operates over varying timespans ranging from momentary tiredness to a lifetime of chronic fatigue.

The essence of a dimensional model is that the multiple attributes of fatigue may be measured on continuous, quantitative scales (approximating to interval scales). Ideally, researchers use an instrument that is universally applicable, just as a portable weather station measures temperature, humidity, wind speed, etc. at any point on the Earth's surface. For example, broad personality scales such as those for the "Big Five" traits (McCrae & Costa, 2008) owe their popularity in part to their demonstrated reliability and validity across a wide range of applied and laboratory settings. Dimensional measures must not just be sensitive to fatiguing influences, but must meet appropriate psychometric criteria.

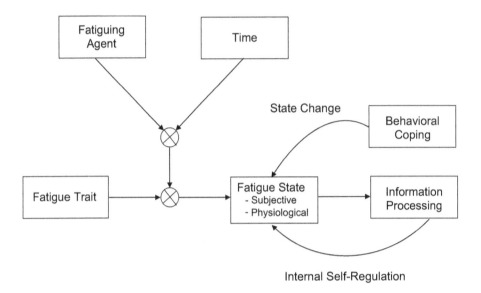

Figure 9.1   A simple trait–state model for fatigue

A simple model for conceptualizing and distinguishing relevant dimensional constructs is shown in Figure 9.1. It differentiates trait and state fatigue, where the trait refers to the person's general vulnerability to fatigue and the state is the immediate experience (and its physiological concomitants). Both traits and states may be considered here as vectors defined by the multiple facets or dimensions of fatigue, broadly defined. The influence of fatigue vulnerability traits is moderated by external agents that promote fatigue. For example, the role of personality differences may become more apparent following extended work periods or sleep loss. Fatigue is defined by the importance of time, which is central to periodic and homeostatic processes in fatigue (Banks, Jackson & Van Dongen, Chapter 11, this volume) and also to temporal decrements in alertness during performance. Figure 9.1 expresses the influence of external agents as time-dependent. The state is assumed to influence information processing and behavior: implications for performance are elaborated by Matthews, Hancock & Desmond (Chapter 10, this volume).

A model of this kind might incorporate multiple feedback loops (Matthews & Desmond, 2002). Only two are shown. It is assumed that the person has some capacity for self-regulation, i.e., monitoring the fatigue state and trying to maintain energy and motivation as appropriate for current goals. In addition, behavioral consequences of fatigue may include "state change" strategies for maintaining alertness (e.g., drinking coffee) or for seeking rest (choosing to sleep). The model does not seek to represent day-to-day cycles of fatigue that may be associated with time periods of weeks or days, e.g., prolonged or chronic fatigue that might be caused by, say, difficulties in sleeping, or excessive daily workloads.

The model differentiates the multiple constructs that may be assessed on a dimensional basis, including personality trait assessments, measures of immediate state of mind, psychophysiological state indices, and use of self-regulative and behavioral fatigue management strategies. There is also some scope for confusing the "stimulus" and "response" sides of fatigue with the fatigue state. For example, an observational measure of sleep disturbance that tracked the times at which the person slept would be informative about a key input, but would only indirectly index the fatigue state itself. Similarly, use of actigraphy provides a measure of the person's level of motor activity, which may be an outcome of fatigue but is not itself a direct index of fatigue (motor activity reflects factors other than the fatigue state).

In the remainder of this chapter, we will use the model as a framework for clarifying some of the different types of fatigue assessment that may be used in research. We will introduce a tentative general taxonomy of fatigue constructs that may be useful in guiding assessment efforts. We will focus especially on the assessment of fatigue vulnerability, the vicissitudes of objective fatigue assessments and a comprehensive state model for mental fatigue dimensions.

A TAXONOMY OF FATIGUE CONSTRUCTS

One way to clarify the wide range of fatigue measures available to the research is to develop a taxonomy or classificatory scheme that differentiates types of measure systematically.

An outline taxonomy for fatigue is shown in Figure 9.2. It incorporates several key distinctions. The first is the timespan for fatigue assessment, corresponding to the trait vs. state distinction of Figure 9.1. We have added an intermediate timespan for chronic fatigue lasting for months or so. The second is whether the assessment is *explicit* – reliant on self-

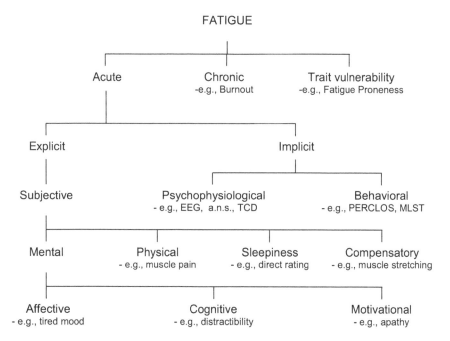

Figure 9.2   A provisional taxonomy of fatigue constructs

report of subjective symptoms, or *implicit* – reliant on behavioral or psychophysiological indices that may reflect unconscious processes. The third distinction refers to qualitative differences between fatigue states, such as mental fatigue, physical discomfort, and sleepiness. We have further subdivided mental fatigue also. We could further extend the taxonomy by differentiating applied contexts for operator fatigue, such as vehicle driving, work and others – examples of fatigue assessments designed for the driving context are discussed below.

In principle, the taxonomy could be presented as a cube or 3-D space in that we can imagine fatigue measures corresponding to each combination of timespan, explicit vs. implicit assessment, and qualitative nature (assuming unconscious mental states are permissible). In fact, it is more convenient to lay out a tree structure, which reflects the preponderance of research being directed toward immediate, conscious fatigue states. Longer-term and implicit measures have received less attention. We will discuss some of the key classes of measure on this basis.

Longer-term Fatigue Measures: Trait and Chronic Fatigue

As is shown in Figure 9.1, some individuals may be especially vulnerable to fatigue, whereas others maintain energy even in fatiguing settings. Michielsen et al. (2004) analyzed data from four scales, such as the Energy and Fatigue subscale from the World Health Organization Quality of Life assessment instrument. Application of factor analysis and item response theory supported a single-factor solution. Their Fatigue Assessment Scale (FAS) comprises 10 items referring to chronic fatigue symptoms, including exhaustion, lack of clarity of thought, and problems in initiating activity (see also Christodoulou, Chapter 8, this volume). If IDs in fatigue show temporal stability, we can examine the place of chronic fatigue within personality trait models, as well as testing associations between personality traits and fatigue states in specific operational contexts.

In terms of the Five Factor Model (FFM) of personality (McCrae & Costa, 2008), the dimension typically linked to fatigue is extraversion–introversion. Extraverts tend to experience more positive affect and subjective energy than introverts, perhaps because of higher levels of activity in a dopaminergic reward system (Corr, 2009). Michielsen, De Vries and Van Heck (2003) found that their FAS correlated negatively with extraversion and positively with neuroticism. However, within specific performance contexts, effect sizes for associations between extraversion and energy states are often small (Matthews & Gilliland, 1999), and, paradoxically, extraverts are often more susceptible to performance decrements in fatiguing environments (Matthews, Deary & Whiteman, 2009). Other FFM factors that may relate modestly to fatigue states include neuroticism (Abdel-Khalek, 2009) and low conscientiousness (Matthews, Emo et al., 2006). More narrowly defined traits may also predict fatigue. Kanfer (2011) implicates various "avoidance" traits in fatigue, including sensitivity to punishment, behavioral inhibition, and anxiety. Recently, Shaw et al. (2010) factor-analyzed a battery of 17 "cognitive–energetic" traits associated with fatigue and lack of alertness. Scales for trait fatigue, including the FAS, loaded on a factor labeled as cognitive disorganization that was also defined by boredom, cognitive symptoms of schizotypy, cognitive failures, and ADD symptoms. Shaw et al. (2010) also found that, although the cognitive disorganization factor was associated with lower conscientiousness and higher neuroticism, it was more predictive of task-induced fatigue than the FFM. A second factor, labeled impulsivity, was also associated with fatigue states. Fatigue vulnerability appears to be linked to cognitive dysfunction as well as to motivational traits.

The trait approach assumes that individuals may show broad vulnerabilities to fatigue that generalize across situations and contexts. This assumption may not be true; a person may consistently find some situations energizing but others fatiguing. In this case, it may be more productive to use "contextualized" fatigue scales that are geared toward a specific setting. An example of this approach is provided by work on driver fatigue vulnerability. Matthews, Desmond et al. (1997) developed a multidimensional instrument for measuring dispositional stress vulnerabilities, the Driver Stress Inventory (DSI). "Fatigue Proneness" emerged as a distinct vulnerability dimension, together with "Aggression," "Sensation-Seeking," "Dislike of Driving" and "Hazard Monitoring." DSI Fatigue Proneness has been validated as a state fatigue predictor in both simulator and field studies (Desmond & Matthews, 2009; Matthews & Desmond, 1998; Neubauer, Matthews & Saxby, Chapter 23, this volume). It predicts both mental fatigue symptoms such as loss of task engagement, and physical symptoms such as muscular discomfort (Matthews & Desmond, 1998). Furthermore, although fatigue-prone drivers tend to be relatively fatigued even prior to vehicle operation, they show larger-magnitude increases in fatigue than fatigue-resistant drivers as a consequence of driving. Thus, the DSI may be useful in selection of individuals who are resilient to fatigue, or in identifying especially fatigue-vulnerable individuals who may not be suitable for employment in occupations such as long-haul trucking. Interestingly, the truck driver sample investigated by Desmond and Matthews (2009) were unusually low in Fatigue Proneness, compared to norms for the DSI, implying these drivers may have been selected for resistance to fatigue.

The trait perspective assumes that genetics and learning "fix" personal characteristics that, while not set in concrete, typically change relatively slowly throughout the lifespan (Matthews, Deary & Whiteman, 2009). However, there are intermediate timespans between trait and state for the chronic fatigue that is linked to life circumstances, but is not a direct expression of stable personality. Instruments may be used to ask about fatigue symptoms over periods such as the last two weeks (Sonnentag & Zijlstra, 2006). For example, fatigue is a common symptom of various health conditions such as chronic fatigue, multiple sclerosis and cancer, but may be alleviated by successful treatment. Fatigue may also be a side effect of treatment, such as chemotherapy in the case of cancer. There are both uni- and multidimensional instruments that are geared toward fatigue in the health context (Dittner, Wessely & Brown, 2004). Another context in which chronic fatigue is common is work, and both unidimensional and multidimensional instruments have been used to determine levels of fatigue in working populations (Beurskens et al., 2000; Sonnentag & Zijlstra, 2006). At the extreme, chronic fatigue shades into burn-out, a syndrome combining severe fatigue (emotional exhaustion), with diminished personal accomplishment, and depersonalization. The gold standard for the assessment of burn-out is the Maslach Burnout Inventory (Maslach, Jackson & Leiter, 1996). Emotional exhaustion – defined as feelings of being overextended and depleted of one's emotional and physical resources – appears as a distinct component of burn-out in confirmatory factor analyses (Worley et al., 2008).

Objective Measures of Fatigue

There is a basic distinction between *explicit* and *implicit* aspects of fatigue. Explicit elements of the fatigue response are consciously accessible, and provide the basis for measuring subjective fatigue through self-reports. However, some aspects of fatigue are likely to be implicit, in that that the operator has little conscious awareness of them. Implicit fatigue must be assessed using objective response measures psychophysiology or behavioral responses (see Figure 9.2). We will not discuss the various measures in detail here. Chapters in Section IV of this book, "The Neuroscience of Fatigue" cover various psychophysiological indices in the context of the neuroscience of fatigue, including electroencephalographic (EEG) measures (Chapter 12, Craig & Tran), biochemical markers (Chapter 14, Watanabe, Kuratsune & Kajimoto), cerebral bloodflow velocity (Chapter 13, Warm, Tripp, Matthews & Helton), and brain-imaging techniques (Chapter 11, Banks, Jackson & Van Dongen).

Christodoulou (Chapter 8, this volume) summarizes the principal behavioral techniques. These measures are primarily directed toward sleepiness, which may be assessed using methods such as the PERCLOS technique that measures eyelid closure and the Multiple Sleep Latency Test (MSLT), which measures the length of time taken to fall asleep in a darkened room. Another behavioral measure is the actigraph, a portable motion-sensing device, which records motor activity and is especially useful for naturalistic research.

Our discussion will be limited to some general remarks on using objective measures to support psychometrically sound dimensional models. First, in principle, both trait and state fatigue might be measured using implicit measures. For example, EEG parameters show some long-term stability (Corsi-Cabrera et al., 2007), and implicit personality assessments based on techniques such as the Implicit Association Test (IAT) have become increasingly popular (Schnabel, Asendorpf & Greenwald, 2008). Thus far, implicit measures of stable fatigue vulnerability have received little attention. Second, it is important to differentiate the different elements of a causal process, as indicated in Figure 9.1. If a fatigue state is the cause, and loss of task performance is the effect, then use of performance measures to assess fatigue risks generating circular arguments. It is also important to differentiate the fatigue state from indices of compensatory efforts to manage fatigue (Christodoulou, Chapter 8, this volume; Hockey, Chapter 3, this volume), labeled as self-regulation and state change in Figure 9.1.

Third, some special problems attach to psychophysiological indices of fatigue states. Typically, recording requires specialized skills, such as processing of artifacts (Craig & Tran, Chapter 12, this volume), which limits universality of application (Belz, Robinson & Casali, 2004). Another well-known issue in stress research is "response specificity" (Lacey, 1967); there are individual differences in the sensitivity to stress of responses, such as electrodermal and electrocardiac responses. Fatigue responses may also be subject to this principle, though the issue has been little researched.

Fourth, explicit psychometric models, such as latent factor models, are rather infrequently applied to psychophysiological data. Indeed, some of the basic assumptions made may not readily translate into psychometrically acceptable scales. Every introductory psychology student knows that as cortical arousal increases, we see in the EEG, successively, theta waves, followed by alpha waves, and finally beta waves. The problem is that measures of the spectral power density of the different frequency bands are not strongly correlated (Matthews & Amelang, 1993), and cannot be used to identify a psychometrically sound arousal dimension. The point is not that the frequency bands are unrelated to fatigue

– they are (Craig & Tran, Chapter 12, this volume). It is that standard psychometric techniques may be ill-suited to analyzing individual differences in power densities so as to assign a quantitative "fatigue state" score to the individual. There may be other means for developing dimensional models from EEG (or other psychophysiological) data, such as developing specific, composite indices of fatigue. For example, two EEG indices of task engagement (i.e., low fatigue) are the ratios (1) [beta/(alpha + theta)] and (2) [frontal midline theta/parietal alpha], as discussed by Hockey et al. (2009). Although promising in experimental studies, such measures await rigorous psychometric analysis.

Fifth, it may be difficult or impossible to interpret objective measures without knowledge of the context in which the data were recorded. Even if fatigue consistently affects an index, we cannot necessarily work back from the index to infer fatigue. For example, a change in an actigraphic measure of motor activity might simply reflect a change in the activities performed by participants in a study, rather than any change in fatigue. Galinsky et al. (1993) showed that during vigilance, participants show greater rather than less motor activity; fatigue leads to fidgeting in this context. The classic arousal indices are also subject to extraneous influences, including physical activity (electrocardiac measures) and intensity of visual stimulation (EEG).

None of these considerations rules out development of general fatigue metrics based on objective indices. Indeed, rapid advancements in psychophysiology and in behavioral measures of implicit mental states are a source of optimism. But it is challenging to develop general factor models for objective indices of the kind commonplace in personality and affective state research.

Dimensions of Subjective Fatigue

Fatigue states may be assessed on a unitary basis (Christodoulou, Chapter 8, this volume). Early studies of subjective fatigue during vehicle operation typically used multi-point rating scales, such as Pearson's Fatigue Scale, which distinguished 13 levels of fatigue ranging, in one version, from "bursting with energy" to "ready to drop" (Pearson 1957). However, use of single-item measurement is not psychometrically satisfactory, and more recent work has developed multi-item fatigue scales, reviewed by Christodoulou (Chapter 8, this volume).

Various authors have proposed multi-dimensional models for fatigue states. Several investigations in Japan (e.g., Yoshitake, 1978), conducted in an occupational health context, discriminated and validated dimensions relating to:

1. drowsiness, dullness and tiredness,
2. physical fatigue and bodily discomfort, and
3. poor concentration and thinking difficulties.

Western studies have also discriminated multiple dimensions. Smets et al. (1995) used confirmatory factor analysis to differentiate five dimensions of general fatigue, physical fatigue, mental fatigue, reduced motivation and reduced activity. Interestingly, the study also showed that some fatigue scales, notably mental fatigue, were not well predicted from a visual-analogue rating scale, as commonly used in applied research. Unfortunately, there is no general consensus on the dimensionality of fatigue. In part, the lack of a universal structural model for fatigue states reflects psychometric issues.

For example, some researchers have used orthogonal varimax rotations (e.g., Shapiro et al., 2002), where a correlated-factor solution would be more appropriate. As well as technical issues, there is lack of consensus on the sampling of fatigue items and scales. Naturally, the output of a factor analysis is critically dependent on sampling of inputs. Researchers differ on the *range* of constructs sampled. To the extent that researchers predominantly collect data on tiredness in its various forms, we may expect a strong general factor to emerge. Similarly, when researchers analyze multiple instruments (e.g., Michielsen et al., 2004) there will be considerable item overlap; almost all measures include items for tiredness. Sampling may then be *unbalanced*, so that tiredness is over-represented relative to other aspects of fatigue. Sampling a wider range of constructs is likely to support multiple factors. Fuller (1983) refers to constructs such as perceived performance ability, risk-taking, time perception and day-dreaming, which are distinct from tiredness.

A taxonomic approach can impose some conceptual order on the multiple constructs that are covered by the broad umbrella of fatigue. Our taxonomy recognizes four broad expressions of fatigue in subjective experience, most of which can be further subdivided. It broadly differentiates mental fatigue (e.g., tiredness), physical fatigue (e.g., muscular discomfort), sleepiness (e.g., difficulty staying awake) and compensatory activities (e.g., pacing back and forth). Mental fatigue is the best understood of these modes of expression, and we will return to dimensional models of mental fatigue shortly.

Dimensions of *physical fatigue* include muscular discomfort, eyestrain (asthenopia) and headache (Matthews & Desmond, 1998). They relate, of course, to physiological processes such as loss of muscle function due to prolonged use, a loss that has a well-defined recovery function. However, the subjective awareness of physical fatigue does not correspond directly to the physiological process (see Noakes, Chapter 7, this volume). People differ in their awareness of physical symptoms and the importance attributed to them. Noakes' Central Governor Model sees subjective fatigue as a functional state that regulates physical exertion (see Chapter 7, this volume). Thus, scales are needed to assess the individual's perceptions of physical symptoms, which may influence behavior somewhat separately from the physiological fatigue process itself.

Sleepiness represents a single, rather special dimension linked to the specific behavior of falling asleep, although subjective sleepiness may not be an accurate predictor of the behavior, as measured objectively by the MLST (Carskadon & Dement, 1982). Sleepiness is commonly assessed by single rating scales, such as the 9-point Karolinska Sleepiness Scale (Åkerstedt & Gillberg, 1990), and the 7-point Stanford Sleepiness Scale (Hoddes et al., 1973). People in performance settings actively resist fatigue, with varying degrees of success (Hanks et al., 1999). However, there has been little systematic work performed on dimensions of coping or *compensatory activities*, again referring to the subjective experience of fighting fatigue rather than the objective behaviors themselves. Hanks et al. (1999) discriminated 10 strategies that vehicle drivers use, including drinking caffeinated beverages, stopping for a nap or a walk, snacking, rolling the window down, talking with a passenger and even chewing ice. However, this and other studies have not aimed to develop systematical dimensional models. More subtly, people also regulate the level of effort allocated to coping with fatigue, and may accept some level of fatigue in order to avoid the strain of resistance (see Hockey, Chapter 3, this volume). Fairclough (2001) proposed that people regulate discomfort by balancing task and comfort-seeking goals, so that fatigue represents a state of motivational conflict.

Table 9.1   Illustrative items for seven dimensions of the Driving Fatigue Scale (Matthews, Saxby & Hitchcock, 2008)

Factorial Scale	Item
Muscular fatigue (α= 0.89)	Having tremors in my limbs Legs and arms feel stiff
Exhaustion–sleepiness (α= 0.94)	Fighting myself to stay awake Overtired
Boredom (α= 0.90)	Find driving repetitive Don't want to do this ever again
Confusion–distractibility (α= 0.90)	I'm easily distracted I catch myself daydreaming
Performance worries (α= 0.85)	I'm finding it hard to control my speed I keep losing track of where I am on the road
Comfort-seeking (α= 0.92)	I just want to take things easy I don't feel like exerting myself
Self-arousal (α= 0.83)	Listening to the radio Talking to somebody else

Table 9.1 illustrates a recent attempt to develop a multidimensional fatigue scale, for the vehicle- driving context, with scales corresponding to all four modes of fatigue experience. Matthews, Saxby and Hitchcock (2008) collected data on a new Driver Fatigue Scale from 288 participants in fatiguing driver simulator studies. Seven factors were extracted; factor correlations ranged from 0.19–0.45. Scales include those for motivational (boredom) and cognitive (performance worries, confusion) elements of mental fatigue. As is discussed below, further, affective symptoms of fatigue may be assessed with general state measures. Muscular fatigue was the most prominent sign of physical fatigue in this study. A sleepiness scale also included items for exhaustion (e.g., wiped out, burned out). Two "compensatory" factors were also found, consistent with Fairclough's (2001) analysis: self-arousal and comfort-seeking. Clearly, such work could be taken further, and different factor sets might emerge in different operational contexts. However, the study illustrates how the taxonomy provides a framework for differentiation of fatigue symptoms.

A COMPREHENSIVE STATE MODEL FOR MENTAL FATIGUE AND STRESS

There are multiple facets of mental fatigue states, including tiredness, loss of task motivation and cognitive dysfunctions. Some fatigue symptoms may overlap with those of stress, including negative emotions and mind-wandering (Matthews & Desmond, 2002). Matthews, Campbell et al. (2002) attempted to develop a comprehensive dimensional model for both stress and fatigue states. They used the classic "trilogy of mind" (Hilgard, 1980) as the basis for sampling systematically the different aspects of subjective states characteristic of task performance settings. They selected items to represent fundamental dimensions of mood (or basic affects), motivation and cognition, as experienced subjectively.

Table 9.2   A summary of the scales of the DSSQ

Domain	Scale	Example item	α
Affect/mood	Energetic arousal	I feel … vigorous	82
	Tension	I feel … nervous	87
	Hedonic Tone	I feel … contented	88
Motivation	Interest	The content of the task is interesting	81
	Success	I want to perform better than most people do	87
Cognition	Self-focus	I am reflecting about myself	87
	Self-esteem	I am worrying about looking foolish (-ve)	89
	Concentration	My mind is wandering a great deal (-ve)	89
	Confidence-control	I feel confident about my abilities	84
	INT-TR	I have thoughts of … The difficulty of the task	77
	INT-TI	I have thoughts of … Personal worries	85

Note: INT-TR = Task-Relevant Cognitive Interference; INT-TI = Task-Irrelevant Cognitive Interference.

Factor analyses of item data distinguished 11 dimensions of subjective state (see Table 9.2). These dimensions may be measured by the Dundee Stress State Questionnaire (DSSQ; see Matthews, Campbell et al., 2002). Each dimension is represented by a 7- or 8-item scale; internal consistencies (αs) are satisfactory and typically greater than 0.8.

The factor analyses employed oblique rotations that extracted correlated factors. Second-order factor analyses of the scale scores consistently demonstrated a three-factor solution delineating broader factors that tended to integrate the three domains of experience (Matthews, Campbell et al., 2002). *Task Engagement* is defined primarily by high energy, motivation and concentration: low scores define a fatigue state relating to tiredness, loss of interest in task performance, and distractibility. *Distress* refers to high tension, unpleasant mood and low confidence and control, whereas *Worry* is a purely cognitive factor defined by self-focused attention, low self-esteem and high levels of cognitive interference, i.e., intrusive thoughts about the task or personal concerns. The DSSQ provides the researcher with the option of measuring broad state constructs, such as task disengagement, or performing a more fine-grained assessment that differentiates specific elements of mood, cognition and motivation that may dissociate. It also allows discrimination of a broad mental fatigue state (low task engagement), which is distinct from stress and worry. Both correlational and experimental studies have contributed to validating the DSSQ for assessment of fatigue and stress states in performance settings, As has previously been discussed, task-induced fatigue (loss of engagement) is associated with cognitive-energetic traits such as cognitive-disorganization and impulsivity (Shaw et al., 2010). Matthews, Hancock and Desmond (Chapter 10, this volume) review correlational studies that link stress states including task engagement and distress to task performance. Next, we will summarize findings from studies relating stress states to psychophysiology and cognitive stress processes. Subjective and psychophysiological states overlap to a modest degree, but each type of index adds unique information about the operator (Matthews, Warm et al., 2010). Psychophysiological research on the DSSQ and its component dimensions supports this conclusion. Several mood studies (e.g., Matthews, 1987; Thayer, 1989) confirmed that both subjective energy and tension relate to indices of autonomic arousal such as heart rate and skin conductance. A study of the DSSQ (Fairclough & Venables, 2004) found that both task engagement and distress related to multiple indicators of autonomic and central nervous system activity. Toward the end of an 80-minute multi-component task, increased engagement was independently predicted by increased respiration rate, lower EEG alpha power, and shorter eyeblink duration. The multiple correlates of the subjective state confirm Thayer's (1989) argument that subjective arousal measures reflect the integration of several distinct systems. As is discussed elsewhere in this volume (Chapter 10, Matthews, Hancock & Desmond); Chapter 13, Warm, Tripp, Matthews & Helton), task engagement correlates also with a physiological index of workload – cerebral bloodflow velocity (CBFV) in the middle cerebral arteries, measured by transcranial Doppler sonography (TCD).

Both stress and fatigue states may be understood in relation to the transactional model of stress, which attributes emotional states to cognitive appraisal and coping processes (Lazarus, 1999; Szalma, Chapter 5, this volume). Studies have investigated how task-induced changes in engagement, distress and worry relate to key aspects of appraisal and coping identified by Lazarus (1999). Tasks include standard laboratory tasks such as rapid information processing (Matthews & Campbell, 2009) and vigilance (Shaw et al., 2010), and simulations of operational skills such as customer service (Matthews & Falconer, 2000). Each state factor appears to be associated with a characteristic pattern of appraisal and coping, as shown in Table 9.3. Low task engagement seems to be generated by lack of challenge, use of avoidance coping, and failure to use task-focused coping (Matthews, Warm et al., 2010). Such findings serve to link operational fatigue and stress to the cognitive theory of stress and emotion (Lazarus, 1999).

Experimental studies have investigated how task performance influences subjective state change. Changes in state from pre- to post-task may be expressed in standardized form, relative to scale distributions in a large normative sample for the DSSQ (Matthews, Campbell et al., 2002), such that normative mean and SD are 0 and 1 respectively. Figure 9.3 illustrates how various tasks elicit differing multivariate profiles of state change. The top part of the figure shows broadly "fatiguing" tasks that elicit large-magnitude

Table 9.3   Patterns of appraisal and coping associated with three DSSQ secondary factors

Task Engagement	Distress	Worry
Primary states		
Energetic arousal	Tense arousal	Self-consciousness
Motivation (interest)	Low hedonic tone	Low self-esteem
Motivation (success)	Low confidence	Cog. interference (task- related)
Concentration		Cog. interference (personal)
Appraisal		
Challenge	Threat	
High mental demands	Low controllability	
High effort	High total workload	
	Failure to reach goals	
Coping		
Task-focus	Emotion-focus	Emotion-focus
Low avoidance		Avoidance

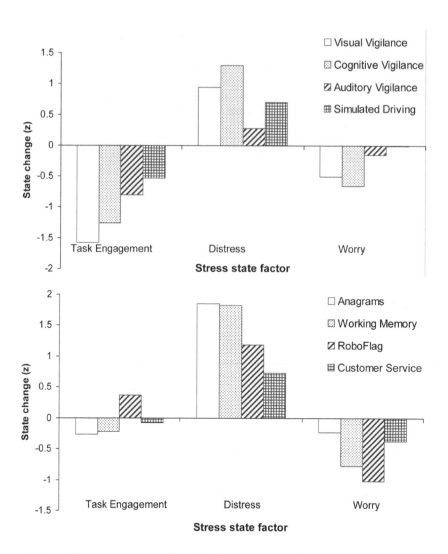

Figure 9.3   **Stress state profiles for (a – top) "fatiguing" and (b - bottom) "stressful" tasks**

declines in task engagement, including 36-minute visual and cognitive vigilance tasks, a 24-minute auditory vigilance task, and monotonous simulated driving. The lower part illustrates tasks that are "stressful" but not fatiguing, including a time-pressured working memory task, solving impossible anagrams, and simulations of unmanned vehicle control (Roboflag task), and customer service work.[1] The characteristic features

---

1   Data sources are as follows: Matthews, Warm et al., 2010 – visual (N=187) and cognitive (N=107) vigilance; Matthews, Campbell et al., 2002 – auditory vigilance (N=114); Neubauer et al., 2011 – simulated driving (N=180); Matthews, Emo et al., 2006 – working memory (N=50) and anagrams (N=50); Guznov, Matthews & Warm, 2010 – Roboflag task (N=150), Matthews & Falconer, 2000 – customer service (N=86).

of these tasks are increased distress, with variable patterning of other components. A similar evaluation may be applied to real-world tasks. For example, Desmond and Matthews (2009) used selected scales from the DSSQ to demonstrate task-induced fatigue in both commercial and noncommercial vehicle drivers. The scale has also been applied to profiling the effects of stress factors including loud noise (Helton, Matthews & Warm, 2009), cold infection (Matthews, Warm et al., 2001) and time pressure (Matthews & Campbell, 2009).

Thus, scales such as the DSSQ are useful in operational practice for characterizing fatigue in relation to both persons and situations, picking up two of the essential inputs to the trait-state conceptual model shown in Figure 9.1. On the one hand, the DSSQ may be used to confirm the fatigue vulnerability of individuals high in relevant traits, and to explore the cognitive stress processes that may mediate associations between traits and fatigue states (Shaw et al., 2010). On the other hand, the pattern of state change at the group level (e.g., Figure 9.3) provides a standard means for describing the general impact of a task environment on fatigue state, and for differentiating effects of external stressors from those of task demands.

CONCLUSIONS

The multifaceted nature of individual differences in fatigue is a challenge for assessment. Most generally, it is essential to distinguish trait and state characteristics of the person, and to separate inputs and outputs of the neurocognitive and self-regulative processes that generate fatigue. The fatigue state is a general quality of the person that has multiple concomitants, both physiological and psychological. The taxonomy we have presented in this chapter may help to differentiate some of the multiple expressions of fatigue states. The implicit components of the fatigue state are best known from psychophysiology. The introduction of objective, performance-based measures of implicit beliefs represents a new frontier for assessment, though not without its challenges. Despite their limitations, it is likely that self-report assessments of fatigue state will continue to be popular. Broadly, such explicit states may be divided into mental, physical, sleepiness-related and compensatory expressions of fatigue. In the realm of mental fatigue, use of self-report requires comprehensive dimensional models supporting state scales validated against objective measures. Theoretical understanding of state dimensions, whether based on cognitive stress theory or neuropsychology, is also essential. The multivariate state model we have advocated in this chapter has several advantages for operational practice. As is discussed in the previous section, stress/fatigue state patterns may be used to characterize both persons and task situations. Measurement of fatigue states also provides a means for separating environmental and personal influences on operator fatigue from the outcomes of the fatigue state. Measurement of state may especially contribute to understanding the impact of fatigue on information processing and performance. An important caveat is that not all elements of the state are accessible to conscious awareness. A challenge for the future is the development of strong psychometric models for "implicit" fatigue, whether psychophysiological or behavioral methods are used. Finally, state measurement is important for evaluating the effectiveness of countermeasures to fatigue.

REFERENCES

Abdel-Khalek, A.M. (2009). The relationship between fatigue and personality in a student population. *Social Behavior and Personality*, 37, 1357–68.

Åkerstedt, T. & Gillberg, M. (1990). Subjective and objective sleepiness in the active individual. *International Journal of Neuroscience*, 52, 29–37.

Belz, S.M., Robinson, G.S. & Casali, J.G. (2004). Temporal separation and self-rating of alertness as indicators of driver fatigue in commercial motor vehicle operators. *Human Factors*, 46, 154–69.

Beurskens, A.J., Bultmann, U., Kant, J., Vercoulen, J.H., Bleijenberg, G. & Swaen, G.M. (2000). Fatigue among working people: Validity of a questionnaire measure. *Occupational and Environmental Medicine*, 57, 353–537.

Carskadon, M.A. & Dement, W.C. (1982). The Multiple Sleep Latency Test: What does it measure? *Sleep: Journal of Sleep Research & Sleep Medicine*, 5(Suppl 2), 67–72.

Corr, P.J. (2009). The reinforcement sensitivity theory of personality, in P.J. Corr & G. Matthews (eds), *The Cambridge Handbook of Personality*. Cambridge: Cambridge University Press, 347–76.

Corsi-Cabrera, M., Galindo-Vilchis, L., del-Ro-Portilla, Y., Arce, C. & Ramos-Loyo, J. (2007). Within-subject reliability and inter-session stability of EEG power and coherent activity in women evaluated monthly over nine months. *Clinical Neurophysiology*, 118, 9–21.

Desmond, P.A. & Matthews, G. (2009). Individual differences in stress and fatigue in two field studies of driving. *Transportation Research. Part F, Traffic Psychology and Behaviour*, 12, 265–76.

Dittner, A.J., Wessely, S.C. & Brown, R.G. (2004). The assessment of fatigue: A practical guide for clinicians and researchers. *Journal of Psychosomatic Research*, 56, 157–70.

Fairclough, S.H. (2001). Mental effort regulation and the functional impairment of the driver, in P.A. Hancock & P.A. Desmond (eds), *Stress, Workload and Fatigue*. Mahwah, NJ: Lawrence Erlbaum, 479–502.

Fairclough, S.H. & Venables, L. (2006). Prediction of subjective states from psychophysiology: A multivariate approach. *Biological Psychology*, 71, 100–110.

Fuller, R.G.C. (1983). Effects of prolonged driving on time headway adopted by HGV (Heavy Goods Vehicle) drivers. Report number A685631. Alexandria, VA: U.S. Army Research Institute for the Behavioral and Social Science.

Galinsky, T.L., Rosa, R.R., Warm, J.S. & Dember, W.N. (1993). Psychophysical determinants of stress in sustained attention. *Human Factors*, 35, 603–14.

Gawron, V.J., French, J. & Funke, D. (2001). An overview of fatigue, in P.A. Hancock & P.A. Desmond (eds), *Stress, Workload and Fatigue*. Mahwah, NJ: Lawrence Erlbaum, 581–95.

Guznov, S., Matthews, G. & Warm, J.S. (2010). Team member personality, performance, and stress in a Roboflag synthetic task environment. *Proceedings of the Human Factors and Ergonomics Society*, 54, 1679–83.

Hanks, W.A., Driggs, X.A., Lindsay, G.B. & Merrill, R.M. (1999). An examination of common coping strategies used to combat driver fatigue. *Journal of American College Health*, 48, 135–7.

Helton, W.S., Matthews, G. & Warm, J.S. (2009). Stress state mediation between environmental variables and performance: The case of noise and vigilance. *Acta Psychologica*, 130, 204–13.

Hilgard, E.R. (1980). The trilogy of mind: Cognition, affection, and conation. *Journal of the History of the Behavioral Sciences*, 16, 107–17.

Hockey, G.R.J., Nickel, P., Roberts, A.C. & Roberts, M.H. (2009). Sensitivity of candidate markers of psychophysiological strain to cyclical changes in manual control load during simulated process control. *Applied Ergonomics*, 40, 1011–18.

Hoddes, E., Zarcone, V., Smythe, H., Phillips, R. & Dement, W.C. (1973). Quantification of sleepiness: A new approach. *Psychophysiology*, 10, 431–6.

Kanfer, R. (2011). Determinants and consequences of subjective cognitive fatigue, in P.L. Ackerman (ed.), *Cognitive Fatigue: The Current Status and Future for Research and Applications*. Washington, D.C.: APA, 189–207.

Lacey, J.I. (1967). Somatic response patterning and stress: Some revisions of activation theory, in M.H. Appley & R. Trumbull (eds), *Psychological Stress: Issues in Research*. New York: Appleton Century-Crofts.

Lazarus, R.S. (1999). *Stress and Emotion: A New Synthesis*. New York: Springer.

Maslach, C., Jackson, S.E. & Leiter, M.P. (eds) (1996). *MBI Manual*. Third edition. Palo Alto, CA: Consulting Psychologists Press.

Matthews, G. (1987). Personality and multidimensional arousal: A study of two dimensions of extraversion. *Personality and Individual Differences*, 8, 9–16.

Matthews, G. & Amelang, M. (1993). Extraversion, arousal theory and performance: A study of individual differences in the EEG. *Personality and Individual Differences*, 14, 347–63.

Matthews, G. & Campbell, S.E. (2009). Sustained performance under overload: Personality and individual differences in stress and coping. *Theoretical Issues in Ergonomics Science*, 10, 417–42.

Matthews, G. & Desmond, P.A. (1998). Personality and multiple dimensions of task-induced fatigue: A study of simulated driving. *Personality and Individual Differences*, 25, 443–58.

Matthews, G. & Desmond, P.A. (2002). Task-induced fatigue states and simulated driving performance. *Quarterly Journal of Experimental Psychology*, 55A, 659–86.

Matthews, G. & Falconer, S. (2000). Individual differences in task-induced stress in customer service personnel. *Proceedings of the Human Factors and Ergonomics Society*, 44, 145–8.

Matthews, G. & Gilliland, K. (1999). The personality theories of H.J. Eysenck and J.A. Gray: A comparative review. *Personality and Individual Differences*, 26, 583–626.

Matthews, G., Campbell, S.E., Falconer, S., Joyner, L., Huggins, J., Gilliland, K., Grier, R. & Warm, J.S. (2002). Fundamental dimensions of subjective state in performance settings: Task engagement, distress and worry. *Emotion*, 2, 315–40.

Matthews, G., Deary, I.J. & Whiteman, M.C. (2009). *Personality Traits*. 3rd ed. Cambridge: Cambridge University Press.

Matthews, G., Desmond, P.A., Joyner, L.A. & Carcary, B. (1997). A comprehensive questionnaire measure of driver stress and affect, in E. Carbonell Vaya & J.A. Rothengatter (eds), *Traffic and Transport Psychology: Theory and Application*. Amsterdam: Pergamon, 317–24.

Matthews, G., Emo, A.K., Funke, G., Zeidner, M., Roberts, R.D., Costa, P.T., Jr. & Schulze, R. (2006). Emotional intelligence, personality, and task-induced stress. *Journal of Experimental Psychology: Applied*, 12, 96–107.

Matthews. G., Saxby, D.J. &. Hitchcock, E.M. (2008). Driver Fatigue Scale: Initial summary report. Unpublished report. Cincinnati, OH: NIOSH.

Matthews, G., Warm, J.S., Dember, W.N., Mizoguchi, H. & Smith, A.P. (2001). The common cold impairs visual attention, psychomotor performance and task engagement. *Proceedings of the Human Factors and Ergonomics Society*, 45, 1377–81.

Matthews, G., Warm, J.S., Reinerman, L.E., Langheim, L, Washburn, D.A. & Tripp, L. (2010). Task engagement, cerebral blood flow velocity, and diagnostic monitoring for sustained attention. *Journal of Experimental Psychology: Applied*, 16, 187–203.

McCrae, R.R. & Costa Jr., P.T. (2008). Empirical and theoretical status of the five-factor model of personality traits, in G.J. Boyle, G.M. Matthews & D.H. Saklofske (eds), *The SAGE Handbook of Personality Theory and Assessment*, Vol. 1: *Personality Theories and Models*. Thousand Oaks, CA: Sage Publications, 273–94.

Michielsen, H.J., De Vries, J. & Van Heck, G.L. (2003). In search of personality and temperament predictors of chronic fatigue: A prospective study. *Personality and Individual Differences*, 35, 1073–87.

Michielsen, H.J., De Vries, J., Van Heck, G.L., Van de Vijver, F.J.R. & Sijtsma, K. (2004). Examination of the dimensionality of fatigue: The construction of the Fatigue Assessment Scale (FAS). *European Journal of Psychological Assessment*, 20, 39–48.

Neubauer, C., Langheim, L.K., Matthews, G. & Saxby, D.J. (2011 [in press]). Fatigue and voluntary utilization of automation in simulated driving. *Human Factors*. DOI: 10.1177/0018720811423261.

Pearson, R.G. (1957). Scale analysis of a fatigue checklist. *Journal of Applied Psychology*, 41, 186–91.

Schnabel, K., Asendorpf, J.B. & Greenwald, A.G. (2008). Assessment of individual differences in implicit cognition: A review of IAT measures. *European Journal of Psychological Assessment*, 24, 210–17.

Shapiro, C.M., Devins, G.M., Flanigan, M., Fleming, J.A.E., Morehouse, R., Moscovitch, A., Plamondon, J. & Reinish, L. (2002). Development of an adjective checklist to measure five FACES of fatigue and sleepiness: Data from a national survey of insomniacs. *Journal of Psychosomatic Research*, 52, 467–73.

Shaw, T.H., Matthews, G., Warm, J.S., Finomore, V., Silverman, L. & Costa, P.T., Jr. (2010). Individual differences in vigilance: Personality, ability and states of stress. *Journal of Research in Personality*, 44, 297–308.

Smets, E.M.A., Garssen, B., Bonke, B. & de Haes, J.C.J.M. (1995). The Multidimensional Fatigue Inventory (MFI): Psychometric qualities of an instrument to assess fatigue. *Journal of Psychosomatic Research*, 39, 315–25.

Sonnentag, S. & Zijlstra, H. (2006). Job characteristics and off-job activities as predictors of need for recovery, well-being, and fatigue. *Journal of Applied Psychology*, 91, 330–50.

Thayer, R.E. (1989). *The Biopsychology of Mood and Arousal*. New York: Oxford University Press.

Worley, J.A., Vassar, M., Wheeler, D.L. & Barnes, L.L.B. (2008). Factor structure of scores from the Maslach Burnout Inventory: A review and meta-analysis of 45 exploratory and confirmatory factor-analytic studies. *Educational and Psychological Measurement*, 68, 797–823.

Yoshitake, H. (1978). Three characteristic patterns of subjective fatigue symptoms. *Ergonomics*, 21, 231–3.

10

Models of Individual Differences in Fatigue for Performance Research

Gerald Matthews, P.A. Hancock and Paula A. Desmond

INTRODUCTION

Experimental studies of fatigue and performance typically reveal substantial individual differences (IDs). Some operators appear much better able than others to sustain effective performance under conditions of sleep loss, monotony or prolonged high workload. These IDs may derive from a variety of different sources. These include susceptibility to fatiguing agents, neurological functioning, motivation, and the utilization of effective strategies for countering fatigue, among others. Thus, efforts at measuring fatigue need to go further than simply trying to identify fatigue-vulnerable and fatigue-resistant operators. This chapter reviews the existing models that may be used to guide research on IDs in performance under fatiguing conditions. We first contrast differing causal models for IDS, before outlining theoretical perspectives from both neuroscience and cognitive psychology, as explanatory accounts of performance variation in fatigue. Finally, we suggest that models focusing on individual differences in fatigue state responses may provide a means for integrating multiple perspectives on IDs in fatigue vulnerability. We discussed psychometric and conceptual aspects of state models of fatigue in Chapter 9 of this volume (Matthews, Desmond & Hitchcock). The present chapter explores their utility in understanding IDs in performance.

CAUSAL MODELS FOR INDIVIDUAL DIFFERENCES IN PERFORMANCE IN STATES OF FATIGUE

Studies of performance demonstrate IDs in a dependent variable or outcome. Typically, this dependent variable is an index of performance expressed in terms of the speed or accuracy of response to task stimuli. The key research question is the nature of the "upstream" causal processes that generate such performance variation. Figure 10.1 expands on some of these causal possibilities. Any causal model must be a *moderator*

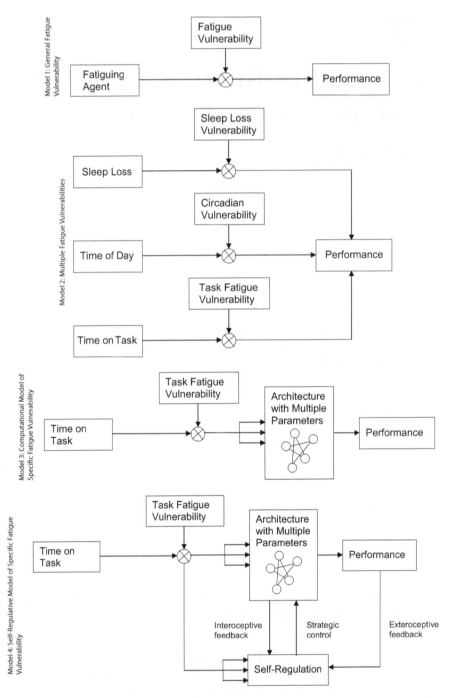

Figure 10.1 Four schematic models for causal influence of dispositional fatigue vulnerability traits

model, in which fatigue-proneness influences a causal path from a fatiguing agent to downstream performance. The simplest moderator model (1a) sees fatigue-proneness as a stable trait that amplifies or buffers the impact of fatiguing agents (e.g., sleep loss, workload, etc.). The idea is evident in models of temperament in children that propose dimensions such as activity and endurance (see Matthews, Deary & Whiteman, 2009).

Unfortunately, this model is largely insufficient as a basis for research, for two reasons. First, fairly extensive research has failed to identify any personality or ability trait that reliably affects performance across the full spectrum of fatiguing environments. The best candidate for such a trait has been extraversion–introversion, which has been linked, for example, to vulnerability to sleep deprivation effects (Matthews, Deary & Whiteman, 2009) and to vigilance decrement (Koelega, 1992), as well as subjective fatigue (Matthews, Desmond & Hitchcock, Chapter 9, this volume). However, it transpires that effects of extraversion on performance are typically contingent on a variety of other moderator factors, including time of day (Humphreys & Revelle, 1984), and information-processing demands (Szalma, 2008). Indeed, extraverts can outperform introverts on certain vigilance tasks (Matthews, Davies & Lees, 1990). Thus, while extraversion is on occasion a moderator of fatigue effects, it cannot be identified with a general fatigue vulnerability. Similar instabilities of prediction are observed with other personality traits (Finomore, Matthews & Warm, 2009).

In response to this initial disappointment, we might propose separate vulnerabilities for different sources of fatigue. Some people may be especially vulnerable to sleep loss, but not to task-induced fatigue, for example. This selective vulnerability is expressed in Model 2. Models of this kind appear more promising, at least for the practical task of predicting performance under fatigue within a given operational domain or context. However, we must be cognizant of the fact that their greater explanatory power may in part reflect the greater number of explanatory degrees of freedom that they possess.

IDs research that is focused on a specific fatiguing agent has also had some success. For example, Van Dongen et al. (2007) showed systematic IDs in vulnerability to sleep loss effects. Similarly, studies of the morningness–eveningness trait confirm that morning and evening types show appropriately synchronized vulnerabilities to circadian phase effects (Waterhouse, Chapter 16, this volume). Fatigue vulnerability may be assessed within specific contexts such as vehicle driving, occupational shift work, and various medical conditions (Matthews, Davies et al., 2000). However, a "specific vulnerabilities" approach is limited by neglect of mediating processes. Like stressors, the effects of fatiguing agents are dependent on task parameters (Hockey, 1984), and, thus, closer attention to information-processing mechanisms is necessary to understand IDs.

Model 3 assumes that models are focused on a specific source of fatigue. It adds an additional mediating process – a computational architecture controlling performance (typically within a limited task domain). Information processing is regulated by parameters of the architecture, such as the probability of a process executing correctly and capacities of memory stores. We can then model the conjoint effects of the fatiguing agent and ID factors on such parameters. Simple versions of this model, like those previously described, can be tested without formal computational modeling. For example, if we suppose that sleep loss lowers the value of a parameter controlling working memory (WM) capacity, and anxiety has a similar effect, we can derive testable predictions, e.g., that sleep loss effects on WM should be more severe for more anxious individuals.

A more powerful approach is to develop a full computational model for the task of interest, and model IDs in parameters (cf., Gunzelman et al., 2011). Models might be either the traditional symbolic-processing models of cognitive psychology, or subsymbolic connectionist models. In either case, observed moderator effects of ID factors might derive from the independent effects of the fatiguing agent and the ID factor on a common parameter. Alternatively, effects might reflect an interactive effect on the parameter of the two factors. For example, Matthews and Harley (1993) modeled the interactive effects of extraversion and subjective fatigue on semantic priming by investigating the impact of IDs in several quantitative parameters of a connectionist model of word recognition. In this case, variation in a parameter controlling levels of random noise provided performance outcomes closest to those observed in real data.

Model 3 has the advantage of incorporating multiple information-processing components (and hence task moderator factors). However, it is limited by its linear progression from independent variables to performance outcome, thus largely neglecting the role of feedback (other than local feedback loops internal to the architecture). Cybernetic models of the fatigue process (Hockey, 2011, see also Hockey, Chapter 3, this volume) emphasize the importance of performance monitoring and effort regulation. In Hockey's (2011) conceptualization, fatigue corresponds to a mode of executive control of performance in which task goals are increasingly threatened by difficulties in recruiting sufficient effort to maintain performance.

Such considerations suggest Model 4 – as an outline model (the models just cited are considerably more sophisticated). It introduces "self-regulation" as an additional causal factor, which is embedded within feedback loops driven by internal and external cues to performance, with its own internal structure (not shown). Self-regulation is supported by executive processes that handle such dilemmas as whether to respond to fatigue and failing effort by striving to increase effort voluntarily, at the cost of increased strain, or by lowering the perceived threshold for acceptable performance (Hockey, 2011; Langner et al., 2010). Such a model suggests two levels of IDs. It continues to represent IDs in the computational architectures supporting processing of task stimuli, as in Model 3. However, it adds IDs in self-regulation. For example, people may vary in their willingness to continue to exert effort as a particular task is appraised as increasingly monotonous. Sources of IDs in self-regulation can be very diverse. They include traits such as optimism–pessimism, personal goals, self-referent knowledge about personal competence, and metacognitions of cognitive and emotional functioning (Wells & Matthews, 1994). This model might also allow us to incorporate "idiographic" sources of variation such as idiosyncratic likes or dislikes for the specific task concerned. Self-regulation is influenced by the person's appraisal of task demands and their personal significance (Szalma, 2008, 2009). Positive appraisals of the task and its importance may protect against fatigue, or generate a state of fatigue that is relatively benign in preserving performance.

The separation between cognitive architectures for self-regulation and for task processing suggests a further differentiation of IDs. Suppose we find that, following a given number of hours of sleep deprivation, some individuals perform worse than others. There are several possible explanations, but we will focus on two. The first is that the fatiguing agent (sleep loss) fatigues some individuals more than others. There may be IDs in subjective state, in self-regulative processes and in their neurological concomitants. A second explanation is that individuals differ in how vulnerable task processing is to a given fatigue state. In this case, we would find performance differences between individuals matched for subjective state or matched for broad psychophysiological measures (e.g., EEG spectral power densities). In relation to Model 4, IDs would be confined to the architecture for self-regulation.

In this chapter, we will focus primarily on the self-regulative perspective on IDs represented by Model 4. An advantage of this model is that it can also accommodate the "state" perspective advanced in our earlier chapter (Matthews, Desmond & Hitchcock, Chapter 9, this volume). That is, fatigue can be characterized as a broad-based, transient characteristic of neuropsychic functioning that can be assessed, at least in part, through subjective measures. States such as task disengagement and distress may correspond, at least imperfectly, to the overall functioning of the self-regulative system. States are often more predictive of performance than stable traits (Finomore et al., 2009; Matthews, Warm et al., 2010a; Shaw et al., 2010), and so it is important to represent these state factors in models of IDs.

THEORETICAL PERSPECTIVES

Thus far, we have discussed causal models for IDs in general terms, without particular reference to specific processes or mechanisms. The next step is to consider theories of fatigue and cognition that may provide a platform for generating hypotheses in IDs research.

The immediate theoretical challenge is that there are many parameters of neural and mental functioning that can differ across individuals. Broadly, we can attribute IDs to brain processes, to parameters of virtual architectures that control information processing, or to high-level motivational and strategic processes (Matthews, 2001). The classic psychobiological model for fatigue, the traditional arousal theory, suggests that individuals differ in some general fatigue vulnerability (Model 1). As has already been discussed, this model is unsatisfactory for IDs. We may also cite the general weaknesses of arousal theory and the Yerkes-Dodson Law as its underlying descriptive principle (Hancock & Ganey, 2003; Hockey, 1984).

In personality research, the predominant psychobiological theory is that extraversion and neuroticism relate to motivational systems controlling response to reward and punishment stimuli. These are referred to as the Behavioral Activation System (BAS) and Behavioral Inhibition System (BIS), respectively (Corr, 2009). The BAS is associated with increased activity in dopaminergic afferents to areas of cortex, and thus can be seen as offsetting fatigue. Animal evidence supports this perspective (Foley et al., 2006). For example, rats with depletion of dopamine in the basal ganglia prefer reduced-effort strategies for seeking food (Salamone et al., 2009). Effects of fatigue on the coherence of the EEG during human performance have been similarly interpreted in terms of dopaminergic mechanisms (Lorist et al., 2009). Note, however, that changes in dopamine may be embedded within a more complex biochemical fatigue process mediated by additional neurotransmitters, especially serotonin (Watanabe, Kuratsune & Kajimoto, Chapter 14, this volume)

Applied to personality, the dopamine hypothesis would appear to predict that extraverts (having a stronger BAS response) should be *more* resistant to fatigue than introverts. While the evidence is mixed, extraverts actually often appear to be more sensitive to fatigue than introverts (Matthews, Deary & Whiteman, 2009). Thus, it may be simplistic to identify extraversion with a general tendency toward stronger approach behavior irrespective of context.

A potentially more promising approach may be to identify and investigate specific neural circuits that vary across individuals. As understanding of the neuroscience of sleep deprivation and circadian rhythms improves, we may be able to attribute IDs in vulnerability to these fatiguing agents to the brain systems concerned (corresponding to Model 2). Whether these more localized IDs have any broader connection with broader, emergent personality traits remains to be determined. Sleep loss may relate to impairment of specific executive processes supported by prefrontal cortex (Killgore, Chapter 15, this volume; Killgore et al., 2009). The increasingly differentiated models of executive functioning adopted by ID researchers may support a similarly differentiated view of the prefrontal circuits maximally sensitive to fatigue, or to specific fatiguing agents.

Although neurological and information-processing models are increasingly integrated within cognitive neuroscience, models of virtual architectures are not obliged to correspond in any direct, isomorphic manner to their neurological foundations (Matthews, 2001). Consonant with Model 3, researchers may then seek the parameters of information processing that support predictive modeling of IDs. In fact, much of the relevant research has referred to rather general qualities of the architecture that may reflect multiple specific parameters. In this context, two relevant constructs are attentional resources (ARs; Wickens, 2002) and working memory (WM; Ilkowska & Engle, 2010).

ARs refer to a metaphorical reservoir of energy or fuel required for successful completion of various component processes. ARs are thought to be required for "effortful" or "controlled" processing (see Hancock, Oron-Gilad & Szalma, 2007). Because more demanding tasks typically require a higher allocation of resources, the concept is appealing in explaining fatigue effects that increase with task demands (Matthews, Davies et al., 2000). Resource theory has been especially influential in research on sustained attention and vigilance (see Hancock & Warm, 1989). More demanding signal detection tasks show greater temporal decrement in perceptual sensitivity, as indexed by the d' measure ("vigilance decrement"; See et al., 1995). Prolonged mental workload, on a monotonous task, is thought to lead to depletion of resources over time. More demanding tasks are more sensitive to this draining of the resource pool (Warm, Matthews & Finomore, 2008).

Individuals are likely to differ in the size of the resource reservoir. If so, those with a pond rather than an ocean of resources should be more sensitive to those fatiguing agents that deplete resources. As we shall see, this idea has proved useful for explaining effects of subjective fatigue (low energy and task disengagement) on sustained attention tasks (see Matthews, Warm et al., 2010b). Specifically, subjective fatigue appears to be more damaging to vigilance tasks that impose higher workloads (Matthews, Davies & Lees, 1990). It is noteworthy that this result is the opposite of the pattern of outcomes which is predicted by the Yerkes-Dodson Law, i.e., that easy, undemanding tasks should be maximally impaired by low arousal (fatigue).

Similarly, there is ample evidence to show that WM is vulnerable to sleep loss (Ilkowska & Engle, 2010), and more demanding WM tasks show larger impairments (Babkoff et al., 1988). In addition, WM load moderates vigilance decrement (Caggiano & Parasuraman, 2004; See et al., 1995), implying some commonality to WM and AR. Indeed, if we define WM in relation to the executive control of attention, it becomes hard to distinguish the WM and AR hypotheses. The WM/attention hypothesis also has the advantage that the neural circuits supporting these cognitive functions are becoming increasingly well understood (Ilkowska & Engle, 2010).

However, this perspective on IDs in vulnerability to fatigue-induced performance impairments has its limitations. First, ARs have proven to be difficult to identify psychometrically (Matthews, Warm et al., 2010b), although latent factor models of batteries of single attentional and WM tasks have had success in isolating broad factors (Schweizer, 2010). Second, the concept of resources has been attacked as being redundant (Navon, 1984; Pashler, 1998). Perhaps, apparent resource limitations may simply reflect a multitude of independent constraints on information processing, without there being any general reservoir of energy per se. Arguments over whether resources are unitary or multiple (Wickens, 2002) also tend to muddy the waters. In future research, it may be preferable to work with more differentiated models of executive processes (e.g., Friedman et al., 2006), than with broader resource models. Third, it may be difficult to differentiate fatigue effects on resource availability from those on strategic allocation or deployment of resources. For example, in a study of simulated driving, Matthews and Desmond (2002) found that fatigue impaired vehicle control and reduced effort (indexed by small-magnitude steering movements) in undemanding driving conditions (straight roads) but not when demands increased (curved roads). They concluded that fatigue induced by their driving task did not affect resource availability, but, rather, the driver's willingness to exert effort when the task appeared routine and undemanding.

The importance of strategic effects of fatigue (c.f., Hockey, 2011; Hockey, Chapter 3, this volume) inspires Model 4, in which self-regulative processes act as the primary moderator of the effects of fatiguing agents. Self-regulation may moderate the impacts of fatiguing agents on both neuropsychic state and information processing. The increasing reluctance of the fatigued operator to apply effort to task performance is well known. This basic observation suggests motivational models for fatigue, which in turn suggests further sources of IDs. Kanfer and her colleagues (e.g., Kanfer, 2011; Kanfer & Ackerman, 1989) proposed that subjective experience of cognitive fatigue signals a motivational conflict between the short-term goal of reducing effort in order to rest, and the longer-term goal of accomplishing performance goals. The behavioral consequences of the fatigue state may depend on personality traits linked to approach, such as extraversion and achievement motivation. According to Kanfer (2011), individuals high in approach-related traits will use self-regulative strategies to maintain effort and a focus on performance goals. An assumption that extraversion relates more strongly to the regulation of fatigue than to the fatigue experience itself might resolve some of the inconsistencies in the experimental findings that we have previously indicated.

Szalma (2009) has provided a contrasting account of IDs in motivation and performance that emphasizes the role of personal autonomy. Deci and Ryan's (2009) self-determination theory distinguishes motivation based on interest and autonomous agency (intrinsic motivation) from motivation-based reinforcement and punishment contingencies (extrinsic motivation). Intrinsic motivation broadly tends to energize and maintain behavior, and thus may be especially important in fatigue states. Evidence supporting these claims comes from studies which show that even highly fatigued operators can maintain performance if the task is of sufficient intrinsic interest (see Matthews, Davies et al., 2000).

A limitation of motivational theories, especially those that are derived from personality rather than human factors (e.g., Deci & Ryan, 2009), is that they are typically vague about precise components of information processing. In relation to Model 4, motivational theories may specify the cognitive architecture for self-regulation but not for its interface with task processes. The Hancock and Warm (1989) dynamic model of stress, fatigue and sustained attention (see Hancock, Desmond & Matthews, Chapter 4, this volume; Szalma,

Chapter 5, this volume) seeks to remedy this limitation. Two of its features are especially relevant to IDs in motivational processes (Szalma, 2008). First, subjective discomfort precedes cognitive and physiological instability. Consistent with classical cognitive emotion theory (cf., Szalma, 2009), the fatigue state itself is becomes a signal of impending performance breakdown that may motivate compensatory efforts. As is discussed by Kanfer (2011), personality may then moderate how much and in what ways the person seeks to achieve this compensation (see also Hancock & Caird, 1993). The model also suggests the importance of metacognitions in the adaptive response, e.g., whether the person interprets fatigue as a sign that the task is not worth bothering with, or as a sign that effort should be increased. Metacognitive variables have important moderating effects in states of anxiety and emotional distress (Wells & Matthews, 2006), but their role in fatigue has yet to be explored. Kustubayeva, Panganiban and Matthews (2010) found that the relationship between positive affect and search for information in a tactical decision-making task is moderated by the affective context (success vs. failure), implying that the motivational impact of fatigue may be sensitive to such contextual factors, and, hence, to personality factors that serve to influence perceptions of context.

A second feature of the Hancock and Warm (1989) model is that it suggests some specific cognitive processes that may mediate fatigue effects on performance (see also Szalma, 2008). Breakdown in adaptation is equated with loss of AR, consistent with the resource model of vigilance decrement (Warm, Matthews & Finomore, 2008). Thus, individuals that compensate more successfully in fatigue states should behave as having functionally more resources. More subtly, Hancock and Warm (1989) draw attention to the task strategies that operators may use to counter loss of adaptability. In the case of fatigue (hypostress), operators are expected to seek greater temporal information flow and reduced spatial structure as compensatory strategies. In principle, such adaptations can be housed in machine interfaces so that the machine component of the human–machine system may contribute to the effort of adaptation (see Hancock & Szalma, 2003). The prevalence of the vigilance decrement in contemporary operational settings may in part reflect the limited scope that highly constrained monitoring tasks provide for such compensatory activities. Thus, while a range of motivational theories provide ID constructs that may serve as independent variables in empirical studies, the Hancock and Warm (1989) model indicates some promising dependent variables, not least perceptual sensitivity (d') during vigilance.

STATE MODELS FOR FATIGUE EFFECTS ON SUSTAINED ATTENTION

Thus far, we have emphasized the complexity of models for IDs in fatigue during sustained performance. Not only can research be directed toward IDs in brain processes, in information processing and in self-regulation, but there are a multiplicity of relevant ID constructs at each level of analysis. Of course, research may investigate specific mechanisms, whether at a neurological or cognitive level of analysis. However, we also need more integrative perspectives on multiple pathways for fatigue effects. We therefore conclude this chapter by suggesting that a focus on IDs in fatigue states may fulfill such an integrative role.

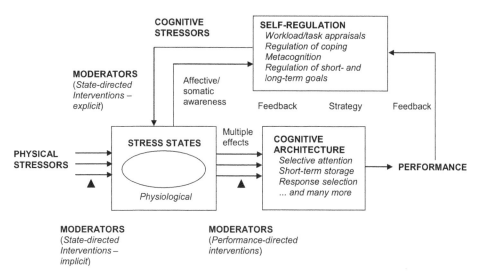

Figure 10.2   A multidimensional model for stress and fatigue effects on self-regulation, cognitive processing, and performance

Figure 10.2 summarizes a conceptual state model for stress and fatigue effects on performance, building on Model 4 in Figure 10.1, and an architecture for cognition and emotion described by Wells and Matthews (1994). We assume here that "fatigue" refers to the subset of stress effects that elicit task disengagement. At the core of the model is a multidimensional state of stress defined by psychological qualities such as subjective fatigue, as well as by indices of physiological functioning such as EEG and CBFV. As has previously been described (Matthews, Desmond & Hitchcock, Chapter 9, this volume), the Matthews, Campbell et al. (2002) model of multidimensional subjective states identifies task engagement, distress, and worry as key components of the model. It is assumed that there is partial overlap between psychological and physiological state dimensions (Thayer, 1996). For example, Fairclough and Venables (2006) found that up to 30 percent of the variance in the stress states measured by the DSSQ (Matthews, Campbell et al., 2002) was explained by measures of autonomic and central nervous system arousal.

Stress state is influenced by physical stressors and by cognitive stressors whose effects are mediated by self-regulative processes, including the person's appraisals of task demands and workload (Hancock & Caird, 1993), and consequent efforts at coping. (We assume that causal effects of stressors on states are controlled by both neural and "virtual" cognitive mechanisms). Stress states then bias the operation of discrete components of information processing, which in turn influence overt performance. These biases are multivariate in nature (Hockey, 1984): each state dimension is associated with a distinctive "cognitive patterning" of multiple changes in processing components. Workload is important within the model in two respects. First, high workload may signal vulnerability to cognitive overload and stress-related impairments such as loss of situational awareness (e.g., Warm & Dember, 1998). Second, the operator's appraisal of workload plays a critical role in cognitive stressor effects such as those of demand transitions (Hancock, Williams et al., 1995; Helton et al., 2008). The conceptual framework also distinguishes several distinct types of moderator effect. Traits for fatigue vulnerability may impact paths between stressors and stress-state response, or they may influence effects of stress states on parameters of the architecture for task processing. It follows that multiple traits, operating via different mechanisms, may play a role in vulnerability.

Recent data on fatigue effects on sustained attention fit the model (see Matthews, Warm et al., 2010a, Matthews, 2011, for reviews). Cognitive mediators of the impact of task stressors on subjective states have been explored in studies based on the transactional theory of stress (Lazarus, 1999). We have already reviewed key findings from this research (Matthews, Desmond & Hitchcock, Chapter 9, this volume). In brief, studies of sustained performance (Matthews & Campbell, 2009; Shaw et al., 2010) have shown that loss of engagement during performance, relative to the pre-task baseline level, relates to appraising the task as less challenging, using less task-focused coping and more avoidance coping. By contrast, the distress response to sustained performance is tied more closely to the overall workload of the task, to appraisals of threat and controllability, and to use of self-critical emotion-focused coping (Matthews & Campbell, 2009; Matthews, Campbell et al., 2002).

The conceptual model allows for the two different kinds of impact of stress and fatigue state on information processing as represented in Model 4: (1) impacts of each dimension of state on multiple parameters of task information processing, and (2) indirect effects mediated by top-down self-regulative processes that are dependent on motivations, goal-setting and effort-regulation (Szalma, 2009). Sustained attention studies suggest that task engagement influences the availability of ARs, i.e., an effect of state on a key parameter(s) for processing. Numerous studies have shown that task engagement measured prior to performance predicts higher perceptual sensitivity on a variety of tasks requiring sustained attention, vigilance, and controlled visual and semantic search (e.g., Matthews, Davies & Lees, 1990, Matthews, Warm et al., 2010a, Shaw et al., 2010). Some studies also show that higher engagement relates to a smaller temporal decrement in performance (Matthews, Warm et al., 2010a). A recent study (Matthews, Zeidner & Zwang, 2012) has shown that engagement correlates with superior executive processing, measured by Fan et al.'s Attentional Network Test (2002), suggesting a more precise account of the resource mechanism.

These findings demonstrate a specific link between subjective task engagement and attention. Other state dimensions map onto other component processes. For example, in a study that used a structural equation modeling (SEM) approach with longitudinal data, Matthews and Campbell (2010) confirmed that day-to-day variation in distress influenced day-to-day variation on a WM task. Engagement showed a similar but separate effect, consistent with the role of executive control of attention in WM previously described.

We can attribute this influence of task engagement to a resource mechanism on several grounds. First, effects of engagement or subjective energy are workload-dependent, and reliable only when the task imposes substantial attentional demands (Matthews, Davies & Lees, 1990). Second, the engagement effect generalizes across a wide variety of tasks that recruit different components of information processing (Matthews, 2011; Matthews, Warm et al., 2010a). Matthews, Warm et al. (2010b) fitted an SEM to data from a *sensory* vigilance task, in which initial task engagement was a causal influence on subsequent attentional efficiency. The causal model exhibited good generalization (high fit) to a second data set derived from a *cognitive* vigilance task which imposed high demands on WM. Third, formal tests of resource theory predictions have been confirmed (Matthews & Margetts, 1991; Matthews, Warm et al., 2010b). Fourth, Matthews, Warm et al. (2010b) showed that task engagement was associated with the CBFV response to high-workload

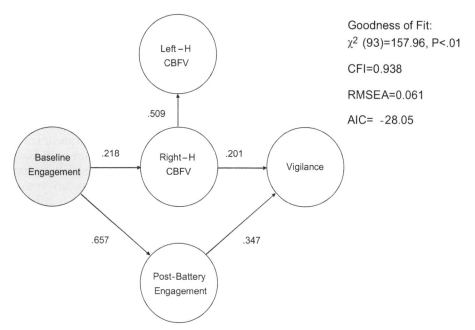

Goodness of Fit:
χ^2 (93)=157.96, P<.01

CFI=0.938

RMSEA=0.061

AIC= -28.05

Figure 10.3   Latent factor model for Matthews, Warm et al. (2010b) data, omitting measurement model and error terms (CBFV = Cerebral bloodflow velocity, H = hemisphere)

tasks; this metabolic index has been linked to resource utilization during attentional task performance in several studies (Warm, Tripp, Matthews & Helton, Chapter 13, this volume). Figure 10.3 shows an outline (latent factor model only) of the causal model that interrelated engagement, CBFV and subsequent performance, on both sensory and cognitive tasks.

An alternative hypothesis for fatigue effects on sustained performance refers to self-regulation and effort-regulation (c.f., Hockey, 2011; Langner et al., 2010). Perhaps vigilant operators simply lose their motivation and subsequently reduce their effort as they become fatigued. This hypothesis gains credibility from findings that task manipulations that increase operator control over the task increase both engagement and performance (Parsons et al., 2007) as well as a spectrum of findings on motivational effects on vigilance (Warm & Dember, 1998). On the other hand, workload remains high even in fatigued operators (Warm & Dember, 1998), and as Helton (e.g., 2009) has argued in a series of recent articles, the vigilance decrement cannot be attributed to simple "mindlessness." Although there is more to this issue, we simply note here that the role of IDs in effort regulation during vigilance performance requires more investigation of this suite of complex, interactive effects. There seems to be some general alignment between loss of resources and loss of task motivation, given that increasing avoidance and decreasing task-focused coping appear to be hallmarks of the fatigued state (Matthews & Campbell, 2009).

The multidimensional state model also explains effects of potentially fatiguing stressors on sustained attention. Matthews et al. (2001) found that infection with the common cold was related to both performance impairment and to task engagement. The effect of the infection on vigilance appeared to be fully mediated by its effects on subjective state. Similarly, Helton, Matthews and Warm (2009) used structural modeling to show that effects of airplane jet engine noise on vigilance were mediated by task engagement. In this case, noise actually served to increase engagement, thus offsetting the fatigue induced by performance of the task.

Thus, the state model suggests various strategies through which countermeasures might effectively operate (see Figure 10.3). First, "state-directed" interventions may seek to reduce the impact of external agents on the fatigue state. Design of system and interface characteristics that increase the interest and challenge of the task can therefore presumably protect against loss of engagement (cf., Hancock, 1997). However, it may be relatively hard to mitigate against the effects of agents that have a strong neurological basis such as sleep loss and circadian phase. Some interventions (e.g., habituation to novelty) may influence implicit paths, whereas others (e.g., stress management) may influence paths mediated by explicit task appraisals. Second, performance-directed interventions may reduce the vulnerability of information-processing to fatigue states that develop. Task design and operator training may both reduce dependence of performance on AR, for example (e.g., Wickens, 2002). The problem here is that tasks designed to be undemanding may also be inherently less challenging, leading to more rapid build-up of fatigue.

SUMMARY AND CONCLUSIONS

Research on IDs in fatigue vulnerability has come a long way from its origins in the simplicities of unitary arousal theory. ID factors such as extraversion–introversion which have been traditionally related to fatigue vulnerability are, in fact, influential only in very specific circumstances. We have identified several key moderator factors of IDs, including the source of fatigue, the cognitive demands of the task and the operator's motivations and strategies for fatigue management. At one level, research on IDs in fatigue vulnerability must thus be highly focused on specific neurocognitive or strategic mechanisms that may produce performance impairment. Fine-grained investigations of such mechanisms may be complemented by a state-based approach. Specifically, the task engagement state may integrate a variety of processes, including brain metabolic support for processing, mobilization of AR, and task-focused coping strategies that maintain effort. The hypothesis that task engagement provides a marker for individual resource availability has been quite successful in explaining IDs in sustained attention. It is likely that other state models, with a greater emphasis on neural processes, are necessary for the understanding of IDs in vulnerability to other precursors of fatigue such as sleep loss and circadian phase. Separating IDs in vulnerability to fatigue states, and in vulnerability to the impact of fatigue on information processing may be a useful step in developing countermeasures that accommodate the individual characteristics of each particular person.

REFERENCES

Babkoff, H., Mikulincer, M., Caspy, T., Kempinski, D. & Sing, H. (1988). The topology of performance curves during 72 hours of sleep loss: A memory and search task. *The Quarterly Journal of Experimental Psychology*, 40A, 737–56.

Caggiano, D.M. & Parasuraman, R. (2004). The role of memory representation in the vigilance decrement. *Psychonomic Bulletin & Review*, 11, 932–7.

Corr, P.J. (2009). The Reinforcement Sensitivity Theory of personality, in P.J. Corr & G. Matthews (eds), *Cambridge Handbook of Personality*. Cambridge: Cambridge University Press, 347–76.

Deci, E.L. & Ryan, R.M. (2009). Self-determination theory: A consideration of human motivational universals, in P.J. Corr & G. Matthews (eds), *Cambridge Handbook of Personality*. Cambridge: Cambridge University Press, 441–56.

Fairclough, S.H. & Venables, L. (2006). Prediction of subjective states from psychophysiology: A multivariate approach. *Biological Psychology*, 71, 100–110.

Fan, J., McCandliss, B.D., Sommer, T., Raz, A. & Posner, M.I. (2002). Testing the efficiency and independence of attentional networks. *Journal of Cognitive Neuroscience*, 14, 340–7.

Finomore, V.S., Matthews, G. & Warm, J.S. (2009). Predicting vigilance: A fresh look at an old problem. *Ergonomics*, 52, 791–808.

Foley, T.E., Greenwood, B.N., Day, H.E.W., Koch, L.G., Britton, S.L. & Fleshner, M. (2006). Elevated central monoamine receptor mRNA in rats bred for high endurance capacity: Implications for central fatigue. *Behavioural Brain Research*, 174, 132–42.

Friedman, N.P., Miyake, A., Corley, R.P., Young, S.E., DeFries, J.C. & Hewitt, J.K. (2006). Not all executive functions are related to intelligence. *Psychological Science*, 17, 172–9.

Gunzelmann, G., Moore, L.R., Gluck, K.A., Van Dongen, H.P.A. & Dinges, D.F. (2011). Fatigue in sustained attention: Generalizing mechanisms for time awake to time on task, in P.L. Ackerman (ed.), *Cognitive Fatigue: The Current Status and Future for Research and Applications*. Washington, DC: APA, 83–101.

Hancock, P.A. (1997). *Essays on the Future of Human–Machine Systems*. Eden Prairie, MN: Banta.

Hancock, P.A. & Caird, J.K. (1993). Experimental evaluation of a model of mental workload. *Human Factors*, 35, 413–29.

Hancock, P.A. & Ganey, N. (2003). From the inverted-U to the extended-U: The evolution of a law of psychology. *Journal of Human Performance in Extreme Environments*, 7, 5–14.

Hancock, P.A. & Szalma, J.L. (2003). The future of neuroergonomics. *Theoretical Issues in Ergonomic Science*, 4, 238–49.

Hancock, P.A. & Warm, J.S. (1989). A dynamic model of stress and sustained attention. *Human Factors*, 31, 519–37.

Hancock, P.A., Oron-Gilad, T. & Szalma, J.L. (2007). Elaborations of the multiple resource theory of attention, in A.F. Kramer, D.A. Wiegmann & A. Kirlik (eds), *Attention: From Theory to Practice*. Oxford: Oxford University Press, 45–56.

Hancock, P.A., Williams, G., Miyake, S. & Manning, C.M. (1995). The influence of task demand characteristics on workload and performance. *International Journal of Aviation Psychology*, 5, 63–85.

Helton, W.S. (2009). Impulsive responding and the sustained attention to response task. *Journal of Clinical and Experimental Neuropsychology*, 31, 39–47.

Helton, W.S., Matthews, G. & Warm, J.S. (2009). Stress state mediation between environmental variables and performance: the case of noise and vigilance. *Acta Psychologica*, 130, 204–13.

Helton, W.S., Shaw, T.H., Warm, J.S., Matthews, G., Dember, W.N. & Hancock, P.A. (2008). Effects of warned and unwarned demand transitions on vigilance performance and stress. *Anxiety, Stress and Coping*, 21, 173–84.

Hockey, G.R.J. (1984). Varieties of attentional state: The effects of the environment, in R. Parasuraman & D.R. Davies (eds), *Varieties of Attention*. New York: Academic Press, 449–83.

Hockey, G.R.J. (2011). A motivational control theory of cognitive fatigue, in P.L. Ackerman (ed.), *Cognitive Fatigue: The Current Status and Future for Research and Applications*. Washington, DC: APA, 167–87.

Humphreys, M.S. & Revelle, W. (1984). Personality, motivation and performance: A theory of the relationship between individual differences and information processing. *Psychological Review*, 91, 153–84.

Ilkowska, M. & Engle, R.W. (2010). Trait and state differences in working memory capacity, in A. Gruszka, G. Matthews & B. Szymura (eds), *Handbook of Individual Differences in Cognition: Attention, Memory and Executive Control*. New York: Springer, 295–320.

Kanfer, R. (2011). Determinants and consequences of subjective cognitive fatigue, in P.L. Ackerman (ed.), *Cognitive Fatigue: The Current Status and Future for Research and Applications*. Washington, DC: APA, 189–207.

Kanfer, R. & Ackerman, P.L. (1989). Motivation and cognitive abilities: An integrative/aptitude treatment interaction approach to skill acquisition. *Journal of Applied Psychology Monograph*, 74, 657–90.

Killgore, W.D.S., Grugle, N.L., Reichardt, R.M., Killgore, D.B. & Balkin, T.J. (2009). Executive functions and the ability to sustain vigilance during sleep loss. *Aviation, Space, and Environmental Medicine*, 80, 81–7.

Koelega, H.S. (1992). Extraversion and vigilance performance: 30 years of inconsistencies. *Psychological Bulletin*, 112, 239–58.

Kustubayeva, A., Panganiban, A.R. & Matthews, G. (2010). Affective biases in information search during tactical decision-making. *Proceedings of the Human Factors and Ergonomics Society*, 54, 1057–61.

Langner, R., Willmes, K., Chatterjee, A., Eickhoff, S.B. & Sturm, W. (2010). Energetic effects of stimulus intensity on prolonged simple reaction-time performance. *Psychological Research/ Psychologische Forschung*, 74, 499–512.

Lazarus, R.S. (1999). *Stress and Emotion: A New Synthesis*. New York: Springer.

Lorist, M.M., Bezdan, E., ten Caat, M., Span, M.M., Roerdink, J.B.T.M. & Maurits, N.M. (2009). The influence of mental fatigue and motivation on neural network dynamics: An EEG coherence study. *Brain Research*, 1270, 95–106.

Matthews, G. (2001). Levels of transaction: A cognitive sciences framework for operator stress, in P.A. Hancock & P.A. Desmond (eds), *Stress, Workload, and Fatigue*. Mahwah, NJ: Erlbaum, 5–33.

Matthews, G. (2011). Personality and individual differences in cognitive fatigue, in P.L. Ackerman (ed.), *Cognitive Fatigue: The Current Status and Future for Research and Applications*. Washington, DC: APA, 209–27.

Matthews, G. & Campbell, S.E. (2009). Sustained performance under overload: Personality and individual differences in stress and coping. *Theoretical Issues in Ergonomics Science*, 10, 417–42.

Matthews, G. & Campbell, S.E. (2010). Dynamic relationships between stress states and working memory. *Cognition and Emotion*, 24, 357–73.

Matthews, G. & Desmond, P.A. (2002). Task-induced fatigue states and simulated driving performance. *Quarterly Journal of Experimental Psychology*, 55A, 659–86.

Matthews, G. & Harley, T.A. (1993). Effects of extraversion and self-report arousal on semantic priming: A connectionist approach. *Journal of Personality and Social Psychology*, 65, 735–56.

Matthews, G. & Margetts, I. (1991). Self-report arousal and divided attention: A study of performance operating characteristics. *Human Performance*, 4, 107–25.

Matthews, G., Campbell, S.E., Falconer, S. Joyner, L., Huggins, J., Gilliland, K. Grier, R. & Warm, J.S. (2002). Fundamental dimensions of subjective state in performance settings: Task engagement, distress and worry. *Emotion*, 2, 315–340.

Matthews, G., Davies, D.R. & Lees, J.L. (1990). Arousal, extraversion, and individual differences in resource availability. *Journal of Personality and Social Psychology*, 59, 150–68.

Matthews, G., Davies, D.R., Westerman, S.J. & Stammers, R.B. (2000). *Human Performance: Cognition, Stress, and Individual Differences*. Hove: Psychology Press.

Matthews, G., Deary, I.J. & Whiteman, M.C. (2009). *Personality Traits*. Third edition. Cambridge: Cambridge University Press.

Matthews, G., Warm, J.S., Dember, W.N., Mizoguchi, H. & Smith, A.P. (2001). The common cold impairs visual attention, psychomotor performance and task engagement. *Proceedings of the Human Factors and Ergonomics Society*, 45, 1377–81.

Matthews, G., Warm, J.S., Reinerman, L.E., Langheim, L.K. & Saxby, D.J. (2010a). Task engagement, attention and executive control, in A. Gruszka, G. Matthews & B. Szymura (eds), *Handbook of Individual Differences in Cognition: Attention, Memory and Executive Control*. New York: Springer, 205–30.

Matthews, G., Warm, J.S., Reinerman, L.E., Langheim, L, Washburn, D.A. & Tripp, L. (2010b). Task engagement, cerebral blood flow velocity, and diagnostic monitoring for sustained attention. *Journal of Experimental Psychology: Applied*, 16, 187–203.

Matthews, G., Zeidner, M. & Zwang, N. (2012). Trait and state correlates of the Attentional Network Test. Paper presented at the Biennial Meeting of the International Society for the Study of Individual Differences, London, July 2011.

Navon, D. (1984). Resources—a theoretical soup stone? *Psychological Review*, 91, 216–34.

Parsons, K.S., Warm, J.S., Nelson, W.T., Riley, M. & Matthews, G. (2007). Detection–action linkage in vigilance: Effects on workload and stress. *Proceedings of the Human Factors and Ergonomics Society*, 51, 1291–95.

Pashler, H.E. (1998). *The Psychology of Attention*. Cambridge, MA: MIT Press.

Salamone, J.D., Correa, M., Farrar, A.M., Nunes, E.J. & Pardo, M. (2009). Dopamine, behavioral economics, and effort. *Frontiers in Behavioral Neuroscience*, 3 [no pagination].

Schweizer, K. (2010). The relationship of attention and intelligence, in A. Gruszka, G. Matthews and B. Szymura (eds), *Handbook of Individual Differences in Cognition: Attention, Memory and Executive Control*. New York: Springer, 247–62.

See, J.E., Howe, S.R., Warm, J.S. & Dember, W.N. (1995). Meta-analysis of the sensitivity decrement in vigilance. *Psychological Bulletin*, 117, 230–49.

Shaw, T.H., Matthews, G., Warm, J.S., Finomore, V., Silverman, L. & Costa, P.T., Jr. (2010). Individual differences in vigilance: Personality, ability and states of stress. *Journal of Research in Personality*, 44, 297–308.

Szalma, J.L. (2008). Individual differences in stress reaction, in P.A. Hancock & J.L. Szalma (eds), *Performance Under Stress*. Aldershot: Ashgate Publishing, 323–57.

Szalma, J.L. (2009). Individual differences: Incorporating human variation into human factors/ ergonomics research and practice. *Theoretical Issues in Ergonomics Science*, 10, 377–9.

Thayer, R.E. (1996). *The Origin of Everyday Moods*. New York: Oxford University Press.

Van Dongen, H.P.A., Mott, C.G., Huang, J., Mollicone, D.J., McKenzie, F.D. & Dinges, D.F. (2007). Optimization of biomathematical model predictions for cognitive performance impairment in individuals: Accounting for unknown traits and uncertain states in homeostatic and circadian processes. *Sleep: Journal of Sleep and Sleep Disorders Research*, 30, 1129–43.

Warm, J.S. & Dember, W.N. (1998). Tests of a vigilance taxonomy, in R.R. Hoffman, M.F. Sherrick & J.S. Warm (eds), *Viewing Psychology as a Whole: The Integrative Science of William N. Dember*. Washington, DC: American Psychological Association, 87–112.

Warm, J.S., Matthews, G. & Finomore, V.S. (2008). Workload and stress in sustained attention, in P.A. Hancock & J.L. Szalma (eds), *Performance Under Stress*. Aldershot: Ashgate Publishing, 115–41.

Wells, A. & Matthews, G. (1994). *Attention and Emotion: A Clinical Perspective*. Hove: Erlbaum.

Wells, A. & Matthews, G. (2006). Cognitive vulnerability to anxiety disorders: An integration, in L.B. Alloy & J.H. Riskind (eds), *Cognitive Vulnerability to Emotional Disorders*. Hillsdale, NJ: Lawrence Erlbaum, 303–25.

Wickens, C.D. (2002). Multiple resources and performance prediction. *Theoretical Issues in Ergonomic Science*, 3, 159–77.

The Neuroscience of Fatigue

11

Neuroscience of Sleep and Circadian Rhythms

Siobhan Banks, Melinda L. Jackson and Hans P.A. Van Dongen

Although the exact function of sleep is still debated, sleep deprivation studies have shown that lack of sleep increases fatigue, produces decrements in neurocognitive performance and degrades physiological functioning (Banks & Dinges, 2007; Jackson & Van Dongen, in press). Our understanding of the neurobiological mechanisms that underlie sleep and sleep regulation has improved in recent years, but there are still many important questions to be answered. Much more is known about circadian rhythms, which are closely associated with sleep regulation and co-determine fatigue, neurocognitive performance and physiological functioning. This chapter outlines current knowledge regarding the neurobiological regulation of sleep and circadian rhythms and their effects on fatigue and performance.

NEUROBIOLOGY OF CIRCADIAN RHYTHMS

Humans, like all organisms, show daily changes in their biology and behavior that are not just controlled by external environmental stimuli, but are driven by internally generated cycles. These (near-)24-hour cycles, known as circadian rhythms, affect many physiological and neurobiological variables, including core body temperature, circulating hormones, sleep patterns and fatigue (Van Dongen et al., 2004).

The circadian system is composed of a central "clock," or circadian pacemaker, which is located in the suprachiasmatic nuclei (SCN) of the hypothalamus in the brain, and peripheral clocks distributed across the body and brain. Other important components include photoreceptors and visual pathways (e.g., retinohypothalamic tract; RHT); and output pathways such as the pineal gland. The SCN is synchronized ("entrained") to a 24-hour cycle by external Zeitgebers (time cues), the most salient of which is light (Wever, 1989).

The recently discovered blue-sensitive photoreceptor melanopsin (Panda et al., 2005), the RHT and the optic chiasm of the brain are the primary pathways through which information regarding the intensity of light at the eye is relayed to the SCN (Golombek & Rosenstein, 2010). The SCN receives feedback from the pineal gland through the hormone melatonin, which is secreted at night, stabilizes the rhythm of the SCN, and signals that it is an opportune time to sleep. Production of melatonin is inhibited by bright light (principally blue light; Lockley, Brainard & Czeisler, 2003).

The timing ("phase") of the circadian rhythm determines whether people are morning types (larks) or evening types (owls; Kerkhof & Van Dongen, 1996). Exogenous administration of melatonin and exposure to light can (transiently) influence the SCN and shift the timing of the circadian rhythm, and thereby the biologically opportune timing for sleep (Rajaratnam et al., 2004). These effects are critically dependent on time of day: bright light in the morning or melatonin administration late in the evening results in an advance of the circadian rhythm, whereas bright light exposure in the evening or melatonin administration in the early morning has the opposite effect (Lewy et al., 1998). In many animal species, melatonin also helps the organism track the seasons (Arendt, 1998), but in humans seasonal rhythms tend to be negligible (Van Dongen, Kerkhof & Souverijn, 1998). The circadian rhythm as driven by the SCN has a profound effect on sleep, fatigue and performance, as described below under "Interaction of Sleep Homeostasis with Circadian Rhythms."

NEUROBIOLOGY OF SLEEP

It is a well-established fact that the longer people are awake, the greater their propensity to fall asleep. Conversely, the longer the sleep period, the more the propensity for sleep is dissipated. The neurobiological process responsible for this is commonly known as the sleep homeostatic process (Achermann & Borbély, 2010). It maintains a balance between wakefulness and sleep, and so is dependent on hours of wake, hours of sleep, and current "sleep debt" from prior days (McCauley et al., 2009).

The neurobiological nature of the sleep homeostatic process is largely unknown. However, it appears to be reflected in increased adenosine production in the extracellular space in the basal forebrain and areas of the cortex (Porkka-Heiskanen et al., 1997). Adenosine is a metabolic by-product of the brain's use of adenosine triphosphate (ATP), the primary source of cellular energy. Adenosine serves as a neuromodulator, signalling sleepiness through binding to the adenosine receptor (for review, see Basheer et al., 2004). Caffeine binds to this receptor and thereby blocks adenosine signaling, which is how caffeine promotes alertness (Nehlig, Daval & Debry, 1992). It is not clear, however, what processes go on during sleep, in relation to adenosine, that reduce the homeostatic propensity for sleep (Benington & Heller, 1995). Most likely, a large number of additional sleep regulatory substances are involved (Krueger, Obál & Fang, 1999), with adenosine being a part of a complicated cascade of compounds and interactions (Van Dongen, Belenky & Krueger, in press).

Waking alertness is sustained by the combined action of various wake-promoting centers in the brain (e.g., reticular activating system, locus coeruleus and tuberomammillary nucleus), which utilize monoaminergic (serotonin, dopamine, histamine, adrenalin, noradrenalin) and cholinergic neurotransmitters to promote neuronal activation (Saper, Chou & Scammell, 2001; see Figure 11.1). A brain area called the ventrolateral preoptic nucleus (VLPO) monitors the sleep propensity state of the brain (possibly via detection of increases in extracellular ATP; Krueger et al., 2010) and triggers whole-brain sleep when sleep propensity becomes excessive. The VLPO blocks the activity of wake-promoting centers in the brain (such as the locus coeruleus and tuberomammillary nucleus) through the inhibitory neurotransmitter GABA, and thereby promotes sleep (Fuller, Gooley & Saper, 2006).

The VLPO works like a bistable ("flip-flop") switch (Saper, Chou & Scammell, 2001); it tends to get stuck in one of its two states, wake pass-through or wake inhibition, and thereby consolidates sleep and wake periods. In this manner, the VLPO prevents awakening

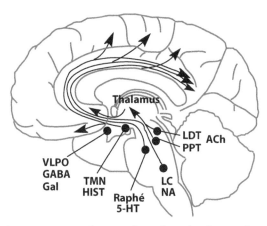

Figure 11.1   Brain systems orchestrating sleep/wake cycles and fatigue

Note: The ascending arousal system sends projections from the brainstem and posterior hypothalamus throughout the forebrain promoting wakefulness (arrows). Cholinergic (ACh; acetylcholine) neurons of the laterodorsal tegmental and pedunculopontine tegmental nuclei (LDT and PPT), serotonergic (5-HT; serotonin) neurons of the dorsal and median raphé, histaminergic (HIST) neurons of the tuberomammillary nucleus (TMN), and noradrenergic (NA) neurons of the locus coeruleus (LC) project to many forebrain targets and the thalamus, promoting wakefulness. Sleep-promoting neurons of the ventrolateral preoptic nucleus (VLPO) containing GABA and galanin (Gal) block these wake-promoting systems to initiate and maintain sleep.

Source: Reprinted from Trends in Neuroscience, 24(12): 726–31: Saper, C.B., Chou, T.C. & Scammell, T.E., "The sleep switch: hypothalamic control of sleep and wakefulness." Copyright © 2001, with permission from Elsevier.

shortly after falling asleep due to the associated dissipation of sleep propensity; rather, sleep propensity has to be dissipated almost completely before the VLPO will no longer inhibit the wake-promoting centers of the brain and allow natural awakening to occur. The thresholds for homeostatic sleep pressure that trigger VLPO state switches are modulated by the circadian rhythm (Saper, Scammell & Lu, 2005). During the day, therefore, it takes more sleep propensity to fall asleep, and less dissipation of sleep propensity to wake up again, which is why sleeping during the day is more difficult than at night.

Neurons in the hypothalamus produce a neurotransmitter called orexin (or hypocretin), which directly stimulates arousal centers as well as the cerebral cortex itself (de Lecea et al., 1998; Sakurai et al., 1998). Mathematical modeling efforts have shown that orexinergic/ hypocretinergic neurons provide a possible mechanism for resisting a VLPO-induced switch-over to the sleep state during periods of sleep deprivation (Fulcher, Phillips & Robinson, 2010; Rempe, Best & Terman, 2010). This could explain how deliberate effort to stay awake may overcome the natural tendency to fall asleep when the homeostatic drive for sleep supersedes the threshold that would otherwise trigger a VLPO switch-over, allowing humans to sleep-deprive themselves.

LOCAL, USE-DEPENDENT SLEEP AND WAKE STATE INSTABILITY

Evidence is accumulating that sleep is not fundamentally a property of the whole brain, but rather of neuronal assemblies (Krueger et al., 2008). These are highly interconnected neuronal groups such as cortical columns, which are believed to be the basic units of information processing in the brain (Koch, 2004). They produce electrical signals that can be measured with electrodes and can be classified as wake-like or sleep-like. Animal experiments have revealed that cortical columns can exhibit the sleep-like state while neighboring cortical columns and the whole organism are awake (Rector et al., 2005). The sleep-like state appears more frequently when the cortical column has been used (i.e., required to process information) more intensively, so that the sleep-like state occurs in a homeostatic fashion. When the cortical column is in the sleep-like state, however, it does not process information accurately, which may result in cognitive failure (Van Dongen, Belenky & Krueger, in press).

Subcortical brain mechanisms like the VLPO switch serve to orchestrate the states of the neuronal assemblies, such that a person would not be awake and interacting with the environment while a significant portion of his/her neuronal assemblies is in the sleep-like state and therefore not processing information properly. However, it has been theorized that when wakefulness is extended (i.e., during sleep deprivation), such an uncoordinated brain condition may actually exist (Van Dongen, Belenky & Krueger, in press). This theory provides a possible explanation for the increased variability in cognitive performance that occurs during sleep deprivation (Doran et al., 2001). This phenomenon, known as (wake) state instability, may be a key underlying mechanism for the link between sleep deprivation and increased risk of errors and accidents (Van Dongen & Hursh, 2010).

INTERACTION OF SLEEP HOMEOSTASIS WITH CIRCADIAN RHYTHMS

Under natural conditions, the timing of sleep and wake is in synchrony with the circadian control of the sleep cycle. As wakefulness continues across the day, homeostatic pressure for sleep builds up, while at the same time, the circadian pressure for wakefulness increases. The effect of this opposition of the sleep homeostatic and circadian processes is to maintain a relatively even level of alertness across approximately 16 hours of waking (Dijk & Czeisler, 1994). Paradoxically, it also makes it difficult to fall asleep in the early evening, because that is when the circadian pressure for wakefulness is greatest (the so-called "wake maintenance zone").

Later in the evening, the circadian pressure for wakefulness gradually falls, along with systematic changes in many other circadian-mediated physiological functions (e.g., decrease in core body temperature, secretion of melatonin), whereas the homeostatic pressure for sleep continues to increase. This causes the VLPO to switch from the wake pass-through to the wake inhibition state, triggering sleep. During sleep, homeostatic pressure steadily dissipates, while the circadian pressure for wakefulness declines further to a nadir (approximately in mid-sleep). Thus, there is little waking pressure through most of the night, which, in healthy individuals, results in a consolidated period of sleep (Dijk & Czeisler, 1994). In the morning, the circadian pressure for wakefulness increases again, which forces the VLPO to switch back to the wake pass-through state, whereupon another wake–sleep cycle begins.

In shift workers, the timing of sleep and wakefulness is not in synchrony with the circadian rhythm. For instance, night shifts involve working during a time when the circadian drive for alert wakefulness decreases, and thus amplifies rather than opposes the increasing homeostatic drive for sleep across the waking hours (Van Dongen, 2006).

This is an important source of fatigue in night workers, especially toward the end of the shift and during the commute home, but it is not the only reason why night work is fatiguing. Working nights also implies that sleep times are placed during the day, when the circadian drive for wakefulness is increasing. This results in early awakening (Van Dongen, 2006) and reduced sleep duration to six hours or less in the average shift worker (Åkerstedt, 1998). Thus night work is fatiguing because of the adverse circadian timing of both the waking period and the sleep period and the sleep loss that is associated with the latter.

Similar problems occur when traveling to a different time zone, which places the circadian rthythm out of synchrony with the external light–dark cycle. It takes a few days for the circadian pacemaker to entrain to the new time zone. During this period, the sleep and wake periods are out of alignment with the circadian rhythm, resulting in a transient state similar to that experienced for night work. This fatiguing state is commonly known as "jet lag" (Monk, 1990).

NEUROIMAGING OF BRAIN MECHANISMS OF FATIGUE

Neuroimaging techniques such as positron emission tomography (PET) and functional magnetic resonance imaging (fMRI) have significantly advanced the ability to study brain function during sleep and waking (Maquet, 2010). Both PET and fMRI are based on the assumption that increased blood volume and/or flow reflects increased neuronal activity, since active neurons require more oxygenated blood. The interpretation of changes in brain activity caused by fatigue, however, tends to be ambiguous, particularly in the context of individual differences (Van Dongen, 2005).

Relative decreases in oxygenated blood signal in specific brain regions have been interpreted as a failure to preserve function and thus vulnerability to sleep loss, or as a beneficial protective mechanism. Similarly, relative increases in oxygenated blood signal in specific brain regions during sleep loss have been interpreted as a beneficial compensatory brain response, or as evidence of a need for greater neuronal effort and thus vulnerability. Further, fMRI investigations of cognitive performance have shown that increases in activation are correlated with less degraded task performance following sleep loss in some (Chee & Choo, 2004; Chee et al., 2006), but not all (Drummond & Brown, 2001; Habeck et al., 2004) cognitive performance paradigms. This inconsistency in the literature is symptomatic of an uncertainty about how directional changes in brain activation patterns in response to sleep loss should be interpreted. In that sense, understanding the neural underpinnings of the fatigued brain and the consequences for behavior and performance is still in its infancy.

Neuroimaging studies have revealed a decrease in metabolic activity in the thalamus following sleep deprivation (Thomas et al., 2000; Wu et al., 1991), indicating that sleep deprivation affects the brain's arousal mechanisms (Wu et al., 1991). This could be one of the ways in which sleep loss causes lapses in attention, cognitive variability, and overall impairments in performance. A functional neuroimaging study recently documented in detail what occurs in the brain during attentional lapses (Chee et al., 2008). Lapses related to sleep deprivation were associated with a significant reduction in frontal and parietal control areas, reflecting lower cognitive control, and reduced thalamic activation, reflecting a period of lowered arousal. These neural changes resulted in the attenuation of sensory processing and a brief apparent shift from wake to sleep, or a "microsleep." In the well-rested state, this sleep effect would normally be countered by top-down control from the frontal and parietal regions of the brain. Performance variability and attentional lapses following sleep loss may reflect the opposing effects of involuntarily progression of the brain toward sleep and voluntary effort of the individual to maintain wakefulness (Doran, Van Dongen & Dinges, 2001).

SLEEP, LEARNING AND MEMORY

There is a growing literature asserting a critical role for sleep in learning and memory (for review, see Walker & Stickgold, 2006). Sleep is made up of two main components: rapid eye movement (REM) sleep, identified by high frequency, low amplitude electroencephalogram (EEG), absence of muscle tone, and bursts of rapid eye movements; and non-REM (NREM) sleep, identified by moderate to low frequency, mixed to high amplitude EEG patterns depending on the non-REM sleep stage (N1 through N3, from light to deep sleep). These two aspects of sleep are believed to be involved in different facets of memory formation. This is supported by evidence from sleep studies suggesting that following learning of new information, changes in the NREM/REM structure of sleep occur.

A recent hypothesis states that sleep is essential for brain plasticity, and posits a specific role for NREM sleep (Tononi & Cirelli, 2005). According to the hypothesis, learning and other experiences during wakefulness result in increases in synaptic strength in many brain areas. NREM sleep (in particular stage N3) would be involved in downscaling the strength of all neuronal connections to a level that is energetically sustainable (and creates space for new neural connections to be formed the next day). During this process of synaptic downscaling, weak (functionally irrelevant) neuronal connections get lost. Strong neuronal connections become weaker in an absolute sense as well, but compared to the already weak connections the strong connections come out relatively stronger. This increases the signal-to-noise ratio of the neuronal connections that have been built up due to learning, and thus helps to consolidate memory. In this context, sleep seems to be critical for the homeostatic control of brain plasticity (Krueger & Obál, 1993; Tononi & Cirelli, 2005).

Consolidation of information learned during the day appears to be dependent on the reactivation of specific regions of the brain during sleep. Reactivation of neuronal assemblies in the hippocampus, a region known to be involved in memory formation, has been observed during post-training NREM sleep, with an improvement in next-day behavioral performance (Peigneux et al., 2004; Walker, 2009). In line with this, PET studies have documented increased glucose metabolism in the hippocampus during NREM sleep (compared to wakefulness), believed to reflect increased metabolic requirements in this region (Nofzinger et al., 2002). Reactivation is thought to strengthen the neural connections between different brain regions, and may help to store newly learned information or skills in long-term memory.

Deprivation of sleep has been repeatedly documented to disrupt memory, not only by disrupting the consolidation process, but also by degrading encoding, that is, the conversion of sensory input into a neural representation (Maquet, 2001; Walker & Stickgold, 2006). Functional neuroimaging studies have revealed that after one night of sleep deprivation, significant reduction in hippocampal activation is observed during encoding of episodic (autobiographical) information. This reduction in activation of the hippocampus, which is particularly critical for learning new episodic memories, is associated with a decline in encoding success (Yoo et al., 2007b). Adequate sleep prior to learning therefore appears to be necessary for the proper functioning of brain regions involved in encoding new memories.

Further to this, sleep deprivation has been shown to have a differential effect on learning depending on the emotional content of the information. Negative emotional stimuli seem to be relatively resistant to encoding impairment from prior sleep loss, as compared to positive and neutral emotional stimuli (Walker & Stickgold, 2006). In support of this, evidence from functional neuroimaging research has revealed that sleep loss causes uncoupling of the amygdala and the prefrontal cortex (Yoo et al., 2007a), which under rested conditions typically act as an integrated inhibitory, top-down control mechanism regulating emotional responses. The uncoupling due to sleep loss may result in inappropriate modulation of the response to negative information, producing a negative learning bias. This may have downstream effects for other cognitive processes, such as decision-making.

FRONTAL DEACTIVATION DURING SLEEP AND SLEEP LOSS

During normal wakefulness, the prefrontal cortex has the highest metabolic rate of all cortical regions, performing higher-order cognitive functions such as judgment, planning and situational awareness, integrating sensory and limbic information, organizing behavioral responses and supporting working memory (Braun et al., 1997; Beebe & Gozal, 2002). As such, this region is believed to have the greatest need for recovery during sleep (Horne, 2000). In line with this, both PET studies (Braun et al., 1997; Nofzinger et al., 2002) and quantitative EEG studies (Werth, Achermann & Borbély, 1997) have found frontal deactivations during NREM sleep. During REM sleep, the brain appears to be functionally active overall, as is characterized by sustained neuronal activity and high cerebral energy requirements and blood flow (Maquet, 2000). However, relative decreases in activity in prefrontal regions of the brain are observed during REM sleep as well (Maquet, 2000). Thus, it appears that the prefrontal cortex is selectively targeted for metabolic slowing throughout sleep (Balkin et al., 2002).

Sleep deprivation studies support the role of sleep in the restoration of the prefrontal cortex. There is slowing of the waking EEG (a physiological marker of sleep pressure) in predominantly frontal EEG derivations (Cajochen et al., 1999; Tinguely et al., 2006; see Figure 11.2). Reduction in prefrontal metabolism as measured by fMRI is observed during sleep deprivation as well (Thomas et al., 2003); the effect is reversed following recovery sleep (Wu et al., 2006). Declines in prefrontal activation have been associated with attentional impairment (Thomas et al., 2003); thus, sleep deprivation studies typically interpret reduced prefrontal activation as detrimental. This interpretation is tenuous, however, given that reduced activation during NREM sleep has been interpreted as a restorative process. This again highlights the uncertainty of how to understand changes in brain activation during sleep deprivation.

Figure 11.2   Topographic distribution of slowing in the waking EEG after one day of total sleep deprivation

Note: Average increases relative to baseline in EEG delta power (2–4 Hz; left panel) and theta power (5–8 Hz; right panel) over 8 subjects are plotted on a planar projection of the scalp, and expressed on a gray scale. White represents the greatest observed slowing (146 percent increase for delta power, 198 percent increase for theta power) and black represents the smallest observed slowing (111 percent increase for delta power, 151 percent increase for theta power). Black dots indicate EEG electrode positions.

Source: Graphs adapted from Figure 4 (lower left panels) in *NeuroImage*, 32(1): 283–92: Tinguely, G., Finelli, L.A., Landolt, H.-P., Borbély, A.A. & Achermann, P., "Functional EEG topography in sleep and waking: state-dependent and state-independent features." Copyright © 2006 with permission from Elsevier.

It has been suggested that sleep-related attentional impairments may be caused by a functional disconnect between frontal brain areas and task-related sensory brain regions, resulting in a reduction in top-down processing strength exerted by the prefrontal cortex to these regions (Thomas et al., 2000; Chee et al., 2008; Lim et al., 2010). However, as discussed earlier, performance on sustained attention tasks does not diminish completely during periods of extended wakefulness, but rather performance levels become more variable, with intermittent and unpredictable periods of impairment or lapses in attention (Doran et al., 2001). Thus, what role the reduced prefrontal activation observed during sleep deprivation may play with regard to degradation of cognitive performance is not yet entirely clear.

CONCLUSION

Sleep homeostasis and circadian rhythmicity interact to produce fatigue during wake extension and nocturnal hours of the day. The underlying neurobiology is relatively well known with respect to the circadian system, and the subcortical mechanisms initiating and terminating sleep are also well understood. However, the mechanisms of sleep homeostasis are still a topic of speculation, and possibly have to do with local use of neuronal assemblies. Fatigue from sleep loss is associated with regional changes in brain activation from which specific cognitive deficits may be inferred, but this does not explain the moment-to-moment variability in performance that is a hallmark of the effects of sleep deprivation on cognition. Thus, the neuroscience of sleep and circadian rhythms as it pertains to fatigue is not yet fully developed. At the behavioral level, however, sleep and circadian rhythms have been characterized in sufficient detail to be able to anticipate their overall effects on fatigue and performance. From an operational perspective, therefore, the scientific basis for understanding, predicting and managing fatigue is relatively solid and of high practical use (Van Dongen & Belenky, Chapter 30, this volume).

REFERENCES

Achermann, P. & Borbély, A.A. (2010). Sleep homeostasis and models of sleep regulation, in M.H. Kryger, T. Roth & W.C. Dement (eds), *Principles and Practice of Sleep Medicine*. Fifth edition. St. Louis, MO: Elsevier Saunders, 431–44.

Åkerstedt, T. (1998). Shift work and disturbed sleep/wakefulness. *Sleep Medicine Reviews*, 2, 117–28.

Arendt, J. (1998). Melatonin and the pineal gland: Influence on mammalian seasonal and circadian physiology. *Reviews of Reproduction*, 3, 13–22.

Balkin, T.J., Braun, A.R., Wesensten, N.J., Jeffries, K., Varga, M., Baldwin, P. et al. (2002). The process of awakening: A PET study of regional brain activity patterns mediating the re-establishment of alertness and consciousness. *Brain*, 125, 2308–19.

Banks, S. & Dinges, D.F. (2007). Behavioral and physiological consequences of sleep restriction. *Journal of Clinical Sleep Medicine*, 3, 519–28.

Basheer, R., Strecker, R.E., Thakkar, M.M. & McCarley, R.W. (2004). Adenosine and sleep–wake regulation. *Progress in Neurobiology*, 73, 379–96.

Beebe, D.W. & Gozal, D. (2002). Obstructive sleep apnea and the prefrontal cortex: Towards a comprehensive model linking nocturnal upper airway obstruction to daytime cognitive and behavioral deficits. *Journal of Sleep Research*, 11, 1–16.

Benington, J.H. & Heller, C.H. (1995). Restoration of brain energy metabolism as the function of sleep. *Progress in Neurobiology*, 45, 347–60.

Braun, A.R., Balkin, T.J., Wesensten, N.J., Carson, R.E., Varga, M., Baldwin, P. et al. (1997). Regional cerebral blood flow throughout the sleep–wake cycle. An H2(15)O PET study. *Brain*, 120, 1173–97.

Cajochen, C., Khalsa, S.B., Wyatt, J.K., Czeisler, C.A. & Dijk, D.J. (1999). EEG and ocular correlates of circadian melatonin phase and human performance decrements during sleep loss. *American Journal of Physiology*, 277, R640–R649.

Chee, M.W.L. & Choo, W.-C. (2004). Functional imaging of working memory after 24 hours of total sleep deprivation. *The Journal of Neuroscience*, 24, 4560–67.

Chee, M.W.L., Chuah, L.Y.M., Venkatraman, V., Chan, W.Y., Philip, P. & Dinges, D.F. (2006). Functional imaging of working memory following normal sleep and after 24 and 35 h of sleep deprivation: Correlations of fronto–parietal activation with performance. *NeuroImage*, 31, 419–28.

Chee, M.W.L., Tan, J.C., Zheng, H., Parimal, S., Weissman, D.H., Zagorodnov, V. et al. (2008). Lapsing during sleep deprivation is associated with distributed changes in brain activation. *Journal of Neuroscience*, 28, 5519–28.

de Lecea, L., Kilduff, T.S., Peyron, C., Gao, X., Foye, P.E., Danielson, P.E. et al. (1998). The hypocretins: Hypothalamus-specific peptides with neuroexcitatory activity. *Proceedings of the National Academy of Sciences*, 95, 322–7.

Dijk, D.-J. & Czeisler, C.A. (1994). Paradoxical timing of the circadian rhythm of sleep propensity serves to consolidate sleep and wakefulness in humans. *Neuroscience Letters*, 166, 63–8.

Doran, S.M., Van Dongen, H.P.A. & Dinges, D.F. (2001). Sustained attention performance during sleep deprivation: Evidence of state instability. *Archives of Italian Biology*, 139, 253–67.

Drummond, S.P.A. & Brown, G.G. (2001). The effects of total sleep deprivation on cerebral responses to cognitive performance. *Neuropsychopharmacology*, 25, S68–S73.

Fulcher, B.D., Phillips, A.J.K. & Robinson P.A. (2010). Quantitative physiologically based modeling of subjective fatigue during sleep deprivation. *Journal of Theoretical Biology*. 264, 407–19.

Fuller, P.M., Gooley, J.J. & Saper, C.B. (2006). Neurobiology of the sleep–wake cycle: Sleep architecture, circadian regulation, and regulatory feedback. *Journal of Biological Rhythms*, 21, 482–93.

Golombek, D.A. & Rosenstein, R.E. (2010). Physiology of circadian entrainment. *Physiological Review*, 90, 1063–102.

Habeck, C., Rakitin, B.C., Moeller, J., Scarmeas, N., Zarahn, E., Brown, T. et al. (2004). An event-related fMRI study of the neurobehavioral impact of sleep deprivation on performance of a delayed match-to-sample task. *Cognitive Brain Research*, 18, 306–21.

Horne, J. (2000). Neuroscience: Images of lost sleep. *Nature*, 403, 605–6.

Jackson, M.L. & Van Dongen, H.P.A. (2011). Cognitive effects of sleepiness, in M.J. Thorpy & M. Billiard (eds), *Sleepiness*. New York: Cambridge University Press.

Kerkhof, G.A. & Van Dongen, H.P.A. (1996). Morning-type and evening–type individuals differ in the phase position of their endogenous circadian oscillator. *Neuroscience Letters*, 218, 153–6.

Koch, C. (2004). *The Quest for Consciousness: A Neurobiological Approach*. Englewood, CO: Roberts & Company.

Krueger, J.M. & Obál, J.F. (1993). A neuronal group theory of sleep function. *Journal of Sleep Research*, 2, 63–69.

Krueger J.M., Obál, J.F. & Fang, J. (1999). Humoral regulation of physiological sleep: Cytokines and GHRH. *Journal of Sleep Research*, 8, 53–59.

Krueger, J.M., Rector, D.M., Roy, S., Van Dongen, H.P.A., Belenky, G. & Panksepp, J. (2008). Sleep as a fundamental property of neuronal assemblies. *Nature Reviews Neuroscience*, 9, 910–19.

Krueger, J.M., Taishi, P., De, A., Davis, C.J., Winters, B.D. Clinton, J., Szentirmai, E. & Zielinski. M.R. (2010). ATP and the purine type 2 X7 receptor affect sleep. *Journal of Applied Physiology*, 109, 1318–27.

Lewy, A.J., Bauer, V.K., Ahmed, S., Thomas, K.H., Cutler, N.L., Singer, C.M. et al. (1998). The human phase response curve (PRC) to melatonin is about 12 hours out of phase with the PRC to light. *Chronobiology International*, 15, 71–83.

Lim, J., Tan, J.C., Parimal, S., Dinges, D.F. & Chee, M.W.L. (2010). Sleep deprivation impairs object-selective attention: A view from the ventral visual cortex. *PLoS One, 5*, e9087.

Lockley, S.W., Brainard, G.C. & Czeisler, C.A. (2003). High sensitivity of the human circadian melatonin rhythm to resetting by short wavelength light. *Journal of Clinical Endocrinology and Metabolism*, 88: 4502–5.

Maquet, P. (2000). Functional neuroimaging of normal human sleep by positron emission tomography. *Journal of Sleep Research*, 9, 207–31.

Maquet, P. (2001). The role of sleep in learning and memory. *Science*, 294, 1048–52.

Maquet, P. (2010). Understanding non rapid eye movement sleep through neuroimaging. *The World Journal of Biological Psychiatry*, 11, 9–15.

McCauley, P., Kalachev, L.V., Smith, A.D., Belenky, G., Dinges, D.F. & Van Dongen, H.P.A. (2009). A new mathematical model for the homeostatic effects of sleep loss on neurobehavioral performance. *Journal of Theoretical Biology*, 256, 227–39.

Monk, T.H. (1990). The relationship of chronobiology to sleep schedules and performance demands. *Work & Stress*, 4, 227–36.

Nehlig, A., Daval, J.L. & Debry, G. (1992). Caffeine and the central nervous system: Mechanisms of action, biochemical, metabolic, and psychostimulant effects. *Brain Research Reviews*, 17, 139–70.

Nofzinger, E.A., Buysse, D.J., Miewald, J.M., Meltzer, C.C., Price, J.C., Sembrat, R.C. et al. (2002). Human regional cerebral glucose metabolism during non-rapid eye movement sleep in relation to waking. *Brain*, 125, 1105–15.

Panda, S., Nayak, S.K., Campo, B., Walker, J.R., Hogenesch, J.B. & Jegla, T. (2005). Illumination of the melanopsin signaling pathway. *Science*, 307, 600–604.

Peigneux, P., Laureys, S., Fuchs, S., Collette, F., Perrin, F., Reggers, J. et al. (2004). Are spatial memories strengthened in the human hippocampus during slow wave sleep? *Neuron*, 44, 535–45.

Porkka-Heiskanen, T., Strecker, R.E., Thakkar, M., Bjorkum, A.A., Greene, R.W. & McCarley, R.W. (1997). Adenosine: A mediator of the sleep-inducing effects of prolonged wakefulness. *Science*, 276, 1265–68.

Rajaratnam, S.M.W., Middleton, B., Stone, B.M., Arendt, J. & Dijk, D.-J. (2004). Melatonin advances the circadian timing of EEG sleep and directly facilitates sleep without altering its duration in extended sleep opportunities in humans. *Journal of Physiology* (London), 561, 339–51.

Rector, D.M., Topchiy, I.A., Carter, K.M. & Rojas, M.J. (2005). Local functional state differences between rat cortical columns. *Brain Research*, 1047, 45–55.

Rempe, M., Best, J. & Terman, D. (2010). A mathematical model of the sleep/wake cycle. *Journal of Mathematical Biology*, 60, 615–44.

Sakurai, T., Amemiya, A., Ishii, M., Matsuzaki, I., Chemelli, R.M., Tanaka, H. et al. (1998). Orexins and orexin receptors: A family of hypothalamic neuropeptides and G protein-coupled receptors that regulate feeding behavior. *Cell*, 92, 573–85.

Saper, C.B., Chou, T.C. & Scammell, T.E. (2001). The sleep switch: Hypothalamic control of sleep and wakefulness. *Trends in Neuroscience*, 24, 726–31.

Saper, C.B., Scammell, T.E. & Lu, J. (2005). Hypothalamic regulation of sleep and circadian rhythms. *Nature*, 437, 1257–63.

Thomas, M., Sing, H., Belenky, G., Holcomb, H., Mayberg, H., Dannals, R. et al. (2000). Neural basis of alertness and cognitive performance impairments during sleepiness. I. Effects of 24 h of sleep deprivation on waking human regional brain activity. *Journal of Sleep Research*, 9, 335–52.

Thomas, M., Sing, H., Belenky, G., Holcomb, H., Mayberg, H., Dannals, R. et al. (2003). Neural basis of alertness and cognitive performance impairments during sleepiness II. Effects of 48 and 72 h of sleep deprivation on waking human regional brain activity. *Thalamus & Related Systems*, 2, 199–229.

Tinguely, G., Finelli, L.A., Landolt, H.-P., Borbély, A.A. & Achermann, P. (2006). Functional EEG topography in sleep and waking: State-dependent and state–independent features. *NeuroImage*, 32, 283–92.

Tononi, G. & Cirelli, C. (2005). Sleep function and synaptic homeostasis. *Sleep Medicine Reviews*, 10, 49–62.

Van Dongen, H.P.A. (2005). Brain activation patterns and individual differences in working memory impairment during sleep deprivation. *Sleep*, 28, 386–8.

Van Dongen, H.P.A. (2006). Shift work and inter-individual differences in sleep and sleepiness. *Chronobiology International*, 23, 1139–47.

Van Dongen, H.P.A. & Hursh, S.R. (2010). Fatigue, performance errors, and accidents, in M.H. Kryger, T. Roth & W.C. Dement (eds), *Principles and Practice of Sleep Medicine*. Fifth edition. St. Louis, MO: Elsevier Saunders, 753–9.

Van Dongen, H.P.A., Belenky, G. & Krueger, J.M. (2011). A local, bottom-up perspective on sleep deprivation and neurobehavioral performance. *Current Topics in Medicinal Chemistry*, 11, 2414–22.

Van Dongen, H.P.A., Kerkhof, G.A. & Dinges, D.F. (2004). Human circadian rhythms, in A. Sehgal (ed.), *Molecular Biology of Circadian Rhythms*. Hoboken, NJ: John Wiley & Sons, 255–69.

Van Dongen, H.P.A., Kerkhof, G.A. & Souverijn, J.H.M. (1998). Absence of seasonal variation in the phase of Walker, M.P. (2009). The role of sleep in cognition and emotion. *Annual New York Academy of Science*, 1156, 168–97.

Walker, M. & Stickgold, R. (2006). Sleep, memory, and plasticity. *Annual Reviews in Psychology*, 57, 139–66.

Werth, E., Achermann, P. & Borbély, A. (1997). Fronto-occipital EEG power gradients in human sleep. *Journal of Sleep Research*, 6, 102–12.

Wever, R.A. (1989). Light effect of human circadian rhythms: a review of recent Andechs experiments. *Journal of Biological Rhythms*, 4, 161–86.

Wu, J.C., Gillin, J.C., Buchsbaum, M.S., Chen, P., Keator, D.B., Khosla Wu, N. et al. (2006). Frontal lobe metabolic decreases with sleep deprivation not totally reversed by recovery sleep. *Neuropsychopharmacology*, 31, 2783–92.

Wu, J.C., Gillin, J.C., Buchsbaum, M.S., Hershey, T., Hazlett, E., Sicotte, N. et al. (1991). The effect of sleep deprivation on cerebral glucose metabolic rate in normal humans assessed with positron emission tomography. *Sleep*, 14, 155–62.

Yoo, S.-S., Gujar, N., Hu, P., Jolesz, F.A. & Walker, M.P. (2007a). The human emotional brain without sleep – a prefrontal amygdala disconnect. *Current Biology*, 17, R877–R878.

Yoo, S.-S., Hu, P., Gujar, N., Jolesz, F.A. & Walker, M.P. (2007b). A deficit in the ability to form new human memories without sleep. *Nature Neuroscience*, 10, 385–92.

12

The Influence of Fatigue on Brain Activity

Ashley Craig and Yvonne Tran

INTRODUCTION

Fatigue has been identified as a major source for accidents and subsequent injury when driving (Åkerstedt & Kecklund, 2001; Connor et al., 2002) and when performing process type work-related tasks (Baker, Olson & Morisseau, 1994; Frank, 2000). Fatigue is a recognized problem in the transport industry with suggestions that fatigue is a factor in at least 40 percent of road crashes (Fletcher et al., 2005). The cost of fatigue-related accidents in financial terms is high, for instance, the Bureau of Transport Economics, Australia, estimated that the annual cost to the Australian community could be as high as A\$3 billion (Parliament of the Commonwealth of Australia, 2000). In Australia, at least 20 percent of road accidents every year are reported to be driver fatigue-related (Roads Traffic Authority, 2001). Research has shown that workers employed in the transport and storage industries have the highest rate of work-related road deaths (15.5 per 100,000), in which fatigue is an important contributing factor (Mitchell, Driscoll & Healey, 2004). More than likely, the frequency of fatigue-related accidents is underestimated due to a lack of a universally accepted definition for fatigue, as well as a lack of methodology that is able to measure validly the level of fatigue immediately after an accident (Dobbie, 2002).Operator fatigue in the process, health, mining, and assembly industries, is also problematic. Fatigue is a risk when people work without having breaks, and risks are increased especially where work is conducted in night shifts. For instance, Frank (2000) showed that shiftwork is associated with a high risk of industrial injuries, while Bourdouxhe et al. (1999) found elevated risks of sleep deficit, chronic fatigue, health problems and disruption of social and family life in shift workers in a Canadian refinery workplace where operators work in rotating 12-hour shifts. Kopardekar and Mital (1994) investigated the effect of different work–rest schedules on fatigue and performance in an operator's task. They found that working in a period of two hours without a break was associated with increased errors in performance. Baker, Olson and Morisseau (1994) studied work practices that possibly contributed to fatigue-induced performance decrements in the commercial nuclear power industry. Results found a relationship between overtime and plant safety performance. Other researchers such as van der Linden, Frese and Meijman (2003) found that fatigue compromised executive brain capacity, such as the ability to regulate perceptual and motor processes for goal-directed behavior. They used a card-sorting test as a measure of executive capacity and fatigue was induced through participants working through a two-hour task. They found fatigued participants performed more poorly (e.g. slower planning times) on the card tests. The authors believed these results are evidence of compromised executive mental control under fatigue conditions, which help explain the risk of errors and poorer performance that occurs when people become fatigued. Finally, Gold et al. (1992) conducted an investigation into accidents in a large sample of nurses in an American hospital. In comparison to those nurses who worked only day or evening shifts, nurses employed on night shifts had increased risk of being fatigued, twice the chance of "nodding off" while driving to or from work and twice the risk of a work-related accident or error. Fatigue is also a common symptom. It is associated with many diseases such as sleep apnoea (George, 2004), chronic fatigue syndrome (Boneva et al., 2007), depressive and anxiety-based disorders (Friedman & Thayer, 1998; Fuhrer & Wessely, 1995), trait anxiety and stress (Craig et al., 2006) and life-threatening diseases such as cancer (Mills et al., 2005). Clearly, the problem of fatigue is widespread and considerable in our communities and thus it is crucial that a better understanding is achieved concerning its causes and impacts. The primary aim of this chapter will be to explore what we know about the neural basis of operator fatigue, specifically, the influence of fatigue on brain activity.

DEFINITION AND DETERMINANTS OF FATIGUE

It is important to note that a universally accepted definition for fatigue remains elusive. What is meant by the term "fatigue?" First, the term as used in this chapter does not refer to muscular fatigue. Muscular fatigue is a physical state caused by prolonged and strong contraction of a muscle, and is related to the rate of muscle glycogen depletion. Muscle fatigue is associated with oxygen lack and an increased level of blood and muscle lactic acid. In contrast to specific muscle fatigue, we are concerned with fatigue that has mental and physical components, which has been described as a change in psychophysiological state due to sustained performance, involving both psychological and physical aspects of tiredness (Desmond & Hancock, 2001; Williamson, Feyer & Friswell, 1996). Fatigue can also be defined in terms of a phase of sleepiness, that is, as the state before the onset of stage 1 microsleep (Broughton & Hasan, 1995). Brown (1994) defined fatigue as a subjective psychological state related to drowsiness (Brown, 1994; Craig et al., 2006). Finally, it has been defined as a stress-related condition, with some suggesting that fatigue and "distress" are the same construct but distinct conditions (Bultmann et al., 2002).In the light of the above definitions, and for the purposes of this chapter, we define fatigue as:

> a neuropsychophysiological state that occurs in a person who is feeling tired or drowsy, to the extent that they have reduced capacity to function, resulting in performance decrement, negative emotions and boredom.

DETERMINANTS OF FATIGUE

Given the variety of definitions used in fatigue research, it is important to clarify the determinants of fatigue. We have found that it is necessary to employ a range of outcome measures to achieve this goal (Craig et al., 2006). Assuming that fatigue has environmental, physiological and psychological determinants (Matthews & Desmond, 1998), then it is essential to employ outcome measures that encompass possible associated factors. For example, assessing physical aspects (e.g. brain wave activity; general health, age), psychological aspects (e.g. mood, anxiety) and environmental aspects (e.g. time of day) will result in an improved understanding of the nature of operator fatigue (Craig et al., 2006; Wijesuriya, Tran & Craig, 2007). The type of outcome measure used to establish how much a person fatigues is crucial (Craig et al., 2006). However, many researchers employ a single fatigue outcome measure such as performance decrement, self-reported fatigue or, perhaps, change in some neurophysiological activity such as brain wave activity (electroencephalography or EEG) and eye activity (electrocculography or EOG). We recommend the use of multiple measures such as performance decrement, as well as physiological and psychological outcome measures (Craig et al., 2006; Wijesuriya et al., 2007).It is interesting to note that psychological/self-report measures of fatigue have been found to be significantly correlated with psychological determinants such as mood (Craig et al., 2006; Wijesuriya et al., 2007). However, psychological factors have been shown to have a non-significant association with performance decrement and physiological fatigue outcome measures (Craig et al., 2006). This lack of association between physiological and psychological factors suggests multiple independent factors contribute to fatigue. A model that supports the notion that multiple determinants contribute to fatigue is the two-level processing model of fatigue (Verwey & Zaidel, 2000). This model predicts that physiological fatigue factors tap into automatic lower level processing, while psychological fatigue factors are associated with higher-level "executive" cognitive or cortical processes (van der Linden, Frese & Meijman, 2003). Studying fatigue using this model can assist in understanding the nature of fatigue. This chapter will therefore describe the relationship between fatigue and neural factors such as brain activity in some detail.

ELECTROENCEPHALOGRAPHY (EEG)

The EEG is considered a promising measure of fatigue (Santamaria & Chiappa, 1987; Wijesuriya et al., 2007). EEG activity reflects the summation of the inhibitory and excitatory postsynaptic potentials of nerve cells (Santamaria & Chiappa, 1987). The EEG measures the *synchronous activity* of cortical neurons that have a similar spatial direction, and is typically described in terms of rhythmic activity. The rhythmic activity is divided into bands by frequency, including delta activity (0.5–4 Hz), theta activity (4–7.5 Hz), alpha activity (8–13 Hz) and beta activity (13.5–30Hz).

Delta waves
These are slow waves between 0.5 and 4 Hz. Delta waves have been shown to be present during transition to drowsiness and are associated with slow-wave sleep (Lal & Craig, 2001a).

Theta waves
These are slow waves within the frequency range of 4–7.5 Hz. Theta rhythms are associated with a variety of psychological states including meditative states, and low levels of alertness during drowsiness and sleep, but not deep sleep, and as such have been associated with decreased information processing (Lal & Craig, 2001a) They replace the alpha components at the onset of sleep (Santamaria & Chiappa, 1987).

Alpha waves

Alpha waves have a frequency range of 8–13 Hz, They occur during wakefulness, particularly over the occipital cortex, appear markedly at eye closure and decrease at eye opening and are highly attenuated during attention (Tran, Craig & McIsaac, 2001). The alpha wave rhythm is known to be responsive to arousal and mental activity (Santamaria & Chiappa, 1987; Tran, Craig & McIsaac, 2001). Alpha wave activity has been traditionally considered to be a thalamocortical rhythm (Knyazev & Slobodskaya, 2003), and cortical arousal is linked to the amplitude and frequency of the alpha rhythm (Golan & Neufield, 1996), with high amplitude, low frequency activity associated with low cortical arousal, and low amplitude high frequency activity associated with high cortical arousal (Tran, Craig & McIsaac, 2001).

Beta waves

Beta waves are fast waves with frequency range between 13.5 and 30 Hz. It is the EEG potentials associated with increased alertness, arousal and excitement (Lal & Craig, 2001a).In their seminal paper, Santamaria and Chiappa (1987) defined the changes in EEG activity associated with drowsiness into four periods. Period 1 (Transitional period) is the period just before the disappearance of alpha waves, determined by the appearance or increase of centrofrontal alpha lasting 1–10 seconds, with bursts of increased amplitudes at frequencies slower that that observed with eyes closed alpha. Temporal alpha also appear with a progressive slowing in EEG activity (delta and theta waves appearing). Period 2 (Transitional and Post-transitional) is defined as the period between Transitional and Post-transitional and has EEG features of either or both periods. Period 3 (Post-transitional) consists of the EEG segments after alpha wave disappearance and comprises the early Stage 1 of sleep, marked by centrofrontal and posterior slowing of EEG. Period 4 (Arousal) is the period where the person emerges from drowsiness and consists of centrofrontal beta and centrofrontal alpha. Figure 12.1a shows 32 channels over the

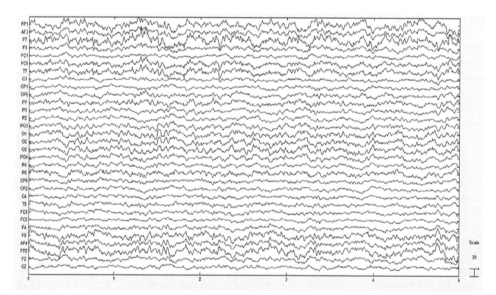

Figure 12.1a   Raw EEG signals across 32 channels over the entire cortex, typical of an alert person

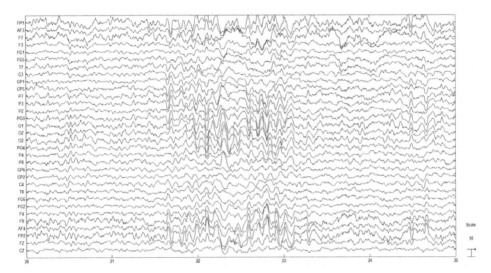

Figure 12.1b **Change in the EEG signals when a person fatigues: typically, the change involves increases in slow wave activity**

entire cortex of typical raw EEG signals from an alert EEG. Figure 12.1b shows the EEG during the onset of fatigue. The figure shows slowing of EEG activity, especially in the centrofrontal and posterior regions, typical of the Transitional–Post-transitional period.

EEG STUDIES AND FATIGUE

Table 12.1 shows details of major studies that have investigated EEG changes that occur as a person becomes fatigued. The information presented in this table demonstrates that brain activity does change in a consistent manner when a person fatigues or becomes drowsy (Santamaria & Chiappa, 1987). The information in Table 12.1 provides some confirmation on the direction of change that occurs in the EEG frequency bands. It has not been clear whether, for instance, alpha wave activity increases or decreases with increasing fatigue levels. The systematic review isolated 17 research papers that investigated the impact on brain activity using EEG measures. The results of these 17 studies are presented in Table 12.1 and most of these studies are further discussed below. Table 12.1 shows details on each study, including the number of participants, the type of fatiguing task, the influence on the four major frequency bands and the electrode sites used (Åkerstedt, Kecklund & Knutsson, 1991; Caldwell et al.,1995; Cajochen et al., 1996; Campagne, Pebayle & Muzet, 2004; Dumont, et al., 1997; Eoh, Chung & Kim, 2005; Kecklund & Åkerstedt, 1993; Lal & Craig, 2002; Macchi et al., 2002; Pal et al., 2008; Papadelis et al., 2006; Schier, 2000; Strijkstra et al., 2003; Tanaka, Hayashi & Hori, 1997; Torsvall & Åkerstedt, 1987; Trejo et al., 2005).

Åkerstedt, Kecklund & Knuttson (1991), using only one bipolar site, studied the influence of shiftwork on delta, theta and alpha activity in 25 shift workers. Alpha activity power was found to increase significantly during the shift (at night) and was found to correlate with subjective ratings of fatigue. The authors failed to find significant changes in delta and theta activity. Caldwell, Hall & Erickson (2002) employed six monopolar

Table 12.1 Details of studies that have investigated EEG changes as a person fatigues, including number of participants, task or fatigue condition, impact on the 4 EEG bands and number of electrode scalp sites

Study	Subjects	Task	delta	theta	alpha	Beta	Sites
Åkerstedt et al., 1991	25 shift workers	Shift work periods	NC	NC	↑	–	C2–O2 bipolar
Caldwell et al., 2002	10 pilots	Flying sessions	↑	↑	↓	NC	6 monopolar
Cajochen et al., 1995	9 female subjects	Up to 34 hrs awake	–	↑	↑	–	2 monopolar C3, C4
Cajochen et al., 1996	8 male students	Dose of melatonin	–	↑	↑	–	2 monopolar C3, C4
Campagne et al., 2004	46 males	Simulated driving	–	↑	↑	–	4 sites in left hemisphere
Dumont et al., 1997	9 subjects	Up to 24 hrs awake	–	↑	↑	–	C3 monopolar site
Eoh et al., 2005	8 subjects	Simulated driving	–	NC	↑	↑	8 monopolar
Kecklund & Åkerstedt, 1993	18 truck drivers	Truck driving	–	↑	↑	–	Cz–Oz bipolar
Lal & Craig, 2002	35 subjects	Simulated driving	↑	↑	↑	↑	19 10–20 sites
Macchi et al., 2002	8 truck drivers	Simulated driving	–	↑	↑	–	8 monopolar
Pal et al., 2008	13 subjects	1 hr car drive	–	↑	↑	–	O2 monopolar
Papadelis et al., 2006	20 subjects	Sleep deprived for 24 hrs	↓	↑	↑	NC	16 10–20 sites
Schier, 2000	2 drivers	Simulated driving	–	–	↑	–	F3, F4, P3, P4
Strijkstra et al., 2003	10 subjects	40 hrs awake	–	↑	↓	–	28 10–20 sites
Tanaka et al., 1997	10 male subjects	Sleep task	↑	↑	↑	↑	12 10–20 sites
Torsvall & Åkerstedt, 1987	11 train drivers	Shift work	↑	↑	↑	–	O2–P4 bipolar
Trejo et al., 2005	16 subjects	Arithmetic tasks up to 3 hrs	–	↑	↑	–	30 10–20 sites

Note: ↑ means significant increase in activity; ↓ means significant decrease in activity in the EEG bands, while NC means no significant change; – means the impact on that band was not reported; definition of EEG frequency bands may differ slightly across studies; however, generally, delta refers to 0.5–3.5Hz EEG activity; theta refers to 4–7.5Hz EEG activity; alpha refers to 8–13Hz activity and beta refers to 14–30Hz activity.

electrode sites to study the neural effects of flying for long periods in 10 helicopter pilots. They found increases in delta and theta wave activity, a decrease in alpha wave activity, but found no significant changes in beta wave activity. Cajochen, et al. (1995, 1996) found significantly increased theta and alpha wave changes in response to long periods of staying awake (1995) and as a result of melatonin doses (1996). The results of the two studies are limited by small numbers of participants and only two electrode sites were used to measure EEG change. In a simulated driving task with 46 males, Campagne, Pebayle & Muzet, (2004) showed that as vigilance level decreased, so performance decrements increased and this was associated with higher levels of theta and alpha wave activity. Eoh, Chung & Kim (2005) studied EEG changes resulting from a simulated continuous driving task in a small number of participants using eight monopolar sites. They found increased alpha and decreased beta wave activity associated with the development of fatigue. However, Eoh, Chung & Kim (2005) did not find significant changes in theta wave activity during the fatiguing task. In a field study by Kecklund and Åkerstedt (1993) using the Cz-Oz bipolar site, EEG was studied at different stages during day and night driving in a group of 18 truck drivers. The higher subjective drowsiness and reduced driving performance observed in the night group was associated with increased theta and alpha activity observed during the last few hours of driving. In a simulated driving task with 35 non-professional drivers, Lal and Craig (2002) assessed EEG change across 19 10–20 electrode sites. They found significantly increased levels of delta and theta wave activity over the entire scalp and significant though smaller change in alpha and beta wave activity associated with fatigue. In another simulated night- shift driving study with eight long-haul truck drivers, Macchi et al. (2002) studied EEG in two frequency bands (theta and alpha wave activity) using eight monopolar electrode sites. They compared EEG activity in long-haul truck drivers who had an afternoon nap, compared to a group of drivers who had not had a nap. The EEG alpha wave activity of both groups was associated with their reaction time, with faster reaction times found in those who had had an afternoon nap. EEG changes found to occur with progressive drowsiness were increased amounts of theta and alpha wave activity. Pal et al. (2008) conducted a study of fatigue with 13 participants who were subjected to a one-hour car drive. They found increased theta and alpha wave activity associated with fatigue, though they only used one posterior electrode site. Papadelis et al. (2006) sleep-deprived 20 participants who were assessed for EEG change using 16 10–20 electrode sites. They found reduced delta activity, increased theta and alpha wave activity and no change in beta activity associated with fatigue. Employing 28 10–20 electrode sites across the entire cortex, Strijkstra et al. (2003) studied the association between drowsiness and EEG change during 40 hours of staying awake in ten participants. They found alpha wave levels decreased as participants fatigued over the 40 hours, while theta wave activity increased over the same period, especially in frontal regions. Tanaka, Hayashi & Hori (1997) employed a sleep task with ten participants and studied EEG change using 12 EEG 10–20 sites. They found increased delta and theta wave activity associated with fatigue. They also found that alpha amplitude increased as drowsiness increased, and that beta wave activity increased with drowsiness (defined as sigma activity 13–14.8 Hz band). Torsvall and Åkerstedt (1987) studied EEG change using the O2–P4 bipolar site with 11 train drivers over a night and day trip. They found increased delta, theta and alpha wave activity associated with fatigue. In summary, the majority of studies investigated theta and alpha amplitude or power change associated with fatigue. Except for Åkerstedt et al., 1991 and Eoh, Chung & Kim, 2005, in which there was no significant change in theta wave activity, all studies found theta wave activity increased as fatigue developed. Similarly, except for Caldwell, Hall & Erickson, (2002) and Strijkstra et al. (2003), all found increased alpha wave activity occurred with fatigue. Clearly, brain wave activity changes do reliably occur when a person fatigues, and these changes include increased slow wave activity (theta) and increased alpha wave activity, both of which are symptomatic of reduced cortical alertness (Santamaria & Chiappa, 1987). The change that occurs in the delta and beta bands is less clear, and additional research needs to be carried out to clarify the situation. It would also be important to determine whether fatigue is differentially associated with alpha wave activity changes in low versus high apha activity. That is, between so-called apha-1 (8.0–9.0 Hz), alpha-2 (9.5–11 Hz) and alpha-3 (11.5–13 Hz) bands. As far as the authors are aware, this has not yet been explored.

CONCLUSIONS

Studies into the psychophysiological determinants of fatigue have isolated factors that are associated with higher risks of developing fatigue (Wijesuriya et al., 2007). Factors that seem to be supported in the research literature include unique contributors, such as a tendency toward extraversion, sensation-seeking behaviour and raised levels of anxiety. Furthermore, the change in brain activity as a person becomes fatigued or drowsy has been studied over many years with some uncertainty as to the relationship (Lal & Craig, 2001a,b). However, the results of our systematic review have helped clarify these changes. Not surprisingly, the EEG changes found to occur reflect reduced cortical arousal. In support of this, Dirnberger et al. (2004) studied the relationship between movement-related cortical potentials and fatigue. They found lower amplitudes in movement potentials were associated with greater levels of fatigue. They suggested that this supports the notion that fatigue reduces cognitive capability, suggesting that fatigue reduces cortical arousal. Lehmann, Grass & Meier (1995) conducted canonical analyses between subjective cognitions and EEG spectral power profiles with four channels of EEG in 20 healthy men and women. They found significant pairs of variables that support the finding that fatigue is associated with reduced cortical arousal. For instance, they found 2–6 Hz activity to be significantly associated with cognitions reflecting reduced cognitive capacity (e.g. lacking orientation, low recall ability). Another significant canonical pair included 4–7 Hz and 10–13 Hz was also associated with reduced cognitions.

Assuming that changes in theta, alpha and perhaps delta spectral frequency bands are reliably associated with fatigue and drowsiness, then the question remains as to the implications for the application of these findings. Importantly, evidence suggests that fatigue- related changes in theta and alpha activity occur before any deterioration in performance occurs (Gevins et al., 1990). That is, performance decrements are preceded in time by these increases in theta and alpha waves, suggesting brain wave activity may be a sensitive indicator of fatigue and its associated performance decrements. The obvious advantage is that EEG change may reliably predict performance decrement before errors or accidents occur. To confirm whether EEG changes precede performance decrement will be a challenge for future research. Further, it will also be important to estimate just how much time before the decrement occurs, does the EEG change. Nevertheless, there is optimism that EEG activity will be useful as a measure for a fatigue countermeasure device. For instance, it could provide a basis for continuous fatigue-monitoring technology that may be successfully employed to reduce fatigue-related accidents. EEG could also provide a useful indicator of the success of countermeasures. For instance, Horne and Reyner (1996) used EEG slow wave measures to confirm the usefulness of employing coffee (150mg caffeine) and taking a break in 10 tired drivers. They found that compared to the control, caffeine and taking a nap significantly reduced driving impairments and subjective sleepiness, and the EEG measure used confirmed that brain activity changed towards increased cortical arousal. A further useful progression in this research could well be the presence of fatigue-related "EEG signatures." For instance, does the change in the EEG reflect fatigue arising from different sources (such as performance-related fatigue in a boring task versus sleep debt-related fatigue)? This will remain a challenge for future research. Limitations of the findings discussed in this chapter need to be raised. First, few if any studies have employed comprehensive artefact removal in the EEG signal. Artefact is especially prominent in low-frequency activity, and the EEG signal with a high noise to signal ratio will produce confounded results. Muscle artefact will be prominent in studies that measure EEG while the subject is performing a task such as simulated driving. Sophisticated methods are being employed to remove artefact successfully (Delorme & Makeig, 2004; Tran, Thuraisingham, Craig & Nguyen, 2009), and the challenge for future research in this area will be to employ such techniques to clean the signal so the impact of fatigue on the brain can be better detected. Also, participant numbers are small in most studies. Issues of statistical power need to be considered in the planning stages of fatigue-related EEG research, so that studies have sufficient numbers of participants and thus sufficient statistical power to reject the null hypothesis of no difference. Clearly, if fatigue does influence brain activity, then the study needs to be able to confirm this difference statistically. Furthermore, most studies employ a small number of EEG sites to detect brain activity change. Future studies need to employ multiple 10–20 cortical sites of at least 20–30 referenced electrode sites. Finally, research needs to explore new methods of analyzing EEG data and signals associated with fatigue. For instance, we have explored the value of employing non-linear strategies (such as sample entropy and second order difference plots) for processing the fatigue-related EEG signal (Tran et al., 2007, 2008). We have also trialled techniques such as measuring microstate segments, that is, microstate intensity methodology, in relation to spontaneous electrical change on the cortex (Thuraisingham et al., 2009). These techniques measure the complexity and variability of the EEG wave, and a great advantage is that they yield a single value per time as opposed to, say, assessing changes in at least three or perhaps four frequency bands in the spectral EEG. This means that these non-linear signal-processing approaches could provide a faster and more convenient approach for detecting fatigue in a countermeasure device in comparison to analyzing spectral bands (Tran et al., 2007, 2008). These innovative signal-processing techniques also provide an alternative picture of the change in the brain related to fatigue, and this is potentially valuable data for understanding the neural basis of fatigue as well as applying the information to fatigue countermeasures (Tran et al., 2007, 2008).

REFERENCES

Åkerstedt, T. & Kecklund, G. (2001). Age, gender and early morning highway accidents. *Journal of Sleep Research*, 10, 105–10.

Åkerstedt, T., Kecklund, G. & Knuttson, A. (1991). Manifest sleepiness and the EEG spectral content during night work. *Sleep*, 14, 221–5.

Baker, K., Olson, J. & Morisseau, D. (1994). Work practices, fatigue, and nuclear power plant safety performance. *Human Factors*, 36, 244–57.

Boneva, R.S., Decker, M.J., Maloney, E.M., Lin, J.M., Jones, J.F., Helgason, H.G. et al. Higher heart rate and reduced heart rate variability persist during sleep in chronic fatigue syndrome: A population-based study. *Autonomic Neuroscience*, 137, 94–101.

Bourdouxhe, M.A., Queinnec, Y., Granger, D., Baril, R.H., Guertin, S.C., Massicotte, P.R. et al. (1999). Aging and shiftwork: The effects of 20 years of rotating 12-hour shifts among petroleum refinery operators. *Experimental Aging Research*, 25, 323–9.

Broughton, R. & Hasan, J. (1995). Quantitative topographic electroencephalographic mapping during drowsiness and sleep onset. *Journal of Clinical Neurophysiology*, 12, 372–86.

Brown, I. (1994). Driver fatigue. *Human Factors*, 36, 298–314.

Bultmann, U., Kant, I., Kasl, S., Beurskens, A.J.H.M. & van den Brandt, P.A. (2002). Fatigue and psychological distress in the working population: Psychometrics, prevalence and correlates. *Journal of Psychosomatic Research*, 52, 445–52.

Cajochen, C., Brunner, D.P., Kräuchi, K., Graw, P. & Wirz-Justice, A. (1995). Power density in theta/alpha frequencies of the waking EEG progressively increases during sustained wakefulness. *Sleep*, 18, 890–94.

Cajochen, C., Krauchi, K., von Arx, M-A., Mori, D., Graw, P. & Wirz–Justice, A. (1996). Daytime melatonin administration enhances sleepiness and theta/alpha activity in the waking EEG. *Neuroscience Letters*, 207, 209–13.

Caldwell, J.A., Hall, K.K. & Erickson, B.S. (2002). EEG data collected from helicopter pilots in flight are sufficiently sensitive to detect increased fatigue from sleep deprivation. *The International Journal of Aviation Psychology*, 12, 19–32.

Campagne, A., Pebayle, T. & Muzet, A. (2004). Correlation between driving errors and vigilance level: Influence of the driver's age. *Physiology and Behavior*, 80, 515–24.

Connor, J., Norton, R., Ameratunga, S., Robinson, E., Civil, I., Dunn, R. et al. (2002). Driver sleepiness and risk of serious injury to car occupants: Population based case control study. *British Medical Journal*, 324, 1125.

Craig, A., Tran, Y., Wijesuriya, N. & Boord, P. (2006). A controlled investigation into the psychological determinants of fatigue. *Biological Psychology*, 72, 78–87.

Delorme, A. & Makeig, S. (2004). EEGLAB: An open source toolbox for analysis of single-trial EEG dynamics. *Journal of Neuroscience Methods*, 134, 9–21.

Desmond, P.A. & Hancock, P.A. (2001). Active and passive fatigue states, in P.A. Hancock & P.A. Desmond (eds), *Stress, Workload, and Fatigue*. Mahwah, NJ: Lawrence Erlbaum Associates, 455–65.

Dirnberger, G., Duregger, C., Trettler, E., Lindinger, G. & Lang, W. (2004). Fatigue in a simple repetitive motor task: A combined electrophysiological and neurophysiological study. *Brain Research*, 1028, 26–30.

Dobbie, K. (2002). Fatigue–related crashes, an analysis of fatigue-related crashes on Australian roads using an operational definition of fatigue. Road safety research report OR 23. Canberra: Australian Transport Safety Bureau.

Dumont, M., Macchi, M., Piche, M.C., Carrier, J. & Hebert, M. (1997). Waking EEG activity during a constant routine: Effects of sleep deprivation. *Sleep Research*, 26, 712.

Eoh, H.J., Chung, M.K. & Kim, S.-H. (2005). Electroencephalographic study of drowsiness in simulated driving with sleep deprivation. *International Journal of Industrial Ergonomics*, 35, 307–20.

Fletcher, A., McCulloch, K., Baulk, S.D. & Dawson, D. (2005). Countermeasures to driver fatigue: A review of public awareness campaigns and legal approaches. *Australian and New Zealand Journal of Public Health*, 29, 471–6.

Frank, A.L. (2000). Injuries related to shiftwork. *American Journal of Preventive Medicine*, 18, 33–6.

Friedman, B.H. & Thayer, J.F. (1998). Autonomic balance revisited: Panic and heart rate variability. *Journal of Psychosomatic Research*, 44, 33–151.

Fuhrer, R. & Wessely, S. (1995). The epidemiology of fatigue and depression: A French primary care study. *Psychological Medicine*, 25, 895–905.

George, C.F.P. (2004). Sleep 5: Driving and automobile crashes in patients with obstructive sleep apnoea/hypopnoea syndrome. *Thorax*, 59, 804–7.

Gevins, A.S., Bressler, S.L., Cutillo, B.A., Illes, J. & Fowler-White, R.M. (1990). Effects of prolonged mental work on functional brain topography. *Electroencephalography and Cinical Neurophysiology*, 76, 39–350.

Golan, Z. & Neufield, M.Y. (1996). Individual differences in alpha rhythm as characterizing temperament related to cognitive performances. *Personality and Individual Differences*, 21, 775–84.

Gold, D.R., Rogacz, S., Bock, N., Tosteson, T.D., Baum, T.M., Speizer, F.E. et al. (1992). Rotating shift work, sleep, and accidents related to sleepiness in hospital nurses. *American Journal of Public Health*, 82, 1011–14.

Horne, J.A. & Reyner, L. (1996). Counteracting driver sleepiness: effects of napping, caffeine, and placebo. *Psychophysiology*, 33, 306–309.

Kecklund, G. & Åkerstedt, T. (1993). Sleepiness in long-distance truck driving: an ambulatory EEG study of night driving. *Ergonomics*, 36, 1007–17.

Knyazev, G.G. & Slobodskaya, H.R. (2003). Personality trait of behavioral inhibition is associated with oscillatory systems reciprocal relationships. *International Journal of Psychophysiology*, 48, 247–61.

Kopardekar, P. & Mital, A. (1994). The effect of different work–rest schedules on fatigue and performance of a simulated directory assistance operator's task. *Ergonomics*, 37, 1697–707.

Lal, S. & Craig, A. (2001a). A critical review of psychophysiology of driver fatigue. *Biological Psychology*, 55, 173–94.

Lal, S. & Craig, A. (2001b). Electroencephalography activity associated with driver fatigue: Implications for a fatigue countermeasure device. *Journal of Psychophysiology*, 15, 183–9.

Lal, S. & Craig, A. (2002). Driver fatigue: Electroencephalography and psychological assessment. *Psychophysiology*, 39, 313–21.

Lehmann, D., Grass, P. & Meier, B. (1995). Spontaneous conscious covert cognition states and brain electric spectral states in canonical correlations. *International Journal of Psychophysiology*, 19, 41–52.

Macchi, M.M., Boulos, Z., Ranney, T., Simmons, L. & Campbell, S.S. (2002). Effects of an afternoon nap on nighttime alertness and performance in long-haul drivers. *Accident Analysis Prevention*, 34, 825–34.

Matthews, G. & Desmond, P.A. (1998). Personality and multiple dimensions of task-induced fatigue: A study of simulated driving. *Personality and Individual Differences*, 25, 443–58.

Mills, P.J., Parker, B., Dimsdale, J.E., Sadler, G.R. & Ancoli-Israel, S. (2005). The relationship between fatigue and quality of life and inflammation during anthracycline-based chemotherapy in breast cancer. *Biological Psychology*, 69, 85–96.

Mitchell, R., Driscoll, T. & Healey, S. (2004). Work-related road fatalities in Australia. *Accident Analysis and Prevention*, 36, 851–60.

Pal, N.R., Chuang, C.-Y., Ko, L.-W., Chao, C.-F., Jung, T.-P., Liang, S.-F. et al. (2008). EEG-based subject and session independent drowsiness detection: An unsupervised approach. *EURASIP Journal on Advances in Signal Processing*, 1–11.

Papadelis, C., Kourtidou-Papadeli, C., Bamidis, P.D., Chouvarda, I., Koufogiannis, D., Bekiaris, E. et al. (2006). Indicators of sleepiness in an ambulatory EEG study of night driving. Proceedings of the 28th IEEE EMBS Annual International Conference, New York City, August 30–September 3.

Parliament of the Commonwealth of Australia, The (2000). *Beyond the Midnight Oil: An Inquiry into Managing Fatigue in Transport*. Canberra: CanPrint Communications Pty Ltd.

Roads Traffic Authority (2001). *Driver Fatigue: Problem Definition and Countermeasure Summary – 2001*. Sydney: Road and Traffic Authority.

Santamaria, J. & Chiappa, K.H. (1987). The EEG of drowsiness in normal adults. *Journal of Clinical Neurophysiology*, 4, 327–82.

Schier, M.A. (2000). Changes in EEG alpha power during simulated driving: A demonstration. *International Journal of Psychophysiology*, 37, 155–62.

Strijkstra, A.M., Beersma, D.G.M., Drayer, B., Halbesma, N. & Daan, S. (2003). Subjective sleepiness correlates negatively with global alpha (8–12 Hz) and positively with central frontal theta (4–8 Hz) frequencies in the human resting awake electroencephalogram. *Neuroscience Letters*, 340, 17–20.

Tanaka, H., Hayashi, M. & Hori, T. (1997). Topographical characteristics and principal component structure of the hypnagogic EEG. *Sleep*, 20, 523–34.

Torsvall, L. & Åkerstedt, T. (1987). Sleepiness on the job: Continuously measured EEG changes in train drivers. *Electroencephalography and Clinical Neurophysiology,* 66, 502–11.

Tran, Y., Craig, A. & McIsaac, P. (2001). Extraversion/Introversion and 8–13 Hz wave in frontal cortical regions. *Personality and Individual Differences*, 30, 205–15.

Tran, Y., Thuraisingham, R.A., Craig, A. & Nguyen, H.T. (2009). Evaluating the efficacy of an automated procedure for EEG artifact removal. 31st Annual International Conference of the IEEE Engineering in Medicine and Biology Society, Minneapolis, MN, September 2–6, 376–9.

Tran, Y., Thuraisingham, R.A., Wijesuriya, N., Nguyen, H. & Craig, A. (2007). Detecting neural changes during stress and fatigue effectively: A comparison of spectral analysis and sample entropy. Third International IEEE EMBS Conference on Neural Engineering, 350–53.

Tran, Y., Wijesuriya, N., Thuraisingham, R., Craig, A. & Nguyen, H. (2008). Increase in regularity and decrease in variability seen in electroencephalography (EEG) signals from alert to fatigue during a driving simulated task. 30th Annual International Conference of the IEEE Engineering in Medicine and Biology Society to be held in Vancouver, British Columbia, August 20–24, 1096–9.

Thuraisingham, R.A., Tran, Y., Craig, A., Wijesuriya, N. & Nguyen, H.T. (2009). Using microstate intensity for the analysis of spontaneous EEG: Tracking changes from alert to the fatigue state. 31tst Annual International Conference of the IEEE Engineering in Medicine and Biology Society, Minneapolis, Minnesota, September 2–6, 4982–85.

Trejo, L.J., Kochavi, R., Kubitz, K., Montgomery, L.D., Rosipal, R. & Matthews, B. (2005). EEG-based estimation of cognitive fatigue. Proceedings of SPIE, 5797, 105–15.

van der Linden, D., Frese, M. &. Meijman, T.F. (2003). Mental fatigue and the control of cognitive processes: Effects on perseveration and planning. *Acta Psychologica*, 113, 45–65.

Verwey, W.B. & Zaidel, D.M. (2000). Predicting drowsiness accidents from personal attributes, eye blinks and ongoing driving behaviour. *Personality and Individual Differences*, 28, 123–42.

Wijesuriya, N., Tran, Y. & Craig, A. (2007). The psychophysiological determinants of fatigue. *International Journal of Psychophysiology*, 63, 77–86.

Williamson, A.M., Feyer, A. & Friswell, R. (1996). The impact of work practices on fatigue in long-distance truck drivers. *Accident Analysis and Prevention*, 28, 709–19.

13

Cerebral Hemodynamic Indices of Operator Fatigue in Vigilance

Joel S. Warm, Lloyd D. Tripp, Gerald Matthews and William S. Helton

INTRODUCTION

Vigilance or sustained attention refers to the ability of organisms to maintain their focus of attention and to remain alert to stimuli over prolonged periods of time. That aspect of behavior is of considerable concern to specialists in applied psychology and human systems engineering because vigilance is a critical component of human performance in many work environments where automated systems are involved. These include military surveillance, air traffic control, cockpit monitoring, unmanned aerial vehicle control, airport, border, and homeland security, industrial process/quality control, and medical functions such as cytological screening and the inspection of anesthesia gauges during surgery (Warm, Parasuraman & Matthews, 2008). Failure to detect untoward events or critical signals in such situations can have disastrous consequences for system productivity and for public safety and health. Thus, it is important to uncover the factors that control vigilance performance. The quintessential finding in studies of vigilance is that sustained attention is fragile, waning quickly over time, a phenomenon known as the *vigilance decrement*. It takes the form of a decline in signal detections or a rise in response time (RT) to signal detections over the course of the vigilance session (Davies & Parasuraman, 1982; Dinges & Powell, 1985; Gunzelmann et al., 2008; Matthews et al., 2000; Warm, 1984). Investigators have frequently turned to physiological measures to understand the nature of the decrement. Consistent with a view proposed by Sir Charles Sherrington many years ago (Roy & Sherrington 1890), a considerable amount of current research on brain imaging indicates that there is a close tie between cerebral blood flow and neural activity in the performance of mental tasks (Raichle, 1998). Along that line, positive emission tomography (PET) and functional magnetic resonance imaging (fMRI) techniques have been successful in identifying several brain regions that are activated in the performance of vigilance tasks. The regions range from high level centers in the frontal cortex and the cingulate gyrus to lower centers including the thalamus and the reticular formation in the brainstem (Parasuraman, Warm & See, 1998). However, as Parasuraman et al. (1998) have noted, these studies are subject to two serious limitations. First, they have generally neglected to link the brain centers they have identified to performance efficiency, perhaps because of the high cost associated with using PET and fMRI during the prolonged running times (30 minutes or more) characteristic of vigilance research. Second, they feature environments that are noisy and restrictive, in which observers must remain nearly motionless during the experiment. Noise is one of several variables that can degrade vigilance performance (Hancock, 1984) and observers in vigilance experiments rarely remain motionless. Instead, they tend to fidget and do so increasingly over the course of the vigil (Galinsky et al., 1993). Consequently, imaging techniques are needed that will avoid these shortcomings and help bridge the gap in our understanding of brain systems and the vigilance decrement. Two such techniques are Transcranial Doppler Sonography (TCD), which measures cerebral blood flow velocity (CBFV) and an index of regional cerebral oxygen saturation (rSO$_2$) secured by means of Functional Near-Infrared Spectroscopy (fNIRS).

OVERVIEW OF MEASUREMENT TECHNIQUES

TCD is a noninvasive neuroimaging technique that employs ultrasound signals to gauge CBFV or hemovelocity in the middle, anterior, and posterior intracranial arteries. These arteries are readily isonated through a cranial transtemporal window and exhibit discernable measurement characteristics that facilitate their identification (Aaslid, 1986). As is illustrated in Figure 13.1, the TCD procedure uses a small 2-Mz pulsed Doppler transducer to gauge arterial blood flow. The transducer is placed just above the zygomatic arch along the temporal bone, a part of the skull that is functionally transparent to ultrasound. The depth of the pulse is adjusted until the desired intracranial artery is isonated. TCD measures the difference in frequency between the outgoing and reflected energy as the outgoing energy strikes moving erythrocytes. The low weight and the small size of the transducer and the ability to embed it conveniently in a headband worn by an observer permit real-time assessment of CBFV in both cerebral hemispheres without limiting or being hampered by body motion. Hence, TCD enables inexpensive, continuous, and prolonged monitoring of CBFV in both cerebral hemispheres concurrent with task performance. Blood flow velocities measured in centimeters per second are typically highest in the middle cerebral artery (MCA) which carries about 80 percent of the blood flow within each cerebral hemisphere (Toole, 1984). Consequently, CBFV in the left and right branches of the MCA was measured in the experiments to be discussed below. In these studies, CBFV was scored as a percentage of the last 60 seconds of a five-minute resting baseline as recommended in the TCD literature (Aaslid, 1986; Tripp & Warm, 2007).

When a particular area of the brain becomes metabolically active, as in the performance of mental tasks, by-products of this activity, such as carbon dioxide (CO$_2$), increase. This increase leads to a dilation of blood vessels serving that area, which in turn results in an increased blood flow to that region (Aaslid, 1986). Consequently, TCD permits measurement of changes in metabolic activity during task performance. Unlike the PET and fMRI procedures, TCD is limited in its spatial resolution. While it provides information about both cerebral hemispheres, it does not do so about specific brain loci. Nevertheless, it offers good temporal resolution (Aaslid, 1986), and compared to PET and fMRI, it can track rapid changes in blood flow in real time under less restrictive and invasive conditions. A wide variety of short-duration cognitive tasks elicit increases in CBFV in the MCA (Duschek & Schandry, 2003; Tripp & Warm, 2007). These TCD measured changes in CBFV are linked to the cognitive demand imposed by the tasks, implying that the CBFV

Transcranial Doppler Technique

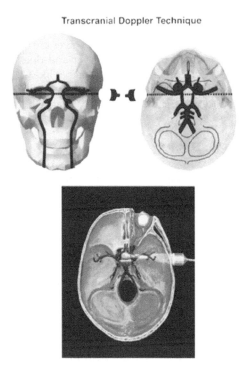

Figure 13.1   The transcranial Doppler technique, showing placement of the ultrasound transducer adjacent to the medial cerebral artery

response may reflect mobilization of information-processing resources (Tripp & Warm, 2007). Responses also tend to be lateralized, with verbal tasks eliciting stronger left-hemisphere responses, and spatial tasks increasing right-hemisphere CBFV (Stroobant & Vingerhoets, 2000). As described by Graton and Fabiani (2007), the fNIRS technique used to measure rSO$_2$ is based upon two major principles. Photons of near-infrared light (NIR) penetrate the skull and diffuse into brain tissue. The diffusion process is influenced by the scattering and absorption of properties of the tissue brought about by physiological events in the brain. By measuring the quantity of returning photons as a function of wavelength, one can infer the spectral absorption of the underlying tissue and draw conclusions about its average level of oxygenation (McCormick et al., 1991)

Human tissue is translucent to NIR photons having wavelengths between 650 and 1000nm. Even small amounts of colored material (chemophores) will cause wavelength-dependent absorption of the photons which produces characteristic signatures in the spectrum of the emerging light (Norris, 1977). The chemophores with the highest absorption in body tissue are the red-colored hemoglobin molecules found within the erythrocytes circulating in the blood. The measure of rSO$_2$ is accomplished by transmitting NIR optical wavelengths from a light-emitting diode through the subcutaneous tissue and the skull approximately 5 mm into cortical tissue (Eggert & Blazek, 1992). Two sensors, positioned 30mm and 40mm from the light source, respectively, collect the returning light that was not absorbed by the hemoglobin in the cortical tissue. Signals from the sensors are channeled to a computer, which calculates the oxygen saturation based upon the wavelengths of the returning photons of NIR light. The procedure is non-invasive and non-restrictive, since the sensors needed to emit and record returning light waves are housed in a headband and the recording equipment is isolated from the observer and does not restrict movement. Previous studies with the fNIRS technique have shown that like CBFV, rSO_2 is linked to the information processing demands of the task being performed (Franceschini & Boas, 2004; Gratton & Fabiani, 2007; Punwani et al., 1998; Toronov et al., 2001).

VIGILANCE AND CBFV

Using the TCD procedure, several experiments have revealed a close linkage between CBFV and vigilance performance. These studies, which are described in detail in Warm and Parasuraman (2007) and Warm et al. (2008), are summarized below. Although short tasks elevate CBFV, the vigilance studies, typically lasting 30–60 minutes, show that a temporal decline in the frequency of signal detections is paralleled by a *decline* in CBFV. This linkage is illustrated in Figure 13.2 from a study by Hitchcock et al. (2003), which examined the influence of the reliability of automation cues to impending signals on vigilance performance in a simulated air traffic control task wherein observers were asked to detect planes flying on a collision course over the center of a city. It is evident in the figure that the detection rate of critical signals was very high and remained stable over time with 100 percent cue-reliability, but declined over time in the remaining cue conditions so that by the end of the task, performance efficiency was best in the 100 percent condition followed in order by conditions in which cueing was only 80 percent or 40 percent reliable or was not available at all (no-cue condition). These performance effects were very closely mirrored by changes in CBFV.

In this study and all of those that have followed, the temporal decline in CBFV was restricted to conditions in which observers had to actively perform a vigilance task.

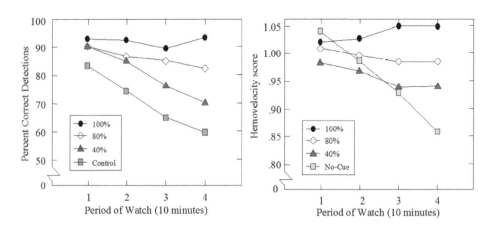

Figure 13.2   Mean percentages of correct detections (left panel) and mean CBFV values (right panel) as a function of periods of watch in four cue-reliability conditions (after Hitchcock et al., 2003)

Cerebral blood flow velocity remained stable over time in control observers who were asked to merely view the vigilance display without a work imperative. Results of this sort indicate that the changes in CBFV are task-determined and are not the result of spontaneous changes in arousal that may appear over time. Moreover, the temporal decline in CBFV occurs in a similar manner with both visual and auditory signals (Shaw et al., 2009) indicating that the effect is general in nature and not restricted to vision alone. The temporal decline in CBFV during the course of a vigil has been used to support a resource model of vigilance performance in which the vigilance decrement is considered to result from a decline in information-processing assets that are not replenished over time (Warm & Parasuraman, 2007; Warm et al., 2008; Warm, Matthews & Parasuraman, 2009). Consequently, the temporal decline in CBFV has important implications for "monitoring the monitor" and determining when an operator is in need of rest or replacement. These implications will be discussed more fully below.

Two additional elements of the CBFV/vigilance relation are important to note. One is that the findings related to CBFV have occurred predominantly in the right cerebral hemisphere. This result is consistent with prior findings with the fMRI and PET procedures (Parasuraman et al., 1998) and points to a right hemispheric system in the control of vigilance (but see Schultz et al., 2009).

The second element is that the time-dependent and hemispheric-dependent changes in CBFV are only observable when vigilance tasks are difficult enough for critical signals to be missed and correct detections are the measure of performance efficiency. A recent study by Funke and his associates (Funke et al., 2010a) has found that tasks in which signals are always detected and RT is the performance index do not show these time and laterality effects. In this study, observers accomplished a simulated air traffic control assignment involving the detection of a plane flying in the wrong direction in a flight of military aircraft. The monitored display featured what the authors designated as dynamic or static conditions. In the former case, the task was of sufficient difficulty perceptually for signals to be missed and performance efficiency was indexed by the percentage of correct detections. In the static case, differential discriminations were not required; observers simply responded whenever the flight of planes appeared on their monitored display and RT was the index of performance. The results of the study are displayed in Figure 13.3.

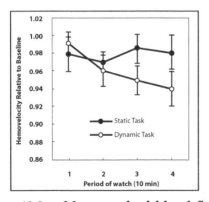 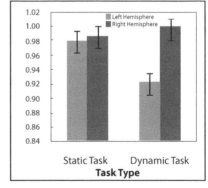

Figure 13.3   Mean cerebral blood flow velocity scores over time in the left and right hemispheres for the static and dynamic tasks – Error bars are standard errors (after Funke et al., 2010a)

Although both conditions showed a time-based reduction in performance efficiency, changes in CBFV were dramatically different in the two conditions. As can be seen in Figure 13.3, CBFV declined over time in the dynamic condition but not in the static condition, and the overall level of CBFV was greater in the right than the left hemisphere in the dynamic condition, while hemispheric disparities in overall CBFV were absent in the static condition. The CBFV findings in this study have two important messages for vigilance researchers. First, they indicate that a long-held belief in the literature that correct detections and RT are interchangeable indices of the same underlying process in vigilance (Buck, 1966) is not correct. Second, they serve to alert researchers to the need to consider the dynamic/static distinction in their models of vigilance performance. While the resource model may be applicable to dynamic-type tasks, other models involving elements such as criterion shifts in decisions to respond (Davies & Parasuraman, 1982), the negative effects of monotony (Larue, Rakotonirainy & Pettitt, 2010), or increased motor fatigue (Funke et al., 2010a) may be more suitable for static-type tasks.

CBFV AND THE DIAGNOSIS OF OPERATOR FATIGUE IN VIGILANCE

The studies described link declines in CBFV to the vigilance decrement in performance in group data. They also raise the possibility that CBFV may be diagnostic of fatigue in the individual operator. That is, CBFV might be used to test the operator's fitness to perform a fatiguing task prior to actual performance. In addition, monitoring CBFV during performance might provide a concurrent index of cognitive fatigue. The capacity to monitor the individual's level of fatigue might be useful in the range of operational environments listed in the introduction to this chapter (Reinerman-Jones et al., 2010).

Two recent TCD studies (Matthews et al., 2010) explored the potential utility of CBFV for diagnostic monitoring of the individual. These studies employed a dual-phase methodology in which psychophysiological responses to a cognitive challenge, the performance of a battery of short, high-workload tasks, were tested as predictors of a subsequently longer 36-min vigilance task consisting of four continuous 9-min watchkeeping periods. One study (N=187) made use of the Hitchcock et al. (2003) "sensory-based" task, whereas the other (N=107) used a more "cognitively based" task requiring the decoding of sequences of alphanumeric characters. It was hypothesized that the CBFV response to the short task battery would provide an index of individual differences in resource availability, which, in turn, would predict both types of vigilance tasks, consistent with the resource theory outlined above (Warm et al., 2008). CBFV was also recorded during vigilance performance itself. At the group level, CBFV responses were as expected. The short tasks elicited appropriately lateralized increases in CBFV, relative to baseline, similar to prior studies (e.g., Stroobant & Vingerhoets, 2000). For example, a line length discrimination task ("sensory" task) elicited a stronger right hemisphere response, whereas a working memory task requiring retention of word sequences produced a stronger left hemisphere response. During vigilance, declines in CBFV were seen in both hemispheres, for both tasks.

The first requirement for the diagnostic use of CBFV is that psychometrically sound indices of individual response are secured. Matthews et al. (2010) found that responses to the three short tasks used in the first phase of the studies were substantially intercorrelated, suggesting that individual differences generalize across different task types, consistent with a resource model. Indeed, left and right hemisphere responses correlated at about .40–.50. Factor analysis suggested that two correlated left- and right-hemisphere factors could be extracted, explaining about 70 percent of the response variance. Individual differences in CBFV during vigilance also appeared to be quite consistent across the four periods of work.

Consistent with prediction, CBFV response to the short battery correlated at about 0.3, with overall perceptual sensitivity on the vigilance task in both studies. The CBFV indices were also related to the temporal decrement in performance. These data suggest that individual differences in resources mobilized to meet a cognitive challenge provide an index of resource availability during vigilance. Furthermore, CBFV responses to the individual short-battery tasks were equally predictive of vigilance, an outcome that is critical on both psychometric and theoretical levels. Suppose that predictive validity depends on a match between the specific processes supporting the short task and those supporting the vigilance task. In that case, the CBFV response to the line length discrimination task should predict sensory vigilance more strongly than does the CBFV response to the working memory task. Conversely, response to the working memory task should be more highly correlated with cognitive vigilance. However, no such differences in prediction were found. This result was taken to support resource theory, in that the CBFV index appears to be diagnostic of the capacity to sustain attention irrespective of the task eliciting the CBFV response.

The Matthews et al. (2010) studies also showed that CBFV is positively correlated with subjective measures indicating low fatigue: task engagement and task-focused coping. As is further discussed by Matthews, Hancock and Desmond (Chapter 10, this volume), these state measures may provide additional indices of resource availability and utilization. One cautionary note should be sounded, though. CBFV was not significantly correlated with concurrent vigilance; individuals showing the largest decline in CBFV were not necessarily those showing the largest performance decrements. We take the concurrent CBFV measure to indicate a general vulnerability to loss of alertness, but it may be necessary to identify additional moderator factors that determine whether that vulnerability translates into actual performance loss.

As Reinerman-Jones et al. (2010) and Matthews et al. (2010) discuss, these findings suggest practical applications for diagnostic testing of the operator's fitness to perform. Should a fatigued soldier be sent on a further mission requiring sustained vigilance? His or her CBFV response to a short, high-workload task might provide a valid diagnostic index. Is a trucker who has slept badly for several nights fit to drive? Again, TCD might provide relevant evidence. The utility of TCD in the transportation context is also supported by data showing bilateral declines in CBFV during a monotonous drive (Reinerman et al., 2008). This study showed correlations between CBFV during the drive and higher task engagement. The robust nature of the CBFV decline in group data suggests that it remains promising as a concurrent index of loss of functional resources, but more research is needed to unravel the nature of individual differences in concurrent measures.

COMPARISON OF TCD AND FNIRS MEASURES IN RELATION TO VIGILANCE

There have been limited studies employing fNIRS during vigilance tasks, making comparisons between the fNIRS and the TCD measurement procedures tenuous. In a direct evaluation of the two hemodynamic indices, Helton and colleagues (Helton et al., 2007) compared them during an abbreviated 12-min vigilance task. Both measures showed higher levels of cerebral metabolic activity in the right than in the left cerebral hemisphere. This result was consistent with previous findings indicating right cerebral dominance during vigilance tasks. Although there was a significant decline in performance over time during the abbreviated vigilance task, there were no significant temporal changes in cerebral metabolic activity as indexed by the TCD and the fNIRS measures. This latter finding did not match the temporal decline in cerebral metabolic activity detected in long-duration vigils via the TCD measure as discussed in earlier sections of this chapter. Nevertheless, the consistent result regarding right hemispheric dominance between the two measures is encouraging.

While not directly comparing TCD and fNIRS, Li et al. (2009) studied fNIRS activity during prolonged vehicle driving. They examined 40 healthy male participants who were divided at random into two groups: an experimental driving group who performed a simulated drive for three hours and a control group who did not perform the drive. Unlike in other studies, Li and colleagues only monitored the left frontal lobe using fNIRS. While they detected significant increases in fNIRS levels at the beginning of the driving task compared with resting baseline values, they also found that the fNIRS level in the driving task group was significantly lower following the three-hour drive compared with that in the control group. Additionally, they reported a significant negative correlation in the experimental group between reaction time to critical signals and the suppressed fNIRS scores following the drive that was not present in the control group. A result of this sort suggests that the rSO_2 levels reflected in the fNIRS measure may be indicative of central fatigue. In another recent study comparing the TCD and fNIRS measurement techniques, Funke and his associates (Funke et al., 2010b) showed divergent results between the two measures during a 40-minute vigilance task. The TCD scores declined with time on-task, whereas the fNIRS scores became elevated over time; and the fNIRS scores were directly related to task difficulty while the TCD scores were unrelated to the difficulty of the task. The two measures were also largely uncorrelated in the study. While this may appear discouraging, there are at least two physiological reasons for the apparent inconsistency in these results. First, there may be differences in the precise brain areas generating the two different signals. Second, the fNIRS signal does not penetrate with much depth into cerebral tissue and thus is much more likely to pick up the metabolic activity in a smaller region of the cortex than would be the case with TCD signals recorded at the MCA, which are picking up most of the blood delivery to the entire brain. Undoubtedly, more studies integrating TCD, fNIRS, and other psychophysiological and imaging techniques should be employed in the future. Nevertheless, it should kept in consideration that both the measures from TCD and fNIRS are actually better physiologically understood than the BOLD signal from fMRI and, indeed, fNIRS is being used increasingly to verify the inferences made using fMRI (Raichle & Mintun, 2006).

CONCLUSIONS

Studies of vigilance provide one of the primary research paradigms for investigating the effects of cognitive fatigue on attention. Although there is a burgeoning cognitive neuroscience of vigilance (Parasuraman et al., 1998; Warm et al., 2009; Warm & Parasuraman, 2007; Warm et al., 2008), neurophysiological methods for tracking the temporal decrement in alertness have proved frustratingly elusive. The hemodynamic indices provided by TCD and fNIRS that we have reviewed appear especially promising as correlates of the performance decrement. Consistent with evidence from brain-imaging studies (Parasuraman et al., 1998), the vigilance decrement in signal detection appears to be reliably associated with declining CBFV in the right hemisphere. Importantly, the decline in CBFV is not simply an expression of reduced cortical arousal as the operator becomes increasingly fatigued. Simply watching task displays without attempting to respond does not affect CBFV (e.g., Hitchcock et al., 2003). Changes in CBFV appear to be more tightly coupled with changes in information-processing than are conventional neurophysiological indices of autonomic and cortical arousal.

Right-hemisphere CBFV depends on the same workload parameters as does performance. That is, the cognitive fatigue induced by high workloads on monotonous tasks depresses both signal detection and CBFV. Both performance and CBFV decrements may be attributed to resource depletion, consistent with vigilance theory (Warm et al., 2008). Resource theory has also proved useful as a framework for understanding how individual differences in CBFV relate to individual differences in performance (Matthews et al., 2010). At the level of the individual, however, the phasic increase in CBFV induced by demanding tasks appears to be more diagnostic of performance than CBFV recorded concurrently with vigilance.

The utility of TCD-determined CBFV as a diagnostic index of alertness suggests that other hemodynamic measures, especially fNIRS measures of rSO_2 should be similarly related to vigilance. In fact, the limited evidence available suggests that fNIRS measures may be indicative of processes that are ancillary to signal detection such as compensatory effort (Funke et al., 2010a). Further research is needed on these techniques, and on their relationships to the regional hemodynamic responses measured by fMRI.

At a practical level, CBFV as measured by the TCD procedure is potentially a useful addition to diagnostic tests for operator fatigue (Matthews et al., 2010; Reinerman-Jones et al., 2010). Its application is most easily justified for tasks for which vigilance is central, such as military surveillance, industrial inspection, or other tasks requiring extended monitoring of displays. However, researchers are beginning to explore the utility of TCD-determined CBFV in additional performance contexts including vehicle operation (Reinerman et al., 2008), making shoot/don't shoot decisions (Schultz et al., 2009), and command and control environments (de Visser et al., 2010). Further progress in the use of CBFV for diagnostic monitoring may depend on a more detailed theoretical articulation of its relationship with resource allocation processes.

REFERENCES

Aaslid, R. (1986). Transcranial Doppler examination techniques, in R. Aaslid (ed.), *Transcranial Doppler Sonography*. New York: Springer-Verlag, 39–59.

Buck, L. (1966). Reaction time as a measure of perceptual vigilance. *Psychological Bulletin*, 65, 291–304.

Davies, D.R. & Parasuraman, R. (1982). *The Psychology of Vigilance*. London: Academic Press.

de Visser, E., Guagliardo, L., Parasuraman, R. & Shaw, T. (2010). Using Transcranial Doppler Sonography to measure cognitive load in a command and control task. *Proceedings of the Human Factors and Ergonomics Society*, 54, 249–53.

Dinges, D.F. & Powell, J.W. (1985). Microcomputer analysis of performance on a portable, simple visual RT task during sustained operations. *Behavior Research Methods, Instruments and Computers*, 17, 652–5.

Duschek, S. & Schandry, R. (2003). Functional transcranial Doppler sonography as a tool in psychophysiological research. *Psychophysiology*, 40, 436–54.

Eggert, H.R. & Blazek, V. (1992). Optical properties of human brain tissue, meninges, and brain tumors in spectral range of 200–900nm. *Neurosurgery*, 76, 315–18.

Franceschini, M.A. & Boas, D.A. (2004). Noninvasive measurement of neuronal activity with near-infrared optical imaging. *NeuroImage*, 21, 372–96.

Funke, M., Warm, J.S., Matthews, G., Finomore, V., Vidulich, M., Knott, B., Helton, W., Shaw, T. & Parasuraman, R. (2010a). Static and dynamic discrimination in vigilance: Effects on cerebral hemodynamics and workload, in M. Tadeusz, W. Karwowski & V. Rice (eds), *Advances in Understanding Human Performance: Neuroergonomics, Human Factors Design, and Special Populations*. Boca Raton, FL: CRC Press, 80–90.

Funke, M.E., Warm, J.S., Matthews, G., Riley M., Finomore, V., Funke, G.J., Knott, B.A. & Vidulich, M.A. (2010b). A comparison of cerebral hemovelocity and blood oxygen saturation levels during vigilance performance. *Proceedings of the Human Factors and Ergonomics Society*, 54 (September), 1345–49.

Galinsky, T.L., Rosa, R.R., Warm, J.S. & Dember, W.N. (1993). Psychophysical determinants of stress in sustained attention. *Human Factors*, 35, 603–14.

Gratton, G. & Fabiani, M. (2007). Optical imaging of brain function, in R. Parasuraman & M. Rizzo (eds), *Neuroergonomics: The Brain at Work*. New York: Oxford University Press, 65–81.

Gunzelmann, G., Moore, R.L., Gluck, K.A., Van Dongen, H.P.A. & Dinges, D.F. (2008). Individual differences in sustained vigilant attention: Insights from conceptual cognitive modeling. Proceedings of the 13th Annual Meeting of the Cognitive Science Society, Austin, TX, 2017–22.

Hancock, P.A. (1984). Environmental stressors, in J.S. Warm (ed.), *Sustained Attention in Human Performance*. Chichester: Wiley, 103–42.

Helton, W.S., Hollander, T.D., Warm, J.S., Tripp, L.D., Parsons, K.S., Matthews, G., Dember, W.N., Parasuraman, R. & Hancock, P.A. (2007). The abbreviated vigilance task and cerebral hemodynamics. *Journal of Clinical and Experimental Neuropsychology*, 29, 545–52.

Hitchcock, E.M., Warm, J.S., Matthews, G., Dember, W.N., Shear, P.K., Tripp, L.D., Mayleben, D.W. & Parasuraman, R. (2003). Automation cueing modulated cerebral blood flow and vigilance in a simulated air traffic control task. *Theoretical Issues in Ergonomic Science*, 4, 89–112.

Larue, G.S., Rakotonirainy, A. & Pettitt, A.N. (2010). Real-time performance modeling of a sustained attention to response task. *Ergonomics*, 53, 1205–16.

Li, Z., Zhang, M., Zhang, X., Dai, S. Yu, X. & Wang, Y. (2009). Assessment of cerebral oxygenation during prolonged simulated driving using near infrared spectroscopy: Its implications for fatigue development. *European Journal of Applied Physiology*, 107, 281–7.

Matthews, G., Davies, D.R., Westerman, S.J. & Stammers, R.B. (2000). *Human Performance: Cognition, Stress and Individual Differences*. Hove: Psychology Press.

Matthews, G., Warm, J.S., Reinerman, L.E., Langheim, L., Washburn, D.A. & Tripp, L. (2010). Task engagement, cerebral blood flow velocity, and diagnostic monitoring for sustained attention. *Journal of Experimental Psychology: Applied*, 16, 187–203.

McCormick, P.W., Stewart, M., Goetting, M.G., Dujovny, M., Lewis, G. & Ausman, J.I. (1991). Noninvasive cerebral optical spectroscopy for monitoring cerebral oxygen delivery and hemodynamics. *Critical Care Medicine*, 19, 89–97.

Norris, K.H. (1977). Light is transmitted through human tissue, in K.C. Smith (ed.), *The Science of Photobiology*. New York: Plenum, 400–409.

Parasuraman, R., Warm, J.S. & See, J.E. (1998). Brain systems of vigilance, in R. Parasuraman (ed.), *The Attentive Brain*. Cambridge, MA: MIT Press, 221–56.

Punwani, S., Ordidge, R.J., Cooper, C.E., Amess, P. & Clemence, M. (1998). MRI measurements of cerebral deoxyhaemoglobin concentration (dhB) – correlation with near-infrared spectroscopy (NIRS). *NMR in Biomedicine*, 11, 281–9.

Raichle, M.E. (1998). Behind the scenes of functional brain imaging: A historical and physiological perspective. *Proceedings of the National Academy of Sciences USA*, 95, 765–72.

Raichle, M.E. & Mintun, M.A. (2006). Brain work and brain imagining. *Annual Review of Neuroscience*, 29, 449–76.

Reinerman, L.E., Warm, J.S., Matthews, G. & Langheim, L.K. (2008). Cerebral blood flow velocity and subjective state as indices of resource utilization during sustained driving. *Proceedings of the Human Factors and Ergonomics Society*, 52, 1252–6.

Reinerman-Jones, L.E., Matthews, G., Warm, J.S. & Langheim, L.K. (2010). Selection for vigilance assignments: A review and proposed new direction. *Theoretical Issues in Ergonomics Science*, 11, 1–24.

Roy, C.S. & Sherrington, C.S. (1890). On the regulation of the blood supply of the brain. *Journal of Physiology* (London), 11, 85–108.

Schultz, N.B., Matthews, G., Warm, J.S. & Washburn, D.A. (2009). A transcranial sonography study of shoot/don't shoot responding. *Behavior Research Methods*, 41, 593–7.

Shaw, T.H., Warm, J.S., Finomore, V.S., Tripp, L., Matthews, G., Weiler, E.M., Dember, W.N. & Parasuraman, R. (2009). Effects of sensory modality on cerebral blood flow velocity during vigilance. *Neuroscience Letters*, 461, 207–11.

Stroobant, N. & Vingerhoets, G. (2000). Transcranial Doppler ultrasonography monitoring of cerebral hemodynamics during performance of cognitive tasks: A review. *Neuropsychology Review*, 10, 213–31.

Toole, J.F. (1984). *Cerebral Vascular Disorders*. Third edition. New York: Raven.

Toronov, V., Webb, A., Choi, J.H., Wolf, M., Michalos, A., Gratton, E. et al. (2001). Investigation of human brain hemodynamics by simultaneous near-infrared spectroscopy and functional magnetic resonance imaging. *Medical Physics*, 28, 521–7.

Tripp, L.D. & Warm, J.S. (2007). Transcranial Doppler sonography, in R. Parasuraman & M. Rizzo (eds), *Neuroergonomics: The Brain at Work*. New York: Oxford University Press, 82–94.

Warm, J.S. (1984). An introduction to vigilance, in J.S. Warm (ed.), *Sustained Attention in Human Performance*. Chichester: Wiley, 1–14.

Warm, J.S. & Parasuraman, R. (2007). Cerebral hemodynamics and vigilance, in R. Parasuraman & M. Rizzo (eds), *Neuroergonomics: The Brain at Work*. New York: Oxford University Press, 146–58.

Warm, J.S., Matthews, G. & Parasuraman, R. (2009). Cerebral hemodynamics and vigilance performance. *Military Psychology*, 21, S75–S100.

Warm, J.S., Parasuraman, R. & Matthews, G. (2008). Vigilance requires hard mental work and is stressful. *Human Factors*, 50, 433–51.

14

Biochemical Indices of Fatigue for Anti-fatigue Strategies and Products

Yasuyoshi Watanabe, Hirohiko Kuratsune and Osami Kajimoto

INTRODUCTION

Although many people are suffering from chronic fatigue, which is characterized as experiencing symptoms of fatigue lasting longer than six months, integrated research on fatigue has not yet been systematically organized. Fatigue is a sensation which all people have experienced and it is therefore a popular phenomenon for us, but its molecular and neural mechanisms have not been elucidated yet, probably because of the complicated nature of their origins. However, we know the decrease in the efficiency of our performance on tasks in studies of fatigue and also the decrease in motivation that accompanies loss of performance. It is of great value in our modern society to extensively analyze the causes of fatigue and motivation loss and to develop quantitative methods for assessment of fatigue and motivation so as to invent methods and therapies for better recovery and avoidance of severe chronic fatigue. The economic gain is really substantial if our chronic fatigue is somehow cured. To our regret and surprise, almost all commercially available goods for recovery from fatigue are not proven by scientific and medical evidence. We therefore organized the integrated research project on "The molecular/neural mechanisms of fatigue and fatigue sensation and the way to overcome chronic fatigue" with a grant from the Ministry of Education, Culture, Sports, Science, and Technology (MEXT) of the Japanese Government from 1999 to 2005. The major contributions are (see Watanabe et al., 2008):

1. Elucidation of the brain regions and their neurotransmitter systems responsible for fatigue sensation and chronic fatigue.
2. Development of a variety of methods and scales to quantitatively evaluate the extent of fatigue.
3. Development of animal models of different causes of fatigue.
4. Elucidation of molecular/neural mechanisms of fatigue in humans and animals.
5. Invention of various methods or therapies for chronic fatigue and chronic fatigue syndrome.

Then, to develop the foods and drugs to overcome fatigue, we have made efforts under the twenty-first century Center of Excellence (COE) program "Establishment of the COE to overcome fatigue" (2004–2009) under the MEXT Japanese Government program. More recently, we included cohort design studies testing motivation, especially learning motivation, in school-age girls and boys. Here, we would like to summarize the biochemical indices or biomarkers for the fatigue and motivation research, and perspectives on remedies of chronic fatigue and chronic fatigue syndrome through evaluations using the biomarkers.

If fatigue is generated by the biochemical pathways of the body, it represents an outcome of an elaborate series of molecular reactions. Thus, a chapter of this kind must necessarily describe the numerous molecules and chemical processes involved, which may be unfamiliar to our readers. As we discuss in section 2 below, a thorough comparison of biochemical markers in fatigued and non-fatigued individuals shows differences in concentrations of a multitude of different molecules. As an introduction, we will here briefly outline the major perspectives on understanding these complex data. The first perspective is to see fatigue as an outcome of the basic metabolic process of catabolism, which generates physical energy from nutrients in food, including carbohydrates and lipids (fats). A key process here is the tricarboxylic acid (TCA) cycle, which occurs in the mitochondria, the energy-generation parts of the cell. The TCA cycle is a series of chemical reactions that converts nutrients into energy (along with water and CO_2). The energy generated by the TCA cycle is used in a subsequent oxidative chemical process (oxidative phosphorylation) that produces adenosine triphosphate (ATP), the molecule that allows energy to be transported around the cell. One hypothesis we will briefly examine (section 2) relates to this oxidative phosphorylation process. It may generate molecules known as reactive oxygen species, including free radicals, which may damage the cell. Prolonged exposure to oxidative stress may cause fatigue. A related hypothesis is that cofactors of metabolism, the enzymes that are required to support processes including energy generation and cell repair, may themselves influence fatigue (section 4). Metabolism may also be affected by the "stress" hormones regulated by the hypothalamic–pituitary–adrenal (APA) axis of the brain. Hormones released from the pituitary may influence enzymes such as carnitine acetyltransferase that participate in energy-generation reactions in the mitochondria (section 5).

A rather different perspective on fatigue comes from studies of the immune system, which protects the body against infections caused by viruses and bacteria. The importance of the immune system is suggested by observations that chronic fatigue may follow viral infection, or a past history of allergies. Substances known as cytokines (e.g., interferon) regulate the activity of the immune system, either stimulating or suppressing immune response. Broadly, prolonged activation of the immune system may be linked to chronic fatigue. In section 6, we examine abnormalities in some of the specific biochemical pathways that have been implicated in fatigue.

We need also to understand dysfunctions in the brain that may follow from metabolic and immune system mechanisms for fatigue (section 7). Evidence may be obtained from studies that image regional brain metabolism and brain anatomy. These studies suggest that serotonergic pathways in frontal regions such as the anterior cingulate may be especially important for fatigue. Multiple neurotransmitters are likely to be involved in fatigue, but the importance of serotonin is suggested by its role in wellbeing and the link between depression and low serotonin levels.

In sum, the biochemistry of fatigue may be understood at different levels of physiology including intracellular metabolic processes, hormonal responses to stress, immunological functioning and neurological pathways. In the sections that follow, we will fill out some details of the different biochemical processes contributing to fatigue, with an emphasis on chronic fatigue syndrome.

1. PAST HYPOTHESES AND PITFALLS

Previously, researchers (Hill & Kupalov, 1929) mentioned the biochemical index for the extent of physical fatigue to be lactate in the plasma, but now we know that lactate is not a fatigue-inducing substance. In fact, lactate level in the plasma and skeletal muscles increases during exercise, especially in anaerobics. However, pH change caused by this elevation is within the range of very slight acidosis (pH 7.4 to pH 6.9~7.2), because of the endogenous buffer action. In experiments, we did the administration of bufferized lactate and citrate in rats at a dose of such an elevation level or 10 times more, but we could not find any harmful effects. We especially found no decrease in spontaneous motor activity (Tanaka & Watanabe, 2008b). Instead, citrate enhanced their spontaneous motor activity or performance of exercise. Citrate is one of the compounds that regulates the generation of energy by the TCA cycle. Lactate is a substantial energy source in neurons and also in muscle cells for glucose, and supply of lactate increases the activity of neurons and muscles. Even a slight acidosis helps the muscle contractile activity (Pedersen et al., 2004). Thus, lactate is rather an indicator of severity of exercise, and an important energy source in the recovery phase from fatigue.

In order to explain the central fatigue during rather severe physical tasks, the amino acids and serotonin hypothesis has been proposed by Newsholme & Blomstrand (1995). A decrease in plasma levels of branched-chain amino acids (BCAA; leucine, valine, and isoleucine) may be led by consumption for protein synthesis and energy production, which results in less competition of L-type amino acid transporter with aromatic amino acids. Free L-tryptophan liberated from serum albumin by increased amount of fatty acids during severe exercise is therefore taken up more into the brain, which stimulates further production of serotonin, resulting in induction of sleep and the demand of rest.

However, in our human study, with a four-hour fatigue load (Nozaki, Tanaka et al., 2009), the plasma level of BCAA increased, probably because of the request by the brain from liver or peripheral organs, or because of the proteolytic degradation of proteins. Such phenomena, i.e., increase in plasma level of BCAA and also other amino acid dynamics, could be observed in a similar pattern even in the exhaustion model used in animals (Jin et al., 2009), as is shown in Figure 14.1.

The discrepancy between the results is most likely explained by the difference of the phase or duration of fatigue, which suggests that the amino acid level is not a proper biomarker for the extent of fatigue. The serotonin elevation hypothesis of fatigue sensation is also not supported, since most of the animal model studies (Tanaka & Watanabe, 2008a), and a human Positron Emission Tomography (PET) study on patients with Chronic Fatigue Syndrome (CFS) (Yamamoto et al., 2004), demonstrated the decrease of serotonin and deterioration of serotonergic systems. Tryptophan elevation theory is now interpreted as the increase in indole ring-cleaved metabolite, neurotoxic metabolite, quinolinate in the brain, which could attack the glutamatergic postsynaptic component and might induce inactivation of neurons.

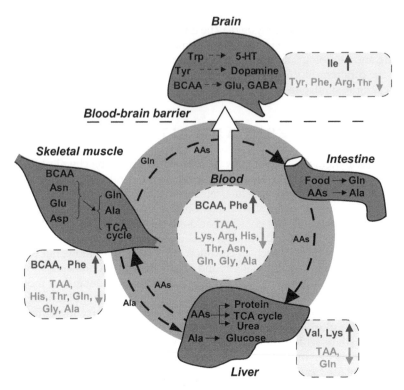

Figure 14.1   Changes in amino acid levels in the plasma and tissues by severe fatigue (TAA, total amino acids; AAs, amino acids; BCAA, branched chain amino acids)

In CFS patients, representing prolonged and severe fatigue cases, several immunological and biochemical biomarkers were reported through a variety of studies, including our own. However, Mathew et al. (2009) demonstrated an increase in lactate signal in lateral cerebral ventricular space of CFS patients as revealed by Magnetic Resonance Spectroscopy (MRS). The raised lactate concentration is characteristic in CFS, as compared to individuals with generalized anxiety disorder and healthy volunteers. They commented that CFS is associated with significantly raised concentrations of ventricular lactate, potentially consistent with recent evidence of decreased cortical blood flow, secondary mitochondrial dysfunction, and/or oxidative stress abnormalities in the disorder. Concerning the concentrations of plasma amino acids in CFS patients, no specific pattern has so far been observed.

2. MENTAL AND PHYSICAL FATIGUE-RELATED BIOCHEMICAL ALTERATIONS

In our research, an acute fatigue load of four hours of mental and physical work was applied to investigate the biomarkers and remedies of fatigue. In order to confirm the fatigue-

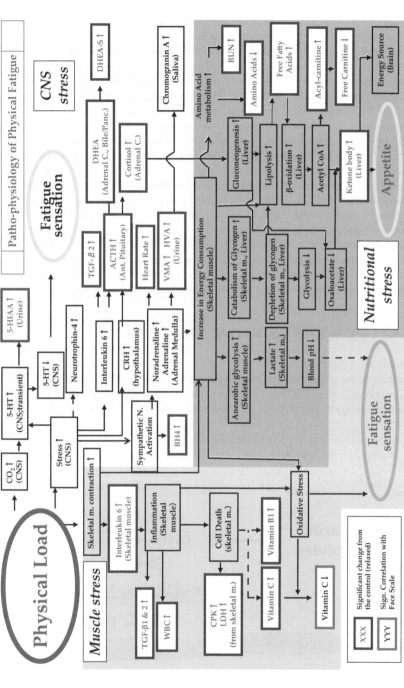

Figure 14.2 One typical example of biochemical alteration map (interaction map) of fatigue after 4-hour physical load as compared with relaxed 4-hour state (see Nozaki et al., 2009) in a randomized crossover fashion

Note: Neuro-immuno-endocrinological perturbation could be demonstrated.

related biochemical alterations, we tested more than 100 biochemical, immunological, and endocrinological factors in a pilot study with two-, four-, and eight-hour loads, and finally selected various parameters for measurement just before and after relaxation and fatigue-inducing mental or physical sessions. Subsequently, 54 healthy volunteers were randomized to perform relaxation and fatigue-inducing mental and physical sessions for four hours in a double-blind, 3-crossover design. Before and after each session, subjects were asked to rate their subjective sensations of fatigue. In addition blood, saliva, and urine samples were taken. After the fatigue-inducing mental and physical sessions, subjective scores of fatigue increased. After the fatigue-inducing mental session, vanillylmandelic acid level in the urine was higher and plasma valine level was lower than after the relaxation session. By contrast, after the fatigue-inducing physical session, serum citric acid, triacylglycerol, free fatty acid, ketone bodies, total carnitine, acylcarnitine, uric acid, creatine kinase, aspartate aminotransferase, lactate dehydrogenase, cortisol, dehydroepiandrosterone, dehydroepiandrosterone sulfate, plasma branched-chain amino acids, transforming growth factor-beta 1 and 2, white blood cell and neutrophil counts, saliva cortisol and amylase, and urine vanillylmandelic acid levels were higher and serum free carnitine and plasma total amino acids and alanine levels were lower than those of the relaxation session. Figure 14.2 shows the summary of the results from the physical fatigue as compared with the relaxation state, as a kind of correlation map. Thus, fatigue provokes a complex pattern of biochemical changes that may reflect a variety of physiological processes.

3. OXIDATIVE STRESS BIOMARKERS

As has previously been mentioned, metabolic processes may produce damaging oxidative molecules such as free radicals. 8-Oxo-deoxy-guanosine (8-OH-dG) (Asami et al., 1998; Shigenaga, Gimeno & Ames, 1989) and 8-isoprostane (Roberts & Morrow, 1994) in the tissues and urine are good biomarkers of moderate to severe exercise fatigue, due to

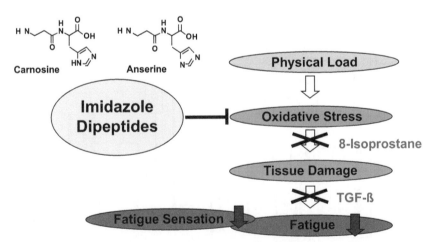

Figure 14.3 **One typical example of the results from the developmental study on anti-fatigue substance and food components**

Note: Imidazole dipeptides showed very good anti-fatigue and anti-fatigue sensation effects with clear evidence on mechanisms, through double-blind, placebo-controlled, crossover trial with 20 healthy volunteers.

DNA damage and lipid peroxidation, respectively. Lipid peroxidation refers to chemical changes in lipids such as fatty acids, which may lead to oxidative cell damage. We utilized both 8-OH-dG and 8-isoprostane for the development of an anti-fatigue food or drink component. For example, imidazole dipeptides, carnosine and anserine showed great anti-fatigue effects if they were taken before exercise, through the mechanisms of reduction of oxidative stress and immunosuppressive cytokine production, which resulted in lowering performance loss and fatigue sensation (Tanaka, Shigihara et al., 2008, as shown in Figure 14.3. (Cytokines are proteins important for communication between cells that modulate the functioning of the immune system, among other functions). Ascorbic acid (Vitamin C) and vitamin E levels also decreased in the peripheral tissues, mostly in the liver and kidney, after the fatigue load in animals, but their concentrations in the plasma and brain were constant, probably due to a compensatory process associated with the peripheral tissues.

4. ENERGY DEMAND FOR REPAIR OF DAMAGED CELLS AND CELLULAR COMPONENTS WITH VITAMINS

Much more energy, supplied by ATP, is required for the recovery from tissue, cell, DNA, membrane phospholipids, and protein damage than for routine "housekeeping" energy. Recently, our metabolome analysis showed less concentration of the metabolites of energy production in an animal model of exhaustion and also in CFS patients. (Metabolome refers to the set of metabolites found within the sample analyzed). Details of the results will be published soon.

There is also a requirement for co-enzymes and co-factors for the TCA cycle and electron transfer system, such as thiamine pyrophosphate (VB1) and Coenzyme Q10, which is evident during the recovery phase of fatigue. Actually, in our study, we demonstrated the effects of such co-factors for better recovery or better protection from fatigue (Nozaki, Mizuma et al., 2009; Mizuno et al., 2008).

We are now routinely using the biochemical/immunological biomarkers related to metabolic processes such as oxidative stress biomarkers in the urine, plasma cytokines, and citrate and TCA cycle intermediates, viral DNA in saliva (mentioned beneath), in combination with physiological biomarkers with autonomic nerve function and behavioral analysis, for evaluation and development of anti-fatigue substances and instruments (Tanaka & Watanabe, 2008a, Tanaka et al. 2008b, & 2009; Ataka et al., 2007 & 2008; Tajima et al., 2008; Mizuma et al., 2009; Mizuno et al., 2010).

5. HYPOTHALAMO–PITUITARY–ADRENAL (HPA) DYSFUNCTION AND RELATED METABOLIC ABNORMALITIES

In 1991, Demitrack et al. (1991) reported the impaired activation of the HPA in patients with CFS, suggesting a possible role for stress hormones. Thereafter, several investigators addressed HPA dysfunction in patients with CFS, including lower basal plasma cortisol levels, reduced salivary cortisol levels, lower ACTH response in insulin tolerance test and psychosocial stress test, reduced ACTH responses to CRH, and prolonged suppression of salivary free cortisol in the low-dose dexamethasone suppression test (Roberts et al., 2004; Gaab, Hüster et al., 2002; Gaab, Engert et al., 2004). Kuratsune, Yamaguti, Sawada et al. (1998) also found that the majority of Japanese patients with CFS had a deficiency in serum dehydro-epiandrosterone-sulfate (DHEA-S). Serum DHEA-S is one of the most abundantly produced hormones secreted from the adrenal glands, and its physiological role is thought to be a precursor of sex steroids. However, DHEA-S itself was recently shown to have physiological properties, acting as a neurosteroid associated with such psychophysiological phenomena as memory, stress, anxiety, sleep, and depression. Therefore, the deficiency in DHEA-S might be related to the neuropsychiatric symptoms in patients with CFS. There is also a possibility that the DHEA-S deficiency is associated with the increased serum level of transforming growth factor beta (TGF-beta).Recently, Kuratsune, Yamaguti, Takahashi et al. (1994) also found that most Japanese patients with CFS showed a low level of serum acetyl-L-carnitine, which is well correlated with the subjective rating score of fatigue, and that a considerable amount of the acetyl moiety of serum acetyl-L-carnitine is taken up into the brain (Yamaguti et al., 1996). Carnitine is necessary for the metabolism of fatty acids, in producing energy via the TCA cycle previously mentioned. As is mentioned earlier DHEA-S is known to regulate the activity of carnitine acetyltransferase (Chiu et al., 1997). Therefore, the decrease in DHEA-S might play an important role in endogenous acetyl-L-carnitine deficiency in serum. Indeed, when Kuratsune et al. (1994) administered DHEA-S to patients with CFS, an apparent increase in serum acetyl-L-carnitine was found. It was also found that the acetyl moiety taken up into the brain through acetyl-L-carnitine is mainly utilized for the biosynthesis of glutamate (Kuratsune, Yamaguti, Lindh et al., 2002). Thus, this metabolic abnormality (i.e., low uptake of acetyl-L-carnitine into the brain) might cause some of the secondary brain dysfunction in CFS. In Figure 14.4 we illustrate our hypothesis of brain dysfunction with chronic fatigue or chronic fatigue syndrome.

More recently, a-MSH was characterized as a biomarker of severe fatigue. a-MSH is one of several peptide hormones produced by the pituitary gland that influence psychological functioning. Ogawa et al. employed an animal model of exhaustion (Tanaka M, Nakamura et al., 2003), and found (Ogawa et al., 2005 & 2009) the atrophy of thymus and spleen, and also found out a dramatic increase in plasma a-MSH level, and then death of cells in the intermediate lobe of the pituitary (melanotroph cells). Furthermore, Shishioh-Ikejima et al. (2010) found an increase of plasma a-MSH level in patients in the early stage of CFS. These increases in a-MSH might be somehow related to the feedback system of the proopiomelanocortin (POMC) gene expression. POMC is a precursor to various hormones produced by the pituitary.

6. IMMUNOLOGICAL ABNORMALITIES

It is well known that the prevalence of having a past history of allergies is high in patients with CFS. Furthermore, CFS patients were reported to have many immunological abnormalities of various types, such as low natural killer cell function, abnormality of T cell population, elevated levels of several kinds of cytokines, the presence of antinuclear antibody, an increased level of immune complexes, and abnormality of the RNase-L pathway (Klimas et al., 1990; Bennet et al., 1997; Moss, Mercandetti & Vojdani, 1999;

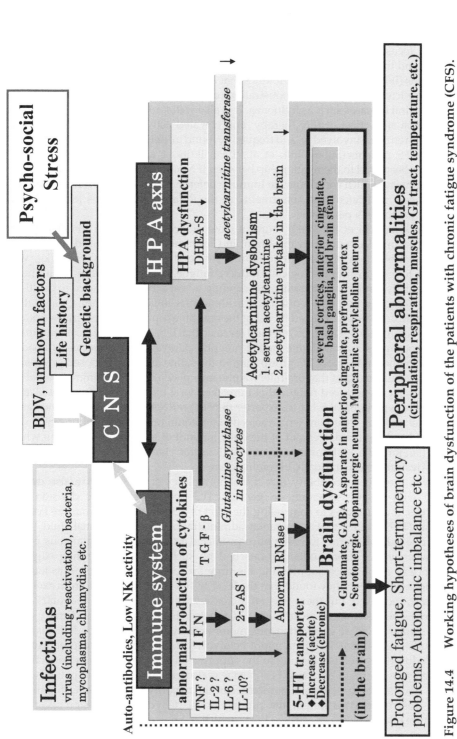

Figure 14.4   Working hypotheses of brain dysfunction of the patients with chronic fatigue syndrome (CFS).

Note: Kuratsune and Watanabe proposed this figure from the data and literatures collected.

Suhadolnik et al., 1994; Ikuta et al., 2003; Demettre et al., 2002). Among these abnormalities, we are now focusing on the elevation of several kinds of cytokines, the presence of auto-antibodies, and the abnormality of the RNase-L pathway.

Interferons (IFNs) are a class of cytokines that are produced by the immune system in response to microbial invasion. Administration of IFN has been used as an experimental therapy for viral diseases such as influenza. It is also well known that flu-like symptoms are a common side effect of IFN therapy; and elevated activity of 2',5'-oligoadenylate synthetase, an enzyme involved in, and frequently found in, peripheral blood mononuclear cells from CFS patients (Suhadolnik et al., 1994; Ikuta et al., 2003). Therefore, much attention has been paid to the relationship between IFN and the pathogenesis of CFS. The RNase-L pathway, which contributes to immune defense by degrading viral and cellular RNA, has been implicated. The abnormality of the RNase-L pathway is located in the downstream of the IFN pathway.

Recently, Katafuchi et al. (2003) found a close association between the changes in the IFN-alpha mRNA content in the brain and immunologically induced fatigue. An intraperitoneal injection of a synthetic double-stranded RNA, poly I:C 3 mg/kg, was given to rats to produce immunologically induced fatigue, i.e., by stimulating the immune system. The daily amounts of spontaneous running wheel activity decreased to ca. 40–60 percent of the preinjection level until day 9, with normal circadian rhythm. Quantitative analysis of mRNA levels conducted by using the real-time capillary reverse transcriptase-polymerase chain reaction (RT-PCR) method revealed that IFN-alpha mRNA contents in the cortex, hippocampus, hypothalamic medial preoptic, paraventricular, and ventromedial nuclei were higher in the poly I:C group than in the saline and heat-exposed groups on day 7. These results suggest that brain IFN-alpha may play a role in the animal model for the immunologically induced fatigue that mimicked viral infection. Katafuchi et al. also found that the expression of 5-HT transporter (5-HTT) mRNA in the brain was increased in this model and that treatment with a selective serotonin reuptake inhibitor (SSRI) was effective for blocking the decrement of the daily amount of spontaneous running wheel activity. Therefore, the relationship among viral infection changes in cytokine production and brain dysfunction is gradually becoming clearer.

Moreover, there is a possibility that the abnormalities of TGF-beta are also deeply concerned with fatigue sensation. In general, TGF-beta regulates the division and death of cells; it contributes to controlling the level of activation of the immune system. Inoue et al. (1999) found that intracranial administration of cerebrospinal fluid (CSF) from exercise-exhausted rats to naïve mice produced a decrease in spontaneous motor activity, whereas CSF from sedentary rats had no such effect. This finding suggests the presence of a substance suppressing the urge for motion as a response to fatigue. Using a bioassay system, they found that the level of TGF-beta, in the CSF from exercise-fatigued rats, increased; but there was no increase in the CSF from the sedentary rats. Furthermore, the injection of recombinant TGF-beta into the brains of sedentary mice elicited a similar decrease in spontaneous motor activity in a dose-dependent manner. These results suggest that TGF-beta might be involved in the molecular mechanism for fatigue produced by exercise, leading to the suppression of spontaneous motor activity. An elevated serum level of bioactive TGF-beta was also frequently found in patients with CFS,[5] and we also confirmed such an increase in the majority of Japanese CFS patients. TGF-beta was reported to inhibit the production of dehydroepiandrosterone sulfate (DHEA-S) (Stankovic, Dion & Parker, 1994), which is known to regulate positively the activity of carnitine acetyltransferase (Chiu et al., 1997), which catalyzes the transfer of free carnitine to acylcarnitine, especially the acetylcarnitine. (We have previously noted the role of carnitine in the TCA cycle). We found that most Japanese CFS patients had a deficiency in DHEA-S (Kuratsune, Yamaguti, Sawada et al., 1998) and in acetylcarnitine (Kuratsune, Yamaguti, Takahashi et al., 1994), and so the increase in TGF-beta would appear to be related to these abnormalities.

Additionally, the presence of auto-antibodies including antinuclear antibody is also thought to be an important key immunological abnormality involved in the pathogenesis of CFS. It is known that antinuclear antibody is frequently found in fatigued patients throughout the world who have various indefinite complaints. However, the role of these auto-antibodies in these patients is unclear. Recently, using a sensitive radioligand assay, requiring the injection of a radioactive biochemical substance, Tanaka S. Kuratsune et al. (2003) examined the sera of CFS patients (n = 60), patients with autoimmune disease (n = 33), and healthy controls (n = 30) for auto-antibodies against various neurotransmitter receptors, i.e., recombinant human muscarinic cholinergic receptor 1 (CHRM1), mu-opioid receptor (OPRM1), 5-hydroxytryptamine receptor 1A (HTR1A), and dopamine receptor D2 (DRD2). The mean anti-CHRM1 antibody index was significantly higher in patients with CFS (p<0.0001) and autoimmune disease (p<0.05), than in healthy controls, and over a half of the patients with CFS (53.3 percent, 32/60) had anti-CHRM1 antibody. Antinuclear antibodies were also found in 56.7 percent (34/60) of the CFS patients, but their titers did not correlate with the activities of the above 4 auto-antibodies. The CFS patients who had positive auto-antibodies against CHRM1 had a significantly higher mean score (1.81) of "feeling of muscle weakness" than those negative for auto-antibodies (1.18) (p<0.01). Higher scores on "painful lymph node," "forgetfulness," and "difficulty in thinking" were also found in CFS patients with anti-CHRM1 antibodies than in those without them, but statistical significance was not reached. Anti-OPRM1 antibodies, anti-HTR1A antibodies and anti-DRD2 antibodies were also found in 15.2, 1.7, and 5.0 percent of patients with CFS, respectively; but no significant relationship was found between the symptoms and existence of these antibodies. Since the anti-CHRM1 antibody is also frequently found in patients with schizophrenic disorders, mood disorders, and other psychiatric disorders, it is not specific for CFS; but the autoimmune abnormalities in neurotransmitter receptors might cause the secondary brain dysfunction including CFS.

7. BRAIN DYSFUNCTION

Recent single-photon emission computed tomography (SPECT) studies (Ichise et al., 1992; Costa, Tannock & Brostoff, 1995; Fischler et al., 1996) using ^{99m}Tc-hexamethyl-propylene-amine oxime revealed that most CFS patients showed cerebral hypoperfusion, i.e., a lower cerebral blood flow rate in a variety of brain regions such as the frontal, temporal, parietal, and occipital cortices; anterior cingulate; basal ganglia; and brain stem. This finding suggested that central nervous system (CNS) dysfunction might be related to the neuropsychiatric symptoms of CFS patients. To confirm these findings, we (Kuratsune, Yamaguti, Lindh et al., 2002) studied the regional cerebral blood flow (rCBF) in 8 CFS patients and 8 age and sex-matched controls by use of ^{15}O-labeled water ($H_2^{15}O$) and positron emission tomography (PET), and found that the rCBF was lower in the CFS patient group than in the control group in the brain regions including the frontal, temporal, and occipital cortices, anterior cingulate, basal ganglia; and brain stem. These brain regions correspond to various neuropsychiatric complaints: autonomic imbalance, sleep disturbance, many kinds of pain, and the loss of concentration, thinking, motivation, and short-term memory. Therefore, our results from the first quantitative rCBF study done on CFS patients with PET are in good agreement with the data from the previous SPECT studies, and indicate that various neuropsychiatric complaints found in CFS patients might be related to dysfunction in these regions of the CNS.

Furthermore, when we (Kuratsune et al., 2002) studied the cerebral uptake of [2-^{11}C] acetyl-L-carnitine in the same 8 CFS patients and 8 age- and sex-matched normal controls by using PET, a significant decrease was found in several brain regions of the patients' group, namely, in the prefrontal (Brodmann's area 9/46d) and temporal (BA21 and 41) cortices, anterior cingulate (BA24 and 33), and cerebellum. These findings suggest that the levels of neurotransmitters biosynthesized through acetylcarnitine might be reduced in some brain regions of chronic-fatigue patients and that this abnormality might be one of the keys to unveil the mechanisms of chronic-fatigue sensation.More recently, using MRI, we found that patients with CFS have reduced gray matter (GM) volume in the bilateral prefrontal cortices (Okada et al., 2004). Furthermore, right-hemisphere GM volume correlated with subjects' fatigue ratings (Okada et al., 2004). This is consistent with the above-mentioned result that described a decrease of uptake of acetylcarnitine, which may indicate a decrease in the biosynthesis of glutamate in the prefrontal cortex. The prefrontal cortex might therefore be part of the neural underpinnings of fatigue. In 2008, de Lange et al. (2008) confirmed our results and reported the recovery from volume reduction in a responder group of cognitive behavioral therapy. The volume reduction might be shrinkage of neuronal dendrites, which might result from reduction of the amount of brain derived neurotrophic factor (BDNF).

We also studied 5-HTT density in 10 patients with CFS and 10 age-matched normal controls by using PET with the radiotracer [^{11}C](+)McN5652. Analysis using a statistical parametric mapping software (SPM99) revealed that the density of 5-HTT in the rostral subdivision of the anterior cingulate was significantly reduced in CFS patients (Yamamoto et al., 2004). In addition, the density of 5-HTT of dorsal anterior cingulate was negatively correlated with the pain score (Yamamoto et al., 2004). Therefore, an alteration in the serotoninergic neurons in the anterior cingulate plays a key role in the pathophysiology of CFS.

These PET results on 5-HTT density seem to be inconsistent with our results (Narita et al., 2003) regarding the 5-HTT gene promoter polymorphism, where CFS patients could have a greater frequency of the L allele, which affords greater transporter efficiency, and hence elevation of serotonin levels. However, it might be the case that the reduction in 5-HTT density in CFS patients with L and XL allelic variants is less than that in CFS patients with S allelic variants. Since 5-HT synthesis in the brain is thought to deteriorate in patients with CFS, 5-HT deficiency in the synapses might be more serious in patients with L and XL allelic variants. If so, it is consistent with the finding that SSRI treatment is effective for some patients with CFS. To clarify, in order to understand the full particulars of brain dysfunction in patients with CFS, we are now studying the 5-hydroxy-L-tryptophan (5-HTP) uptake, L-DOPA uptake, and muscarinic acetylcholine receptor density by using PET. We will further report the results from these studies concerning brain dysfunction found in patients with CFS in the near future.

Thus far, we propose the working hypothesis on the dysfunction in chronic fatigue, as shown in Figure 14.5. First, the activation of a part of orbitofrontal cortex [Brodmann's area 11 (BA11)] occurs with fatigue sensation, then the serotonergic system originating from the Raphe nucleus in the brain stem is activated by the signal from BA11 to activate anterior

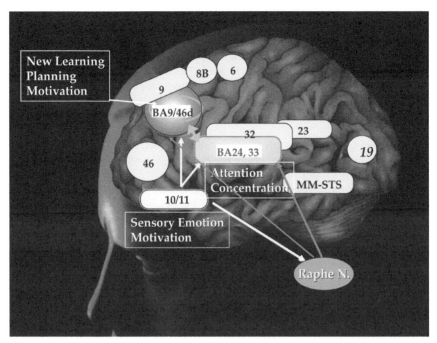

Figure 14.5   Hypothesis of neural circuits responsible for central fatigue and fatigue sensation

cingulate cortex (BA24) and prefrontal cortex (BA9/46d) to overcome the deterioration of the autonomic nerve function and executive function, respectively, by fatigue. Then, after prolongation of serotonergic activation, the system would be exhausted and results in deactivation of anterior cingulate cortex (BA24) and prefrontal cortex (BA9/46d). Also, the other regions in light blue in Figure 14.5, which showed a lower blood flow rate in the PET study, would demonstrate other brain dysfunction along with chronic fatigue. The brain dysfunctions illustrated in Figure 14.5 may be placed within the broader physiological model of Figure 14.4. Effects of psychosocial stressors and genetic background are mediated by multiple processes associated with metabolic and immune dysfunction, producing multiple biochemical markers for fatigue.

REFERENCES

Asami, S., Hirano, T., Yamaguchi, R., Tsurudome, Y., Itoh, H. & Kasai, H. (1998). Effects of forced and spontaneous exercise on 8-hydroxydeoxyguanosine levels in rat organs. *Biochemical and Biophysical Research Communications*, 243, 678–82.

Ataka, S., Tanaka, M., Nozaki, S., Mizuma, H., Mizuno, K., Tahara, T., Sugino, T., Shirai, T., Kajimoto, Y., Kuratsune, H., Kajimoto, O. & Watanabe, Y. (2007). Effects of Applephenon((R)) and ascorbic acid on physical fatigue. *Nutrition*, 23, 419–23.

Ataka, S., Tanaka, M., Nozaki, S., Mizuma, H., Mizuno, K., Tahara, T., Sugino, T., Shirai, T., Kajimoto, Y., Kuratsune, H., Kajimoto, O. & Watanabe, Y. (2008). Effects of oral administration of caffeine and D-ribose on mental fatigue. *Nutrition*, 24, 233–8.

Bennett, A.L., Chao, C.C., Hu, S., Buchwald, D., Fagioli, L.R., Schur, P.H., Peterson, P.K. & Komaroff, A.L. (1997). Elevation of bioactive transforming growth factor-beta in serum from patients with chronic fatigue syndrome. *Journal of Clinical Immunology*, 17, 160–1667.

Chiu, K.M., Schmidt, M.J., Shug, A.L., Binkley, N. & Gravenstein, S. (1997). Effect of dehydroepiandrosterone sulfate on carnitine acetyl transferase activity and L-carnitine levels in oophorectomized rats. *Biochimica et Biophysica Acta*, 1344, 201–9.

Costa, D.C., Tannock, C. & Brostoff, J. (1995). Brainstem perfusion is impaired in chronic fatigue syndrome. *QJM*, 88, 767–73.

de Lange, F.P., Koers, A., Kalkman, J.S., Bleijenberg, G., Hagoort, P., van der Meer, J.W. & Toni, I. (2008). Increase in prefrontal cortical volume following cognitive behavioural therapy in patients with chronic fatigue syndrome. *Brain*, 131, 2172–80.

Demitrack, M.A., Dale, J.K., Straus, S.E., Laue, L., Listwak, S.J., Kruesi, M.J., Chrousos, G.P. & Gold, P.W. (1991). Evidence for impaired activation of the hypothalamic≠pituitary–adrenal axis in patients with chronic fatigue syndrome. *Journal of Clinical Endocrinology & Metabolism*, 73, 1224–34.

Demettre, E., Bastide, L., D'Haese, A., De Smet, K., De Meirleir, K., Tiev, K.P., Englebienne, P. & Lebleu, B. (2002). Ribonuclease L proteolysis in peripheral blood mononuclear cells of chronic fatigue syndrome patients. *Journal of Biological Chemistry*, 277, 35746–51.

Fischler, B., D'Haenen, H., Cluydts, R., Michiels, V., Demets, K., Bossuyt, A., Kaufman, L. & De Meirleir, K. (1996). Comparison of 99mTc HMPAO SPECT scan between chronic fatigue syndrome, major depression and healthy controls: An exploratory study of clinical correlates of regional cerebral blood flow. *Neuropsychobiology*, 34, 175–83.

Gaab, J., Engert, V., Heitz, V., Schad, T., Schürmeyer, T.H. & Ehlert, U. (2004). Associations between neuroendocrine responses to the insulin tolerance test and patient characteristics in chronic fatigue syndrome. *Journal of Psychosomatic Research*, 56, 419–24.

Gaab, J., Hüster, D., Peisen, R., Engert, V., Schad, T., Schürmeyer, T.H. & Ehlert, U. (2002). Low-dose dexamethasone suppression test in chronic fatigue syndrome and health. *Psychosomatic Medicine*, 64, 311–18.

Hill, A.V. & Kupalov, P. (1929). Proceedings of the Royal Society of London. Series B, Biological sciences, 105, 313.

Ichise, M., Salit, I.E., Abbey, S.E., Chung, D.G., Gray, B., Kirsh, J.C. & Freedman, M. (1992). Assessment of regional cerebral perfusion by 99Tcm-HMPAO SPECT in chronic fatigue syndrome. *Nuclear Medicine Communications*, 13, 767–72.

Ikuta, I., Yamada, T., Shimomura, T., Kuratsune, H., Kawahara, R., Ikawa, S., Ohnishi, E., Sokawa, Y., Fukushi, H., Hirai, K., Watanabe, Y., Kurata, T., Kitani, T. & Sairenji, T. (2003). Diagnostic evaluation for 2',5'-oligoadenylate synthetase activities and antibodies against Epstein-Barr virus and Coxiella Burnetii in patients with chronic fatigue syndrome in Japan. *Microbes and Infection*, 5, 1096–102.

Inoue, K., Yamazaki, H., Manabe, Y., Fukuda, C., Hanai, K. & Fushiki, T. (1999). Transforming growth factor-beta activated during exercise in brain depresses spontaneous motor activity of animals: Relevance to central fatigue. *Brain Research*, 846, 145–53.

Jin, G., Kataoka, Y., Tanaka, M., Mizuma, H., Nozaki, S., Tahara, T., Mizuno, K., Yamato, M. & Watanabe, Y. (2009). Changes in plasma and tissue amino acid levels in an animal model of complex fatigue. *Nutrition*, 25, 597–607.

Katafuchi, T., Kondo, T., Yasaka, T., Kubo, K., Take, S. & Yoshimura, M. (2003). Prolonged effects of polyriboinosinic: Polyribocytidylic acid on spontaneous running wheel activity and brain interferon-alpha mRNA in rats: a model for immunologically induced fatigue. *Neuroscience*, 120, 837–45.

Klimas, N.G., Salvato, F.R., Morgan, R. & Fletcher, M.A. (1990). Immunologic abnormalities in chronic fatigue syndrome. *Journal of Clinical Microbiology*, 28, 1403–10.

Kuratsune, H., Yamaguti, K., Lindh, G., Evengård, B., Hagberg, G., Matsumura, K., Iwase, M., Onoe, H., Takahashi, M., Machii, T., Kanakura, Y., Kitani, T., Långström, B. & Watanabe, Y. (2002). Brain regions involved in fatigue sensation: Reduced acetylcarnitine uptake into the brain. *NeuroImage*, 17, 1256–65.

Kuratsune, H., Yamaguti, K., Sawada, M., Kodate, S., Machii, T., Kanakura, Y. & Kitani, T. (1998). Dehydroepiandrosterone sulfate deficiency in chronic fatigue syndrome. *International Journal of Molecular Medicine*, 1, 143–6.

Kuratsune, H., Yamaguti, K., Takahashi, M., Misaki, H., Tagawa, S. & Kitani, T. (1994). Acylcarnitine deficiency in chronic fatigue syndrome. *Clinical Infectious Diseases*, 18, S62–67.

Mathew, S.J., Mao, X., Keegan, K.A., Levine, S.M., Smith, E.L., Heier, L.A., Otcheretko, V., Coplan, J.D. & Shungu, D.C. (2009). Ventricular cerebrospinal fluid lactate is increased in chronic fatigue syndrome compared with generalized anxiety disorder: an in vivo 3.0 T (1)H MRS imaging study. *NMR in Biomedicine*, 22, 251–8.

Mizuma, H., Tanaka, M., Nozaki, S., Mizuno, K., Tahara, T., Ataka, S., Sugino, T., Shirai, T., Kajimoto, Y., Kuratsune, H., Kajimoto, O. & Watanabe, Y. (2009). Daily oral administration of crocetin attenuates physical fatigue in human subjects. *Nutrition Research*, 29, 145–50.

Mizuno, K., Tanaka, M., Nozaki, S., Mizuma, H., Ataka, S., Tahara, T., Sugino, T., Shirai, T., Kajimoto, Y., Kuratsune, H., Kajimoto, O. & Watanabe, Y. (2008). Antifatigue effects of coenzyme Q10 during physical fatigue. *Nutrition*, 24, 293–9.

Mizuno, K., Tanaka, M., Tajima, K., Okada, N., Rokushima, K. & Watanabe, Y. (2010). Effects of mild-stream bathing on recovery from mental fatigue. *Medical Science Monitor*, 16, 8–14.

Moss, R.B., Mercandetti, A. & Vojdani, A. (1999). TNF-alpha and chronic fatigue syndrome. *Journal of Clinical Immunology*, 19, 314–16.

Narita, M., Nishigami, N., Narita, N., Yamaguti, K., Okado, N., Watanabe, Y. & Kuratsune, H. (2003). Association between serotonin transporter gene polymorphism and chronic fatigue syndrome. *Biochemical and Biophysical Research Communications*, 311, 264–6.

Newsholme, E.A. & Blomstrand, E. (1995), in Gandevia et al. (eds), *Fatigue*. New York: Plenum Press.

Nozaki, S., Mizuma, H., Tanaka, M., Jin, G., Tahara, T., Mizuno, K., Yamato, M., Okuyama, K., Eguchi, A., Akimoto, K., Kitayoshi, T., Mochizuki-Oda, N., Kataoka, Y. & Watanabe, Y. (2009). Thiamine tetrahydrofurfuryl disulfide improves energy metabolism and physical performance during physical-fatigue loading in rats. *Nutrition Research*, 29, 867–72.

Nozaki, S., Tanaka, M., Mizuno, K., Ataka, S., Mizuma, H., Tahara, T., Sugino, T., Shirai, T., Eguchi, A., Okuyama, K., Yoshida, K., Kajimoto, Y., Kuratsune, H., Kajimoto, O. & Watanabe, Y. (2009). Mental and physical fatigue-related biochemical alterations. *Nutrition*, 25, 51–7.

Ogawa, T., Kiryu-Seo, S., Tanaka, M., Konishi, H., Iwata, N., Saido, T., Watanabe, Y. & Kiyama, H. (2005). Altered expression of neprilysin family members in pituitary gland of sleep-disturbed rat, an animal model of severe fatigue. *Journal of Neurochemistry*, 95, 1156–66.

Ogawa, T., Shishioh-Ikejima, N., Konishi, H., Makino, T., Sei, H., Kiryu-Seo, S., Tanaka, M., Watanabe, Y. & Kiyama, H. (2009). Chronic stress elicits prolonged activation of α-MSH secretion and subsequent degeneration of melanotroph. *Journal of Neurochemistry*, 109, 1389–99.

Okada, T., Tanaka, M., Kuratsune, H., Watanabe, Y. & Sadato, N. (2004). Mechanisms underlying fatigue: a voxel-based morphometric study of chronic fatigue syndrome. *BMC Neurology*, 4, 14.

Pedersen, T.H., Nielsen, O.B., Lamb, G.D. & Stephenson, D.G. (2004). Intracellular acidosis enhances the excitability of working muscle. *Science*, 305, 1144–7.

Roberts, L.J. 2nd. & Morrow, J.D. (1994). Isoprostanes: Novel markers of endogenous lipid peroxidation and potential mediators of oxidant injury. *Annals of the New York Academy of Science*, 744, 237–42.

Roberts, A.D., Wessely, S., Chalder, T., Papadopoulos, A., Cleare, A.J. (2004). Salivary cortisol response to awakening in chronic fatigue syndrome. *The British Journal of Psychiatry*, 184, 136–41.

Shigenaga, M.K., Gimeno, C.J. & Ames, B.N. (1989). Urinary 8-hydroxy-2'-deoxyguanosine as a biological marker of in vivo oxidative DNA damage. Proceedings of the National Academy of Sciences of the United States of America, 86, 9697–701.

Shishioh-Ikejima, N., Ogawa, T., Yamaguti, K., Watanabe, Y., Kuratsune, H. & Kiyama, H. (2010). The increase of alpha-melanocyte-stimulating hormone in the plasma of chronic fatigue syndrome patients. *BMC Neurology*, 10, 73.

Stankovic, A.K., Dion, L.D., Parker, C.R. Jr. (1994). Effects of transforming growth factor-beta on human fetal adrenal steroid production. *Molecular and Cellular Endocrinology*, 99, 145–51.

Suhadolnik, R.J., Reichenbach, N.L., Hitzges, P., Sobol, R.W., Peterson, D.L., Henry, B., Ablashi, D.V., Müller, W.E., Schröder, H.C., Carter, W.A. & Strayer, D.R. (1994). Upregulation of the 2-5A synthetase/RNase L antiviral pathway associated with chronic fatigue syndrome. *Clinical Infectious Diseases*, 18, 96–104.

Tajima, K., Tanaka, M., Mizuno, K., Okada, N., Rokushima, K. & Watanabe, Y. (2008). Effects of bathing in micro-bubbles on recovery from moderate mental fatigue. *Journal of Ergonomia IJE&HF*, 30, 135–45.

Tanaka, M. & Watanabe, Y. (2008a). Mechanism of fatigue studied in a newly developed animal model of combined (mental and physical) fatigue, in Y. Watanabe et al. (eds), *Fatigue Science for Human Health*. New York: Springer, 203–12.

Tanaka, M. & Watanabe, Y. (2008b). Lactate is not a cause of fatigue, in Y. Watanabe (ed.), *Fatigue Science for Human Health*. New York: Springer.

Tanaka, M., Baba, Y., Kataoka, Y., Kinbara, N., Sagesaka, Y.M., Kakuda, T. & Watanabe, Y. (2008). Effects of (-)-epigallocatechin gallate in liver of an animal model of combined (physical and mental) fatigue. *Nutrition*, 24, 599–603.

Tanaka, M., Mizuno, K., Tajima, S., Sasabe, T. & Watanabe, Y. (2009). Central nervous system fatigue alters autonomic nerve activity. *Life Sciences*, 84, 235–9.

Tanaka, M., Nakamura, F., Mizokawa, S., Matsumura, A., Nozaki, S. & Watanabe, Y. (2003). Establishment and assessment of a rat model of fatigue. *Neuroscience Letters*, 352, 159–62.

Tanaka, M., Shigihara, Y., Fujii, H., Hirayama Y. & Watanabe, Y. (2008). Effect of CBEX-Dr-containing drink on physical fatigue in healthy volunteers. *Japanese Pharmacology & Therapeutics* (abstract in English, text in Japanese), 36, 199–212.

Tanaka, S., Kuratsune, H., Hidaka, Y., Hakariya, Y., Tatsumi, K.I., Takano, T., Kanakura, Y. & Amino, N. (2003). Autoantibodies against muscarinic cholinergic receptor in chronic fatigue syndrome. *International Journal of Molecular Medicine*, 12, 225–30.

Watanabe, Y., Evengård, B., Natelson, B.H., Jason, LA. & Kuratsune, H. (eds) (2008). *Fatigue Science for Human Health*. New York: Springer.

Yamaguti, K., Kuratsune, H., Watanabe, Y., Takahashi, M., Nakamoto, I., Machii, T., Jacobsson, G., Onoe, H., Matsumura, K., Valind, S., Långström, B. & Kitani, T. (1996). Acylcarnitine metabolism during fasting and after refeeding. *Biochemical and Biophysical Research Communications*, 225, 740–46.

Yamamoto, S., Ouchi, Y., Onoe, H., Yoshikawa, E., Tsukada, H., Takahashi, H., Iwase, M., Yamaguti, K., Kuratsune, H. & Watanabe, Y. (2004). Reduction of serotonin transporters of patients with chronic fatigue syndrome. *Neuroreport*, 15, 2571–4.

PART V

Performance Effects of Sleep Loss and Circadian Rhythms

15

Socio-emotional and Neurocognitive Effects of Sleep Loss

William D.S. Killgore

INTRODUCTION

Fatigue can affect both physical and mental performance. Physical fatigue emerges when an individual has overtaxed and depleted the basic physiological resources necessary to keep muscles and bodily systems functioning at full capacity. Similarly, mental fatigue occurs when the physiological systems supporting cognitive and emotional processes have been exhausted faster than they can be replenished and refreshed. While physical fatigue primarily involves cardiovascular systems and large muscle groups, mental fatigue predominantly affects the brain and supporting neural systems. Prolonged waking cognitive activity is associated with regional changes in many aspects of the brain, including glucose metabolism, glycogen stores, accumulation of a sleep-promoting chemical known as adenosine, and changes in synaptic receptor density, just to name a few. As anyone who has played an intense sporting event knows, the effects of physical fatigue can often be rapidly mitigated by a brief period of physical rest. Mental fatigue, on the other hand, appears to require more than simple rest to fully restore cognitive capacities to normal functional levels. Simply put, the body needs rest but the brain needs sleep.

Because the neurocognitive effects of sleep loss are complex and multifaceted, it is not possible to comprehensively review the extensive literature on this topic in an introductory chapter of this scope. Instead, the present chapter presents a selective review of the effects of sleep deprivation on brain and cognitive functioning at the macro level, with a particular focus on those processes most relevant to making errors in operational settings, such as simple inattention, mood changes, and alterations in judgment and decision-making. The chapter will build primarily on a theoretical model positing that one of several major effects of sleep loss is altered functioning of the prefrontal cortex, which can adversely affect a variety of cognitive processes ranging from simple alertness and vigilance to mood and emotional processing and, ultimately, higher-level judgment and decision-making.

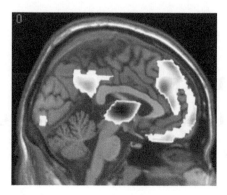

Figure 15.1   The figure shows a mid-sagittal brain image from a positron
               emission tomography (PET) study of the effects of sleep
               deprivation on regional cerebral glucose metabolism (adapted
               from an unpublished image from Thomas et al., 2000)

Source: This image was graciously provided by Maria Thomas, with special thanks to Gregory Belenky of the Walter Reed Army Institute of Research and Henry Halcomb of Johns Hopkins University.

THE BRAIN WITHOUT SLEEP

Sleep deprivation has a profound effect on brain functioning. While this is obvious from its effects on behavior and cognition, only within the last couple of decades has it actually become possible to observe and measure subtle changes in brain functioning that are brought about by sleep loss. Neuroimaging methods, such as positron emission tomography (PET) have been particularly important in this regard. As is shown in Figure 15.1, PET scan technology has revealed that sleep deprivation is associated with significant global reductions in the amount of energy consumed by the brain (Thomas et al., 2000). Since cognitive processing and neural activity require energy metabolism, the finding that the brain is less metabolically active during sleep loss provides a partial neural explanation for the mental sluggishness we experience without sufficient sleep. Moreover, some regions appear more significantly affected by sleep loss than others. Figure 15.1 shows that the greatest reductions are seen in the thalamus, a subcortical structure involved in the transmission of information throughout the brain and in basic alertness and attention (Thomas et al., 2000). Sleep loss is also associated with significant relative declines in energy use within the prefrontal and parietal cortices, regions that are involved in selective allocation of attention, integration of multiple sources of information, emotional regulation, and higher-level cognitive processes such as judgment and decision-making. In light of these alterations in brain activity, it is not surprising that sleep loss is often associated with impairments of the cognitive abilities that rely heavily upon the processing capacities of these regions.

ALERTNESS, VIGILANCE, AND SIMPLE ATTENTION

Alertness and attention serve as the foundation for higher cognitive processing. In well-rested individuals, alertness levels fluctuate only modestly throughout the day and sleepiness is generally not an issue. However, after about 16 hours of continuous wakefulness, even a normally rested person will usually begin to show declines in alertness and vigilance, as is evidenced by slowed reaction times, reduced accuracy of responses, and variability in performance (Goel, et al., 2009). In fact, of all cognitive capacities studied, alertness and vigilance appear to be the most reliably impaired by sleep loss (Lim & Dinges, 2008). These deficits are apparent in several ways.

First, as might be expected, sleep deprivation results in a significant slowing of simple reaction time. It has been suggested that speed of responses may slow by about 25 percent for each night of total sleep loss (Belenky et al., 1994). Response slowing also emerges more insidiously when sleep is chronically restricted to less than adequate levels over several nights. For instance, Van Dongen and colleagues showed that even restricting sleep to six hours per night for two weeks was sufficient to reduce vigilance performance to the same level as someone totally sleep deprived for two full nights (Van Dongen, Maislin et al., 2003).

Sleep loss also increases the frequency and duration of attentional lapses (i.e., periods of non-responsiveness lasting a half second or longer). Lapsing can be particularly serious in occupational environments involving heavy machinery or moving vehicles, where even a brief moment of inattentiveness can have catastrophic effects. Prolonged wakefulness increases the magnitude of homeostatic sleep pressure (i.e., the biological drive for sleep). This biological pressure to sleep is only part of the story, however. The effect of sleep loss on alertness is also affected by the time of day (i.e., circadian phase). Normally, the circadian fluctuation follows a sinusoidal pattern throughout the day. Together, these two influences form the basis of the two-process model of alertness (Borbély, 1982), which asserts that alertness levels and the propensity for sleep are determined by a mathematical combination of two interacting biological processes: 1) a homeostatic drive for sleep that accumulates during wakefulness and is dissipated with sleep (process S), and 2) an oscillating pattern of alertness that fluctuates throughout the day according to a sinusoidal circadian rhythm (process C) (see Figure 15.2). Thus, the level of alertness and vigilance results from the combined product of these interacting processes at any particular moment throughout the day. Over time, insufficient sleep leads to accumulated homeostatic sleep pressure. When the circadian rhythm is at its peak, this sleep pressure may have only modest effect on performance. However, when

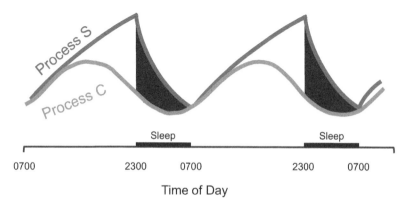

Figure 15.2   The two-process model of sleep homeostasis (Borbély, 1982)

the circadian rhythm is at its nadir, generally in the early morning hours, the effects of sleep deprivation on alertness and vigilance performance may be quite profound (Babkoff et al., 1991).

WAKE STATE INSTABILITY

The net effect of attentional lapses on cognition is instability of performance. When rested, an individual is generally alert and performance remains relatively stable. However, as the homeostatic pressure for sleep increases, it eventually reaches a tipping point where the individual's motivation to remain awake becomes insufficient to consistently suppress the drive for sleep. For most people, the tension between these two systems typically becomes apparent after about 16 to 18 hours of wakefulness. After this point, the opposing forces between the drive for sleep and the motivation to remain awake are in a sort of tug of war, with brief moments where one force suddenly and intermittently overtakes the other. This leads to increased variability in alertness and progressively unreliable performance, a phenomenon known as "wake state instability" (Doran, Van Dongen & Dinges, 2001). In practical terms, a sleep-deprived person may appear to be performing normally at one moment, only to show slowed response time or even a brief period of non-responsiveness the next. This makes the effects of sleep deprivation particularly dangerous in the work environment, as an operator may appear functional and alert when spoken to, but may be at risk of a sudden and unexpected lapse only moments later. This variability in performance can have catastrophic consequences in operational settings where constant vigilance and sustained performance are required.

FUNCTIONAL BRAIN IMAGING OF SLEEP LOSS

As was mentioned earlier, the brain becomes less metabolically active during extended wakefulness, but some regions appear to be particularly affected (Thomas et al., 2000). Notably, metabolic activity declines in regions that are important for allocating attention, processing information, and controlling thought and behavior (Thomas et al., 2000). Using functional magnetic resonance imaging (fMRI), Drummond and colleagues found that the response-slowing characteristic of sleep deprivation was associated with increased activation of the brain's "default mode network," a system of brain regions that are typically more active during daydreaming, self-referential processing, and when cognitive demands are minimal (Drummond, Bischoff-Grethe et al., 2005). Moreover, when a sleep-deprived individual actually experiences an attentional lapse, there appears to be s significantly reduced level of activation of visual sensory processing regions, including parietal and occipital cortices, and the thalamus (Chee et al., 2008). These findings suggest that when lapses occur during the sleep-deprived state, they affect more than simple attention systems; rather, such lapses are also associated with significantly reduced visual sensory processing. This sensory disengagement from the environment during sleep deprivation could be particularly worrisome for sleep-deprived workers who are attempting to sustain visual vigilance, as it suggests that there may be brief moments where visual processing is essentially disengaged.

INDIVIDUAL DIFFERENCES

While it is appealing to formulate general rules about how sleep loss affects performance for most people, emerging evidence suggests that there is wide variation across individuals with regard to their vulnerability to sleep derivation (Van Dongen, Baynard et al., 2004). Some people seem to be able to function amazingly well on relatively little sleep, while others are dramatically impaired by even mild sleep deprivation. Van Dongen and colleagues found that individuals show highly consistent patterns of PVT performance degradation and sleepiness ratings across multiple sessions of sleep deprivation, suggesting that resistance to sleep loss is a stable trait-like phenomenon (Van Dongen et al., 2004). Recent evidence now suggests that these trait-like individual differences may have some genetic (King, Belenky & Van Dongen, 2009) or neuroanatomical basis (Rocklage et al., 2009).

EMOTIONAL PROCESSING

While the declines in vigilance and attention are the most obvious effects of sleep loss, changes in mood and emotional functioning are also commonly observed (Dinges et al., 1997). Furthermore, emotional processes are intimately tied with cognitive processes and, thus, it is likely that changes in cognition during sleep loss may be mediated to some extent by altered emotional processing. Some of these effects will be reviewed briefly.

Self-Reported Mood, Coping, and Affective State

It is common knowledge that people tend to get cranky and irritable when they are lacking sleep, and this is supported by numerous studies of self-reported mood. When compared to normally rested individuals, participants deprived of even a single night of sleep show a significant worsening of negative self-rated mood scores (Tempesta et al., 2010). Sleep deprivation also leads to negative perceptions regarding social and interpersonal situations. In one study, subjects were presented with semi-projective cartoon scenarios portraying frustrating situations. Compared to their baseline responses, these same subjects were more likely to give responses that shifted blame toward others and were less likely to provide responses suggestive of a willingness to smooth over problems or make apologies after they had been sleep deprived for two nights (Kahn-Greene, Lipizzi et al., 2006).

Sleep deprived individuals also show reductions in their self-reported coping capacities and general emotional intelligence. For instance, after two nights without sleep, individuals show a decline in their tendency to think positively, reduced willingness to take active steps to solve problems, and are more likely to rely on fruitless coping strategies such as superstitious thinking and magical beliefs (Killgore, Kahn-Greene et al., 2008). In the same study, subjects also showed significant declines on several facets of emotional intelligence, including self-esteem, ability to empathize with others, comprehension of interpersonal dynamics, and behavioral impulse control (Killgore, Kahn-Greene et al., 2008). In short, sleep deprived individuals are more likely to be easily frustrated, less tolerant, forgiving, or caring, and generally more self-focused than when normally rested.

Even more severe changes in mood and affective functioning can be seen following sleep loss. For example, after 56 hours of sleep deprivation, subjects in one study showed significant elevations on several clinical psychopathology scales, including depression,

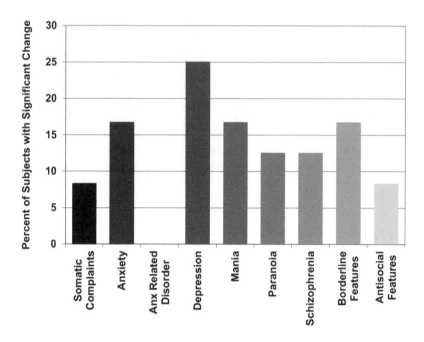

Figure 15.3   The percentage of subjects showing clinically significant increases in symptoms of psychopathology following 56 hours of sleep deprivation (Kahn-Greene, Killgore et al., 2007)

anxiety, paranoia, and somatic problems (Kahn-Greene, Killgore et al., 2007). Compared to their baseline rested scores, this group of healthy individuals was more prone to endorse greater feelings of worthlessness, powerlessness, inadequacy, failure, low self-esteem, and overall reduced life satisfaction. A particularly striking finding was that the change was of sufficient magnitude to qualify as a "clinically significant" increase in depression scores for as many as 25 percent of the participants (see Figure 15.3) (Kahn-Greene, Killgore et al., 2007). These basic mood and emotional changes, while not pathological in absolute severity, were potentially large enough to have a negative operational impact on workers functioning within close quarters for extended periods of time. Thus, within operational settings, the adverse effects of sleep loss on mood and interpersonal functioning need to be considered above and beyond its effects on simple alertness and attention.

Emotional Perception

Not only does sleep loss affect mood and the experience of emotions but it also appears to alter basic perception of emotional stimuli. Generally speaking, sleep deprivation leads to a negative bias in emotional perception, particularly for neutral stimuli. For example, subjects rated photographs with neutral content much more negatively following a night of total sleep deprivation than following a night of normal sleep (Tempesta et al., 2010). Interestingly, sleep deprived subjects are also mildly impaired in their ability to perceive or appreciate humor, a subtle but complex cognitive–emotive capacity (Killgore, McBride et al., 2006). Deficits in the ability to correctly identify emotion in facial expressions have also been reported in sleep-deprived volunteers, particularly for moderately intense emotions of anger and happiness (van der Helm, Gujar & Walker, 2010), and complex blends of facial emotion (Huck et al., 2008). These findings support the hypothesis that discrete affective processing systems, probably involving ventromedial prefrontal and amygdala circuits, appear to be adversely impaired by sleep loss.

Neural Correlates

As is discussed throughout this chapter, sleep loss leads to altered metabolic activity across a variety of regions within the brain, including the prefrontal cortex (Thomas et al., 2000). This diverse area of the brain not only regulates attentional resources, but is intimately involved in the regulation of emotional processes and the expression of behavioral traits that are often described as "personality." A prominent theory of emotional functioning suggests that the ventral and medial aspects of the prefrontal cortex – regions particularly affected by sleep deprivation – are critical for integrating lower-order somatic, visceral, and emotional inputs with higher-order abstract reasoning and decision-making (Damasio, 1994). Furthermore, the medial prefrontal cortex also plays a significant role in regulating emotions by inhibiting the activity of primitive emotional response areas

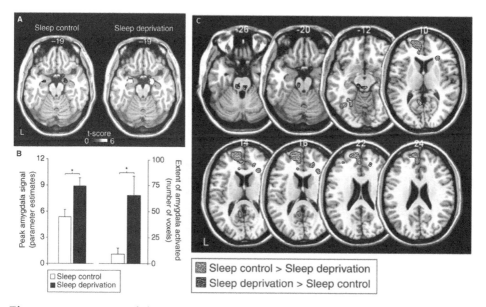

Figure 15.4   Amygdala responses in rested and sleep deprived groups

Note: (A) Amygdala response to emotionally negative stimuli following normal rest and sleep deprivation. (B) Corresponding differences in intensity and volumetric extent of amygdala activation between the rested and sleep deprived conditions. (C) Group level differences in amygdala functional connectivity (combined for left and right), including significantly greater connectivity in the medial-prefrontal cortex (circled areas) for the Sleep control group, yet significantly stronger connectivity with autonomic brainstem regions in the Sleep deprivation group (rectangle areas).

Source: Reprinted with permission from Yoo et al. (2007).

such as the amygdala (Yoo et al., 2007). Reductions in medial prefrontal metabolic activity brought about by sleep deprivation would therefore be expected to impair the normal regulation of amygdala activity. In theory, this might be expected to increase negative affective processing.

A recent neuroimaging study provided support for the hypothesis that sleep deprivation would alter functioning of cortico–limbic emotion regulation circuits. Yoo and colleagues collected fMRI brain scans while presenting emotionally evocative pictures to a group of participants when normally rested and again following 35 hours of sleep deprivation (Yoo et al., 2007). Relative to baseline performance, when sleep deprived, subjects showed significantly greater magnitude and spatial extent of activation within the amygdala to the unpleasant stimuli (see Figure 15.4). Furthermore, the authors examined the strength of functional connectivity between the medial prefrontal cortex and the amygdala at both time points and found that sleep deprivation weakened these connections. Within this context, it is easy to see how sleep deprivation leads to increased negative mood, lower frustration tolerance, increased feelings of depression, and negative perceptual biases for neutral stimuli.

HIGHER ORDER COGNITION

The prefrontal cortex is also critical for carrying out many higher-order cognitive processes that are often loosely classified under the term "executive functions." These capacities generally involve control and coordination of willful action to achieve future goal states (Goel et al., 2009). Although not the only region involved in carrying out these complex cognitive activities, the prefrontal cortex can generally be thought of as the primary engine driving these capabilities.

The prefrontal cortex has been suggested to be particularly vulnerable to the effects of sleep loss because its monitoring and control functions are constantly taxed throughout the waking period (Harrison, Horne & Rothwell, 2000). Just as a muscle is fatigued by overuse, the unceasing demands placed upon the prefrontal cortex during waking activity may degrade executive function systems more rapidly than other cognitive systems that are only intermittently utilized (Harrison & Horne, 2000; Harrison, Horne & Rothwell, 2000). Building on the emotion-related deficits described in the previous section, we will focus here on how the decline in function within the prefrontal cortex, particularly the ventral and medial regions, may impact specific emotionally guided or "hot cognition" aspects of decision-making. Interestingly, many other executive function abilities do not show consistent degradation by sleep loss (Pace-Schott et al., 2009; Tucker, et al., 2010). However, many of the executive function capacities that are resistant to sleep deprivation appear to have a low involvement of emotional brain systems (i.e., they are "cold" executive functions).

HYPOTHESIZED MODEL

We propose that the discrepancies in findings for executive function tasks observed across sleep deprivation studies may be partially explained by the degree to which the tasks rely upon affectively guided decision-making (i.e., hot) systems and the potential for compensatory recruitment of other brain regions by "cold" (i.e., non-emotional) executive function systems. Brain imaging studies by Drummond and colleagues have shown that many tasks that normally depend on prefrontal functioning during the rested state appear to be sustained by compensatory recruitment of other regions such as the parietal cortex when these same processes are engaged during the sleep deprived state (Drummond, Brown et al., 2004; Drummond, Meloy et al., 2005). Despite sleep deprivation, the compensatory recruitment of parietal regions allows many of these "cold" executive capacities to be sustained at near-rested levels of performance. Thus, we propose the speculative hypothesis that the effect of sleep deprivation on executive functions may differ for tasks that rely mostly upon non-emotional dorsolateral capacities versus ventromedial prefrontal regions that are more involved in monitoring and control over more primitive limbic emotional systems. This hypothesized model

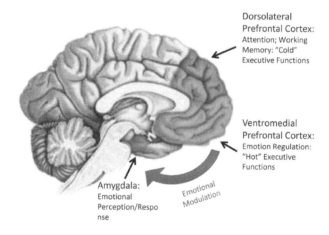

Figure 15.5   A hypothesized model of the effects of sleep deprivation on higher order executive functions

posits that the capacities mediated by the dorsolateral prefrontal systems may be more easily sustained by compensation in other heteromodal cortical regions, whereas the processes carried out by orbital and ventromedial regions of the prefrontal cortex may be less easily compensated for by other regions. Thus, sleep deprivation would be expected to have a more consistently degrading effect on executive function tasks that rely on input, modulation, or integration of emotional information (see Figure 15.5).

When normally rested (Top Panel), the dorsolateral prefrontal regions are generally capable of carrying out complex executive function tasks involving working memory, mental manipulation, allocation of attention, and other "cold" (i.e., nonemotional) cognitive processes, whereas the ventromedial regions are more involved in controlling and modulating emotional processes by inhibiting limbic system (e.g., amygdala) activation (i.e., "hot" cognitive processes). Sleep deprivation (Bottom Panel) reduces metabolic activation within the prefrontal cortex, which affects both the dorsolateral and ventromedial compartments. During sleep loss, many capacities normally mediated by the dorsolateral prefrontal capacities appear to be sustained via compensatory recruitment of other multimodal regions such as the parietal cortex. Consequently, studies of "cold" (i.e., non-emotion related) executive functions often fail to find consistent deficits from sleep loss. In contrast, we hypothesize that degradation of capacities normally mediated by ventromedial regions may not result in comparable compensatory recruitment during sleep loss, leading to sustained deficits in executive function capacities that rely on emotion (e.g., gambling tasks, moral judgment tasks), and difficulties regulating mood and affective responses.

In contrast, executive tasks that depend mostly on working memory or mental manipulation in the dorsolateral prefrontal cortex may be more easily sustained via recruitment of other cortical regions (e.g., parietal cortex) and, therefore, less consistently impaired by sleep loss. The viability of this hypothesis remains to be fully explored, but preliminary evidence from studies of risk-taking and emotional decision-making, which depend heavily on ventromedial prefrontal systems, appears to show relatively consistent deficits during sleep loss.

RISK-TAKING AND DECISION-MAKING

There have been relatively few published studies examining the effects of sleep loss on decision-making, but the available evidence suggests that the relationship is complex. Humans are not unfeeling machines that make decisions based purely on a logical and mathematical weighting of the probable outcomes. Rather, for better or worse, emotions play a major role in our decision-making processes and the willingness to take risks. According to some perspectives, emotions can serve to facilitate decision-making by streamlining the process so that decision-makers are not bogged down by sorting through the details and comparing all possible courses of action (Damasio, 1994; Killgore, 2010). Instead, decision-makers are frequently guided by emotional information that biases decisions based on prior experience in similar situations. A positive experience in the past will emotionally bias future decisions toward similar options that have a high probability of bringing about the same outcome. Likewise, a prior bad experience with something will lead to an affective reaction that biases future decisions away from similar options. While not always accurate, such emotional biasing helps decision-makers skip over many details and focus on the bottom line. These emotional biasing signals, described as "somatic markers" by Antonio Damasio, are integrated with cognition within the ventromedial prefrontal cortex (Damasio, 1994). In fact, Damasio and colleagues found that patients with lesions to the ventromedial prefrontal cortex were impaired on a test specially designed to measure deficits in the ability to use somatic markers to guide decision-making (Bechara, Tranel & Damasio, 2000), a test known as the Iowa Gambling Task (IGT). Furthermore, in healthy normal individuals, IGT performance correlates with ventromedial prefrontal cortex activation during fMRI (Northoff et al., 2006).

Because sleep deprivation reduces regional metabolic activity in the ventromedial prefrontal cortex (as well as other prefrontal regions), Killgore and colleagues hypothesized that it may also lead to deficits in emotional decision-making on the IGT, similar to the deficits seen in patients with actual lesions to that brain region (Killgore, Balkin & Wesensten, 2006). Healthy participants completed the IGT when normally rested and again following 49 hours of sleep deprivation. Participants were told they would win actual money for selecting from the most advantageous decks and should try to avoid the "bad" decks, which would rapidly negate any profits they may have earned. As expected, normally rested participants learned to avoid the more exciting high-risk "bad" decks in favor of "good" decks providing modest but relatively consistent payoffs. Following two nights without sleep, however, these same subjects changed their strategy and shifted toward selecting more frequently from riskier "bad" decks, despite continued losses from doing so (see Figure 15.6). Notably, the pattern of performance expressed

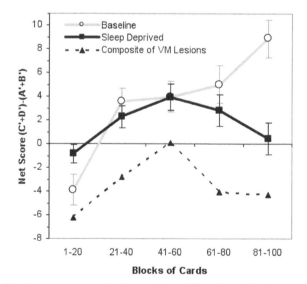

Figure 15.6   **Performance on the Iowa Gambling Task (IGT) at rested baseline and again following two nights of sleep deprivation**

Note: Data reflect the net score difference between the number of selections from advantageous decks minus the number of selections from disadvantageous decks for each block of 20 trials. At rested baseline (grey line with open circles), participants gradually learned to avoid disadvantageous decks and chose more frequently from advantageous decks. In contrast, following 49.5 hours of sleep deprivation (black line with filled squares), the same subjects showed a significantly different pattern of performance that was best characterized by a quadratic function, demonstrating a shift in behavior toward advantageous decks in the first half of the game followed by a reversal of behavior toward progressively more choices from disadvantageous decks in the latter half of the game. For comparison, the figure also shows a composite line calculated by extrapolating and summarizing IGT data reported from previously published reports of patients with damage to the ventromedial prefrontal cortex (dashed line with filled triangles).

Source: Reprinted with permission from Killgore, Balkin and Wesensten, (2006).

among the sleep-deprived volunteers was strikingly similar to (though less severe than) that typically reported in other studies of patients with lesions to the ventromedial prefrontal cortex (Killgore, Balkin & Wesensten., 2006). An essentially identical pattern was replicated in a second study of 75 hours of sleep deprivation (Killgore, Lipizzi et al., 2007). These findings suggest that sleep deprivation impairs the ability to use emotional feedback from prior experience to guide decision-making. Sleep-deprived subjects appear to be short-sighted in their decisions, preferring immediate gratification over longer-term success, a pattern that would be expected from reduced functioning of the ventromedial prefrontal cortex.

The tendency to make riskier decisions during sleep deprivation also depends on the manner in which the risks are presented to the individual. McKenna and colleagues found that if the possible outcome of a risky choice is framed in terms of a potential gain, sleep-deprived subjects tend to make riskier decisions than they ordinarily would when rested. If, on the other hand, the possible outcome is framed as a potential loss, sleep deprivation appears to reduce risk-taking relative to normal rested behavior (McKenna et al., 2007). Again, altered brain functioning within emotion and reward circuitry may play a role in the outcome of risk assessments made by sleep-deprived individuals. Neuroimaging data suggest that sleep deprivation leads to altered activation patterns within regions of the brain that are involved in the evaluation of rewards and punishments, leading to increases in regions involved in the valuation of rewards and reduction in regions associated with the valuation of punishments (Venkatraman et al., 2007). These findings are consistent with the notion that sleep loss may lead to an increased and perhaps unrealistic expectation that risky options will lead to greater rewards relative to losses.

SOCIAL DECISIONS AND THE PERCEPTION OF FAIRNESS

We have discussed how sleep loss affects brain regions that are critical for emotional processes and how these changes appear to be manifested as worsened mood, negatively biased perceptions, feelings of persecution, and reduced behavioral coping and frustration tolerance. It is not inconceivable that these alterations in emotional functioning could also affect decisions within the realm of human social interactions, a possibility that would have clear implications for occupational settings that rely on teamwork and close co-worker interactions. In fact, the evidence suggests that sleep deprivation leads to decreased trust of others and greater social aggression. In one study, participants engaged in a series of "bargaining" and "trust" games that had tangible financial consequences (Anderson & Dickinson, 2010). Participants had to strike financial deals with an unknown partner when rested and again after 36 hours of sleep deprivation. During each bargain, subjects had to either accept or reject the terms proposed by their partner. When sleep deprived, participants were less trusting of the intentions of their unknown partner and were more likely to reject bargaining offers made by their partner if the offers were perceived as "unfair." Of particular interest, sleep-deprived subjects were still more likely to reject unfair offers even if that decision would clearly lead to financial losses for both parties. Apparently, sleep deprivation leads to increased sensitivity to the perception of inequity and feelings that one is being slighted or persecuted. In such cases, sleep-deprived individuals appear less willing to cooperate and would rather respond aggressively to an offending party, even if it is to their own detriment (i.e., that they would also lose money themselves). Thus, sleep deprivation may lead to problems with social judgment, cooperation, and aggression, a finding that may be particularly relevant in operational settings where workers must function effectively for long periods in close proximity.

MORAL JUDGMENT

Neurocognitive changes in the affect regulation system due to sleep loss may also have subtle effects on moral reasoning (Killgore, Killgore et al., 2007). Functional neuroimaging studies have suggested that moral decisions, particularly those involving considerable emotional conflict, appear to activate regions of the medial prefrontal cortex (Greene et al., 2001). Some of the medial prefrontal regions most involved in emotionally charged moral judgments are also among the same prefrontal regions that show profoundly reduced metabolic activity during sleep deprivation (Thomas et al., 2000). In one study, healthy subjects contemplated solutions to moral dilemmas when normally rested and again after two nights of sleep deprivation (Killgore, Killgore et al., 2007). Sleep deprivation had little effect on the speed of decisions about non-moral control dilemmas or even about moral dilemmas that were not very emotionally arousing. Thus, sleep loss did not adversely affect this type of decision-making. However, when the dilemmas actively involved highly emotional choices, sleep-deprived subjects were significantly slower in making their decisions and were more likely to choose options that violated the moral positions they

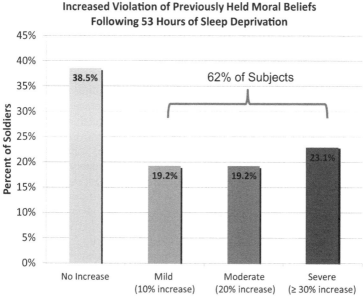

Figure 15.7   Following 53 hours of sleep deprivation, approximately 62 percent of participants showed an increased willingness to select more utilitarian solutions to highly emotionally charged moral dilemmas, compared to their rested baseline answers

Source: Adapted from Killgore, Killgore et al., (2007).

previously expressed when normally rested. In fact, about 62 percent of the participants showed an increase in their willingness to violate their previously held moral positions after two nights without sleep, and almost a quarter showed severe increases in this willingness (see Figure 15.7). Again, these findings suggest that while sleep deprivation may affect a variety of neurocognitive processes, it appears to have particularly disruptive effects on tasks that rely on emotional information (Killgore, Killgore et al., 2007).

CONCLUSIONS

The effects of sleep deprivation on neurocognitive functioning are extensive, ranging from deficits in simple alertness, vigilance, and attention, to emotional processes such as mood, affective perception, and interpersonal responses to frustration, and even higher-order cognitive processes such as risk-taking, decision-making, and moral judgment. A common theme of the present chapter has been that sleep loss affects the normal functioning of prefrontal brain systems, which are critical for motivated and goal- directed behavior. While many other neurocognitive systems are undoubtedly affected by sleep loss as well, we have focused on the prefrontal attention and emotional–cognitive systems because these are likely to have particular relevance within the workplace and other operational environments.

The findings reviewed here suggest that sustained wakefulness is associated with significant declines in global metabolic activity within the brain, particularly within prefrontal inhibitory and thalamic information-processing systems. Without sufficient restorative sleep, simple alertness and attention decline and the probability of brief attentional lapses increases. Insufficient sleep is particularly dangerous during the early morning hours, when the natural rhythm of alertness is lowest, and may lead sleep-deprived workers to experience "microsleeps," or involuntary functional "sleep attacks," where their motivation to remain awake is temporarily overpowered by the biological pressure to sleep.

The prefrontal cortex is also critical for emotion regulation, and reductions in the metabolic activity of this region, particularly the medial cortical systems, appear to be associated with a mild dysregulation of more primitive affective systems deep within the limbic regions of the brain. Lacking sufficient modulation from the prefrontal cortex, these emotional systems appear to worsen general mood, bias some aspects of emotional perception, increase feelings of dysphoria, persecution, and anxiety, and reduce emotional coping capacities, empathy, and frustration tolerance. Without sufficient sleep, workers may be more prone to misinterpret the intentions of others, find less satisfaction in their work, and cope less effectively with the stresses of the job, all of which could have an impact on safety and productivity. Moreover, the neurocognitive effects of sleep loss extend to the realm of judgment and decision-making. Sleep loss appears to disrupt normal functioning of these emotional–cognitive integration systems, leading to increased risk-taking, impaired decision-making, and altered moral reasoning.

Just as muscles need rest, so does the brain need sleep. While there are large differences among individuals in the amount of sleep required for optimal performance, the preponderance of the evidence suggests that workers generally function best on about 7 to 8 hours of regular sleep each night. As sleep durations are dropped below this level, cognitive, social, and emotional processes can be expected to degrade accordingly.

REFERENCES

Anderson, C. & Dickinson, D.L. (2010). Bargaining and trust: The effects of 36-h total sleep deprivation on socially interactive decisions. *Journal of Sleep Research*, 19, 54–63.

Babkoff, H., Caspy, T., Mikulincer, M. & Sing, H.C. (1991). Monotonic and rhythmic influences: A challenge for sleep deprivation research. *Psychological Bulletin*, 109, 411–28.

Bechara, A., Tranel, D. & Damasio, H. (2000). Characterization of the decision-making deficit of patients with ventromedial prefrontal cortex lesions. *Brain*, 123, 2189–202.

Belenky, G., Penetar, D., Thorne, D., Popp, K., Leu, J., Thomas, M., Sing, H., Balkin, T., Wesensten, N. & Redmond, D. (1994). The effects of sleep deprivation on performance during continuous combat operations, in B.M. Marriott (ed.), *Food Components to Enhance Performance*. Washington, DC: National Academy Press, 127–35.

Borbély, A.A. (1982). A two process model of sleep regulation. *Human Neurobiology*, 1, 195–204.

Chee, M.W., Tan, J.C., Zheng, H., Parimal, S., Weissman, D.H., Zagorodnov, V. & Dinges, D.F. (2008). Lapsing during sleep deprivation is associated with distributed changes in brain activation. *Journal of Neuroscience*, 28, 5519–28.

Damasio, A.R. (1994). *Descartes' Error: Emotion, Reason and the Human Brain*. New York: Grosset/Putnam.

Dinges, D.F., Pack, F., Williams, K., Gillen, K.A., Powell, J.W., Ott, G.E., Aptowicz, C. & Pack, A.I. (1997). Cumulative sleepiness, mood disturbance, and psychomotor vigilance performance decrements during a week of sleep restricted to 4–5 hours per night. *Sleep*, 20, 267–77.

Doran, S.M., Van Dongen, H.P. & Dinges, D.F. (2001). Sustained attention performance during sleep deprivation: Evidence of state instability. *Archives Italiennes de Biologie* (Pisa), 139, 253–67.

Drummond, S.P., Bischoff-Grethe, A., Dinges, D.F., Ayalon, L., Mednick, S.C. & Meloy, M.J. (2005). The neural basis of the psychomotor vigilance task. *Sleep*, 28, 1059–68.

Drummond, S.P., Brown, G.G., Salamat, J.S. & Gillin, J.C. (2004). Increasing task difficulty facilitates the cerebral compensatory response to total sleep deprivation. *Sleep*, 27, 445–51.

Drummond, S.P., Meloy, M.J., Yanagi, M.A., Orff, H.J. & Brown, G.G. (2005). Compensatory recruitment after sleep deprivation and the relationship with performance. *Psychiatry Research*, 140, 211–23.

Goel, N., Rao, H., Durmer, J.S. & Dinges, D.F. (2009). Neurocognitive consequences of sleep deprivation. *Seminars in Neurology*, 29, 320–39.

Greene, J.D., Sommerville, R.B., Nystrom, L.E., Darley, J.M. & Cohen, J.D. (2001). An fMRI investigation of emotional engagement in moral judgment. *Science*, 293, 2105–8.

Harrison, Y. & Horne, J.A. (2000). The impact of sleep deprivation on decision making: A review. *Journal of Experimental Psychology, Applied*, 6, 236–49.

Harrison, Y., Horne, J.A. & Rothwell, A. (2000). Prefrontal neuropsychological effects of sleep deprivation in young adults – a model for healthy aging? *Sleep*, 23, 1067–73.

Huck, N.O., McBride, S.A., Kendall, A.P., Grugle, N.L. & Killgore, W.D. (2008). The effects of modafinil, caffeine, and dextroamphetamine on judgments of simple versus complex emotional expressions following sleep deprivation. *International Journal of Neuroscience*, 118, 487–502.

Kahn-Greene, E.T., Killgore, D.B., Kamimori, G.H., Balkin, T.J. & Killgore, W.D.S. (2007). The effects of sleep deprivation on symptoms of psychopathology in healthy adults. *Sleep Med*, 8, 215–21.

Kahn-Greene, E.T., Lipizzi, E.L., Conrad, A.K., Kamimori, G.H. & Killgore, W.D.S. (2006). Sleep deprivation adversely affects interpersonal responses to frustration. *Personality & Individual Differences*, 41, 1433–43.

Killgore, W.D.S. (2010). Asleep at the trigger: Warfighter judgment and decisionmaking during prolonged wakefulness, in P.T. Bartone, R.H. Pastel & M.A. Vaitkus (eds), *The 71F advantage: Applying Army Research Psychology for Health and Performance Gains*. Washington, D.C.: National Defense University Press, 59–77.

Killgore, W.D.S., Balkin, T.J. & Wesensten, N.J. (2006). Impaired decision-making following 49 hours of sleep deprivation. *Journal of Sleep Research*, 15, 7–13.

Killgore, W.D.S., Kahn-Greene, E.T., Lipizzi, E.L., Newman, R.A., Kamimori, G.H. & Balkin, T.J. (2008). Sleep deprivation reduces perceived emotional intelligence and constructive thinking skills. *Sleep Med*, 9, 517–26.

Killgore, W.D.S., Killgore, D.B., Day, L.M., Li, C., Kamimori, G.H. & Balkin, T.J. (2007). The effects of 53 hours of sleep deprivation on moral judgment. *Sleep*, 30, 345–52.

Killgore, W.D.S., Lipizzi, E.L., Kamimori, G.H. & Balkin, T.J. (2007). Caffeine effects on risky decision-making after 75 hours of sleep deprivation. *Aviation Space and Environmental Medicine*, 78, 957–62.

Killgore, W.D.S., McBride, S.A., Killgore, D.B. & Balkin, T.J. (2006). The effects of caffeine, dextroamphetamine, and modafinil on humor appreciation during sleep deprivation. *Sleep*, 29, 841–7.

King, A.C., Belenky, G. & Van Dongen, H.P. (2009). Performance impairment consequent to sleep loss: Determinants of resistance and susceptibility. *Current Opinion in Pulmonary Medicine*. 15, 559–64.

Lim, J. & Dinges, D.F. (2008). Sleep deprivation and vigilant attention. *Annals of the New York Academy of Sciences*, 1129, 305–22.

McKenna, B.S., Dickinson, D.L., Orff, H.J. & Drummond, S.P. (2007). The effects of one night of sleep deprivation on known-risk and ambiguous-risk decisions. *Journal of Sleep Research*, 16, 245–52.

Northoff, G., Grimm, S., Boeker, H., Schmidt, C., Bermpohl, F., Heinzel, A., Hell, D. & Boesiger, P. (2006). Affective judgment and beneficial decision making: Ventromedial prefrontal activity correlates with performance in the Iowa Gambling Task. *Human Brain Mapping*, 27, 572–87.

Pace-Schott, E.F., Hutcherson, C.A., Bemporad, B., Morgan, A., Kumar, A., Hobson, J.A. & Stickgold, R. (2009). Failure to find executive function deficits following one night's total sleep deprivation in university students under naturalistic conditions. *Behavioral Sleep Medicine*, 7, 136–63.

Rocklage, M., Williams, V., Pacheco, J. & Schnyer, D.M. (2009). White matter differences predict cognitive vulnerability to sleep deprivation. *Sleep*, 32, 1100–103.

Tempesta, D., Couyoumdjian, A., Curcio, G., Moroni, F., Marzano, C., De Gennaro, L. & Ferrara, M. (2010). Lack of sleep affects the evaluation of emotional stimuli. *Brain Research Bulletin*, 82, 104–8.

Thomas, M., Sing, H., Belenky, G., Holcomb, H., Mayberg, H., Dannals, R., Wagner, H., Thorne, D., Popp, K., Rowland, L., Welsh, A., Balwinski, S. & Redmond, D. (2000). Neural basis of alertness and cognitive performance impairments during sleepiness. I. Effects of 24 h of sleep deprivation on waking human regional brain activity. *Journal of Sleep Research*, 9, 335–52.

Tucker, A.M., Whitney, P., Belenky, G., Hinson, J.M. & Van Dongen, H.P. (2010). Effects of sleep deprivation on dissociated components of executive functioning. *Sleep*, 33, 47–57.

van der Helm, E., Gujar, N. & Walker, M.P. (2010). Sleep deprivation impairs the accurate recognition of human emotions. *Sleep*, 33, 335–42.

Van Dongen, H.P., Baynard, M.D., Maislin, G. & Dinges, D.F. (2004). Systematic interindividual differences in neurobehavioral impairment from sleep loss: Evidence of trait-like differential vulnerability. *Sleep*, 27, 423–33.

Van Dongen, H.P., Maislin, G., Mullington, J.M. & Dinges, D.F. (2003). The cumulative cost of additional wakefulness: Dose-response effects on neurobehavioral functions and sleep physiology from chronic sleep restriction and total sleep deprivation. *Sleep*, 26, 117–26.

Venkatraman, V., Chuah, Y.M., Huettel, S.A. & Chee, M.W. (2007). Sleep deprivation elevates expectation of gains and attenuates response to losses following risky decisions. *Sleep*, 30, 603–9.

Yoo, S.S., Gujar, N., Hu, P., Jolesz, F.A. & Walker, M.P. (2007). The human emotional brain without sleep – a prefrontal amygdala disconnect. *Current Biology*, 17(20), R877–878.

16

Circadian Rhythms and Mental Performance

Jim Waterhouse

INTRODUCTION

In order to understand the origin of rhythmic changes in mental performance, and how these change in individuals on altered sleep–wake cycles, it is necessary to understand body rhythms in general.

Homeostasis and Rhythmic Changes in the Body

Homeostasis, keeping the internal environment within narrow limits, is a key concept in biology, and failure in this process can lead to death. This stability is normally achieved by reflex mechanisms acting via negative feedback loops. In spite of the importance of homeostasis, however, repeated measurements during the course of a day have indicated that daily rhythms are superimposed upon this stability. For example, in humans, normally active in the daytime and asleep at night, core temperature is 1–2°C higher in the daytime; day–night changes are observed in physiological variables in general (Minors & Waterhouse, 1981; Reilly, Atkinson & Waterhouse, 1997; Waterhouse & DeCoursey, 2004a).

The environment in which we live and the lifestyle we lead both show daily rhythmicity and these factors will contribute to the observed rhythms. Other factors are involved also, and a protocol to illustrate this is the "constant routine." In this protocol, subjects are required: to stay awake and sedentary or lying down for at least 24 hours in an environment of constant temperature, humidity and lighting; to engage in little physical activity (generally reading or listening to music); and to eat identical meals at regular intervals. Rhythmicity due to an individual's environment or lifestyle is removed by this protocol, and so any residual rhythmicity must be internal in origin.

Results from this protocol are illustrated for core (rectal) temperature in Figure 16.1. The daily rhythm shows a peak around the late afternoon and a minimum during sleep at around 05:00 hours. During the constant routine (dashed line), the rhythm of core temperature persists but with a diminished amplitude. Three deductions can be made from these results:

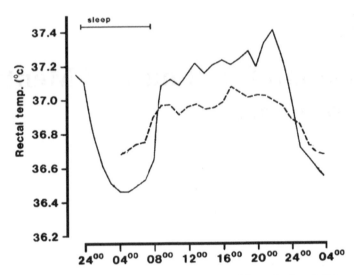

Figure 16.1   Core (rectal) temperature measured hourly in eight subjects

Note: Solid line: living normally and sleeping from 24:00 to 08:00 hours; dashed line: woken at 04:00 hours and spending the next 24 hours on a "constant routine".

Source: From Minors & Waterhouse, 1981.

1. The rhythm observed during the constant routine arises internally; it is the "endogenous" component of the rhythm and its generation is attributed to a body clock.
2. Since the two rhythms are not identical, effects due to the environment and lifestyle are also present when the subjects live conventionally, the difference between the two rhythms being the "exogenous" component of the rhythm.
3. These two components are in phase. During the daytime, body temperature is raised by the body clock acting in synchrony with a more "dynamic" environment and increased physical and mental activities; during the night, the clock, a more restful environment and relaxed lifestyle all act to reduce core temperature.

Although all biological rhythms show endogenous and exogenous components, there are differences in detail – mainly the timing of the rhythm, the nature of the exogenous component, and the relative strengths of the endogenous and exogenous components.

1. For variables associated with activity (e.g. cardiovascular function and plasma adrenaline), the timing is similar to that of core temperature whereas, for variables associated growth or repair (e.g. growth hormone, which occurs more during sleep), the timing tends to be the opposite.
2. The exogenous component for core temperature (as for heart rate and blood pressure) consists of increases during waking caused by light and activity (including mental, physical and social activity) and decreases at night caused by darkness, change in posture, sleep and inactivity. For insulin, the exogenous component is mainly food intake; for growth hormone, sleep onset; for melatonin, strong light (which suppresses its secretion by the pineal gland). For mental performance, the exogenous component

is the ambient conditions (light, noise and temperature) under which the task is being performed.

3. For many variables (e.g. heart rate, blood pressure and growth hormone), the exogenous component is far stronger than the effect of the body clock, as a result of which these rhythms are poor markers of the phase of the body clock. By contrast, melatonin secretion and cortisol secretion have stronger endogenous components and so their rhythms can be used as "clock markers." In subjects with normal amounts of activity (that is, without being confined to their bed or excessively active), the two components for core body temperature are similar in size (see Figure 16.1). Mental performance rhythms are poor markers of the body clock because, regardless of the relative strengths of the endogenous and exogenous components, the endogenous component is not due to the body clock only (see below).

The Body Clock and Some of its Properties

The body clock consists of paired suprachiasmatic nuclei at the base of the hypothalamus; many details of its genetics and molecular biochemistry are now known (Clayton, Kyriacou & Reppert, 2001; Reppert & Weaver, 2001). Being sited close to areas that exert widespread effects upon the body (e.g. temperature regulation, hormone secretion and the sleep–wake and feeding cycles), the body clock produces rhythmicity throughout the body.

When individuals are studied in time-free environments (such as an underground cave), the body clock and rhythms produced by it continue to be manifest but with a period closer to 25 than 24 hours. Such rhythms are called circadian (Latin: "about a day") and the timing system is described as "free-running." This period of about 25 hours does not reflect the intrinsic period of the body clock, due to effects of light exposure during waking. Current evidence (from protocols using very dim light during waking or from blind subjects) indicates that the true intrinsic period of the body clock averages about 24.3 hours (Shanahan & Czeisler, 2000).

Whatever its exact period, the body clock will be of value only if it and the rhythms it drives are synchronized to a solar (24-hour) day. This adjustment of the body clock is achieved by zeitgebers (German: "time-giver"), rhythms with a 24-hour period resulting directly or indirectly from the environment. In humans, the most important zeitgeber is the light–dark cycle, coupled to the regular secretion of melatonin from the pineal gland during darkness. Other zeitgebers (exercise, social factors and food intake) play a minor role only (Mrosovsky, 1999; Edwards, Reilly & Waterhouse, 2009). The suitable use of zeitgebers, particularly the light–dark cycle, can be used to promote adjustment of the body clock to a new time zone or during night work (Waterhouse et al., 2007). Under normal circumstances, all zeitgebers indicate the same external time and so act harmoniously to adjust the body clock.

When light acts upon the body clock, its effect depends on the time of presentation relative to the temperature minimum (normally around 05:00 hours, see Figure 16.1). Light presented in the 6 hours after this minimum advances the body clock to an earlier time, in the 6 hours before, delays it, and at other times exerts no effect (Khalsa, Jewett & Cajochen, 2003). The size of the phase shifts depends upon the intensity of the light, and domestic lighting is normally sufficiently strong to adjust the body clock.

Melatonin ingestion also adjusts the body clock; ingestion in the afternoon and early evening advances it and, in the second half of sleep and during the early morning, delays it. Since bright light inhibits endogenous melatonin secretion, the clock-shifting effects of these two zeitgebers reinforce each other, bright light in the hours immediately after the temperature minimum advancing the body clock not only directly but also indirectly (by suppressing melatonin secretion and so preventing the phase-delaying effect that melatonin would have exerted at this time) (Lewy, Bauer & Ahmed, 1998).

The body clock promotes diurnality in humans, separating daytime activities from nocturnal sleep (when restitution occurs). It also *prepares* individuals for sleep in the evening and waking in the morning, changes that require an ordered sequence of biochemical and physiological events (Waterhouse & DeCoursey, 2004a). The rhythmicity that is produced reflects integration between the body clock and the sleep–wake cycle – core temperature and melatonin secretion being important links in this (Murphy & Campbell, 1997; Shochat, Luboshitsky & Lavie, 1997).

The body clock and the rhythms it produces are robust, as a result of which irregularities in lifestyle (taking a daytime night or waking at night) or the environment (flashes of lightning at night or dark thunder clouds blacking out the daylight) do not alter their phase. This robust property has strong ecological value, but it causes problems when the individual's sleep–wake cycle is changed, as after a time-zone transition or during night work (below).

CIRCADIAN RHYTHMS OF MENTAL PERFORMANCE

Measuring Rhythms

Mental performance tasks, like core temperature, show rhythms with endogenous and exogenous components, but added complexities exist. These include: the nature of the performance task itself, effects of time awake and sleep loss, the environmental factors that affect performance, and the role of practice.

It is possible to consider mental performance tasks as falling on a continuum between "simple" tasks (e.g. simple reaction time, where the role of cognition, due to the CNS, is limited) and "complex" tasks (where cognition is dominant) (Folkard, 1990; Waterhouse et al., 2001). As will be considered later, the effects of time awake and sleep loss upon task performance as well as the time of peak of the rhythm depend upon the type of task.

Environmental conditions affecting performance include lighting, temperature and noise (Wojtczak-Jaroszowa & Jarosz, 1987). Performance deteriorates if individuals feel uncomfortable, are distracted, or the conditions are non-ideal (e.g. poor traffic conditions, or weather when working outdoors). Controlling the last of these is not possible, of course, and so laboratory-based studies have advantages. In the laboratory, controlling ambient temperature at a comfortable level is an obvious improvement, as is decreasing noise levels – but levels that are too low might become soporific. Brightening ambient lighting is also advantageous (producing general activation of the CNS and improving visual acuity) but light that is too bright can be disturbing and make visual display units difficult to read.

The effects of practice need to be incorporated into an experimental protocol. As individuals become familiar with any task, the speed and accuracy with which it can be performed increase. Particularly for cognitive-based tasks, it is also necessary to develop a suitable strategy, and this requires practice. Three methods have been used to deal with effects of practice (Blatter & Cajochen, 2007):

1. The task is performed often enough before the main investigation for practice effects to have disappeared. Such a familiarization process can be rather time-consuming, however.
2. Subjects are divided into several subgroups, each subgroup starting its sequence of 24-hour measurements at a different time of day (e.g. 24:00, 04:00 ... 20:00 hours). The mean performance at each time of day *for a group as a whole* will contain the same mixture of effects due to practice, and so the performance rhythm will be superimposed upon "noise" due to this mixture.
3. A control group, identical to the experimental group except for the intervention or effect being investigated, is incorporated into the experiment, but this design doubles the number of subjects required.

Circadian Rhythms in Mental Performance

Changes in self-rated subjective feelings and many mental performance tasks have been measured in the daytime (diurnally) in healthy subjects living a conventional sleep–activity schedule (Folkard, 1990; Blatter & Cajochen, 2007). Rhythmic changes are present, with worse performance in the early morning and late evening and best performance somewhere near the middle of the day. Measuring performance during the night (unless subjects are working night shifts) is less common because this requires subjects to remain awake or to be woken; both procedures present a problem since mental performance deteriorates in the presence of extended time awake and is transiently depressed immediately after subjects are woken (see below). Nevertheless, when subjective feelings and mental performance are measured at night (Monk et al., 1997), values are lower than during the daytime.

Typical diurnal changes in a simple and complex performance task are illustrated diagrammatically in Figure 16.2. The tasks differ in their time-courses, simpler tasks showing a closer parallelism with core temperature and peaking in the late afternoon, complex tasks peaking earlier, around midday. Differences in time-course are due to variations in the rate at which decrement occurs due to time elapsed since waking; as the task becomes more complex (that is, its cognitive element increases), this rate of deterioration increases. The decline in performance is often referred to as "fatigue" and is also found in field studies (Wojtczak-Jaroszowa & Jarosz, 1987; Åkerstedt, 2007). When performance rhythms are assessed mathematically (cosinor analysis), the acrophase (time of peak of the fitted cosine curve) is around midday for a complex task but in the late afternoon (as is the acrophase for core temperature) for a simple task.

The parallelism between performance at simple tasks and core temperature has long been known (Kleitman, 1963), and the possibility of a causal link has often been suggested. However, such parallelism is also observed when adrenaline and other rhythms are considered (Reilly, Atkinson & Waterhouse, 1997). The current consensus is that temperature and adrenaline are, like performance at simple tasks when time awake is comparatively unimportant, measures of the degree of "activation" of the body, and the parallelism results from independent manifestations of the widespread rhythmicity produced by the body clock.

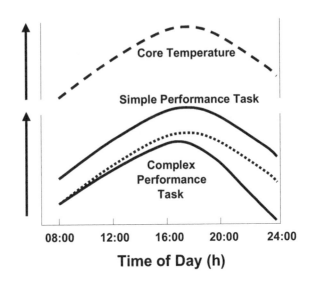

Figure 16.2   Diagrammatic illustration of the daytime rhythms of, top, core temperature and, bottom (full lines), simple and complex performance tasks

Note: The dotted line (bottom) shows the time-course of a complex performance test in the absence of the effects of fatigue. Higher core temperature and better mental performance are plotted upwards

Source: From Valdez, Reilly & Waterhouse, 2008.

Separating Circadian and Time-awake Effects in Mental Performance

Because time awake affects mental performance rhythms, the constant routine protocol has a fundamental flaw insofar as it requires subjects to remain awake for at least 24 hours and so accentuates time-awake effects. (This problem is not seen in other variables such as core temperature, where time awake has a negligible effect). When mental performance tasks have been measured during such a protocol (Johnson et al., 1992; Carrier & Monk, 2000), rhythmicity has been retained but superimposed upon a general decline in performance due to time awake (Figure 16.3). Over the course of a daily cycle, this decline can be measured by comparing mental performance at points separated by 24 hours (see "X," at 08:00 h, in Figure 16.3).

The circadian and time-awake components of mental performance can be separated by a "forced desynchronization" protocol. This protocol is based upon early work by Kleitman (1963). Subjects are required to live on "days" of abnormal length, equal to 21, 27 or 28 solar hours, for example. With a 28-hour "day," the subjects' times of retiring, rising and eating meals, as well as all daily activities and the imposed light–dark cycle, become 4 hours later each solar day. Since the body clock cannot adjust to such periods (which are too far removed from its intrinsic period of 24.3 hours, see above), it "free-runs" with a period just over 24 hours. After about six imposed "days," the sleep–wake cycle and body clock coincide once again, because [6 x 28 h] is equal to [7 x 24 h]; this length of time is called a beat cycle.

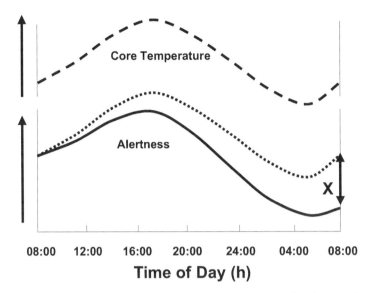

Figure 16.3   Diagrammatic illustration of the daily rhythms of, top, core temperature and, bottom (full line), alertness during a constant routine

Note: The dotted line (bottom) shows the time-course of alertness in the absence of effects due to fatigue; X, the fall in alertness after 24 h due to effects of time awake; Increased temperature and alertness plotted upwards.

Source: From Valdez et al., 2008.

Performance data are collected at regular intervals during the waking periods throughout a whole beat cycle. The measured performance at any point during the beat cycle is due to a combination of two factors: the amount of time awake (often described in 3-hour blocks: 0–3 hours after waking, 3–6 hours after waking, etc.) and the circadian rhythm, due to the body clock and estimated from the phase of the circadian rhythm of core temperature (often described in 4-hour blocks: 2–6 hours after the acrophase, 6–10 hours after the acrophase, etc.). All the results obtained from a beat cycle can then be averaged in two ways (for details, see Harrison, Jones & Waterhouse, 2007). In the first, all results where the amount of time awake is the same (e.g. 0–3 hours) are averaged. This process cancels out the effects of the circadian rhythm and so the value after 0–3 hours time awake can be shown. This averaging is repeated for the other blocks of time awake. In the second way, all results where the circadian phase is the same (e.g. 2–6 hours after the acrophase) are averaged. This process cancels out the effects of time awake and so the value 2–6 hours after the acrophase can be shown. This averaging is repeated for the other blocks of circadian phase.

Figure 16.4 illustrates results using this method (Harrison, Jones & Waterhouse, 2007). Assessments of sleepiness were made 1, 4, 7, 10, 13 and 16 h after waking (total waking time being 18.67 h per "day"). The top part of the Figure shows sleepiness at six circadian phases of core temperature, where: p1 indicates a 4-h "bin" centred on the time of the temperature acrophase; p2, a bin 2-6 h after the acrophase; p3, a bin 6-10 h

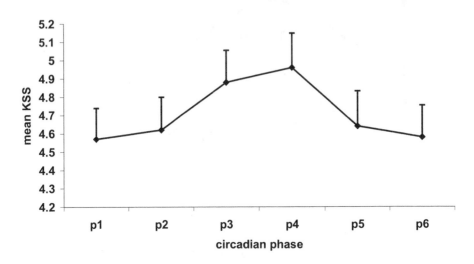

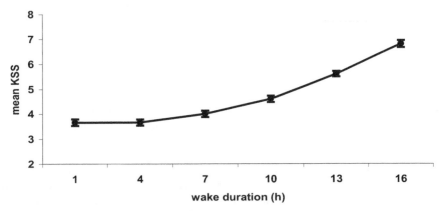

Figure 16.4   Mean (+SE) scores for the Karolinska Sleepiness Scale (KSS) taken from subjects undergoing a forced desynchronization protocol (imposed "days" of 28 hours)

Note: For further details, see text.
Source: From Harrison, Jones & Waterhouse, 2007.

after the acrophase; p4, a bin 10-14 h after the acrophase (centred on the time of the temperature minimum); p5, a bin 6-10 h before the acrophase; and p6, a bin 2-6 h before the acrophase. The bottom part of the Figure shows sleepiness after different times awake. Independent circadian (parallel to core temperature) and time-awake effects are clearly evident (compare with Figures 16.2 and 16.3). This method has been used to investigate several other rhythms of subjective feelings and cognitive tasks (Boivin et al., 1997; Wright, Hull & Czeisler, 2002). It seems that the time-awake effect should be more marked as the degree of cognition involved in the task increases, but this does not seem to have been studied systematically.

Even though this protocol is currently the best way to separate circadian and time-awake effects in mental performance tasks, it is time-consuming and effects due to practice and/or boredom might intrude. Also, if performance changes in the hours immediately after waking are investigated in detail, the general decline in performance due to time awake often shows a transient trend in the *opposite* direction (that is, performance improves) for about the first hour after waking. In other words, the benefits of a sleep (see below) are not apparent immediately after waking from it. This temporarily poorer performance is "sleep inertia" (Naitoh, Kelly & Babkoff, 1993); it tends to be more marked after sleeps taken during the night than the daytime, if the subject wakes from slow wave sleep, or the environment is soporific rather than alerting (Stampi, 1992). Effects of sleep inertia have substantially worn off about 10 minutes after waking, though they can take at least 1 hour to disappear completely.

Modeling Rhythms of Mental Performance and Effects of Sleep Loss

Several models of the rhythm of mental performance have been developed that are based upon the concept that subjective feelings and mental performance are determined by summed effects of circadian (parallel to core temperature) and time-awake (homeostatic) components – that is, are based on a two-process model of sleep regulation (Borbély, 1982). The "two-process model of alertness regulation" has been validated against subjective and objective ratings of performance and sleepiness (lack of alertness) in the laboratory and field; it also predicts sleep length and latency, and the risk of falling asleep (Åkerstedt & Folkard, 1997; Åkerstedt, 2007; Beersma & Gordijn, 2007). Other models also incorporate effects due to prior sleep loss (see review by McCauley et al., 2009) and sleep inertia. A workshop compared their effectiveness in a variety of scenarios that might be expected in field conditions, including night work (Van Dongen, 2004). It was concluded that such models need refinement to become versatile enough to describe all possibilities found in field conditions.

Many studies indicate that sleep loss of as little as two hours leads to a decline in mental performance (Dinges, 1995; Van Dongen & Dinges, 2005; Åkerstedt, 2007). Complex and vigilance-requiring tasks are affected most, partly due to loss of motivation that accompanies fatigue. Effects of sleep loss are lessened if the task does not involve continuous testing, there is opportunity to recover, or short sleeps (naps) can be taken. Naps can counteract performance decrement in adults who have lost sleep or are working at night (Stampi, 1992), though their recuperative value varies with the type of mental task (Åkerstedt, 2007). Current investigations are concentrating on the value of naps of different length and taken at different circadian phases, as well as the composition of the nap with regard to sleep stages. The recuperative role of a nap seems to be equivalent to that of a sleep (when differences in length have been taken into account), but there is also evidence that alertness can be increased by the process of sleep onset itself (Lack & Tiezel, 2000; Saper, Chou & Scammell, 2001).

CHANGED SLEEP–WAKE SCHEDULES AND MENTAL PERFORMANCE

The robustness of the body clock and the ecological advantage this brings have already been mentioned. However, this slow adjustment of the body clock means that circumstances when an individual's sleep–wake cycle is changed less transiently (such as flying to a new time zone or working at night) will cause problems because the normal synchrony (see Figure 16.1) between an individual's environment, lifestyle and body clock (the exogenous and endogenous components of a rhythm) is lost.

TIME-ZONE TRANSITIONS

This loss of the normal synchrony between endogenous and exogenous components of a rhythm causes the symptoms of "jet lag," important components of which are poor sleep at night, daytime fatigue, loss of motivation, and decrements in mental performance (Waterhouse et al., 2007). The wide range of difficulties reflects the many variables that show circadian rhythmicity. Travellers might try to adjust their lifestyle to the new environment, but this will no longer accord with the rhythms being promoted by the unadjusted body clock. To take the difficulty of sleeping during the new nighttime as an example: after a westward flight across eight time zones, the individual will feel tired at 16:00 hours local time (equivalent to 24:00 hours in the time zone just left, the time zone to which the body clock is still adjusted), and will then begin to feel more alert at local midnight (08:00 hours by "body time"). By contrast, after an eastward flight across eight time zones, the individual does not feel tired at midnight by local time (16:00 by body time), but is ready to go to sleep as the new day dawns at 08:00 hours (24:00 hours by body time). Other components of jet lag arise for similar reasons.

Adjustment of the body clock is by exposure to zeitgebers in the new time zone and is equivalent to 1–2 time zones per day, symptoms of jet lag abating as this adjustment takes place. In practice, some symptoms disappear more quickly than others, due to the relative strengths of the exogenous and endogenous components of the circadian rhythm for each variable. For variables which have a strong exogenous component, adjustment will appear to be more rapid (as the traveller's lifestyle is adjusted to the new time zone); for variables with a strong endogenous component, the rate of adjustment will be slower, in line with that of the body clock.

Deterioration in mental performance is expected in the daytimes immediately after the time-zone transition for two reasons. First, after a flight across eight time zones to the east, core temperature will not peak at 17:00 hours by new local time but rather at 01:00 hours and, after a flight across eight time zones to the west, at 09:00 hours. The component of the rhythm of mental performance that parallels core temperature will be phased in the same way, the individual trying to perform mental tasks at the wrong "body time." Second, since sleep is more difficult to initiate and sustain during the new nighttime, some sleep loss will be incurred, and this will independently produce performance decrement.

Results from mental performance tests carried in the field are comparatively rare; not only are such tests difficult to perform under these conditions but also most travellers prefer work or sightseeing to mental performance testing! Even so, there are recent studies upon long-haul passengers (Vanttinen et al., 2008). Aircrew also have been studied: Graeber et al. (1986) objectively measured crews' sleep during rest days before they returned to their home time zone; Samel, Wegmann & Vejvoda (1995) measured the electroencephalogram of crews during long-haul flights; and measurements of eye-blinking and wrist activity have been made in subjects studied in simulators (French et al., 1994; Rosekind et al., 1995). Nevertheless, most studies have used sleep diaries and subjective assessments of fatigue/alertness, measurements that are quick to perform and intrude minimally upon other activities. Important reviews include those by Graeber (1989), Åkerstedt (1995), Dinges (1995) and Srinivasan et al. (2008).

Long-haul flights to the south or north, few time zones being crossed, do not produce changes in sleep pattern or fatigue (unless they involve a night flight, see below); by contrast, long-haul flights to the east or west lead to changed sleep–wake patterns, increased fatigue, sleepiness and incidence of "microsleeps" during the cruise phase, and slower mental performance with more errors (Green, 1985; Graeber et al., 1986; Samel, Wegmann & Vejvoda, 1995; Lemmer et al., 2002; Takahashi, Nakata & Arito, 2002; Carvalho-Bos et al., 2003; Richmond et al., 2004).

Westward and eastward flights do not produce identical changes to the sleep–wake cycle, amount of fatigue or frequency of microsleeps (Graeber et al., 1986; Samel, Wegmann & Vejvoda, 1995; Takahashi, Nakata & Arito, 2002; Carvalho-Bos et al., 2003); westward time-zone transitions are more often followed by an appropriate shift of the sleep–wake cycle (that is, a delay equal to the number of time zones crossed), by less fatigue and more rapid recuperation than are eastward flights (which require an advance of the sleep–wake cycle). The explanation generally offered (Waterhouse et al., 2007) is that sleep after a westward flight is delayed (and so easier to achieve as the traveler is more tired), that the body clock tends to run "slow" (and so can adjust more rapidly to the required delay), and that the zeitgeber effects of the natural light–dark cycle at the destination tend to adjust the body clock appropriately (cause a delay). It is common after westward flights for sleep of about the normal length to be taken as a single episode during the new nighttime; after eastward flights, by contrast, sleep is often split into two sessions (one in phase with home time and the other with destination time), or even for all sleep to remain in accord with home time zone (Graeber et al., 1986; Carvalho-Bos et al., 2003). Retaining home time for hours of sleep is common if the return flight takes place soon after the outward flight (Lowden & Åkerstedt, 1998). These differences (in sleep patterns and fatigue) due to flight direction are likely to result in analogous differences in decrements of mental performance. An added difficulty is that long-haul flights often include overnight travel – that is, effects due to night work exist in addition to those due to the time-zone transition.

MENTAL PERFORMANCE DURING NIGHT WORK

During night work, problems similar to those after a time-zone transition are found, and for similar reasons; the requirement to work at night and sleep during the daytime does not accord with the normal timing of the body clock. Unsurprisingly, symptoms similar to jet lag are present and are known as "shift worker's malaise." However, unlike the symptoms of jet lag, those suffered by night workers are not transient and, in those who are particularly susceptible (see below), their persistence can lead to individuals leaving night work. The problems are more persistent because the body clock is slower to adjust (possibly never adjusting fully) than is the case after a time-zone transition (where adjustment can be complete after a week or so). The reason for the poorer adjustment is that different zeitgebers during night work no longer give unambiguous information with regard to the appropriate timing of circadian rhythms; even if an individual attempts to adjust to the demands of night work (adjusting sleep times, altering meal times and being active during free time), some zeitgebers, including the natural light–dark cycle and rhythms associated with the rest of the population, do not adjust. The problem is even more intransigent because, during rest days, there is then a strong tendency for the individual to conform with the rest of society and so ALL zeitgebers now act to retain the body clock on "normal" time. That is, and any partial adjustment of the body clock obtained during night work is rapidly lost on rest days. These general principles are reviewed in Waterhouse et al. (2001) and Waterhouse & DeCoursey (2004b).

Deterioration in mental performance is to be expected, for several reasons:

1. Individuals will be working at a time closer to the temperature trough than its peak, so performance will be poorer and individuals will feel sleepy.
2. Daytime sleep will be more difficult to initiate and sustain (even if the bedroom is darkened, the phone is switched off, the rest of the household is quiet) due to the inappropriately timed temperature rhythm. Partial sleep loss will occur, adding to the difficulties described in 1. above.
3. There is a more subtle problem. During day work, individuals generally work in the first part of their waking day and relax with friends and family during the second part. By contrast, in order to spend some time with the family, night workers tend to sleep in the morning after work and meet their family in the evening before going to work. That is, night workers tend to work in the second part of their waking span, as a result of which mental performance will decrease due to increased time awake.

Reviews and studies of the problems associated with shift work (e.g. Graeber, 1989; Dinges, 1995; Signal & Gander, 2007; Balkin et al., 2008; Walker, 2008) all stress the negative effects of night work upon mood, sleepiness, alertness and mental performance (particularly upon cognitive tasks), and the critical role played by sleep loss.

INTER-INDIVIDUAL DIFFERENCES

Individuals do not suffer equally the negative effects associated with changes to the sleep–wake cycle and some are more susceptible to longer-term effects of time-zone transitions or night work. There is some evidence that chronic exposure to disruption produced by time-zone transitions leads to long-term cognitive problems and psychiatric morbidity (Cho, 2001; Katz et al., 2002), as well as to menstrual disorders (Grajewski et al., 2003). Also, in night workers, it is known that menstrual disorders and miscarriages are more common, as are eating disorders, obesity, and the incidence of cardiovascular morbidity. Much research has been concerned with devising ways to predict those individuals who will be more susceptible, either to warn them or to emphasize the value of palliatives.

Age and various personality traits have been considered (Barton et al., 1993). Age is often associated with decreased tolerance to shift work, decreasing flexibility in sleep times and physical fitness no doubt being contributory factors (Harma, 1995; Fullick et al., 2009). Against this, however, must be set the value of increased experience, particularly when coping strategies are considered. These strategies include acceptance of a non-conventional lifestyle (Adams, Folkard & Young, 1986), adoption of methods to promote sufficient sleep (Tepas, 1982) and devising ways of overcoming tiredness (Wedderburn, 1987). These strategies have been considered to indicate "commitment" to night work (Adams, Folkard & Young, 1986).

Certain personality traits (see Kerkhof, 1985; Flower, Irvine & Folkard, 2003; Paine, Gander & Travier, 2006) have also been considered as predictors for poor responses to time-zone transitions and night work, including rigidity vs. flexibility of sleeping habits and inability vs. ability to throw off fatigue (Wedderburn, 1987). Little supporting evidence has been published, but more extensive reports exist on chronotype (Horne & Ostberg, 1976), which refers to inter-individual differences in preferred time of sleep, times when individuals feel most or least able to perform mentally or physically demanding tasks, and times when individuals prefer to relax. Chronotypes can vary from "extreme morning types" or "larks" (whose preferred sleep–wake cycle is phased earlier than average), through "intermediate types" (which represents the middle 80 percent of a population) to "extreme evening types" or "owls" (whose preferred sleep–wake cycle is phased later than average). Evening types show a greater sleep propensity in the morning and a rise in sleep propensity later in the evening (Lavie & Segal, 1989), a later phasing of circadian rhythms of temperature and performance (Vidacek et al., 1988; Kerkhof & Van Dongen, 1986) and a smaller incidence of health disorders when on night shifts (Hildebrandt & Strattman, 1979). Therefore, as would be predicted from these findings, morning types are less suited to night work (but more suited to the morning shift) than are evening types.

Inter-individual differences exist also in the degree to which coping mechanisms are employed successfully (Costa et al., 1989; Harma, 1995; Monk & Folkard, 1989). It is unclear whether those who tolerate night work do so because of their innate traits and/or development of coping mechanisms (needing to be more marked in those lacking the appropriate traits). By contrast, it is clear that those who need to develop coping mechanisms but cannot, or do not wish to do so, are intolerant of night work.

Interest has concentrated recently on individuals' resilience to performance decrement in the face of sleep loss (Czeisler, 2009; King, Belenky & Van Dongen, 2009; Van Dongen & Belenky, 2009). The term "trototype" has been coined to describe differences in this susceptibility and it appears that genetic polymorphism is a contributory factor to these differences. For example, polymorphism exists in one of the human clock genes, PER3, and changes in part of this gene are associated with being an extreme morning type or evening type, extreme morning types also suffering more mental decrement after sleep deprivation or disruption.

CONCLUSIONS

Even though homeostatic control of biological variables remains of paramount importance, superimposed upon this stability are rhythms with a period of 24 hours (circadian rhythms). These rhythms, including those of mental performance, have endogenous (due to a body clock) and exogenous (due to an individual's lifestyle and environment) components. Understanding the origins of these components, together with effects of time awake upon mental performance, enables rhythms of mental performance in subjects living conventionally (sleeping at night and active during the daytime) to be described mathematically Moreover, predictions can be made about changes in sleep propensity and mental performance that will be expected when an individual's sleep–wake cycle is changed – as in the days after a time-zone transition or during night work, for example.

Individuals differ in detail in their responses to such changed sleep–wake schedules, but the extent to which these differences are due to their individual traits and/or coping mechanisms is less clear. Even so, details of links between individuals' rhythms in mental performance, various traits and genetic mechanisms are beginning to be elucidated, the outcome of which will be a firmer rationale for ameliorating any problems that might arise in those undergoing time-zone transitions and working at night.

REFERENCES

Adams, J., Folkard, S. & Young, M. (1986). Coping strategies used by nurses on night duty. *Ergonomics*, 29, 185–96.

Åkerstedt, T. (1995). Work hours, sleepiness and accidents. *Journal of Sleep Research*, 4, 1–3.

Åkerstedt, T. (2007). Altered sleep/wake patterns and mental performance. *Physiology and Behavior*, 90, 209–18.

Åkerstedt, T. & Folkard, S. (1997). The three-process model of alertness and its extension to performance, sleep latency, and sleep length. *Chronobiology International*, 14, 115–23.

Balkin, T., Rupp, T., Picchioni, D. & Wesensten, N. (2008). Sleep loss and sleepiness – current issues, *Chest*, 134, 653–60.

Barton, J., Smith, L., Totterdell, P., Spelten, E. & Folkard, S. (1993). Does individual choice determine shift system acceptability? *Ergonomics*, 36, 93–99.

Beersma, D. & Gordijn, M. (2007). Circadian control of the sleep–wake cycle. *Physiology and Behavior*, 90, 190–95.

Blatter, K. & Cajochen, C. (2007). Circadian rhythms in cognitive performance: Methodological constraints, protocols, theoretical underpinnings. *Physiology and Behavior*, 90, 196–208.

Boivin, D., Czeisler, C., Dijk, D., Duffy, J., Folkard, S., Minors, D., Totterdell, P. & Waterhouse, J. (1997). Complex interaction of the sleep–wake cycle and circadian phase modulates mood in healthy subjects. *Archives of General Psychiatry*, 54, 145–52.

Borbély, A. (1982). Sleep regulation: Circadian rhythm and homeostasis, in D. Ganten & D. Pfaff (eds), *Sleep. Clinical and Experimental Aspects*. Berlin; Springer Verlag, 83–104.

Carrier, J. & Monk, T. (2000). Circadian rhythms of performance: New trends. *Chronobiology International*, 17, 719–32.

Carvalho-Bos, S., Waterhouse, J., Edwards, B., Simons, R. & Reilly, T. (2003). The use of actimetry to assess changes to the rest–activity cycle. *Chronobiology International*, 20, 1039–59.

Cho, K. (2001). Chronic "jet lag" produces temporal lobe atrophy and spatial cognitive deficits. *Nature Neuroscience*, 4, 567–8.

Clayton, J., Kyriacou, C. & Reppert, S. (2001). Keeping time with the human genome. *Nature*, 409, 829–31.

Costa, G., Lievore, F., Casaletti, G., Gaffuri, E. & Folkard, S. (1989). Circadian characteristics influencing interindividual differences in tolerance and adjustment to shiftwork. *Ergonomics*, 32, 373–85.

Czeisler, C. (2009). Medical and genetic differences in the adverse impact of sleep loss on performance: Ethical considerations for the medical profession. *Transactions of the American Clinical and Climatological Association*, 120, 249–85.

Dinges, D. (1995). An overview of sleepiness and accidents. *Journal of Sleep Research*, 4, 4–14.

Edwards, B., Reilly, T. & Waterhouse, J. (2009). Zeitgeber-effects of exercise on human circadian rhythms: What are alternative approaches to investigating the existence of a phase–response curve to exercise? *Biological Rhythms Research*, 40, 53–69.

Flower, D., Irvine, D. & Folkard, S. (2003). Perception and predictability of travel fatigue after long-haul flights: A retrospective study. *Aviation, Space and Environmental Medicine,* 74, 173–9.

Folkard, S. (1990). Circadian performance rhythms: Some practical and theoretical implications. *Philosophical Transactions of the Royal Society, London, B,* 327, 543–53.

French, J., Bisson, R., Neville, K., Mitcha, J. & Storm, W. (1994). Crew fatigue during simulated, long duration B-1B bomber missions. *Aviation, Space and Environmental Medicine,* 65, A1–A6.

Fullick, S., Grindey, C., Edwards, B., Morris, C., Reilly, T., Richardson, D., Waterhouse, J. & Atkinson, G. (2009). Relationships between leisure-time energy expenditure and individual coping strategies for shift-work. *Ergonomics,* 52, 448–55.

Graeber, R. (1989). Jet lag and sleep disruption, in M. Krugger, T. Roth & C. Dement (eds), *Principles and Practice of Sleep Medicine.* London and Philadelphia, PA: W.B. Saunders, 324–31.

Graeber, R., Dement, W., Nicholson, A., Sasaki, M. & Wegmann, H. (1986). International cooperative study of aircrew layover sleep: Operational summary. *Aviation Space and Environmental Medicine,* 57, B10–B13.

Grajewski, B., Nguyen, M., Whelan, E., Cole, R. & Hein, M. (2003). Measuring and identifying large-study metrics for circadian rhythm disruption in female flight attendants. *Scandinavian Journal of Work and Environmental Health,* 29, 337–46.

Green, R.G. (1985). Stress and accidents. *Aviation, Space and Environmental Medicine,* 56, 638–41.

Harma, M. (1995). Sleepiness and shiftwork: Individual differences. *Journal of Sleep Research,* 4, 57–61.

Harrison, Y., Jones, K. & Waterhouse, J. (2007). The influence of time awake and circadian rhythm upon performance on a frontal lobe task. *Neuropsychobiologia,* 45, 1996–72.

Hildebrandt, G. & Strattman, I. (1979). Circadian system response to night work in relation to the individual circadian phase position. *International Archives of Occupational and Environmental Health,* 3, 73–83.

Horne, J. & Ostberg, O. (1976). A self-assessment questionnaire to determine morning–eveningness. *International Journal of Chronobiology,* 4, 97–110.

Johnson, M., Duffy, J., Dijk, D., Ronda, J., Dyal, C. & Czeisler, C. (1992). Short-term memory, alertness and performance: A reappraisal of their relationship to body temperature. *Journal of Sleep Research,* 1, 24–29.

Katz, G., Knobler, H., Laibel, Z., Strauss, Z. & Durst, R. (2002). Time zone change and major psychiatric morbidity: The results of a 6-year study in Jerusalem. *Comprehensive Psychiatry,* 43, 37–40.

Kerkhof, G. (1985). Inter-individual differences in the human circadian system: A review. *Biological Psychology,* 20, 83–112.

Kerkhof, G. & Van Dongen, H. (1996). Morning-type and evening-type individuals differ in the phase position of their endogenous circadian oscillator. *Neuroscience Letters,* 218, 153–6.

Khalsa, S., Jewett, M. & Cajochen, C. (2003). A phase response curve to single bright light pulses in human subjects. *Journal of Physiology,* 549, 945–52.

King, A., Belenky, G. & Van Dongen, H. (2009). Performance impairment consequent to sleep loss: Determinants of resistance and susceptibility. *Current Opinion in Pulmonary Medicine,* 15, 559–64.

Kleitman, N. (1963). *Sleep and Wakefulness.* Chicago, IL: University of Chicago Press.

Lack, L. & Tietzel, A. (2000). Do the benefits of brief naps suggest a fourth biological process determining sleepiness? *Sleep,* 23, A56.

Lavie, P. & Segal, S. (1989). Twenty-four-hour structure of sleepiness in morning and evening persons investigated by ultrashort sleep–wake cycle. *Sleep,* 12, 522–8.

Lemmer, B., Kern, R., Nold, G. & Lohrer, H. (2002). Jet lag in athletes after eastward and westward time-zone transition. *Chronobiology International,* 19, 743–64.

Lewy, A., Bauer, V. & Ahmed, S. (1998). The human phase response curve (PRC) to melatonin is about 12 hours out of phase with the PRC to light. *Chronobiology International*, 15, 71–83.

Lowden, A. & Åkerstedt, T. (1998). Retaining home-base sleep hours to prevent jet lag in connection with a westward flight across nine time zones. *Chronobiology International*, 15, 365–76.

McCauley, P., Kalachev, L., Smith, A., Belenky, G., Dinges, D. & Van Dongen, H. (2009). A new mathematical model for the homeostatic effects of sleep loss on neurobehavioral performance. *Journal of Theoretical Biology*, 256, 227–39.

Minors, D. & Waterhouse, J. (1981). *Circadian Rhythms and the Human*. Bristol: John Wright.

Monk, T.H. & Folkard, S. (1985). Individual differences in shift work adjustment, in S. Folkard & T.H. Monk (eds), *Hours of Work. Temporal Factors in Work Scheduling*. Chichester: John Wiley, 227–37.

Monk, T., Buysse, D., Reynolds, C., Berga, S., Jarrett, D., Begley, A. & Kupfer, D. (1997). Circadian rhythms in human performance and self-rated subjective feelings under constant conditions. *Journal of Sleep Research*, 6, 9–18.

Mrosovsky, N. (1999). Critical assessment of methods and concepts in nonphotic phase shifting. *Biological Rhythms Research*, 30, 135–48.

Murphy, P. & Campbell, S. (1997). Nighttime drop in body temperature: A physiological trigger for sleep onset? *Sleep*, 20, 505–511.

Naitoh, P., Kelly, T. & Babkoff, H. (1993). Sleep inertia: Best time not to wake up? *Chronobiology International*, 10, 109–18.

Paine, S., Gander, P. & Travier, N. (2006). The epidemiology of morningness/eveningness: Influence of age, gender, ethnicity, and socioeconomic factors in adults (30–49 years). *Journal of Biological Rhythms*, 21, 68–76.

Reilly, T., Atkinson, G. & Waterhouse, J. (1997). *Biological Rhythms and Exercise*. Oxford: Oxford University Press.

Reppert, S. & Weaver, D. (2001). Molecular analysis of mammalian circadian rhythms. *Annual Review of Physiology*, 63, 647–78.

Richmond, L., Dawson, B., Hillman, D. & Eastwood, P. (2004). The effect of interstate travel on sleep patterns of elite Australian Rules footballers. *Journal of Science and Medicine in Sport*, 7, 186–96.

Rosekind, M., Smith, R., Miller, D., Co, E., Gregory, K., Webbon, L., Gander, P. & Lebacqz, J. (1995). Alertness management: Strategic naps in operational settings. *Journal of Sleep Research*, 4 (Suppl. 2), 62–6.

Samel, A., Wegmann, H. & Vejvoda, M. (1995). Jet-lag and sleepiness in aircrew. *Journal of Sleep Research*, 4(Suppl. 2), 30–36.

Saper, C.B., Chou, T.C. & Scammell, T.E. (2001). The sleep switch: Hypothalamic control of sleep and wakefulness. *Trends in Neuroscience*, 24, 726–31.

Shanahan, T. & Czeisler, C. (2000). Physiological effects of light on the human circadian pacemaker. *Seminars in Perinatology*, 24, 299–320.

Shochat, T., Luboshitzky, R. & Lavie, P. (1997). Nocturnal melatonin onset is phase locked to the primary sleep gate. *American Journal of Physiology (Regulatory and Integrative Physiology)*, 273, R364–R370.

Signal, T. & Gander, P. (2007). Rapid counterclockwise shift rotation in air traffic control: Effects on sleep and night work. *Aviation, Space and Environmental Medicine*, 78, 878– 85.

Srinivasan, V., Spence, D., Pandi-Perumal, S., Trakht, I. & Cardinali, D. (2008). Jet lag: Therapeutic use of melatonin and possible application of melatonin analogs. *Travel Medicine and Infectious Diseases*, 6, 17–28.

Stampi, C. (1992). The effects of polyphasic and ultrashort sleep schedules, in C. Stampi (ed.), *Why We Nap: Evolution, Chronobiology, and Functions of Polyphasic and Ultrashort Sleep*. Boston, MA: Birkhauser, 137–79.

Takahashi, M., Nakata, A. & Arito, H. (2002). Disturbed sleep–wake patterns during and after short-term international travel among academics attending conferences. *International Archives of Occupational and Environmental Health*, 75, 435–40.

Tepas, D.I. (1982). Shift worker sleep strategies. *Journal of Human Ergology*, 11(Suppl.), 325–36.

Valdez, P., Reilly, T. & Waterhouse, J. (2008). Rhythms of mental performance. *Mind, Brain and Education*, 2, 7–16.

Van Dongen, H. (2004). Comparison of mathematical model predictions to experimental data of fatigue and performance. *Aviation, Space and Environmental Medicine*, 75(Suppl.), A15–A36.

Van Dongen, H. & Belenky, G. (2009). Individual differences in vulnerability to sleep loss in the work environment. *Industrial Health*, 47, 518–26.

Van Dongen, H. & Dinges, D. (2005). Sleep, circadian rhythms, and psychomotor vigilance. *Clinics in Sports Medicine*, 24, 237–52.

Vanttinen, T., Jiang, Y., Baca, A. & Zhang, H. (2008). Athletes' jet-lag as a consequence of traveling across six time zones. Proceedings of the First Joint International Pre-Olympic Conference in Sports Science and Sports Engineering, 1, 402–5.

Vidacek, S., Kaliterna, L., Radosevic-Vidacek B. & Folkard, S. (1988). Personality differences in the phase of circadian rhythms: A comparison of morningness and extraversion. *Ergonomics*, 31, 873–88.

Walker, M. (2008). Cognitive consequences of sleep and sleep loss. *Sleep Medicine*, 9, S29–S34.

Waterhouse, J. & DeCoursey, P. (2004a). Human circadian organization, in J. Dunlap, J. Loros & P. DeCoursey (eds), *Chronobiology. Biological Timekeeping*. Sunderland, MA: Sinauer, 291–323.

Waterhouse, J. & DeCoursey, P. (2004b). The relevance of circadian rhythms for human welfare, in J. Dunlap, J., Loros & P. DeCoursey (eds), *Chronobiology. Biological Timekeeping*. Sunderland, MA: Sinauer, 325–56.

Waterhouse, J., Minors, D., Åkerstedt, T., Reilly, T. & Atkinson, G. (2001). Rhythms of human performance, in J. Takahashi, F. Turek, & R. Moore (eds), *Handbook of Behavioral Neurobiology: Circadian Clocks*. New York: Kluver Academic/Plenum Publishers, 571–601.

Waterhouse, J., Reilly, T., Atkinson, G. & Edwards, B. (2007). Jet lag: Trends and coping strategies. *The Lancet*, 369, 1117–29.

Wedderburn, A.A.I. (1987). Sleeping on the job: The use of anecdotes for recording rare but serious events. *Ergonomics*, 30, 1229–33.

Wojtczak-Jaroszowa, J. & Jarosz, D. (1987). Chronohygenic and chronosocial aspects of industrial accidents, in J.E. Pauly & L.E. Scheving (eds), *Advances in Chronobiology B*. New York: Alan R. Liss, 415–26.

Wright, K., Hull, J. & Czeisler, C. (2002). Relationship between alertness, performance, and body temperature in humans. *American Journal of Physiology (Regulatory, Integrative and Comparative Physiology)*, 283, R1370–R1377.

17
Sleep Loss and Performance

Valerie J. Gawron

INTRODUCTION

The causes of sleep loss include injury associated with physical fatigue (see Chapter 7), psychological stress (see Chapter 5), mental illness (see Chapter 18), disease (see Chapter's 19 and 20), sleep disorders (see Chapter 11), desynchronosis with circadian and other biological rhythms (see Chapter's 16, 28 and 30), and sustained operations (Chapter 21). Each of these has different effects on performance since each of these affects how quickly a person can recover the sleep debt, the quality of that sleep, and the duration of exposure to that cause of sleep loss. For injury and disease, performance decrements may be due to sleep loss but more likely to the injury or the disease itself. An example of comorbid conditions, depression, and sleep quality is presented by Hayashino et al. (2010). The focus of this chapter is on the effect of sleep loss, without the attendant causes, on performance.

The effects of sleep loss can be summarized in common sayings that start with "I am so tired that ...." These include:

1. "I can't see straight" (degradation in visual performance).
2. "I can't think straight" (degradation in cognitive performance).
3. "I can't do anything right" (degradation in routine tasks).
4. "I can't seem to pay attention" (degradation in vigilance performance).
5. "I can't lift a finger" (degradation in physical performance).

The studies supporting or refuting these effects are described in separate sections in this chapter. In each section the details of the relevant study (e.g., number and type of subjects, length of sleep loss, type task performed) are summarized in a separate table for easy reference.

VISUAL PERFORMANCE

The effect of sleep loss on vision has real-world implications in security and defense. Eight studies on the effects of sleep loss on visual performance are summarized here with details presented in Table 17.1. Although not all studies have shown a decrement in visual performance associated with sleep loss, the preponderance of evidence suggests that sleep loss causes a degradation in visual performance, i.e., "I can't see straight."

Table 17.1 Effects of sleep loss on visual performance

Study	Subjects	Task	Sleep Loss	Result
Babkoff et al. (1988)	14 men	cross out previously memorized sets of letters (2, 4, or 6 letters) presented in a 16 by 20 matrix	72-hour sleep deprivation period	number attempted decreased over time, lowest for largest letter set, worst 02:00 to 06:00 hours on second and third days; accuracy lower in larger letter sets and lower on third day
Basner et al. (2008)	24 non-professionals	simulated luggage-screening	after a night without sleep	accuracy and hit rate decreased
DeJohn, Reams, & Hochhaus (1992)	military personnel	cognitive, visual, and auditory tasks	extended overwater training mission	no significant differences
Fröberg et al. (1975)	29 military personnel	shooting range performance	severely restricted sleep	number of shots taken decreased
Hovis & Ramaswamy (2007)	a protanope (unable to distinguish green–yellow–red portion of spectrum) and a color-normal person	Hardy, Rand, Rittler Pseudoisochromatic Plates (HRR), the Farnsworth Munsell D-15 (D-15), the Adams Desaturated D-15 (ADesat D-15), the Medmont C-100 (C-100), and the CN Lantern	1 night of sleep deprivation	color normal person confusing blue and blue-green, blue and purple, blue and green, red and orange, green and light blue, and green and dark green; protanope confused orange and red.
Neri, Dinges & Rosekind (1997)	summary of literature	military surge operations	sleep loss	decrements in visual performance
Tyagi, et al. (2009)	8 male students	auditory working-memory vigilance task and PVT	25-hour sleep deprivation study	significant effect
Wilkinson, Edwards & Haines (1966)	6 enlisted men	auditory discrimination task and addition	0, 1, 2, 3, 5, or 7.5 hours sleep	number of signals detected and number of sums per hour increased and number of errors decreased as the number of hours slept increased

Further, the one nonsupporting study, DeJohn, Reams & Hochhaus (1992) concluded that the improvement may have been due to practice.

COGNITIVE PERFORMANCE

There has been a considerable amount of basic research on the effects of sleep loss on cognitive performance. There has also been applied research in the defense, public safety, and industry domains. Each of these is described in a separate section below.

Basic research has included assessing the effects of sleep loss on memory, tracking, speech, word associations, search, math, reading, and decision-making. The studies are summarized in Table 17.2. Not all studies have shown a decrement in visual performance associated with sleep loss, however. DeJohn, Reams & Hochhaus (1992) found in the performance of the cognitive or visual tasks that there was an improvement in understanding of noise-degraded speech. The authors concluded that the improvement was due to practice. The basic research in cognitive performance shows mixed results based on type of task and attributes of the subjects. The preponderance of evidence suggests, however, that sleep loss causes a degradation in cognitive performance such that it is true that "I am so tired that 'I can't think straight.'"

Defense-related research has applied both simple cognitive tasks and actual defense tasks performed during simulated sustained operations. Studies that are defense-related are summarized in Table 17.3. Decrements occurred in all mission-critical tasks such as target processing, reaction time, math, and memory recall.

Public safety research has focused on aviation (Table 17.4), driving (Table 17.5), industry (Table 17.6), and medicine (Table 17.7). For suggestions on identifying and treating fatigue see Rosenthal (2008).

ROUTINE TASK PERFORMANCE

As part of the National Sleep Foundation's 2010 "Sleep in America" poll, participants were asked how much not getting enough sleep affects their job performance. The results showed that participants agreed that sleep impacts their ability to do everyday tasks (65 percent to 70 percent), perform at work (57 percent to 72 percent), perform household tasks (56 percent to 68 percent), and care for family (51 percent to 57 percent). Caldwell, Prazinko & Caldwell (2002) warn against using naps as a countermeasure for fatigue without consideration of sleep inertia, which can also result in degraded performance.

VIGILANCE PERFORMANCE

Vigilance performance has long been used as an indicator of fatigue in both laboratory and operational settings. Summaries of laboratory studies are presented in Table 17.8. Vigilance studies in operational settings are summarized in Table 17.9 (medical), Table 17.10 (military), Table 17.11 (aviation), and Table 17.12 (industry). Finally effects of sleep loss on physical performance are summarized in Table 17.13.

Table 17.2   Effects of sleep loss on cognitive performance (basic research)

Study	Subjects	Task	Sleep Loss	Result
Angus & Heslegrave (1985)	6 highly physically fit females and 6 females who were not so fit	communications, - choice serial reaction time task, simple iterative subtraction, encoding/decoding, complex interactive subtraction, logical reasoning, short-term memory, paired-associate learning task, plotting, map coordinates	54-hour period without sleep	decrease in number of correct responses on serial reaction task, logical reasoning, and encoding/decoding tasks; decrease in percent correct responses in auditory signal detection task. For message processing task, initial improvement followed by increase in processing time
Binks, Waters & Hurry (1999)	29 sleep-deprived undergraduate students, 32 non-sleep-deprived undergraduate students	verbal production of words beginning with specified letter, card-sorting, categorization, Stroop, paced auditory serial addition, Wechsler Adult Intelligence Scale Revised	34 to 36 hours of sleep deprivation	no differences between sleep-deprived and non-sleep-deprived groups
Blagrove, Alexander & Horne (1995)	63 university students total over 4 experiments	logical reasoning, auditory discrimination task, finding embedded figures	sleep deprivation and sleep reduction(limit of 5 hours of sleep)	no effect of sleep reductions on logical reasoning task; sleep-reduced groups did as well as control on auditory discrimination except for first test after sleep reduction started; no significant effects on other 2 tests
Casagrande et al. (1997)	20 males	letter cancellation task	1 night sleep deprivation	decrement in last 3 (17:30, 19:30, 21:30 hours) of 6 trials (every 2 hours beginning at 11:30)
Couyoumdjian et al. (2010)	54 sleep-deprived university students, 54 non-sleep-deprived students	determine if visually presented digit was 1) odd or even and 2) larger or smaller than 5	night of sleep deprivation	higher switching cost; reaction times shorter for repetition trials than for switch trials especially after sleep deprivation; sleep-deprived made more errors
Dawson & Reid (1997)	40 participants	tracking task	28 hours being awake	between hours 10 and 26 of sleep- deprivation performance decreased 0.74% per hour
Dorrian, Lamond & Dawson (2000)	18 participants	visual comparison, tracking, vigilance, and grammatical reasoning	28 hours being awake	decrements in all but visual comparison and grammatical reasoning accuracy

Table 17.2 Continued

Study	Subjects	Task	Sleep Loss	Result
Ferri et al. (2010)	15 subjects with sleep fragmentation, 8 subjects with no sleep fragmentation	reaction time	sleep fragmentation	longer reaction times on day 2 than on day 4 when figures not rotated or reversed or not rotated and mirror imaged; no differences in spatial attention, inhibition of return, and Stroop task
Harrison & Horne (1997)	9 participants	speech	36 hours' sleep deprivation	sleep-deprived had fewer correctly vocalized words and higher proportion of words with semantic similarity and higher number of adjectives and nouns at the end of 36 hour period versus 6 or 30 hours into sleep-deprivation period
Harrison & Horne, 1998	first study 20 subjects; second 30 subjects	speech	34 hours without sleep	fewer novel associations among words in sleep-deprived group
Harrison & Horne (1999)	10 graduate students	computer-based simulation of marketing decision-making game	36 hours' sleep deprivation	no differences in profitability or production errors; decrease in profitability on 12 hours in sleep-deprived condition; interaction for production errors with more errors for sleep-deprived in last two sessions
Heuer et al. (1998)	11 men	responding to current visual stimulus or predicting 1 every 4 hours across 3 days	first night of 3 was sleep-deprived	reaction times on day 3 shorter than days 1 or 2; variability increased after night without sleep; effect of time on-task largest after night without sleep
Horne (1988)	12 college students	Torrance tests of creative thinking	32 hours without sleep	increases in planning time and decreases in scores for elaboration, originality, flexibility, and fluency for sleep-deprived students
Ikegami et al. (2009)	10 men	simple addition, paced auditory serial addition, simple words memory, reading span, continuous performance, card-sorting	an adaptation day with 7 hours, time in bed followed by a day with sleep deprivation and then four days with 7 hours' time in bed	scores on paced auditory serial addition and reading pan lower and omission and commission errors on continuous performance higher on sleep deprivation day
Jaskowski & Weodarczyk (1997)	13 students	simple and choice reaction time	night of sleep deprivation	longer reaction times from 20:00 hours through 06:00 hours night of sleep deprivation; no effect on response force
Killgore, Balkin & Wesensten (2006)	48 participants aged 19 to 39 years old	Iowa Gambling Task	2 days of sleep deprivation	more risky behavior in older participants after sleep deprivation than younger participants

Table 17.2 Continued

Study	Subjects	Task	Sleep Loss	Result
Kim et al. (2001)	18 males	Luria-Nebraska Neuropsychological Battery and calculation and digit span tests of Korean version of Wechsler Adult Intelligence Scale (K-WAIS)	24 hours of sleep deprivation	"no differences in freedom from distractibility, tactile function, visual function, reading, writing, arithmetic and intellectual process function. However, the cognitive functions such as motor, rhythm, receptive and expressive speech, memory and complex verbal arithmetic function were decreased after sleep deprivation" (p. 127)
Lamond & Dawson (1999)	22 students	identifying block in which visual stimulus appears, unpredictable tracking, choice reaction time, grammatical reasoning	28 hours of sleep deprivation	decreased 4 of the 6 measures (choice reaction time and accuracy, grammatical reasoning latency, and tracking)
Linde, Edland & Bergstrom (1999)	12 male sleep-deprived students and 12 non-sleep-deprived	auditory attention, coding, completing complex numeric patterns, and decision-making; last completed either with or without time pressure	33 hours of sleep deprivation	more errors in auditory task for sleep-deprived participants; no effects on coding, completing numeric patterns, or decision-making
McCarthy & Waters (1997)	71 male undergraduates	reaction time	Sleep-deprived	increases in reaction time for a sleep-deprived group
Mikulincer et al. (1989)	12 persons aged 22 to 32	visual search for memorized letters	72 hours of sleep loss	performance decreased as a function of days without sleep; performance higher in late afternoon than during midnight to early morning hours
Mikulincer et al. (1990)	11 males	visual shape discrimination and logical reasoning	72 hours of sleep loss	decrements for persons in off-task cognitions; increases in lapses in visual shape discrimination over days of sleep loss and from 16:00 to 20:00 hours and 02:00 to 06:00 hours but only for persons in high off-task cognitions group; accuracy increased over days on visual discrimination task; increase in response time of 0.44 second associated with sleep loss; no effect of sleep deprivation on lapses in logical reasoning task; less accurate at 02:00 to 06:00 hours and in 4-letter task than at 10:00 to 14:00 hours in 2-letter task; response time was longer between first and third days at 02:00 and 06:00 hours for 4- letter condition

Table 17.2   Continued

Study	Subjects	Task	Sleep Loss	Result
Miro et al. (2003)	30 volunteers aged 18 to 24	time estimation	60 hours of sleep deprivation	lengthening of time estimations over the 60-hour period
Nakano et al. (2000)	10 men aged 20 to 25 and 10 men aged 52 to 63	tracking and reaction time tasks	19 hour period of sleep deprivation	midrange of performance variability between 01:00 and 05:00 hours while median in self-rated sleepiness and alertness occurred around 12:00 hours
Nesthus, Scarborough & Schroeder (1998a)	27 participants	synthetic work, letter search, 4-choice reaction time	weekend sleep deprivation	older participants had stable performance while younger participants had better overall performance but declined over time
Nesthus, Scarborough & Schroeder (1998b)	27 participants	4-choice reaction time and letter search	4, 10, 16, 22, 28, and 34 hours after waking	correct reaction time and errors increased and percent correct and throughput decreased
Philip et al. (2004)	10 men	reaction time	full night of sleep deprivation	older subjects slower, especially in later testing sessions; had more lapses, especially during later testing sessions
Pilcher & Walters (1997)	44 undergraduate students	Watson-Glaser Critical Thinking Appraisal	24 hours of sleep deprivation	decreases in performance
Polzella (1975)	4 men and author	state if probe had been among stimuli previously presented	rest on seven of the days and stayed awake on two nights	decrement in d' and recognition memory with sleep loss; more lapses with sleep loss than with rest
Van Dongen et al. (2003)	48 participants	PVT, digit substitution, and serial addition or subtraction task	sleep restrictions: 4 hours per night for 14 days, 6 hours per night for 14 days, 8 hours per night for 14 days, or 0 hours per night over 3 days	cumulative performance decrements in 4- and 6-hour sleep conditions compared to 8-hour sleep condition; performance in 4-hour sleep restriction condition equivalent to that without sleep for 2 nights
Williamson & Feyer (2000)	30 transport employees and 9 army personnel	Mackworth clock, simple reaction time, tracking, dual task symbol digit, spatial memory, memory and search, and grammatical reasoning	28 hours of sleep deprivation	16.91 (accuracy symbol digit task) to 18.55 (speed symbol digit task) hours without sleep equivalent to 0.05% BAC; 16.51 (reaction time accuracy) to 18.91 (tracking task) hours were equivalent to 0.1% BAC

Table 17.3   Effects of sleep loss on cognitive performance (defense-related)

Study	Subjects	Task	Sleep Loss	Result
Baranski et al. (2007)	32 military, 32 civilians	Team and Individual Threat Assessment Task (TITAN)	30 hours' sleep loss	2% difference in percent errors and 14-second difference in target processing time; teams better performance than individuals
Dinges et al. (1997)	16 young adults	PVT, probe memory, serial addition	7 consecutive nights of restricted sleep (4 to 5 hours)	poorest PVT performance at 10:00 and best at 22:00 hours; reaction times slowed linearly while lapse frequency increased linearly; best performance at 16:00 hours for probe memory and serial addition
Harville, Harrison & Scott (2006)	30 U.S. Air Force officers	math, a two-back working memory	36 hours of sleep deprivation	decrease accuracy in both tasks
Lieberman et al. (2006)	13 male soldiers	matching to a sample, repeat series of 12 key strokes	4-day sustained operations mission	more time out errors on days 3 and 4; no difference in number of correct matches or response time; no effect on keystroke task
Neri, Shappell & DeJohn (1992)	12 male U.S. Marines	4-choice reaction time, grammatical reasoning, serial add/subtraction, manikin, pattern recognition, time estimation	9 hours' planning, 4 hours' rest, 14-hour daytime mission, 6 hours' rest	linear increases in reaction time for correct responses and in response rate for pattern recognition and manikin tests; linear increase in error rate for pattern recognition test; decrease in reaction time for grammatical reasoning test
Pilcher et al. (2007)	24 college students	WOMBAT (tracking, 3-D figure rotation, quadrant location, digit cancelling, autopilot failure), Automated Performance Test System ((APTS) code substitution, grammatical reasoning, math processing), monitoring chemical plant, PVT, vigilance	night without sleep	WOMBAT scores significantly increased over time attributed to learning; no change APTS, decrease in performance on chemical plant task; increase in PVT reaction time but not lapses; increase in reaction time and decrease in accuracy on vigilance task
Rowe et al. (1992)	9 males	Walter Reed Performance Assessment Battery and Complex Cognitive Assessment Battery	2, 30-hour sleep deprivation periods: one in bright light (3,000 lux) and one in dim light (100 lux)	increase in reaction time and variability in accuracy
Vrijkotte et al. (2004)	17 soldiers	recall of 0, 1, or 2 back letters, Tower of Hanoi pattern transformation, vigilance and tracking task	3-week sustained operations mission	decrements in all tests; linear in weeks 1 and 3 pushed to physical limits; irregular in week 2 pushed to mental and physical limits

Table 17.4   Effects of sleep loss on public safety (aviation)

Study	Subjects	Task	Sleep Loss	Result
Beh & McLaughlin (1991)	46 flight attendants	logical reasoning, vertical addition, horizontal addition, letter cancellation, and Stroop card-sorting	1 hour prior to departure and during each layover versus nonflying	decrease grammatical reasoning, vertical addition task, and one card-sorting task; worse during layovers than during the pre-test; letter cancellation on last layover
Caldwell et al. (2009)	review of literature relevant to aviation		sleep loss	thinking and moving slowly, making mistakes, having memory lapses
Cruz et al. (1995)	28 participants ranging from 20 to 55 years of age	Bakan Vigilance Task at the beginning and end of each shift and three 1.5 hour sessions of the Multiple Task Performance Battery	clockwise rotation (24 hours off at each shift rotation and a 48-hour weekend) versus a counter-clockwise rotation (only 8 hours off at each shift rotation and an 80-hour weekend)	Bakan Vigilance Task more correct responses during the first than during the second week, more correct responses at beginning than at end of workday, "number of correct responses on the first evening shift was significantly higher for the counter-clockwise rotation condition than the clockwise rotation condition"
Foushee et al. (1986)	10 pilots rested, 10 not	routine flight operations and response to a mechanical failure	within 3 hours of completing a 3-day, high-density, short-haul duty cycle	post-duty captains rated selves as more fatigued but performance in flight simulator was rated as better than that of rested crews; post-duty had more stable aircraft handling and exchanged more commands; post-duty captains repeated instructions more often than pre-duty captains or first officers in either condition
Gander (2001)	review of fatigue related to Air Traffic Control	rosters	sleep loss	rapid backward-rotating rosters resulted in significant sleep loss, as did early start times
Heslegrave et al. (1995)	921 controllers	estimate the effects of overtime and shift work on performance	12-hour versus 8-hour shifts, midnight shift	degradation of performance; midnight shifts versus 1 to 4 consecutive midnight shifts, 16% versus 27% reported falling asleep while commuting to work; controllers working 5, 6, and 7 midnight shifts 32%, 40%, and 50% reported falling asleep, respectively

Table 17.4 Continued

Study	Subjects	Task	Sleep Loss	Result
Roach et al. (2006)	134 pilots	respond to locked out engine, load sheet error, runway change, level off altitude change, request to expedite climb, request to maximize speed, respond to erroneous pressure setting, no clearance to land during 70 minutes in B747-400 simulator	immediately after final landing at end of international flight	no effect on threat detection; more errors (40.3%) if had less than 5 hours of sleep in last 24 hours than if had more than 5 hours (26.0%)
Russo et al. (2005)	8 Air Force pilots	simulated 12.5 hour air refueling mission with choice reaction and PVT	19 hours awake	impairments in choice reaction time omissions, azimuth deviations, PVT speed and lapses
Thompson et al. (2006)	28 pilots, sensor operators, and intelligence personnel	unmanned aerial vehicle synthetic task environment, self-paced mental arithmetic task; PVT, boredom proneness scale; Profile of Mood States	Pre- versus post-shift days 1 through 6	over half of the operators reported symptoms associated with Shift Work Sleep Disorder; math test decreased 5.1%; root mean square error deviation increased from pre- to post-shift in altitude (-15.75%), heading (-5.32%), and airspeed (184.56%)

Table 17.5   Effects of sleep loss on public safety (driving)

Study	Subjects	Task	Sleep Loss	Result
Åkerstedt (2007)	review of literature	psychomotor performance, vigilance, cognitive performance, complex decision making	altered sleep/wake patterns	degrade performance
Åkerstedt, Peters, Anund & Kecklund (2005)	5 male and 5 female shift workers	2, 2-hour drives in a high-fidelity, moving–base automobile simulator	after a night shift	4 participants did not complete 2-hour drive after a night shift due to excessive sleepiness; more accidents after the night shift versus after normal sleep
Balkin et al. (2004)	66 commercial motor vehicle drivers (aged 24 to 62)	math performance	9, 7, 5, or 3 hours' time in bed for 4 consecutive days	increase in time to complete a serial addition/ subtraction; larger standard deviations in lane tracking and larger relative lane position in sleep restriction
Barger et al. (2005)	2,737 interns	driving	extended work shifts	"every extended work shift that was scheduled in a month increased the monthly risk of a motor vehicle crash by 9.1 percent … increased the monthly risk of a crash during the commute … by 16.2 percent" (p. 125); working ≥5 extended shifts within one month increased risk of falling asleep while driving or stopped in traffic
Connor et al. (2002)	Drivers admitted to hospital after crash versus non-crash drivers	driving vehicles excluding drivers of heavy vehicles, taxis, and emergency vehicles	sleepiness, using the Stanford sleepiness scale, and alertness (on a rating scale of 1–7) immediately before the crash or survey	drivers with scores of 4 or above on Stanford scale had 11-fold risk of injury crash versus those with scores below 4; increased risk if ≤ 5 hours sleep in last 24; highest risk is ≤ 3 hours sleep; greatest risk driving between 02:00 and 05:00 hours
Danner & Phillips (2008)	Survey data from c. 10,000 teenagers	driving before and after 1 hour delay in school start times	self-reported sleep	crash rate for delay teenagers decreased 16.5% while rate for teenagers without delay increased 7.8%
Gillberg, Kecklund & Åkerstedt (1996)	9 professional drivers	3, 30-minute segments driving truck simulator	day, night, night with a 30 minute rest, and night with a 30 minute nap	last segment speed higher, variation in speed lower; standard deviation in lane position lower during day than for any of 3 night drives
Horne & Reyner (1995)	679 sleep-related accidents on British roads	driving	Sleep loss	sleep-related accidents occurred at three peaks: 02:00–02:59 hours; 06:00– 06:59 hours; and 16:00 – 16:59 hours; 82% of the drivers were men

Table 17.6 Effects of sleep loss on public safety (industry)

Study	Subjects	Task	Sleep Loss	Result
Ansiau et al. (2008)	2,337 workers	immediate free recall from a list of 16 words, delayed free recall, digit symbol substitution, and selective attention	Sleep length	decrease in sleep length associated with being 52 (rather than 32), having worked more than 10 hours previous day (rather than less than 8 hours), with atypical schedule and having high workload; working prior to 06:00 hours or after 22:00 hours on previous day associated with poorer performance on immediate and delayed recall, selective attention; higher sleep lengths associated with poorer performance on selective attention task
Barnes & Wagner (2009)	miners from 1983 to 2006	injury rates, reported sleep	Daylight Savings to standard time	increase with an hour lost due to converting to Daylight Savings; people slept 40 minutes less after Daylight Savings Time than on days without the shift; loss of sleep greater than gain of sleep with return to Standard Time
Mitchell & Williamson (2000)	15 employees working 8-hour shifts versus 12 employees working 12-hour shifts	5-minute simple reaction time task, a 3-minute grammatical reasoning task, a 10-minute vigilance task, and a 5-minute critical tracking task	8- versus 12-hour shift	faster reaction time at end than at beginning of shift on simple reaction time task, 12-hour shift more variance in reaction time for correct responses to grammatical reasoning task, also more errors in vigilance task at end of shift

Table 17.7 Effects of sleep loss on public safety (medicine)

Study	Subjects	Task	Sleep Loss	Result
Andreyka & Tell (1996)	24 obstetrical residents	task performance	being on call for 36 hours (sleep-deprived state)	increases in errors of omission and errors of commission
Bartel et al. (2004)	33 resident anesthetists	simple, choice, sequential one back, sequential two back reaction time	preceding night duty (7.04 hours' sleep) versus 24 hours after night duty (1.66 hours' sleep)	longer after night duty for simple (12%), choice (6%), one back tests (7%); accuracy decreased after night duty for both one (2%) and two back (4%) tests

Table 17.7   Continued

Study	Subjects	Task	Sleep Loss	Result
Beatty, Ahern & Katz (1976)	6 anesthesiologists	simulated surgical task, letter-search task, grammatical reasoning task, rotated letters task	after night on duty with no more than 2 hours of sleep in last 24 hours	surgical task degraded for 4 anesthesiologists; no decrement in the letter-search; 23% increase in the time to complete the grammatical reasoning
Christensen et al. (1997)	7 rested and 7 fatigued residents	nodule detection in radiographs	worked at least 15 hours	no difference
Deaconson et al. (1988)	26 surgical residents	paced auditory serial addition, connecting randomly marked points, grammatical reasoning, mentally assembling geometric figure, Purdue Pegboard	less than 4 hours of sleep in 4 last 24 hours	no significant differences between sleep-deprived and non-sleep-deprived
Deary & Tait (1987)	12 residents	digit span, counting backwards, recall immediately after reading passage, recall delayed after reading passage, sort patient reports normal and abnormal, electrocardiogram assessments	a night spent off duty, a night spent on call, and a night spent admitting emergency cases	impairment of immediate recall associated with waiting while on-call
Dorrian et al. (2008)	41 full-time nurses	errors (frequency, type, and severity), drowsiness while driving home from work	self-reported work hours, estimated sleep length and quality, fatigue, sleepiness, stress	38 errors, majority occurring in morning and day shift; 38 near errors distributed through working hours; 65 reports observing someone else's error majority at night; across 34 nurses who drove to work, 70 occurrences of extreme drowsiness and 7 near accidents with almost half of these occurring end of night shift; 40% of extreme drowsiness and near accidents between 14:00 and 19:00 hours peaking at end of morning shift (15:00 hours).
Dula et al. (2001)	8 emergency medicine residents day shift, 8 night shift	Kaufman Adolescent and Adult Intelligence Test	night shift	higher scores on for those working day shifts
Eastridge et al. (2003)	35 surgical residents	simulated laparoscopic surgery	pre-call, on call, and post-call	more errors and increase in time to complete during post-call
Ellman et al. (2004)	surgeons 6,751 cases collected over a 9-year period	mortality rates, operative, pulmonary, renal, neurological, or infectious complications	sleep deprivation – operating between 23:00 and 07:00 hours	no difference

Table 17.7   Continued

Study	Subjects	Task	Sleep Loss	Result
Engel et al. (1987)	7 interns	clinical performance	being on call	no decrement
Friedman, Bigger, and Kornfeld (1971)	13 male and 1 female intern	detecting arrhythmias in an electrocardiogram over a 20-minute period	1.8 hours sleep during previous 32 hours versus an average of 7 hours' sleep	sleep-deprived participants less able to detect arrhythmias; difficulties thinking, memory loss, depression, irritability, sensitivity to criticism, depersonalization, and black humor
Friedman, Kornfeld & Bigger (1973)	13 male and 1 female interns	detecting arrhythmias in an electrocardiogram over a 20-minute period	1.8 hours' sleep during previous 32 hours versus an average of 7 hours' sleep	sleep-deprived participants less able to detect arrhythmias; difficulties thinking, memory loss, depression, irritability, sensitivity to criticism, depersonalization, and black humor
Gottlieb et al. (1991)	32 internal medicine residents	errors	4-day rotation consisting of long call and short call days followed by one non-admitting day; 7 day rotation with 1 long call, 3 short call, and 3 non-admitting days	16.9 errors per 100 patients in 4day rotation and 12.0 errors in 7 day rotation
Hart et al. (1987)	16 sleep-deprived and 14 normal first year residents	recall from stories, Sternberg short-term memory, paced auditory serial addition	Sleep-deprived	sleep-deprived residents recalled less information and tended to have longer response latencies on the Sternberg task (p < 0.10)
Haynes et al. (1995)	resident surgeons: data on 6,371 surgical cases	postsurgical complications	operated the day after a 24-hour on-call period	no difference to those who had not
Jacques, Lynch & Samkoff (1990)	353 family practice residents	examination	number of hours slept night before, number of hours worked each week, average on-call	decrease in scores for decreased sleep night before and for each year of training with first year residents showing greatest decrement

Table 17.7   Continued

Study	Subjects	Task	Sleep Loss	Result
Jensen et al. (2004)	40 surgical residents	time to complete and number of errors for pegboard, cup drop, rope pass, pattern cutting, endoscopic clip application, endoscopic loop application	self-reported amount of sleep the night before	no effect
Joffe (2006)	review of 37 articles	medical error	sleep loss	fatigue reduction is critical in reducing medical error
Klose, Wallace-Barnhill & Craythorne (1985)	14 anesthesiology residents	digit symbol, card sort, pegboard, Stroop	over 5 days	increase in score over time
Landrigan et al. (2004)	interns	serious medical errors under continuous direct observation over 2,203 patient days	24-hour or more work shifts every other night versus without extended hours but with restricted hours per week	traditional schedule more serious medical errors; more errors in critical care units, more medication errors, more serious diagnostic errors
Leonard et al. (1998)	16 interns	trail-making; Stroop Color Word; delayed story recall, critical flicker fusion, 3-minute grammatical reasoning	end of either normal (8- to 10-hour) shift or a 32-hour shift	degradation on trail making and Stroop for those who worked 32 hour shift
Light et al. (1989)	21 rested and 21 sleep deprived surgical residents	pegboard test	sleep deprivation	decrements for both the dominant and non-dominant hand when sleep-deprived
Lingenfelser et al. (1994)	40 young doctors	connecting numbers, recall of list of items, choice reaction time, Stroop, and electrocardiogram simulation	after night off (at least 6 hours' sleep), after night on call (in hospital for 24 hours)	decrements in all 5 tasks
Mak & Spurgeon (2004)	21 residents	complex numeric pattern	at beginning and at end of a call day	no difference

Table 17.7   Continued

Study	Subjects	Task	Sleep Loss	Result
Muecke (2005)	review of 29 studies	performance	sleep deprivation	older nurses (over 40) more than younger nurses
Orton & Gruzelier (1989)	20 house officers	choice reaction time, simple reaction time, haptic sorting of letters and numbers while blindfolded	night shift	no difference choice reaction times, simple reaction times slower and had greater variability, haptic sorting slower after the night duty but no significant difference in number of errors. Choice reaction times slower for participants after less than 5 hours of sleep in last of 3 blocks of trials
Poulton et al. (1978)	30 junior hospital doctors	grammatical reasoning	sleep deficit of 3 hours or more	reduction in efficiency (i.e., number completed per 3-minute trial)
Reznick & Folse (1987)	21 surgery residents	recall of basic surgical science knowledge, identification of abnormal results in laboratory reports, manual dexterity, Purdue Pegboard test	sleep-deprived	no difference recall, identification, manual dexterity; better on Purdue Pegboard in non-sleep-deprived condition using dominant hand
Richardson et al. (1996)	26 physicians	simultaneous tracking with visual reaction time	4 hours of protected sleep time during a 36-hour call day versus 0 hours protected sleep time	no effect of sleep loss
Rollinson et al. (2003)	12 interns in emergency department	delayed recognition, vigilance, Santa Ana Form Board test	working 12-hour consecutive night shifts	decrease (18.5%) in number correct before first error on delayed recognition test
Samkoff & Jacques (1991)	studies of sleep deprivation on residents	vigilance; manual dexterity, reaction time, short-term recall	sleep-deprivation	vigilance degraded; manual dexterity, reaction time, and short-term recall were not
Scott et al. (2006)	502 critical care nurses	errors, near errors	hours duty	27% reported making at least one error and 38% near errors; risk of making error greater when worked 12.5 or more consecutive hours (odds ratio 1.94) or more than 40 hours per week (odds ratio 1.93)

Table 17.7   Continued

Study	Subjects	Task	Sleep Loss	Result
Smith-Coggins et al. (2006)	49 residents and nurses	PVT, memory, computer-based intravenous insertion simulation, drive car simulator post-shift for about 40 minutes	without nap during night shift	more PVT lapses; more correct responses at 04:00 hours than those who did not nap, no difference driving
Smith-Coggins et al. (1997)	6 attending physicians	PVT, ECG interpretation, intubation of mannequin	after primary sleep period, beginning, middle, and end of shift	PVT reaction time increased across night shift; no effect ECG interpretation; longer to intubate during night than day shift
Stone et al. (2000)	424 surgical residents	American Board of Surgery In-Training Examination (ABSITE) scores	on-call the night before the exam	no effect

Table 17.8   Effects of sleep loss on vigilance performance (laboratory)

Study	Subjects	Task	Sleep Loss	Result
Balkin et al. (2004)	66 commercial motor vehicle drivers	PVT, 10-choice reaction time, Wilkinson 4-choice reaction time	9, 7, or 3 hours' time in bed for four consecutive days	Degraded with decreasing sleep
Belenky et al. (2003)	66 participants	PVT	sleep deprivation	recovery of mean speed took longest in 3 hour sleep condition
Gillberg, Kecklund & Åkerstedt (1994)	6 subjects	28 minute visual vigilance, 11 minute reaction time	at 22:00, 02:00, 04:00, and 06:00 hours during night awake	vigilance percent hits at 22:00, 24:00, and 02:00 hours were high with decreases at 04:00 and 06:00 hours; reaction time increased steadily from 22:00 to 24:00 to 00:20 hours then spiked at 04:00 hours and remained at that level at 06:00 hours
Jewett et al. (1999)	61 participants	PVT	0, 2, 5, or 8 hours' sleep the previous night	degraded as amount of sleep decreased; largest decrement between 2 hours and 0 hours sleep

Table 17.9   Effects of sleep loss on operational vigilance performance (medical)

Study	Subjects	Task	Sleep Loss	Result
Denisco, Drummand & Gravenstein (1987)	21 resident anesthesiologists	detect deviations in heart rate and blood pressure outside defined limits	after 24 hours of in-hospital service	scored lower after service than after rest

Table 17.10   Effects of sleep loss on operational vigilance performance (military)

Study	Subjects	Task	Sleep Loss	Result
Lieberman et al. (2006)	13 male soldiers	visual vigilance	4-day sustained operations mission	decreased number of hits days 3 and 4 but no effect in the reaction time; decreases in number of correct responses and increase in reaction time day 3
McLellan et al. (2007)	20 male New Zealand Special Operations personnel	vigilance task to record where, when, and what activity occurred in building under observation	5-day sustained operations	group using caffeine gum higher vigilance scores on days 3 to 5 and night 3 than placebo group; night vigilance degraded both groups on day 4 compared to day 3

Table 17.11   Effects of sleep loss on operational vigilance performance (aviation)

Study	Subjects	Task	Sleep Loss	Result
Neri et al. (2002)	14 pilot/ copilot crews	PVT during flight in Boeing 747-400 flight simulator	18 to 20 hours without sleep	worst performance between 05:50 and 06:50 hours
Van Dongen, Caldwell & Caldwell (2006)	10 F-117 pilots	F-117A high fidelity flight simulator	38 hours of sustained wakefulness	individual differences in performance degradation as a function of time

Table 17.12   Effects of sleep loss on operational vigilance performance (industry)

Study	Subjects	Task	Sleep Loss	Result
Baulk et al. (2009)	20 male smelter workers	PVT at beginning, middle, and end of each shift	self-reported sleep	PVT slower for night shift and degraded from first to second night of shift work

Table 17.13   Effects of sleep loss on physical performance

Study	Subjects	Task	Sleep Loss	Result
Castellani et al. (2006)	13 male soldiers	maximal lifts in squat and bench press, repetitive bench press throw and squat jump test, repetitive lift of 20.5kg box, traversing obstacle course, grenade throw, marksmanship, wall building	49 hours of military tasks	shallower descent before initiating squat jump, decreased number of boxes lifted, time to complete course highest on day 3, degraded wall building days 3 and 4 as compared to day 1
Welsh et al. (2004)	29 United States Marines	peak jump height and power of 1, 5, and 30 repetition unloaded counter movement squat jumps	after 8 day sustained operations exercise	degraded

Table 17.14   Effects of sleep loss on performance

Study	Visual	Cognitive	Routine	Vigilance	Physical
Åkerstedt, Peters, Anund & Kecklund (2005)		More accidents in driving simulator			
Ansiau et al. (2008)		Poorer recall, selective attention			
Balkin et al. (2004)		Increased time to complete serial addition/subtraction task. Increased deviation lane tracking in driving simulator			

Table 17.14 Continued

Study	Visual	Cognitive	Routine	Vigilance	Physical
Balkin et al. (2004)				Degraded PVT, 10-choice reaction time, Wilkinson four-choice reaction time	
Baranski et al. (2007)		Increased assessment errors and target processing time			
Barnes & Wagner (2009)		Injury rate			
Basner et al. (2008)	Decreased accuracy. Decreased hit rate				
Baulk et al. (2009)				Degraded PVT	
Beatty, Ahern & Katz (1976)		Degraded simulated surgical test. Increased time to complete grammatical reasoning. No effect letter search			
Beh & McLaughlin (1991)		Degraded grammatical reasoning, vertical addition, and card sorting			
Belenky et al. (2003)				Recovery on PVT. Increased number of lapses	
Castellani et al. (2006)		Decreased number of correct responses four- choice reaction time. Increased number of errors. Increased time out errors on matching task			Degraded squat jump (shallower descent before initiating the jump), repetitive lift (number of boxes lifted decreased), obstacle course (time to complete), and wall
Connor et al. (2002)		Increased risk of traffic injury crash			
Couyoumdjian et al. (2010)		Higher switch cost. More errors			

Table 17.14 Continued

Study	Visual	Cognitive	Routine	Vigilance	Physical
Danner & Phillips (2008)		Increased crash rate			
Dawson & Reid (1997)		Degraded tracking			
DeJohn, Reams & Hochhaus (1992)	No effect	Improved understanding of speech			
Denisco, Drummand & Gravenstein (1987)				Degraded detection of deviations	
Dorrian et al. (2008)		More medical errors			
Ferri et al. (2010)		Longer reaction time. No effect spatial attention, Stroop task			
Foushee et al. (1986)					
Friedman, Kornfeld, and & Bigger (1973)		Improved performance flight simulator, more stable in aircraft handling			
Fröberg et al. (1975)	Decreased number	Degraded detection			
Gillberg, Kecklund & Åkerstedt (1996)				No effect percent hits on vigilance or mean reaction time	
Harrison & Horne (1997)		Higher speed in driving simulator			
Harrison & Horne (1998)		Lower speech word count, increased speech semantic similarity			
Harrison & Horne (1999)		Fewer novel speech associations			

Table 17.14   Continued

Study	Visual	Cognitive	Routine	Vigilance	Physical
Hart et al. (1987)		Decreased profitability in marketing game. Increased production errors over time			
Harville, Harrison & Scott (2006)		Less recall. Longer response latencies			
Hovis & Ramaswamy (2007)		Degraded accuracy math and two-back memory tasks			
Ikegami et al. (2009)	Degraded color vision				
Jacques, Lynch & Samkoff (1990)		Degraded simple addition, paced auditory serial addition, simple words memory, reading span, continuous performance. No effect card sort			
Jewett et al. (1999)		Degraded score medical test			
Killgore, Balkin & Wesensten (2006)				Degraded PVT	
Landrigan et al. (2004)		More risky behavior on Iowa Gambling Task			
Leonard et al. (1998)		Increased number of serious medical errors			
Lieberman et al. (2006)		Degraded trail making. Degraded Stroop. No effect recall, critical flicker fusion, or grammatical reasoning			
Lieberman et al. (2006)		More time out errors on matching task. No effect on correct matches or response time			
Lieberman et al. (2006)		No difference repeating key strokes			

Table 17.14 Continued

Study	Visual	Cognitive	Routine	Vigilance	Physical
Mitchell & Williamson (2000)		Decreased number of correct responses reaction time		Degraded visual vigilance	
National Sleep Foundation's 2010 Sleep in America poll		More errors vigilance task			
Neri et al. (2002)			Degraded everyday tasks, perform at work, perform household tasks, care for family		
Neri, Shappell & DeJohn (1992)				Worst PVT between 05:50 and 06:50	
Nesthus, Scarborough & Schroeder (1998a)		Increase reaction time for correct responses on pattern recognition and manikin tests, Increase in error rate for the pattern recognition test. No effect time estimation			
Nesthus, Scarborough & Schroeder (1998b)		Younger participants decline over time in synthetic work, letter search, and four-choice reaction time			
Pilcher et al. (2007)		Correct reaction time and errors increased and percent correct decreased on choice reaction time. Throughput decreased on reaction time and letter search			
Scott et al. (2006)		Wombat scores increased over time. No changes Automated Performance Test System. Degraded chemical plant task, increase reaction time and decrease accuracy			

Table 17.14   Continued

Study	Visual	Cognitive	Routine	Vigilance	Physical
Smith-Coggins et al. (2006)		Self-reported medical errors			
Thompson et al. (2006)		More lapses PVT. Quicker catheter insertion. No difference driving			
Van Dongen, Caldwell & Caldwell (2006)		Decreased throughput reaction time task. Increased error piloting performance			
Vrijkotte et al. (2004)				Degraded pilot performance in simulator	
Welsh et al. (2004)		Degraded recall, reasoning, and tracking			Degraded peak jump Height and power of 1, 5, and 30 repetition unloaded counter movement squat jumps

CONCLUSIONS

Table 17.14 summarizes the effects of sleep loss on performance. As can be seen, the majority of the research has been to identify the effects of sleep loss on cognitive tasks. Less has been done on the effects on physical task performance. Also evident is the diversity of tasks which are degraded by sleep loss. These degradations include both increases in errors and time to compete tasks. But this is not universal. For example, Beatty, Ahem & Katz (1976) reported no difference in letter search performance. Lieberman et al. (2006) reported no effect on correct matches. Ferri et al. (2010) found no effect on spatial attention. These three studies suggest that spatial memory may be more difficult to disrupt with sleep loss.

DeJohn, Reams & Hochhaus (1992) found no effect in visual performance and an improved understanding of speech – although the latter could have been associated with learning. The improvement in flight simulator performance seen in Foushee et al., 1986 may also be due to practice. This may also apply to Wombat score results in Pilcher et al. (2007). Another case may be the no effect on card sort in Ikegami et al., 2009. Time estimation has also been reported not to be affected by sleep loss (Neri, Shappell & DeJohn, 1992). The effects on driving are also not universal (e.g., Connor et al., 2002 versus Smith-Coggins et al., 2006); nor are the effects on the Stroop test: Ferri et al. (2010) reported no effect on the Stroop test, while others have found this test to be degraded (Leonard et al., 1998). By far the most consistent effect and one of the most studied is the effect of sleep loss on vigilance with a clear preference for the use of the PVT. But vigilance performance is not always degraded (Gillberg, Kecklund & Åkerstedt, 1994).

So research supports that at least in most cases of sleep loss, humans "can't 'see straight,'" "can't 'think straight,'" "can't 'do anything right,'" "can't 'seem to pay attention,'" and "can't 'lift a finger.'"

REFERENCES

Åkerstedt, T. (2007). Altered sleep/wake patterns and mental performance. *Physiology & Behavior*, 90, 209–18.

Åkerstedt, T., Peters, B., Anund, A. & Kecklund, G. (2005). Impaired alertness and performance driving home from the night shift: A driving simulator study. *Journal of Sleep Research*, 14, 17–20.

Andreyka, K. & Tell, P. (1996). An analysis of continuous performance test scores before and after sleep deprivation in obstetrical resident physicians. *Archives of Clinical Neuropsychology*, 11, 362–63.

Angus, R.G. & Heslegrave, R.J. (1985). Effects of sleep loss on sustained cognitive performance during a command and control simulation. *Behavior Research Methods, Instruments, and Computers*, 17, 55–67.

Ansiau, D., Wild, P., Niezborala, M., Rouch, I. & Marquie, J.C. (2008). Effects of working conditions and sleep of the previous day on cognitive performance. *Applied Ergonomics*, 39, 99.

Babkoff, H., Mikulincer, M., Caspy, T., Kempinski, D. & Sing, H. (1988). The topology of performance curves during 72 hours of sleep loss: A memory and search task. *The Quarterly Journal of Experimental Psychology*, 40A(4), 737–56.

Balkin, T.J. (2004). Comparative utility of instruments for monitoring sleepiness related performance decrements in the operational environment. *Journal of Sleep Research*. 13(3), 219–27

Balkin, T.J., Bliese, P.D., Belenky, G., Sing, H., Thorne, D.R., Thomas, M., Redmond, D.P., Russo, M. & Wesensten, N.J. (2004). Comparative utility of instrument for monitoring sleepiness-related performance decrements in the operational environment. *Journal of Sleep Research*, 13, 219–27.

Baranski, J.V., Thompson, M.M., Lichacz, F.M.J., McCann, C., Gil, V., Pastò, L. & Pigeau, R.A. (2007). Effects of sleep loss on team decision making: Motivational loss of motivational gain? *Human Factors*, (4), 646–60.

Barger, L.K., Cade, B.E., Ayas, N.T., Cronin, J.W., Rosner, B., Speizer, F.E. & Czeisler, C.A. (2005). Extended work shifts and the risk of motor vehicle crashes among interns. *The New England Journal of Medicine*, 352, 125–34.

Barnes, C.M. & Wagner, D.T. (2009). Changing to Daylight Saving Time cuts into sleep and increases workplace injuries. *Journal of Applied Psychology*, 94, 1305–17.

Bartel, P., Offermeier, W., Smith, F. & Becker, P. (2004). Attention and working memory in resident anaesthetists after night duty: Group and individual effects. *Occupational and Environmental Medicine*, 61, 167–70.

Basner, M., Rubinstein, J., Fomberstein, K.M., Coble, M.C., Ecker, A., Avinash, D. & Dinges, D.F. (2008). Effects of night work, sleep loss and time on task on simulated threat detection performance. *Sleep: Journal of Sleep and Sleep Disorders Research*, 31, 1251–9.

Baulk, S.D., Fletcher, A., Kandelaars, K.J., Dawson, D., and Roach, G.D. (2009). A field study of sleep and fatigue in a regular rotating 12-h shift system. *Applied Ergonomics*, 40, 694–8.

Beatty, J., Ahern, S.K., and Katz, R. (1976). Sleep deprivation and the vigilance of anesthesiologists during simulated surgery, in R.R. Mackie (ed.), *Vigilance Theory: Operational Performance and Physiological Correlates*. New York: Plenum.

Beh, H.C. & McLaughlin, P.J. (1991). Mental performance of air crews following layovers on transzonal flights. *Ergonomics*, 34, 123–35.

Belenky, G., Wesensten, N.J., Thorne, D.R., Thomas, M.L., Sing, H.C., Redmond, D.P., Russo, M.B. & Balkin, T.J. (2003). Patterns of performance degradation and restoration during sleep restriction and subsequent recovery: A sleep dose–response study. *Journal of Sleep Research*, 12, 1–12.

Binks, P.G., Waters, W.F. & Hurry, M. (1999). Short-term total sleep deprivations does not selectively impair higher cortical functioning. *Sleep*, 22, 328 –34.

Blagrove, M., Alexander, C., & Horne, J.A. (1995). The effects of chronic sleep reduction on the performance of cognitive tasks sensitive to sleep deprivation. *Applied Cognitive Psychology*, 9, 21–40.

Caldwell J.A., Mallis, M.M., Caldwell, J.L., Paul, M.A., Miller, J.C. & Neri, D.F. (2009). Fatigue countermeasures in aviation. *Aviation, Space, and Environmental Medicine*, 80, 29–59.

Caldwell, J.A., Prazinko, B.F. & Caldwell, J.L. (2002). Fatigue in aviation sustained operations, the utility of napping, and the problem of sleep inertia. Fort Rucker, AL: Army Aeromedical Research Laboratory, November.

Casagrande, M., Violani, C., Curcio, G. & Bertini, M. (1997). Assessing vigilance through a brief pencil and paper letter cancellation task (LCT): Effects of one night of sleep deprivation and the time of day. *Ergonomics*, 40, 613–30.

Castellani, J.W., Nindl, B.C., Lieberman, H.R. & Montain, S.J. (2006). Decrements in human performance during 72–84 hours of sustained operations. Natick, MA: US Army Research Institute of Environmental Medicine, 1 November.

Christensen, E.E., Dietz, G.W., Murry, R.C. & Moore, J.G. (1997). The effect of fatigue on resident performance. *Diagnostic Radiology*, 125, 103–105.

Connor, J., Norton, R., Ameratunga, S., Robinson, E., Civil, I., Dunn, R., Bailey, J., and Jackson, R. (2002). Driver sleepiness and risk of serious injury to car occupants: Population based case control study. *British Medical Journal*, 324, 1–5.

Couyoumdjian, A., Stefano Sdoia, S., Tempesta, D., Curcio, G., Rastellini, E., De Gennaro, L. & Ferrara, M. (2010). The effects of sleep and sleep deprivation on task-switching performance. *Journal of Sleep Research*, 19, 64–70.

Cruz, C., Detwiler, C., Nesthus, T., Boquet, A. & Della Rocco, P.A. (1995). Laboratory comparison of clockwise and counter-clockwise rapidly rotating shift schedules: Effects on performance, sleep, and subjective ratings. Oklahoma City, OK: Federal Aviation Administration, Civil Aerospace Medical Institute.

Danner, F. & Philips, B. (2008). Adolescent sleep, school start times, and teen motor vehicle crashes. *Journal of Clinical Sleep Medicine*, 4, 533–35.

Darwent, D., Lamond, N. & Dawson, D. (2008). The sleep and performance of train drivers during an extended freight-haul operation. *Applied Ergonomics*, 39, 614–22.

Darwent, D., Dawson, D. & Roach, G.D. (2010). Prediction of probabilistic sleep distributions following travel across multiple time zones. *Sleep: Journal of Sleep and Sleep Disorders Research*, 33, 185–95.

Dawson, D. & Reid, K. (1997). Fatigue, alcohol and performance impairment. *Nature*, 388, 235.

Deary, I.J. and Tait, R. (1987). Effects of sleep deprivation on cognitive performance and mood in medical house officers. *British Medical Journal*, 295, 1513–16.

Deaconson, T.F., O'Hair, D.P., Levy, M.F., Lee, M.B., Schueneman, A.L. & Condon, R.E. (1988). Sleep deprivation and resident performance. *Journal of the American Medical Association*, 260, 1721–27.

DeJohn, C.A., Reams, G.C. & Hochhaus, L.W. (1992). Effects of sustained flight operations on naval aircrew. Pensacola, FL: Naval Aerospace Medical research Laboratory.

Denisco, R.A., Drummand, J.N. & Gravenstein, J.S. (1987). The effect of fatigue on the performance of a simulated anesthetic monitoring task. *Journal of Clinical Monitoring*, 3, 22–24.

Dinges, D.F., Pack, F., Williams, K., Gillen, K.A., Powell, J.W., Ott, G.E., Aptowicz, C. & Pack, A.I. (1997). Cumulative sleepiness, mood disturbance, and psychomotor vigilance performance decrements during a week of sleep restricted to 4–5 hours per night. *Sleep*, 20, 267–77.

Dorrian, J., Lamond, N. & Dawson, D. (2000). The ability to self-monitor performance when fatigued. *Journal of Sleep Research*, 9, 137–44.

Dorrian, J., Tolley, C.T., Lamond, N., van den Heuvela, C., Pincombec, J., Rogers, A.E. & Drew, D. (2008). Sleep and errors in a group of Australian hospital nurses at work and during the commute. *Applied Ergonomics*, 39, 605–13.

Dula, D.J., Dula, N.L., Hamrick, C. & Wood, G.C. (2001). The effect of working serial night shifts on the cognitive functioning of emergency physicians. *Annals of Emergency Medicine*, 38, 152–5.

Eastridge, B.J., Hamilton, E.C., O'Keefe, G.E., Rege, R.V., Valentine, R.J., Jones, D.J., Tesfay, S. & Thal, E.R. (2003). Effect of sleep deprivation on the performance of simulated laparoscopic surgical skill. *The American Journal of Surgery*, 186, 169–74.

Ellman, P.I., Law, M.G., Tache-Leon, C., Reece, T.B., Maxey, T.S., Peeler, B.B., Kern, J.A., Tribble, C.G. & Kron, I.L. (2004). Sleep deprivation does not affect operative results in cardiac surgery. *The Annals of Thoracic Surgery*, 78, 906–11.

Engel, W., Seime, R., Powell, V. & D'Allessandri, R. (1987). Clinical performance of interns after being on call. *Southern Medical Journal*, 80, 761–3.

Ferri, R., Drage, V., Arico, D., Bruni, O., Remington, R.W., Stamatakis, K. & Punjabi, N.M. (2010). The effect of experimental sleep fragmentation on cognitive processing. *Sleep Medicine*, 11, 378–85.

Foushee, H.C., Lauber, J.K., Baetge, M.M. & Acomb, D.B. (1986). Crew factors in flight operations III: the operational significance of exposure to short-haul transport operations. *NASA Technical Memorandum, 88322.*

Friedman, R.C., Bigger, T.J. & Kornfeld, D.S. (1971). The intern and sleep loss. *New England Journal of Medicine*, 283, 201–3.

Friedman, R.C., Kornfeld, D.S. & Bigger, T.J. (1973). Psychological problems associated with sleep deprivation in interns. *Journal of Medical Education*, 48, 436–41.

Fröberg, J.E., Karlsson, C., Levi, L. & Lidberg, L. (1975). Circadian rhythms of catecholamine excretion, shooting range performance and self-ratings of fatigue during sleep deprivation. *Biological Psychology*, 2, 175–88.

Gander, P. (2001). Fatigue management in air traffic control: The New Zealand approach. *Transportation Research Part F* 4, 49–62.

Gillberg, M., Kecklund, G. & Åkerstedt, T. (1994). Relations between performance and subjective ratings of sleepiness during a night awake. *Sleep*, 17, 236–41.

Gillberg, M., Kecklund, G. & Åkerstedt, T. (1996). Sleepiness and performance of professional drivers in a truck simulator – comparisons between day and night driving. *Journal of Sleep Research*, 5, 12–15.

Gottlieb, D.J., Parenti, C.M., Peterson, C.A. & Lofgren, R.P. (1991). Effect of a change in house staff work schedule on resource utilization and patient care. *Archives of Internal Medicine*, 151, 2065–70.

Harrison, Y. & Horne, J.A. (1997). Sleep deprivation affects speech. *Sleep*, 20, 871–7.

Harrison, Y. & Horne, J.A. (1998). Sleep loss impairs short and novel language tasks having a prefrontal focus. *Journal of Sleep Research*, 7, 95–100.

Harrison, Y. & Horne, J.A. (1999). One night of sleep loss impairs innovative thinking and flexible decision-making. *Organizational Behavior and Human Decision Processes*, 78, 128–45.

Hart, R.P., Buchsbaum, D.G., Wade, J.B., Hammer, R.M. & Kwentus, J.A. (1987). Effect of sleep deprivation on first-year residents' response times, memory, and mood. *Journal of Medical Education*, 19, 940 –42.

Harville, D.L., Harrison, R. & Scott, C. (2006). The fatigue equivalent of job experience and performance in sustained operations (AFRL-HE-BR-TR-2006-0042). Brooks City-Base, TX: Air Force Research Laboratory, June.

Hayashino, Y., Yamazaki, S., Takegami, M., Nakayama, T., Sokejima, S. & Fukuhara, S. (2010). Association between number of comorbid conditions, depression, and sleep quality using the Pittsburgh Sleep Quality Index: Results from a population-based survey. *Sleep Medicine*, 11, 366–71.

Haynes, D.F., Schwedler, M., Dyslin, D.C., Rice, J.C. & Kersteain, M.D. (1995). Are postoperative complications related to resident sleep deprivation? *Southern Medical Journal*, 88, 283–9.

Heslegrave, R.J., Rhodes, W., Szlapetis, I., Ujimoto, V., Han, K. & Moldofsky, H. (1995). Impact of shiftwork and overtime on Air Traffic Controllers: Subjective estimates of performance deficits as a function of shift, chronobiological typology, and age. *Sleep Research*, 24, 101.

Heuer, H., Spijkers, W., Kiesswetter, E., and Schmidtke, V. (1998). Effects of sleep loss, time of day, and extended mental work on implicit and explicit learning of sequences. *Journal of Experimental Psychology: Applied*, 4, 139–62.

Horne, J.A. (1988). Sleep loss and "divergent" thinking ability. *Sleep*, 11, 528–36.

Horne, J.A. & Reyner, L.A. (1995). Sleep related vehicle accidents. *British Medical Journal*, 310, March 1995, 565–67.

Hovis, J.K. & Ramaswamy, S. (2007). Color vision and fatigue: An incidental finding. *Aviation, Space, and Environmental Medicine*, 78, 1068–71.

Ikegami, K., Ogyu, S., Arakomo, Y., Nagata, S., Hiro H., Suzuki, K. & Mafune, K. (2009). Recovery of cognitive performance and fatigue after one night of sleep deprivation, *Journal of Occupational Health*, 51, 412–22.

Jacques, C.H.M., Lynch, J.C. & Samkoff, J.S. (1990). The effects of sleep loss on cognitive performance of resident physicians. *Journal of Family Practice*, 30, 223.

Jaskowski, P. & Weodarczyk, D. (1997). Effect of sleep deficit, knowledge of results, and stimulus quality on reaction time and response force. *Perceptual and Motor Skills*, 84, 563–72.

Jensen, A., Milner, R., Fisher, C., Gaughan, J., Rolandelli, R. & Grewal, H. (2004). Short-term sleep deficits do not adversely affect acquisition of laparoscopic skills in a laboratory setting. *Surgical Endoscopy*, 18, 948–53.

Jewett, M.E., Dijk, D., Kronauer, R.E. & Dinges, D.F. (1999). Dose-response relationship between sleep duration and human psychomotor vigilance and subjective alertness. *Sleep*, 22, 171–9.

Joffe, M.D. (2006). Emergency department provider fatigue and shift concerns. *Clinical Pediatric Emergency Medicine*, 7, 248–54.

Killgore, W.D., Balkin, T.J. & Wesensten, N.J. (2006). Impaired decision making following 49 h of sleep deprivation. *Journal of Sleep Research*, 15, 7–13.

Kim, D., Lee, H., Kim, M., Park, Y., Go, H., Kim, K., Lee, S., Chae, J. & Lee, C. (2001). The effect of total sleep deprivation on cognitive functions in normal male adult subjects. *International Journal of Neuroscience*, 109, 127–37.

Klose, K.J., Wallace-Barnhill, G.L. & Craythorne, N.W.B (1985). Performance test results for anesthesia residents over a five day week including on-call duty. *Anesthesiology*, 63, A495.

Lamond, N. & Dawson, D. (1999). Quantifying the performance impairment associated with fatigue. *Journal of Sleep Research*, 8, 255–62.

Landrigan. C.P., Rothschild, J.M., Cronin, J.W., Kaushal, R., Burdick, E., Katz, J.T., Lilly, G.M., Stone, P.H., Lockley, S.W., Bates, D.W. & Czeisler, C.A. (2004). Effect of reducing interns' work hours on serious medical errors in intensive care units. *New England Journal of Medicine*, 1838–48.

Leonard, C., Fanning, N., Attwood, J., and Buckley, M. (1998). The effect of fatigue, sleep deprivation, and onerous working hours on the physical and mental wellbeing of pre-registration house officers. *Irish Journal of Medicine Science*, 167(1), 22–5.

Lieberman, H.R., Niro, P., Tharion, W.J., Nindl, B.C., Castellani, J.W. & Montain, S.J. (2006). Cognition during sustained operations: Comparison of a laboratory simulation to field studies. *Aviation, Space, and Environmental Medicine*, 77, 929–35.

Light, A.I., Sun, J.H., McCool, C., Thompson, L., Heaton, S. & Bartle, E.J. (1989). The effects of acute sleep deprivation on level of resident training. *Current Surgery*, 29–30.

Linde, L., Edland, A. & Bergstrom, M. (1999). Auditory attention and multiattribute decision-making during a 33 h sleep-deprivation period: Mean performance and between-subject dispersions. *Ergonomics*, 42, 696–713.

Lingenfelser, T., Kaschel, R., Weber, A., Zaiser-Kaschels, H. Jakober, B. & Kuper, J. (1994). Young hospital doctors after night duty: Their task-specific cognitive status and emotional condition. *Medical Education*, 28, 566–72.

Mak, S.K. & Spurgeon, P. (2004). The effects of acute sleep deprivation on performance of medical residents in a regional hospital: Prospective study. *Hong Kong Medical Journal*, 10, 14–20.

McCarthy, M.E. & Waters, W.F. (1997). Decreased attentional responsivity during sleep deprivation: Orienting response latency, amplitude, and habituation. *Sleep*, 20, 115–23.

McLellan, T.M., Kamimori, G.H., Voss, D.M., Tate, C. & Smith, S.J.R. (2007). Caffeine effects on physical and cognitive performance during sustained operations. *Aviation, Space, and Environmental Medicine*, 78, 871–7.

Mikulincer, M., Babkoff, H., Caspy, T. & Sing, H. (1989). The effects of 72 hours of sleep loss on psychological variables. *British Journal of Psychology*, 80, 145–62.

Mikulincer, M., Babkoff, H., Caspy, T. & Sing, H. (1990). The impact of cognitive interference on performance during prolonged sleep loss. *Psychological Research*, 52, 80–86.

Miro, E., Cano, M.C., Espinosa-Fernandez, L. & Buela-Casal, G. (2003). Time estimation during prolonged sleep deprivation and its relation to activation measures. *Human Factors*, 45, 148–59.

Mitchell, R.J. & Williamson, A.M. (2000). Evaluation of an 8 hour versus a 12 hour shift roster on employees at a power station. *Applied Ergonomics*, 31, 83–93.

Muecke, S. (2005). Effects of rotating night shifts: Literature review. *Journal of Advanced Nursing*, 50, 433–9.

Nakano, T., Araki, K., Michimori, A., Inbe, H., Hagiwara, H. & Koyama, E. (2000). Temporal order of sleepiness, performance and physiological indices during 19-h sleep deprivation. *Psychiatry and Clinical Neuroscience*, 4, 280–82.

National Sleep Foundation (2010). Sleep in America poll summary of findings. Online. Available at: http://www.sleepfoundation.org

Neri, D.F., Dinges, D.F. & Rosekind, M.R. (1977). Sustained carrier operations: Sleep loss, performance, and fatigue countermeasures. Moffett Field, CA: NASA Ames Research Center, June.

Neri, D.F., Oyung, R.L., Colletti, L.M., Mallis, M.M., Tam, P.Y. & Dinges, D.F. (2002). Controlled breaks as a fatigue countermeasure on the flight deck. *Aviation, Space, and Environmental Medicine*, 73, 654–64.

Neri, D.F., Shappell, S.A. & DeJohn, C.A. (1992). Simulated sustained flight operations and performance, Part 1: Effects of fatigue. *Military Psychology*, 4, 137–55.

Nesthus, T.E., Scarborough, A.L. & Schroeder, D.J. (1998a). Changes in the performance of a synthetic work task as a function of age, gender, and sleep deprivation. *Proceedings of the Human Factors and Ergonomics Society 42nd Annual Meeting*, 181–85.

Nesthus, T.E., Scarborough, A.L. & Schroeder, D.J. (1998b). Four-choice serial reaction time and visual search performance during 34 hours of sleep loss. Poster Presentation Aerospace Medical Association, 69th Annual Scientific Meeting.

Orton, D.J. & Gruzelier, J.H. (1989). Adverse changes in mood and cognitive performance of house officers after night duty. *British Medical Journal*, 298, 21–3.

Philip, P., Taillard, J., Sagaspe, P., Valtat, C., Sanchez-Ortuno, M., Moore, N., Charles, A. & Bioulac, B. (2004). Age, performance, and sleep deprivation. *Journal of Sleep Research*, 13, 105–10.

Pilcher, J.J. & Walters, A.S. (1997). How sleep deprivation affects psychological variables related to college students' cognitive performance. *Journal of American College Health*, 46, 121–6.

Pilcher, J.J., Band, D., Odle-Dusseau, H.N. & Muth, E.R. (2007). Human performance under sustained operations and acute sleep deprivation conditions: Toward a model of controlled attention. *Aviation, Space, and Environmental Medicine*, 78, B15–24.

Pilcher, J.J., McClelland, L.E., Moore, D.D., Haarmann, H., Baron, J., Wallsten, T.S. & McCubbin, J.A. (2007). Language performance under sustained work and sleep deprivation conditions. *Aviation, Space, and Environmental Medicine*, 78(5, Suppl.), B15–24.

Polzella, D.J. (1975). Effects of sleep deprivation on short-term recognition memory. *Journal of Experimental Psychology, Human Learning and Memory*, 104, 194–200.

Poulton, E.C., Hunt, G.M., Carpenter, A. & Edwards, R.S. (1978). The performance of junior hospital doctors following reduced sleep and long hours of work. *Ergonomics*, 21, 279–95.

Reznick, R.K. & Folse, J.R. (1987). Effect of sleep deprivation on the performance of surgical residents. *The American Journal of Surgery*, 154, 520–25.

Richardson, G.S., Wyatt, J.K., Sullivan, J.P., Orav, E.J., Ward, A.E. Wolf, M.A. & Czeisler, C.A. (1996). Objective assessment of sleep and alertness in medical house staff and the impact of protected tie for sleep. *Sleep*, 19, 718–26.

Roach, G.D., Petrilli, R.M., Dawson, D. & Thomas, M.J.W. (2006). The effect of fatigue on the operational performance of flight crew in a B747-400 simulator. *Seventh International AAvPA Symposium*. Manly, NSW.

Rollinson, D.C., Rathlev, N.K., Moss, M., Killiany, R., Sassower, K.C., Auerbach, S. & Fish, S.S. (2003). The effects of consecutive night shifts on neuropsychological performance of interns in the emergency department: A pilot study. *Annals of Emergency Medicine*, 41, 400–406.

Rosenthal, T.C., Majeroni, B.A., Pretorius, R. & Malik, K. (2008). Fatigue: An overview. *American Family Physician*, 78, 1173–80.

Rowe, A.L., French J., Neville, K.J. & Eddy, D.R. (1992). The prediction of cognitive performance degradations during sustained operations. *Proceedings of the Human Factors Society 36th Annual Meeting*, 111–15.

Russo, M.B., Kendall, A.P., Johnson, D.E., Sing, H.C., Thorne, D.R., Escolas, S.M., Santiago, S., Holland, D.A., Hall, T.W. & Redmond, D.P. (2005). Visual perception, psychomotor performance, and complex motor performance during an overnight air refueling simulated flight. *Aviation, Space, and Environmental Medicine*, 76, C92–C103.

Samkoff, J.S. & Jacques, C.H.M. (1991). A review of studies concerning effects of sleep deprivation and fatigue on residents' performance. *Academic Medicine*, 66, 687–93.

Scott, L.D., Rogers, A.E., Hwang, W.T. & Zhang, Y. (2006). Effects of critical care nurses' work hours on vigilance and patients' safety. *American Journal of Critical Care*, 15, 30–37.

Smith-Coggins, R., Howard, S.K., Mac, D.T., Wang, C., Kwan, S., Rosekind, M.R., Sowb, Y., Balise, R., Levis, J. & Gaba, D.M. (2006). Improving alertness and performance in emergency department physicians and nurses: The use of planned naps. *Annals of Emergency Medicine*, 48, 596–604.

Smith-Coggins, R., Rosekind, M.R., Buccino, K.R., Dinges, D.F. & Moser, R.P. (1997). Rotating shiftwork schedules: Can we enhance physician adaptation to night shift? *Academic Emergency Medicine*, 4, 951–61.

Stone, M.D., Doyle, J., Bosch, R.J., Bothe, A. & Steele, G. (2000). Effect of resident call status on ABSITE performance. *Surgery*, 465–84.

Thompson, W.T., Lopez, N., Hickey, P., DaLuz, C. & Caldwell, J.L., 2006. Effects of shift work and sustained operations: Operator performance in remotely piloted aircraft (OP-Repair). (HSW-PE-BR-TR-2006-OOO1). Brooks City-Base, TX: Air Force Research Laboratory 311 Performance Enhancement Directorate, 4 January.

Tyagi, R., Shen, K., Shao, S. & Li, X. (2009). A novel auditory working-memory vigilance task for mental fatigue assessment. *Safety Science*, 47, 967–72.

Van Dongen, H.P.A., Caldwell, J.A. & Caldwell, J.L. (2006). Investigating systematic individual differences in sleep-deprived performance on a high-fidelity flight simulator. *Behavior Research Methods* 38, 333–43.

Van Dongen, H.P.A., Maislin, G., Mullington, J.M. & Dinges, D.F. (2003). The cumulative cost of additional wakefulness: Dose-response effects of neurobehavioral functions and sleep physiology from chronic sleep restriction and total sleep deprivation. *Sleep*, 26, 117–26.

Vrijkotte, S., Valk, P.J.L., Vennstra, B.J. & Visser, T. (2004). Monitoring physical and cognitive performance during sustained military operations. 24th Army Science Conference Proceedings, Poster 14.

Welsh, T.T., Alemany, J.A., Nindl, B.C., Frykman, P.N., Tuckow, A.P. & Montain, S.J. (2004). Monitoring warfighter's physical performance during sustained operations using a field expedient jumping test. 24th Army Science Conference Proceedings.

Wilkinson, R.T., Edwards, R.S. & Haines, E. (1966). Performance following a night of reduced sleep. *Psychonomic Science*, 5, 471–2.

Williamson, A.M. & Feyer, A. (2000). Moderate sleep deprivation produces impairments in cognitive and motor performance equivalent to legally prescribed levels of alcohol intoxication. *Occupational and Environmental Medicine*, 57, 649–55.

PART VI

Fatigue and Health

18
Differentiating Fatigue in Chronic Fatigue Syndrome and Psychiatric Disorders

Leonard A. Jason, Molly Brown, Meredyth Evans and Abigail Brown

INTRODUCTION

Fatigue is a prevalent symptom in the general population, with an average of approximately 25 percent of participants in community-based studies reporting brief periods of fatigue lasting between a day and a month (Lewis & Wessely, 1992). Temporary fatigue is typically attributed to lifestyle or situational factors, including poor sleep quality or stressful life events. In contrast, a smaller proportion of the population experiences fatigue of greater intensity and longer duration that causes severe impairment in functional activity and quality of life (Jason & Choi, 2008). For example, prolonged fatigue is considered to last between one and five months, and was found in approximately 7.1 percent of a community-based sample (Jason et al., 1999). Chronic fatigue defined as lasting six or more months is less common yet, and was found to occur approximately 4.2 percent of a community-based sample (Jason et al., 1999). Women are more vulnerable to prolonged and chronic fatigue than men (Furberg et al., 2005; Steele et al., 1998). Long-term fatigue is often due to medical or psychiatric illness, and treatment of the underlying condition may alleviate the fatigue. It is evident that fatigue is a symptom caused by a range of life circumstances and health factors.

FATIGUE IN PSYCHIATRIC CONDITIONS

Prolonged and chronic fatigue are common complaints reported in primary care settings, and many of those reporting prolonged fatigue meet criteria for a psychiatric disorder (Hickie et al., 1996). Individuals with anxiety and mood disorders report particularly high rates of fatigue, as one primary care study found that 26 percent of patients with an anxiety disorder reported fatigue and 40 percent of patients with a mood disorder had the symptom (Kroenke et al., 1994). The relationship between fatigue and psychiatric conditions is highlighted in psychodiagnostic classification systems such as the American Psychiatric Association's *Diagnostic and Statistical Manual of Mental Disorders* (2000; DSM-

IV-TR). According to the DSM-IV-TR, longstanding fatigue is one of the diagnostic criteria for generalized anxiety disorder (GAD), major depressive disorder (MDD), and dysthymic disorder. Fatigue is also a potential feature of schizoaffective and bipolar disorders involving depressive episodes, as well as undifferentiated somatoform disorder. The directional relationship between chronic fatigue and psychiatric disorders is unclear. It is possible for some individuals to experience a period of fatigue that leads to the development of a psychiatric condition due to the physical misery of fatigue or reduced ability to meet life goals. On the other hand, some individuals may experience fatigue as a direct symptom of a psychiatric disorder. It is also possible that fatigue and psychiatric disorders are due to different etiologies. Several studies have attempted to clarify the directional relationship between psychiatric disorders and fatigue, and some research suggests that fatigue most often precedes the development of a psychiatric disorder (Addington et al., 2001; Dryman & Eaton, 1991) while other research does not support this directional relationship (van der Linden et al., 1999). An additional study found that approximately half of patients with chronic fatigue had a premorbid psychiatric condition (Katon et al., 1991). Given these inconsistent findings, it appears that the temporal relationship between fatigue and psychiatric disorders is bidirectional and differs across individuals.

Fatigue is a troubling symptom for patients with psychiatric conditions. For example, one study found that nearly 42 percent of patients with GAD sought medical consultation for fatigue (Bélanger, Ladouceur & Morin, 2005). In addition, the severity of fatigue is correlated with level of psychological distress (Pawlikowska et al., 1994), and this relationship has been found across cultures (Simon et al., 1996). The majority of individuals in the community who experience chronic fatigue were found to also have multiple medically unexplained somatic symptoms indicative of somatization disorder (Martin et al., 2007). Both somatic symptoms and chronic fatigue were related to poor quality of life outcomes (Martin et al., 2007). Moreover, research suggests that individuals with depression who also experienced symptoms of somatization had higher levels of fatigue than those without multiple somatic symptoms (Lavidor, Weller & Babkoff, 2002). In sum, the consequences of chronic fatigue among people with psychiatric conditions are serious as it is often accompanied by other physical symptoms and increased distress.

While many cognitive or behavioral interventions for depression and anxiety focus primarily on changes in cognition or mood, some research suggests that interventions can also be useful in reducing fatigue severity. For example, exercise regimens have been found to reduce fatigue among people with depression (Lane & Lovejoy, 2001). Cognitive behavior therapy has also been successfully applied to those with depression, and some positive findings have occurred in fatigue levels (Friedberg & Krupp, 1994). Pharmacotherapy has also been used to target symptoms of fatigue among patients with depression. The severity of fatigue has been found to substantially decrease after one month of antidepressant medication, but after one month fatigue levels do not continue to show much reduction (Greco, Eckert & Kroenke, 2004). One study found that among people with depression, the addition of modafinil (i.e., a medication used to improve wakefulness among people with sleep disorders) to an antidepressant regimen significantly improved fatigue compared to placebo, although the improvements were not sustained at a six-week follow-up (DeBattista et al., 2003). Finally, given that fatigue is often a psychiatric symptom, most individuals with depression are able to return to their premorbid level of energy upon recovery from a major depressive episode (Buist-Bouwman et al., 2004).

PSYCHIATRIC COMORBIDITY IN FATIGUING MEDICAL CONDITIONS SUCH AS CHRONIC FATIGUE SYNDROME

Both fatigue and psychiatric conditions, particularly depression and anxiety, are common in many medical conditions. For example, patients with cancer report high rates of fatigue as well as symptoms of depression and anxiety (Tchekmedyian et al., 2003). There are conflicting findings regarding the causal relationship between psychological disorders and fatigue in this population (Brown & Kroenke, 2009). Multiple sclerosis (MS) is another illness commonly associated with mood or anxiety disorders and fatigue (Siegert & Abernethy, 2005). The issue of overlapping symptomatology of illness and psychiatric disorders is particularly salient for patients with chronic fatigue syndrome (CFS), an illness characterized by debilitating fatigue.

Controversy has arisen concerning the actual case definition of CFS, and whether it differentiates those with CFS from other psychiatric conditions. Reeves et al. (2005) published a clinically empiric definition of CFS that sought to operationalize the older Fukuda et al. (1994) CFS case definition. Reeves et al. feel that the specification of instruments and cut-off points would result in a more reliable and valid approach for the assessment of CFS. However, using these new empiric criteria, the estimated rates of CFS have increased to 2.54 percent (Reeves et al., 2007), rates that are about ten times higher than prevalence estimates of other investigators (Jason et al., 1999). It is at least possible that the CFS increases in the United States are due to a broadening of the case definition and possible inclusion of cases with primary psychiatric conditions. In support of this proposition, Jason, Najar, Porter and Reh (2009) recently investigated this CFS empiric case definition with participants with a diagnosis of CFS and participants with a diagnosis of MDD. Jason et al. found that 38 percent of those with a diagnosis of MDD were misclassified as having CFS using the new CDC empiric case definition. Reliable and valid ways of classifying patients into diagnostic categories is essential in order to enable investigators to better understand etiology, pathophysiology, and treatment approaches for CFS and other disorders.

CFS has never been classified as a psychiatric condition in either the American Psychiatric Association's Diagnostic and Statistical Manual of Mental Disorders (DSM) or the World Health Organization's International Classification of Diseases (ICD). Individuals with CFS, originally called the "yuppie flu," were initially characterized as primarily white women, of middle to upper class socioeconomic conditions, with psychogenic explanations of their fatigue. Many of these myths have been subsequently discredited, as CFS is less likely to occur among those of higher socioeconomic status, and this illness is more prevalent among minorities (Jason et al., 1999). According to Richman and Jason (2001), following the failure of scientists to establish a viral etiology of CFS, some researchers adopted a psychiatric explanation for the maintenance of this illness. A number of CFS research studies have focused on the role of illness attributions in the maintenance of somatic symptoms and impairment (Cope, David & Mann, 1994; Deale, Chalder & Wessely, 1998), and adoption of a "sick role" (Abbey, 1993; Ferrari & Kwan, 2001).

Because the exact pathogenesis of CFS remains unclear and it is likely a heterogeneous disorder, a clinical diagnosis is based on self-report of symptoms with no true consensus in the scientific community concerning expected laboratory findings or abnormalities (Afari & Buchwald, 2003). As is mentioned above, due to the symptoms-based nature of this illness, several researchers have posited that CFS is primarily a psychogenic illness

(Abbey & Garfinkel, 1991; Harvey et al., 2007; Surawy et al., 1995). Patients with CFS have uniformly rejected such psychiatric explanations for their illness, and often relationships between patients and health care providers have been strained. Deale and Wessely (2001) found that two-thirds of patients with CFS in a hospital fatigue clinic were dissatisfied with their interactions with physicians, and those who were dissatisfied were more likely to have received a psychiatric diagnosis than those who were satisfied. Likewise, Ax, Gregg and Jones (1997) found that when patients with CFS and physicians did not share the same illness attributions (in all cases the physician considered it psychogenic), the quality of the patient–doctor relationship was quite poor.

Psychiatric comorbidity is found to occur in approximately 45 percent to 50 percent of patients with CFS (Johnson, DeLuca & Natelson, 1996a; Wessely et al., 1996), and up to 82 percent of patients have been diagnosed with a psychiatric disorder during their lifetime (Buchwald et al., 1997), with mood and anxiety disorders occurring most often. However, there is little support for a single directional relationship between CFS and psychiatric disorders. CFS can occur in the absence of psychiatric comorbidity (van der Linden et al., 1999) and some patients have psychiatric diagnoses prior to the onset of CFS, while others develop psychiatric disorders after illness onset (Brown et al., 2010; Tiersky et al., 2003). Thus, it is unlikely that CFS can be considered the result of a psychiatric condition.

DIFFERENTIATING CFS FROM PSYCHIATRIC CONDITIONS

One of the difficulties in differentiating CFS from psychiatric disorders is the overlapping symptomatology found among these conditions. Some of the classic symptoms of CFS used for diagnosis include fatigue, problems with memory or concentration, muscle or joint pain, and unrefreshing sleep (Fukuda et al., 1994). These symptoms are also diagnostic criteria for MDD or GAD. Consequently, determining the source of overlapping symptoms can be challenging, and the same difficulty has been observed in other illnesses (Siegert & Abernethy, 2005; Visser & Smets, 1998). The accurate differentiation of symptoms of CFS from symptoms of psychiatric conditions can have significant implications for perceptions of the cause and maintenance of CFS.

Even though many symptoms of CFS overlap with symptoms of other psychiatric disorders, CFS can be differentiated from other psychiatric illnesses. For example, CFS and GAD share symptoms such as fatigue, difficulty concentrating, sleep disturbance, irritability, restlessness, and rapid heartbeat. But these two conditions can be differentiated by the most prominent symptom, which for GAD is excessive, persistent worry whereas for CFS, it is severe, debilitating fatigue. Another psychiatric illness, somatization disorder, is defined by having four pain symptoms, two gastrointestinal symptoms, one sexual symptom, and one pseudoneurological symptom, and as is evident, there is considerable overlap with CFS symptoms. However, in CFS, fatigue is the primary feature of the illness, but this is not so for somatization disorder. In addition, CFS has a sudden onset symptom complex often occurring in a patient's thirties or forties, whereas symptoms of somatization escalate over several years to full blown disorder by age 30.

One of the more difficult diagnostic challenges involves differentiating CFS from depression. It is important to differentiate those with a principal diagnosis of MDD from those with CFS only because it is possible that some patients with MDD also have chronic fatigue and four minor symptoms that can occur with depression (e.g., unrefreshing sleep, joint pain, muscle pain, impairment in concentration). Fatigue and these four minor symptoms are also defining criteria for CFS.

While fatigue is the principal feature of CFS, fatigue does not assume equal prominence in depression (Friedberg & Jason, 1998). Yet, many of the symptoms experienced by patients with CFS, such as prolonged fatigue after physical exertion, night sweats, sore throats, and swollen lymph nodes, are not commonly found among individuals suffering from MDD. Moreover, illness onset with CFS is often sudden, occurring over a few hours or days, whereas primary depression generally shows a more gradual onset. Individuals with CFS also can be differentiated from those with depression by recordings of skin temperature levels and electrodermal activity (Pazderka-Robinson, Morrison & Flor-Henry, 2004). Patients with CFS show more alpha EEG activity during NREM sleep, but this is not seen in dysthymic disorder or MDD (Whelton, Salit & Moldofsky, 1992).

In addition, Johnson, DeLuca, and Natelson (1995) found that individuals with CFS are unlikely to report having self-reproach, which is a key feature of depression. Hawk, Jason, and Torres-Harding (2006) found that self-reproach was a good discriminator between CFS and MDD, but they also found that five additional predictors (percentage of time fatigue is experienced, post-exertional malaise, unrefreshing sleep, confusion–disorientation, and shortness of breath) were more highly endorsed by individuals with CFS. In fact, when using the above six predictors, it was possible to differentiate those with MDD from those with CFS 100 percent of the time (Hawk et al., 2006). Overall, these findings suggest that it is possible to differentiate between somatic symptoms due to MDD and those due to CFS.

Some individuals with CFS might have had psychiatric problems before and/or after CFS onset, yet other individuals may only have primary psychiatric disorders with prominent somatic features. Including the latter type of patients in the current CFS case definition could confound the interpretation of epidemiologic studies, biological markers for the illness, and treatment studies.

ASSESSING PSYCHIATRIC COMORBIDITY

Unfortunately, physicians are not reliable sources of identifying psychiatric conditions among patients with severe fatigue. For example, Torres-Harding et al., (2002) found under-recognition of psychiatric illness by primary care physicians and physicians had particular difficulty assessing psychiatric disorder in patients with chronic fatigue fully explained by a psychiatric disorder. When a psychiatric diagnosis was present, physicians were able to correctly identify the psychiatric condition in only 40 percent of the psychiatrically explained group, 50 percent in the control group, and 64.3 percent in the CFS group. It is of interest that participants were more accurate than physicians in reporting the presence of a psychiatric disorder and in accurately reporting the presence of a mood or anxiety disorder. This investigation highlights the importance of a thorough psychiatric evaluation, including assessment of the natural history, clinical course, and treatment response of the psychiatric problem as part of the standard evaluation for patients with severe fatigue.

It is important to select appropriate psychodiagnostic instruments for different fatiguing illnesses (Taylor & Jason, 1998). For example, many investigators have used the Diagnostic Interview Schedule (DIS) with patients with CFS (Kruesi, Dale & Strauss,

1989; Katon et al., 1991; Katon & Russo, 1992; Lane, Manu & Matthews, 1991; Demitrack et al., 1991; Johnson, DeLuca & Natelson 1996a). The DIS was not designed to be used for those with medical illnesses, and coding rules can mistakenly have physical symptoms counted as psychiatric symptoms. When Johnson, DeLuca and Natelson (1996b) recoded symptoms reported as present following the onset of CFS as physical, the percentage of patients receiving a diagnosis of somatization disorder dropped to 0 percent. Furthermore, when CFS symptoms were not interpreted as symptoms of lifetime major depression, the percentage of those diagnosed with lifetime depression dropped from 40 percent to 20 percent (Demitrack et al., 1991).

In contrast to the DIS, the Structured Clinical Interview for the DSM-III-R (SCID) (Spitzer et al., 1990) has a trained master's level interviewer, and the ability to differentiate symptoms of CFS from psychiatric illnesses (Johnson, DeLuca & Natelson, 1996a). Taylor and Jason (1998) directly compared the DIS and the SCID by using the scoring methods that have been reported in the CFS literature. Participants with CFS received a significantly greater number of current and total lifetime psychiatric diagnoses resulting from the DIS interview, as opposed to the SCID interview. Fifty percent of participants received at least one current psychiatric diagnosis on the DIS, whereas only 22 percent received a current psychiatric diagnosis on the SCID. Johnson, DeLuca and Natelson (1996b) also found rates for psychiatric comorbidity with CFS to be substantially lower when the SCID is used compared to the DIS. These findings suggest that psychiatric comorbidity rates in individuals with CFS are influenced by the type of psychiatric instrument used. These results help explain the large discrepancies in findings for psychiatric illness in individuals with CFS across investigative studies.

CONCLUSION

Fatigue is a prevalent symptom in the general population, and prolonged and chronic fatigue are common complaints reported in primary care settings. In addition, fatigue is a symptom commonly experienced in psychiatric conditions such as depression and anxiety and is a hallmark symptom of CFS. Research has revealed several characteristics that are different among patients with CFS compared to those with psychiatric disorders. In this chapter, we highlighted differences among patients with CFS compared to individuals with psychiatric conditions.

As the introduction explains, chronic fatigue occurs in about 4 to 5 percent of the population, and about half of this is due to clear medical or psychiatric reasons. The critical question is how many of the remaining 2.5 percent have CFS. The Reeves et al. (2005) empiric CFS case definition estimates that 2.54 percent do have this illness, suggesting that almost all of the remaining 2.5 percent would fall within the CFS category. However, within this 2.54 percent are mood disorders, which are one of the most prevalent psychiatric disorders (one month prevalence rate of major depressive episode is 2.2 percent; Regier, Boyd & Burke, 1988). As is mentioned in this chapter, fatigue and these four minor symptoms are also defining criteria for CFS, so it is possible that some patients with a primary affective disorder could be misdiagnosed as having CFS. Including the latter type of patients in the current CFS case definition could confound the interpretation of epidemiologic and treatment studies, and further complicate efforts to identify biological markers for this illness.

It is evident that fatigue is commonly experienced among individuals with psychiatric conditions. An understanding of each patient's unique symptom experience might best be gained through the use of a thorough psychiatric evaluation that encompasses each patient's natural history, clinical course, and response to treatment (Katon, 1982). In order to differentiate the unique symptom experience among individuals with CFS from those with primary psychiatric disorders, it is important to conduct thorough examinations that utilize appropriate psychodiagnostic instruments. Although there may be a certain degree of psychiatric comorbidity in CFS populations, the over-diagnosis of psychiatric conditions among patients with this illness and the inclusion of those with a pure affective diagnosis within CFS case definitions would have significant negative consequences for understanding and differentiating these conditions.

REFERENCES

Abbey, S.E. (1993). Somatization, illness attribution and the sociocultural psychiatry of chronic fatigue syndrome. *Ciba Foundation Symposium*, 173, 238–52.

Abbey, S.E. & Garfinkel, P.E. (1991). Neurasthenia and chronic fatigue syndrome: The role of culture in the making of a diagnosis. *American Journal of Psychiatry*, 148, 1638–46.

Addington, A.M., Gallo, J.J., Ford, D.E. & Eaton, W.W. (2001). Epidemiology of unexplained fatigue and major depression in the community: The Baltimore ECA follow-up, 1981–1994. *Psychological Medicine*, 31, 1037–44.

Afari, N. & Buchwald, D. (2003). Chronic fatigue syndrome: A review. *American Journal of Psychiatry*, 160, 221–36.

American Psychiatric Association (2000). Diagnostic and Statistical Manual of Mental Disorders. Fourth edition, text revision). Washington, DC: APA.

Ax, S., Gregg, V.H. & Jones, D. (1997). Coping and illness cognitions: Chronic fatigue syndrome. *Clinical Psychology Review*, 21, 161–82.

Bélanger, L., Ladouceur, R. & Morin, C.M. (2005). Generalized anxiety disorder and health care use. *Canadian Family Physician*, 51, 1362–63.

Brown, L.F. & Kroenke, K. (2009). Cancer-related fatigue and its associations with depression and anxiety: A systematic review. *Psychosomatics*, 50, 440–47.

Brown, M.M., Kaplan, C., Jason, L.A. & Keys, C.B. (2010). Subgroups of chronic fatigue syndrome based on psychiatric disorder onset and current psychiatric status. *Health*, 2, 90–96.

Buchwald, D., Pearlman, T., Kith, P., Katon, W. & Schmaling, K. (1997). Screening for psychiatric disorders in chronic fatigue and chronic fatigue syndrome. *Journal of Psychosomatic Research*, 42, 87–94.

Buist-Bowman, M.A., Ormel, J., de Graaf, R. & Vollebergh, W.A.M. (2004). Functioning after a major depressive episode: Complete or incomplete recovery? *Journal of Affective Disorders*, 82, 363–71.

Cope, H., David, A. & Mann, A. (1994). "Maybe it's a virus?": Beliefs about viruses, symptom attributional style and psychological health. *Journal of Psychosomatic Research*, 38, 89–98.

Deale, A., Chalder, T. & Wessely, S. (1998). Illness beliefs and the treatment outcome in chronic fatigue syndrome. *Journal of Psychosomatic Research*, 45, 77–83.

Deale, A. & Wessely, S. (2001). Patients' perceptions of medical care in chronic fatigue syndrome. *Social Science & Medicine*, 52, 1859–64.

DeBattista, C., Doghramji, K., Menza, M.A., Rosenthal, M.H. & Fieve, R.R. (2003). Adjunct modafinil for the short-term treatment of fatigue and sleepiness in patients with major depressive disorder: A preliminary double-blind, placebo-controlled study. *Journal of Clinical Psychiatry*, 64, 1057–64.

Demitrack, M.A., Dale, J.K., Strauss, S.E., Laue, L., Listwak, S.J., Kruesi, M.J.P. & Gold, P.W. (1991). Evidence for the impaired activation of the hypothalamic–pituitary–adrenal axis in patients with chronic fatigue syndrome. *Journal of Clinical Endocrinology and Metabolism*, 73, 1224–34.

Dryman, A. & Eaton, W. (1991). Affective symptoms associated with the onset of major depression in the community. Findings from the US National Institute of Mental Health Epidemiologic Catchment Area Program. *Acta Psychiatric Scandinavia*, 84, 1–5.

Ferrari, R. & Kwan, O. (2001). The no-fault flavor of disability syndromes. *Medical Hypotheses*, 56, 77–84.

Friedberg, F. & Jason, L.A. (1998). *Understanding Chronic Fatigue Syndrome: An Empirical Guide to Assessment and Treatment*. Washington, DC: American Psychological Association.

Friedberg, F. & Krupp, L.B. (1994). A comparison of cognitive behavioral treatment for chronic fatigue syndrome and primary depression. *Clinical Infectious Diseases*, 18, S105–S110.

Fukuda, K., Straus, S.E., Hickie, I., Sharpe, M.C., Dobbins, J.G & Komaroff, A. (1994). The chronic fatigue syndrome: A comprehensive approach to its definition and study. *Annals of Internal Medicine*, 121, 953–59.

Furberg, H., Olarte, M., Afari, N., Goldberg, J., Buchwald, D. & Sullivan, P.F. (2005). The prevalence of self-reported chronic fatigue in a US twin registry. *Journal of Psychosomatic Research*, 59, 283–90.

Greco, T., Eckert, G. & Kroenke, K. (2004). The outcome of physical symptoms with treatment of depression. *Journal of General Internal Medicine*, 19, 813–18.

Harvey, S.B., Wadsworth, M., Wessely, S. & Hotopf, M. (2007). The relationship between prior psychiatric disorder and chronic fatigue: Evidence from a national birth cohort study. *Psychological Medicine*, 38, 933–40.

Hawk, C., Jason, L.A. & Torres-Harding, S. (2006). Differential diagnosis of chronic fatigue syndrome and major depressive disorder. *International Journal of Behavioral Medicine*, 13, 244–51.

Hickie, I.B., Hooker, A.W., Hadzi-Pavlovic, D., Bennett, B.K., Wilson, A.J. & Lloyd, A.R. (1996). Fatigue in selected primary care settings: Sociodemographic and psychiatric correlates. *Medical Journal of Australia*, 20, 585–88.

Jason, L.A. & Choi, M. (2008). Dimensions and assessment of fatigue. In Y. Yatanabe, B. Evengard, B.H. Natelson, L.A. Jason & H. Kuratsune, *Fatigue Science for Human Health*. Tokyo: Springer, 1–16.

Jason, L.A., Jordan, K.M., Richman, J.A., Rademaker, A.W., Huang, C., McCready, W. & Plioplys, S. (1999). A community-based study of prolonged fatigue and chronic fatigue. *Journal of Health Psychology*, 4, 9–26.

Jason, L.A., Najar, N., Porter, N. & Reh, C. (2009). Evaluating the Centers for Disease Control's empirical chronic fatigue syndrome case definition. *Journal of Disability Policy Studies*, 20, 93–100.

Johnson, S.K., DeLuca, J. & Natelson, B. (1995). Depression in fatiguing illness: Comparing patients with chronic fatigue syndrome, multiple sclerosis and depression. *Journal of Affective Disorders*, 39, 21–30.

Johnson, S.K., DeLuca, J. & Natelson, B. (1996a). Personality dimensions in the chronic fatigue syndrome: A comparison with multiple sclerosis and depression. *Journal of Psychosomatic Research*, 30, 9–20.

Johnson, S.K., DeLuca, J. & Natelson, B.H. (1996b). Assessing somatization disorder in the chronic fatigue syndrome. *Psychosomatic Medicine*, 58, 50–57.

Katon, W. (1982). Depression: Somatic symptoms and medical disorders in primary care. *Comprehensive Psychiatry*, 23, 274–87.

Katon, W. & Russo, J. (1992). Chronic fatigue syndrome criteria: A critique of the requirement for multiple physical complaints. *Archives of Internal Medicine*, 152, 1604–9.

Katon, W.J., Buchwald, D.S., Simon, G.E., Russo, J.E. & Mease, P.J. (1991). Psychiatric illness in patients with chronic fatigue and those with rheumatoid arthritis. *Journal of General Internal Medicine*, 6, 277–85.

Kroenke, K., Spitzer, R.L., Williams, J.B.W., Linzer, M., Hahn, S.R., deGruy III, F.V. & Brody, D. (1994). Physical symptoms in primary care: Predictors of psychiatric disorders and functional impairment. *Archives of Family Medicine*, 3, 774–9.

Kruesi, M.J., Dale, J. & Strauss, S.E. (1989). Psychiatric diagnoses in patients who have chronic fatigue syndrome. *Journal of Clinical Psychiatry*, 50, 53–6.

Lane, A.M. & Lovejoy, D.J. (2001). The effects of exercise on mood changes: The moderating effect of depressed mood. *Journal of Sports Medicine and Physical Fitness*, 41, 539–45.

Lane, T.J., Manu, P. & Matthews, D.A. (1991). Depression and somatization in the chronic fatigue syndrome. *American Journal of Medicine*, 91, 335–44.

Lavidor, M., Weller, A. & Babkoff, H. (2002). Multidimensional fatigue, somatic symptoms and depression. *British Journal of Health Psychology*, 7, 67–75.

Lewis, G. & Wessely, S. (1992). The epidemiology of fatigue: More questions than answers. *Journal of Epidemiology and Community Health*, 46, 92–7.

Martin, A., Chalder, T., Rief, W. & Braehler, E. (2007). The relationship between chronic fatigue and somatization syndrome: A general population survey. *Journal of Psychosomatic Research*, 63, 147–56.

Pawlikowska, T., Chalder, T., Hirsch, S.R., Wallace, P., Wright, D.J.M. & Wessely, S.C. (1994). Population based study of fatigue and psychological distress. *British Medical Journal*, 308, 763–6.

Pazderka-Robinson, H., Morrison, J.W. & Flor-Henry, P. (2004). Electrodermal dissociation of chronic fatigue and depression: Evidence for distinct physiological mechanisms. *International Journal of Psychophysiology*, 53, 171–82.

Regier, D.A., Boyd, J.H. &. Burke Jr, J.D. (1988). One month prevalence of mental disorders in the United States: Based on five Epidemiological Catchment area sites. *Archives of General Psychiatry*, 45, 977–86.

Reeves, W.C., Jones, J.J. Maloney, E. Heim, C. Hoaglin, D.C., Boneva, R. & Devlin, R. (2007). New study on the prevalence of CFS in metro, urban and rural Georgia populations. *Population Health Metrics*, 5, 5.

Reeves, W.C., Wagner, D., Nisenbaum, R., Jones, J.F., Gurbaxani, B., Solomon, L. & Heim, C. (2005). Chronic fatigue syndrome: A clinically empirical approach to its definition and study. *BMC Medicine*, 3, 19.

Richman, J.A. & Jason, L.A. (2001). Gender biases underlying the social construction of illness states: The case of chronic fatigue syndrome. *Current Sociology*, 49, 15–29.

Siegert, R.J. & Abernethy, D.A. (2005). Depression in multiple sclerosis: A review. *Journal of Neurology, Neurosurgery and Psychiatry*, 76, 469–75.

Simon, G., Gater, R., Kisely, S. & Piccinelli, M. (1996). Somatic symptoms of distress: An international primary care study. *Psychosomatic Medicine*, 58, 481–8.

Spitzer, R.L., Williams, J.B.W., Gibbon, M. & First, M.B. (1990). Structured Clinical Interview for DSM-III-R – Non-Patient Edition (SCID-NP, Version 1.0). Washington, DC: American Psychiatric Press.

Steele, L., Dobbins, J.G., Fukuda, K., Reyes, M., Randall, B., Koppelman, M. & Reeves, W.C. (1998). The epidemiology of chronic fatigue in San Francisco. *American Journal of Medicine*, 105, 83–90.

Surawy, C., Hackmann, A., Hawton, K. & Sharpe, M. (1995). Chronic fatigue syndrome: A cognitive approach. *Behaviour Research and Therapy*, 33, 535–44.

Taylor, R.R. & Jason, L.A. (1998). Comparing the DIS with the SCID: Chronic fatigue syndrome and psychiatric comorbidity. *Psychology and Health*, 13, 1087–104.

Tchekmedyian, N.S., Kallich, J., McDermott, A., Fayers, P. & Erder, M.H. (2003). The relationship between psychologic distress and cancer-related fatigue. *Cancer*, 98, 198–203.

Tiersky, L.A., Matheis, R.J., DeLuca, J., Lange, G. & Natelson, B.H. (2003). Functional status, neuropsychological functioning, and mood in chronic fatigue syndrome (CFS). *Journal of Nervous and Mental Disease*, 191, 324–31.

Torres-Harding, S.R., Jason, L.A., Cane, V., Carrico, A. & Taylor, R.R. (2002). Physicians' diagnoses of psychiatric disorders for persons with chronic fatigue syndrome. *The International Journal of Psychiatry in Medicine*, 32, 109–24.

van der Linden, G., Chalder, T., Hickie, I., Koschera, A., Sham, P. & Wessely, S. (1999). Fatigue and psychiatric disorder: Different or the same? *Psychological Medicine*, 29, 863–8.

Visser, M.R.M. & Smets, E.M.A. (1998). Fatigue, depression and quality of life in cancer patients: How are they related? *Supportive Care in Cancer*, 6, 101–8.

Wessely, S., Chalder, T., Hirsch, S., Wallace, P. & Wright, D. (1996). Psychological symptoms, somatic symptoms, and psychiatric disorder in chronic fatigue and chronic fatigue syndrome: A prospective study in the primary care setting. *American Journal of Psychiatry*, 153, 1050–59.

Whelton, C.L., Salit, I. & Moldofsky, H. (1992). Sleep, Epstein-Barr virus infection, musculoskeletal pain, and depressive symptoms in chronic fatigue syndrome. *The Journal of Rheumatology*, 19, 939–43.

19
Chronic Fatigue Syndrome

Diane L. Cox

INTRODUCTION

What is Chronic Fatigue Syndrome? Chronic fatigue syndrome (CFS) has been described as a relatively common, complex, clinically defined condition characterized by persistent or relapsing debilitating fatigue which affects both physical and mental functioning (Fukuda et al., 1994). The disabling fatigue, the principal symptom, arises in combination with other accompanying symptoms which may include difficulty with memory and concentration, muscle and/or joint pain, unrefreshing sleep, new headaches, tender lymph nodes, sore throats and post-exertion malaise. The condition usually presents in a person who has no previous history of similar symptoms. The term "myalgic encephalomyelitis" (ME) is often used by patients and patient organizations, particularly in the UK, so the acronym "CFS/ME" is used here. Post-infectious fatigue syndrome (PIFS) has been suggested as a subtype of CFS/ME where the illness is associated with infection corroborated by laboratory evidence (Sharpe et al., 1991).

Chronic fatigue syndrome has been shown to share a common set of symptom domains, which can be readily identified in the community and at all levels of health care across countries and cultures (Hickie et al., 2009). The international case definition (Fukuda et al., 1994) is still the most widely accepted and requires a substantial reduction in previous levels of occupational, educational, social and personal activities from new fatigue in excess of six months. CFS/ME is currently recognized on clinical grounds alone, therefore diagnosis is made by clinical evaluation to identify and exclude all other recognized causes of fatigue and any pre-existing underlying or contributing physical or mental health conditions that may require treatment (Fukuda et al., 1994, NICE, 2007).

The clinical evaluation by a suitably qualified healthcare professional, physician or general practitioner (GP) should include;

1. A full history that covers medical and psychosocial circumstances.
2. An assessment of psychological well-being.
3. A thorough physical examination.
4. A minimum battery of laboratory screening tests.

The identifiable symptom domains of illness that are common to CFS/ME across settings are; prolonged fatigue, musculoskeletal pain, neurocognitive difficulties, sleep disturbance and mood disturbance (Hickie et al., 2009).

Although it is generally accepted that a diagnosis of CFS/ME cannot be made until symptoms have persisted for six months, it has been suggested that the diagnosis in adults in most cases should be considered when symptoms have lasted for four months, to ensure the most proactive approach to maintain fitness for work and education and minimize the impact on daily life and activities (NICE, 2007). NICE (2007) also suggests that advice on symptom management should not be delayed until a diagnosis is established and that this advice should be tailored to the person's specific symptoms.

The pattern and intensity of the symptoms can vary between people and during the course of each person's illness (Cox, 2000, NICE, 2007). The categories of mild, moderate, severe and very severe have been previously proposed to indicate the spectrum of severity and functional impact of CFS/ME on daily life (Cox, 1998, Cox & Findley, 1998). People with CFS/ME:

- in the *mild* category will be mobile and self-caring and able to manage light domestic and work tasks with difficulty;
- in the *moderate* category will have reduced mobility and be restricted in all activities of daily living; will usually have stopped work and will require many rest periods;
- in the *severe* category will be able to carry out minimal daily tasks, such as face-washing and cleaning teeth; will have severe cognitive difficulties such as retention of information, short-term memory difficulties and word-finding difficulties, and depend on a wheelchair for mobility;
- in the *very severe* category will be unable to mobilize or carry out any daily task for themselves and will be in bed the majority of the time.

Most people with CFS/ME appear to be within the mild to moderate categories (NICE, 2007). To date, the majority of the research on treatment and management has been carried out with outpatients in the mild to moderate range (Larun et al., 2009, Price et al., 2008, Whiting et al., 2001). However, the illness can have a significant impact on work regardless of the level of ability, as people with CFS/ME may have difficulty undertaking their usual working hours and performing to their usual standard. They may also have recurrent or prolonged sickness absence. In addition, uncertainty about the diagnosis may lead to friction with employers or colleagues (NHS Plus, 2006). It has been suggested that CFS/ME is associated with considerable work-related disability, with a proportion of people losing their job through their ill-health (Taylor & Kielhofner, 2005). The work-related limitations include difficulties with getting to work, learning and remembering new material, communicating, making more work errors, and difficulty in harnessing the energy to sustain task-related effort and concentration. The economic impact of CFS/ME through direct and indirect costs has been suggested as high for people with CFS/ME, their families and society as a whole (Jason et al., 2008).

The NHS Plus review (2006) found that concurrent depression in individuals with CFS is associated with poorer work outcomes. Therefore, comorbid conditions such as depression should be treated concurrently. Other factors found by NHS Plus which predicted poor work outcomes included a greater number of physical symptoms and signs, and older age at presentation.

EVIDENCE-BASED MANAGEMENT OF CFS/ME

The current best evidence suggests that cognitive behaviour therapy (Price et al., 2008), graded exercise therapy (Larun et al., 2009) and activity management (NICE, 2007) are the most appropriate and effective approaches in the management of CFS/ME. Cognitive behaviour therapy and graded exercise therapy have been shown to be effective in restoring the ability to work in those who are currently absent from work (NHS Plus, 2006).

Cognitive Behaviour Therapy

(CBT) is a psychological talking therapy that facilitates the identification of unhelpful, anxiety-provoking thoughts. Once identified, these thoughts and assumptions are then challenged through collaborative discussion, diary-keeping, and testing of beliefs between sessions, with skills training in sessions. CBT therefore involves the achievement of mutually agreed clear goals, the use of homework tasks and a consideration between the links of cognitive, behaviour, social, environmental and physiological effects (Price et al., 2008, NICE, 2007). CBT is usually delivered by cognitive behaviour therapists who may have a psychology or nursing professional background.

CBT for CFS/ME involves a more rehabilitative approach where balanced levels of activity and rest are established, unhelpful thoughts and beliefs about CFS/ME that may impair recovery are addressed, and specific lifestyle changes are encouraged (Scheeres et al., 2008, Godfrey et al., 2007). The essential elements are the assessment of illness beliefs and coping strategies, structuring of daily rest, sleep and activity, to establish a stable baseline of general activities, with a graduated return to normal activity, and collaborative challenging of unhelpful attributions and beliefs about symptoms and activity. The cognitive behavioural model is therefore based on the understanding that thoughts, feelings and actions interlink with each other; what we do influences thoughts and feelings and equally, the way we think affects actions and feelings. A course of CBT is usually 12–16 sessions. CBT is effective in adults with CFS/ME and has been shown to reduce symptoms and improve function and quality of life (NICE, 2007).

Graded Exercise Therapy

(GET) is a structured, mutually developed and monitored programme that plans gradual increments of exercise or physical activity. GET is different to a general exercise programme that may involve simply "going to the gym" or "just walking more" as it is structured and delivered by suitably trained therapists, usually physiotherapists or exercise physiologists (NICE, 2007).

GET for CFS/ME is based on the illness model of deconditioning. The approach involves a physical assessment of the person with CFS/ME, mutually negotiated goal-setting, and education. The person with CFS/ME is encouraged to gradually increase physical exercise initially in duration and then intensity to improve physical functioning, muscular strength and endurance (Clark & White, 2005, Larun et al., 2009). Following the physical assessment, sessions usually involve establishing a stable baseline level of physical activity and negotiation of an individually designed home exercise programme with set target heart rates and times. A course of GET is usually delivered over 12 weeks and most studies have suggested exercise is carried out at home 3–5 times a week. GET for adults with CFS/ME is effective in improving fatigue, physical function, global improvement, sleep, mood and cognition (Larun et al., 2009, NICE, 2007).

Activity Management

Activity management is a person-centered approach to managing a person's symptoms by using activity with the aim of increasing function by introducing a balance of activity, rest and sleep throughout the day and week. Activity management is usually delivered by occupational therapists and or physiotherapists within rehabilitation teams (NICE, 2007, Cox, 2000). It is goal-directed and uses activity analysis and graded activity to enable people to improve, evaluate, restore and/or maintain their function and well-being in self-care, work and leisure. The goals are mutually agreed and individual. Activity [Lifestyle] management programmes need to be adapted to consider the spectrum of ability from mild to very severe seen in CFS/ME.

MANAGEMENT OF CFS/ME IN PRACTICE

All aspects of treatment need to be based on improving the people's ability to manage their daily life and increase their overall functioning. Pacing and activity management are seen as an integral component in a combined Cognitive Behaviour Therapy (CBT) and/or Graded Exercise Therapy (GET) approach often termed "lifestyle management" (Cox, 2000). This combined approach to the management of CFS/ME has been used in community (Taylor, 2004), outpatient (McDermott et al., 2004) and inpatient (Cox, 2002) settings. Lifestyle management is frequently delivered by CFS/ME services as part of a rehabilitation programme and aims at preventing the "boom and bust" cycle of overactivity and relapse (discussed in more detail later). A daily routine can be established with the help of daily diaries and the use of realistic negotiated goals. The daily routine should include a balance of mental and physical activity with regular rest periods, ideally using daily relaxation techniques. The intention is to establish a "baseline of activity" – a level of activity which can be repeated consistently without worsening symptoms. Once this is obtained, it is possible to gradually increase activity in a structured way with regular review of activity and goals. Many people find that once they are able to manage their activity in this way, they are more able to manage their illness, stabilize symptoms and experience improvements (Cox, 2000, 2002).

The key elements in management of CFS/ME regardless of philosophical approach include:

- Making an accurate early diagnosis, which considers comorbidities and alternative diagnoses;
- Identifying and managing symptoms, which includes listening to warning signs related to patterns of activity;

- Establishing a collaborative working partnership with the person with CFS/ME, with mutual agreement of all aspects of treatment and changes

Assisting a person with CFS/ME to understand their condition and the impact it can have on daily life is essential in the therapeutic process. The use of techniques such as diaries and explanation on avoiding "boom and bust" in activity are important as part of this process and can often be a useful starting point in engaging a person in therapy. All treatment and care should take into account people's individual needs and preferences.

Assessment Measures

A number of standardized assessment measures are commonly used in practice and research to assist in evaluating fatigue and other symptoms in CFS/ME (Cox, 2000, Hjollund, Anderson & Bech, 2007, White et al., 2007). These are;

- The 11-item Chalder Fatigue Scale (Chalder et al., 1993), designed specifically for chronic fatigue populations, to measure severity of fatigue, detection of fatigue, and as a valid estimator of change in hospital and community settings;
- The Medical Short Form-36 (SF-36) (Ware & Sherborne, 1992), designed to survey differences in physical and mental health status and the effect of treatment on general health status. It has eight subscales: physical functioning; social functioning; role functioning: physical; body pain; mental health; role functioning: emotional; vitality; general health. The physical functioning subscale of the SF-36 is most commonly used with CFS/ME populations to measure physical function in daily life;
- The Work and Social Adjustment Scale (Mundt et al., 2002), designed to assess disability and functional impairment. It has five items which measure participation in occupational and domestic activities;
- The Hospital Anxiety and Depression Scale (HADS) (Zigmond & Snaith, 1983), designed to assess anxiety and depression.

In addition, a Multidimensional Fatigue Inventory (MFI-26) has recently been validated with a US adult population (Lin et al., 2009). The MFI-26 is designed to measure fatigue severity in fatiguing illnesses such as CFS/ME. It comprises five subscales; general fatigue; physical fatigue; mental fatigue; reduced activity; and reduced motivation (Smets, Garssen & Bonke, 1995). Each subscale includes four items with five point Likert scales.

Lifestyle Activity Management

A lifestyle activity management programme needs to include a number of elements, as outlined in Table 19.1. In addition to these elements treatment is usually divided into 3 phases:

- Assessment, engagement and education;
- Practising self-management and control of symptoms (active treatment);
- Ending treatment and planning for future self-management.

Table 19.1   Elements of a lifestyle (activity) management programme for CFS/ME

- History taking,
  - allowing the person with CFS/ME to tell their story (Gray & Fossey 2003, Hughes 2002)
- Exploration of Daily Activity Levels
  - impact on daily life/ occupational performance, disruption and adaptation
- Understanding of Rest and Relaxation
  - what it is/ what it isn't
- Developing knowledge and understanding of Graded Activity/ Exercise
  - activity analysis & modification
- Energy conservation strategies
  - task analysis and energy expenditure
  - examining diurnal patterns (Taylor & Kielhofner 2003)
- Re-introduction of activity over time.
  - this includes prioritising and appropriate goal setting
- Establishing a daily balance and structure
  - rest/ activity, wants/ needs, mental/ physical tasks, work/ leisure
- Impact and re-adjustment of sleep patterns
- Impact of feelings and thoughts on activity performance & balance
- Strategies for coping with others & change
- Planning ahead & ending active treatment
- Involvement of Family & Friends

ASSESSMENT, ENGAGEMENT AND EDUCATION

During this assessment phase a detailed picture of the person with CFS/ME identifying the pattern of symptoms, activity and rest is established. The assessment, engagement and education phase includes taking a full history that allows the person with CFS/ME to tell their story (Gray & Fossey, 2003; Hughes, 2002). Giving time to hear a person's story and the impact of the condition on their daily life is crucial in this process.

Daily Activity Levels

During and before this narrative discussion it is useful to explore a person's current daily activity levels to consider the impact on daily life, on their domestic and occupational performance, and any disruption and adaptation they have made. The person's understanding of rest and relaxation is also explored, as rest can mean different things to different people; how a person relaxed or rested prior to being ill may not be how they need to rest in order to gain a balance in their daily routines (Cox, 1999). Table 19.2 shows a typical daily routine for a moderate presentation of CFS/ME prior to intervention.

A person's routine and pattern of activity can usually be seen if that person is asked to complete a daily diary. The most common pattern seen in the diaries of people with CFS/ME is a "boom and bust" pattern – that is, peaks and troughs of activity during the day and the week (Cox, 2000). Activity logs have also been used to record a person's daily

Table 19.2   CFS/ME daily diary (example)

Time	Activity	Fatigue level 0 – none/10 – maximum)
10am	Got up, washed, dressed, fed dog, prepared lunches, did some light chores	6
12pm	Had lunch	7
	Watched TV, read, listened to radio	7
2pm	Rest, often sleep	4
4pm	Did some housework, prepared dinner	5
6pm	Feed children, had dinner, talked	7
7pm	Watched TV	6
9pm	Went to bed (got to sleep about 11pm)	8

Source: Adapted from Cox, D.L. (2000).

physical activity and explore how each activity is associated with fatigue or pain (Jason et al., 2009).

The boom and bust cycle, also known as "peak and trough," "overactivity/ underactivity," or the "see-saw cycle," is a natural way of managing limited energy and coping with the demands of life when a person has CFS/ME. When a person feels able, they do more in an attempt to catch up ("boom," a good day); this usually pushes them over their available energy level, which is then followed by a trough ("bust," a bad day). The diagram below (Figure 19.1) attempts to show this. Once this concept is described to people they can often see in their diaries where the "boom and bust" patterns are occurring. Most people at the moderate CFS/ME level of ability, prior to intervention, will show this boom and bust pattern in their daily routine by doing most activity in the morning (after waking), resting/sleeping in the afternoon and doing a lower level of activity in the evening before going to bed. Their day is usually substantially shorter than when they were working.

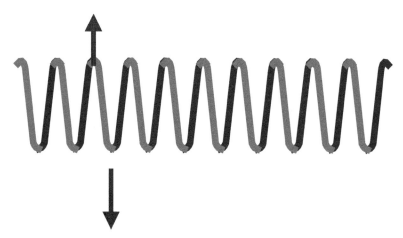

Figure 19.1   Boom and bust cycle

Use and Understanding of Energy

All activities use energy. Human energy is understood to have cognitive, physical and emotional components. Metaphors such as "topping up," or "emptying bank accounts," "batteries," "envelopes," and "jugs" are used to raise awareness of how activity uses energy. These metaphors all introduce energy as a dynamic, which the person can control through recharging, making deposits, budgeting, turning off lights, etc. People with CFS/ME learn to work with this dynamic by learning to analyze what is taking and what is giving energy and to what extent (Cox, 2000; Taylor, 2004; Jason, 2008). It is, however, sometimes hard for people to remember or identify what it is they can actually do to rest and relax. It can therefore be useful for them to record separately what they do to rest/relax over the course of a week. The activities that helped the person to relax before they had CFS/ME might now not be something that "tops up their battery" or their energy level. It is therefore important that they think about what they are doing that might be an aid to expanding their energy. Everyone does different things to rest or relax. For some people this is sitting listening to relaxing music, having a bath, or watching TV. Other people might find reading or meditating relaxing. Some people with CFS/ME will describe sleep as rest.

Understanding and Using Baselines

A baseline is the amount and type of activity an individual can do on a regular basis, this assists the person in knowing what they can manage and when to prioritize activities (Cox, 2000). It can also be used to stabilize and stop the boom and bust cycle. Baselines are what a person can do, despite having a number of symptoms; the focus is on activity levels, not symptom levels. The initial baseline of activity should be seen as a starting point, as the person stabilizes, they can increase and adapt this baseline to include further activities. Appropriate aims and priorities need to be mutually negotiated and agreed and then activity can be built up as the person feels able to do so.

PRACTICING SELF-MANAGEMENT AND CONTROL OF SYMPTOMS (ACTIVE TREATMENT)

The aim of this phase is to implement strategies, to monitor the effect on symptoms, encouraging the person with CFS/ME to take control of symptoms and proactively manage their daily activity levels. Energy conservation strategies considering task/activity analysis and energy expenditure, and examining diurnal patterns (Taylor & Kielhofner, 2003) are an important part of enabling self-management and control of symptoms.

During the process of therapy, each person with CFS/ME needs to develop their knowledge and understanding of graded activity and exercise. This includes practising activity analysis and modification of activities to enable participation in all aspects of daily life (Cox, 1999; 2000). The person with CFS/ME is encouraged to reintroduce activity over time, which involves prioritizing activities and appropriate goal-setting. A daily and weekly balance and structure need to be established. There are a number of elements in daily life that need to be balanced: rest/activity, wants/needs, mental/physical tasks, and work/leisure. In addition, the impact and readjustment of sleep patterns, and feelings and thoughts on occupational performance and daily activity balance, need to be considered.

Structure and Balancing of Daily and Weekly Life

For some people it is helpful to initially structure each day to establish a routine to reduce the boom and bust pattern of activity and balance rest and activity. The new structured day needs to be based on the person's current daily diaries, only changing their routines slowly over time. It is useful to set a regular get up time, a go to bed time, which reinforces healthy sleep habits at night, and set times for rest/relaxation at least three times a day, spread throughout the day (Cox, 1999, 2000). In practice, reducing a long rest period in the afternoon or an afternoon sleep may take time. However, once a person starts to introduce other rest periods into their day, they usually find their need for a sleep in the afternoon is less, and eventually is not required.

The introduction of the importance of balance in daily and weekly plans is crucial if the person is to achieve success in managing their available energy. It is therefore important to address the issue of balance not just in terms of establishing a balance between activity and rest but also between activities they feel they "must do," the self-care and domestic activities, and those that they "would like to do," the enjoyable activities. Due to the nature of CFS/ME and people experiencing reduced energy levels it is often the enjoyable activities, the "would like to do" activities, which get missed out. This can be due to people feeling they have to focus their available energy on the "must do" activities to get through the day. This technique is helpful in getting each person to focus on the activities they enjoy or would like to do in their life, which can then be considered when setting goals.

Energy Expenditure

Table 19.3 gives an example of how activities can be divided into three energy levels; high, medium and low. These categories are often colour-coded; high energy: red, medium energy: green, low energy: blue (rest periods as yellow), or in colours the person chooses

Table 19.3   Energy requirements (example)

How much does the ACTIVITY cost you in terms of energy needed?	
Low	Eating, Dressing, Cleaning teeth, TV
Medium	Showering, Watching a film, Preparing food, Making phone calls
High	Driving, Working, Going out, Seeing friends

Table 19.4   Chronic fatigue syndrome: Weekly plan (example)

	8am				9:30	10:30	11:00	12:30	2:00	2:30	4:00		6:30	7:00	8:00	9:00-9:30
Mon	H	M	M	L	Rest	M	H	M	Rest	Low	M	L	Rest	M	Low	Wind down Bed
Tues	H	M	L	M	Rest	M	H	M	Rest	Low	L	M	Rest	M	Low	Wind down Bed
Wed	H	M	M	L	Rest	M	H	M	Rest	Low	L	H	Rest	M	Low	Wind down Bed
Thurs	H	M	L	M	Rest	M	H	M	Rest	Low	M	L	Rest	M	Low	Wind down Bed
Fri	H	M	M	L	Rest	M	H	M	Rest	Low	M	H	Rest	M	Low	Wind down Bed
Sat	H	M	L	M	Rest	M	H	M	Rest	Low	L	M	Rest	M	Low	Wind down Bed
Sun	H	M	M	L	Rest	M	H	M	Rest	Low	M	L	Rest	M	Low	Wind down Bed

Source: Adapted from Cox, D.L. (2000).

(Cox, 2000). It should be noted that the demands an activity takes in terms of energy is individual, i.e. that driving may be a high-energy activity for one person and a medium- or even low-energy activity for another.

The purpose of introducing this technique is for people to see whether their activities are balanced in terms of energy demands and rest in their day and week. Table 19.4 is an example of a balanced, colour- coded weekly plan.

The colour coding also enables a visual representation of where a person's high-energy activities are and to see if they could be moved to another part of the day, or broken down into components to make the activity less energy-consuming. The daily or weekly plan is then redesigned, spacing out blocks of high-energy activity where possible.

Activity Analysis & Modification

Activity analysis enables people to think about how to break an activity down into its component parts and how to manage the associated energy demands. Activity modification follows on from activity analysis when a person considers the changes they could make to an activity to save energy. This concept of analyzing activities to break the activity down into achievable components can be the key to the success of an activity management programme.

Implementing Activity Management Strategies in Daily Life

For activity management to be incorporated into the person's daily life they need to use their knowledge about themselves and their energy to manage their daily routine within their baseline and available energy. The focus is on balance and there is a hierarchy, starting with sleep management, setting up a daily routine that encourages sleep at night, prioritizing one or two high-level activities in a day, and appropriate rest points. The planning can become more refined and complex as the process develops, e.g. improving the balance of high-, medium- and low-energy activities, mental and physical, social, work tasks etc. A diary sheet, daily and weekly plans, and priority exercises are often used.

ENDING TREATMENT AND PLANNING FOR FUTURE SELF-MANAGEMENT

Ending treatment and planning for future self-management starts from the initial session. However, there are a number of considerations that are needed as sessions are coming to the end of a planned treatment phase, usually 8–14 sessions. The person will need strategies for coping with others and change and there is a need to educate and inform family and friends of the activity management techniques.

Considering Obstacles to Action

Priorities and standards, conflicting demands, pressures to deviate, and dealing with exacerbations of CFS/ME symptoms can be obstacles to managing daily routines. This part of the process includes raising awareness of and exploring the pressures that maintain the boom and bust pattern of symptoms and activity levels and enhancing each person's understanding of the patterns of their illness. Obstacles and pressures are further examples of mental/emotional energy use and a number of strategies and skills can be used that include role play, problem-solving and the application of relaxation techniques.

Setbacks and Relapses

People with CFS/ME need to be advised that setbacks and relapses in activity and symptoms do occur. They can be triggered by an unexpected activity/event, infection, poor sleep or stress. A plan that considers possible setbacks should be discussed. If a setback occurs it needs to be discussed, considering the possible cause, the nature, severity and duration and a review of the current management plan should be included. Strategies that can help in a setback are maintaining activity levels by alternating activities with more rest, using relaxation techniques, and reducing high-energy activities to stabilize symptoms and re-establish a baseline of activity (Cox, 1999, 2000; NICE, 2007).

Monitoring Progress and Moving On

The person with CFS/ME can monitor their progress and move their activity level on by reviewing their baseline regularly, knowing when and how to increase their activities, using problem-solving skills and considering priorities for now and priorities for the future. At this point in the therapy, people are encouraged to develop their own language, model and descriptions of what will enable them to stay in control and to manage their setbacks and progressions on their own.

Return to Work

In order to help people with CFS/ME to return to work when they are ready and fit enough, all healthcare professionals should advise about fitness for work and recommend flexible adjustments or adaptations to work through mutual agreement and collaboration (NICE, 2007). There should be liaison between the employee, employer, healthcare professional, occupational health and human resources, and a suitable individual return to work plan should be developed (Rimes & Chalder, 2005). Work adjustments may include working from home, reducing or limiting hours of work, reducing workload or type of work, limiting or reducing mental or physical tasks, and changing location of work (NHS Plus, 2006).

CONCLUSIONS

Chronic fatigue syndrome is an identifiable, clinically defined condition for which there are evidence- based treatments that can aid recovery. The current best evidence suggests that cognitive behaviour therapy (Price et al., 2008), graded exercise therapy (Larun et al., 2009) and activity management (NICE, 2007) are the most appropriate and effective approaches in the management of CFS/ME. People with CFS/ME can often remain in work or return to work over time with an appropriately time-managed individual programme.

REFERENCES

Chalder, T., Berelowitz, G., Hirsch, S., Pawlikowska, T., Wallace, P., Wessely, S. & Wright, D. (1993). Development of a fatigue scale. *Journal of Psychosomatic Research*, 37(2), 147–53.

Clark, L.V. & White, P.D. (2005). The role of deconditioning and therapeutic exercise in chronic fatigue syndrome (CFS). *Journal of Mental Health*, 14(3), 237–52.

Cox, D.L. (1998). Management of CFS: Development and evaluation of a service. *British Journal of Therapy and Rehabilitation*, 5(4), 205–9.

Cox, D.L. (1999). Chronic fatigue syndrome – an occupational therapy programme. *Occupational Therapy International*, 6, 52–64.

Cox, D.L. (2000). *Occupational Therapy and Chronic Fatigue Syndrome*. London: Whurr (Wiley).

Cox, D.L. (2002). Chronic fatigue syndrome: An evaluation of an occupational therapy inpatient intervention. *British Journal of Occupational Therapy*, 65(10), 461–8.

Cox, D.L. & Findley, L.J. (1998). The management of chronic fatigue syndrome in an inpatient setting: Presentation of an approach and perceived outcome. *British Journal of Occupational Therapy*, 61(9), 405–9.

Fukuda, K., Straus, S.E., Hickie, I., Sharpe, M., Dobbins, J.G. & Komaroff, A. & the International Chronic Fatigue Syndrome Study Group (1994). The chronic fatigue syndrome: a comprehensive approach to its definition and study. *Annals of Internal Medicine*, 121, 953–9.

Godfrey, E., Chalder, T., Ridsdale, L., Seed, P. & Ogden, J. (2007). Investigating the "active ingredients" of cognitive behaviour therapy and counselling for patients with chronic fatigue in primary care: Developing a new process measure to assess treatment fidelity and predict outcomes. *British Journal of Clinical Psychology*, 46, 253–72.

Gray, M.L. & Fossey, E.M. (2003). Illness experience and occupations of people with Chronic Fatigue Syndrome. *Australian Occupational Therapy Journal*, 50(3), 127–36.

Hickie, I., Davenport, T., Vernon, S.D., Nisenbaum, R., Reeves, W.C., Hadzi-Pavlovic, D., Lloyd, A. & the International Chronic Fatigue Syndrome Study Group (2009). Are chronic fatigue and chronic fatigue syndrome valid clinical entities across countries and health-care settings? *Australian and New Zealand Journal of Psychiatry*, 43, 25–35.

Hjollund, N.H., Anderson, J.H. & Bech, P. (2007). Assessment of fatigue in chronic disease: A bibliographic study of fatigue measurement scales. *Health and Quality of Life Outcomes*, 5(12).

Hughes, J.L. (2002). Illness narrative and chronic fatigue syndrome/myalgic encephalomyelitis: A review. *British Journal of Occupational Therapy*, 65(1), 9–14.

Jason, L. (2008). The energy envelope theory and myalgic encephalomyelitis/chronic fatigue syndrome. *American Association of Occupational Health Nursing*, 56(5), 189–95.

Jason, L.A., Benton, M.C., Valentine, L., Johnson, A., Torres-Harding, S. (2008). The economic impact of ME/CFS: Individual and societal costs. *Dynamic Medicine*, 7(6).

Jason, L.A., Timpo, P., Porter, N., Herrington, N., Brown, M., Torres-Harding, S. & Friedberg F. (2009). Activity logs as a measure of daily activity among patients with chronic fatigue syndrome. *Journal of Mental Health*, 18(6), 549–56.

Larun, L., McGuire, H., Edmonds, M., Odgaard-Jensen, J. & Price, J.R. (2009). Exercise therapy for chronic fatigue syndrome (Review). *The Cochrane Library*, Issue 1. Chichester: Wiley.

Lin, J.M.S., Brimmer, D.J., Maloney, E.M., Nyarko, E., Belue, R. & Reeves, W.C. (2009). Further validation of the Multidimensional Fatigue Inventory in a US adult population sample. *Population Health Metrics*, 7(18), doi:10.1186/1478-7954-7-18.

McDermott, C., Richards, S.C.M., Ankers, S., Selby, M., Harmer, J., & Moran, C.J. (2004). An evaluation of a chronic fatigue lifestyle management programme focusing on the outcome of return to work or training. *British Journal of Occupational Therapy*, 67(6), 269–73.

Mundt, J.C., Marks, I.M., Shear, K. & Griest, J.H. (2002). The work and social adjustment scale: A simple measure of impairment in functioning. *British Journal of Psychiatry*, 180, 461–4.

NHS Plus (2006). *Occupational Aspects of the Management of Chronic Fatigue Syndrome: A National Guideline*. London: NHS Plus.

NICE (National Institute for Health and Clinical Excellence) (2007). Chronic fatigue syndrome/ myalgic encephalomyelitis (or encephalopathy): Diagnosis and management of chronic fatigue syndrome/myalgic encephalomyelitis (or encephalopathy) in adults and children. London: NICE. Available at: http://www.nice.org.uk/page.aspx?o=111636.

Price, J.R., Mitchell, E., Tidy, E. & Hunot, V. (2008). Cognitive behaviour therapy for chronic fatigue syndrome in adults (Review). *The Cochrane Library*. Issue 3. Chichester: Wiley.

Rimes, K.A. & Chalder, T. (2005). Treatments for chronic fatigue syndrome. *Occupational Medicine*, 55(1), 32–9.

Scheeres, K., Wensing, M., Knoop, H. & Bleijenberg, G. (2008). Implementing cognitive behavioral therapy for chronic fatigue syndrome in a mental health center: A benchmarking evaluation. *Journal of Consulting & Clinical Psychology*, 76(1), 163–71.

Sharpe, M.C., Archard, L.C., Banatvala, J.E., Borysiewicz, J.E., Clare, A.W., David, A., Edwards, R.H.T., Hawton, K.E.H., Lambert, H.P., Lane, R.J.M., McDonald, E.M., Mowbray, J.F., Pearson, D.J., Peto, T.E.A., Preddy, V.R., Smith, A.P., Smith, D.G., Taylor, D.J., Tyrell, D.A.J., Wessely, S. & White, P.D. (1991). A report – chronic fatigue syndrome: guidelines for research. *Journal of the Royal Society of Medicine*, 84, 118–21.

Smets, E.M., Garssen, B. & Bonke, B. (1995). *Manual: Multidimensional Fatigue Inventory*. Amsterdam: Medical psychology, Academic Medical Centre.

Taylor, R.R. (2004). Quality of life and symptom severity for individuals with chronic fatigue syndrome: Findings from a randomised controlled trial. *American Journal of Occupational Therapy*, 58(1), 35–43.

Taylor, R.R. & Kielhofner, G.W. (2003). An approach to persons with chronic fatigue syndrome based on a model of human occupation. Part Two: Assessment and intervention *Occupational Therapy in Health Care*, 17(2), 63–88.

Taylor, R.R. & Kielhofner, G.W. (2005). Work-related impairment and employment-focused rehabilitation options for individuals with chronic fatigue syndrome: A review. *Journal of Mental Health*, 14(3), 253–67.

Ware, J.E. & Sherborne, C.D. (1992). The MOS 36-item short form health survey (SF-36): Conceptual framework and item selection. *Medical Care*, 30(6), 473–83.

White, P., Sharpe, M.C., Chalder, T., DeCesare, J., Walwyn, R. & The PACE Trial Group (2007). Protocol for the PACE trial: A randomised controlled trial of adaptive pacing, cognitive behaviour therapy, and graded exercise as supplements to standardised specialist medical care versus standardised specialist medical care alone for patients with the chronic fatigue syndrome/myalgic encephalomyelitis or encephalopathy. *BMC Neurology*, 7(6), doi:10.1186/1471-2377-7-6.

Whiting, P., Bagnall, A.M., Sowden, A.J., Cornell, J.E., Mulrow, C.D. & Ramirez, G. (2001). Interventions for the treatment and management of Chronic Fatigue Syndrome. *Journal of the American Medical Association*, 286(11), 1360–68.

Zigmond, A.S. & Snaith, R.P. (1983). The hospital anxiety and depression scale. *Acta Psychiatrica Scandinavica*, (67), 361–70.

20
Upper Respiratory Tract Illnesses and Fatigue

Andrew P. Smith

INTRODUCTION

Upper respiratory tract illnesses (URTIs), such as the common cold and influenza, are frequent and widespread. They have a large impact on healthcare costs and are a major cause of absenteeism from work and education (see Smith, 1992a). In addition, quality of life is reduced by such illnesses and the malaise associated with them has been the subject of recent research. As well as the specific local (e.g. increased nasal secretion; nasal stuffiness) and systemic symptoms (e.g. fever; sore throat) these illnesses are associated with increased fatigue. The research described in this chapter aimed to describe the behavioural changes associated with URTIs, elucidate underlying mechanisms, and consider the practical implications of such effects. The next section of the chapter considers some early anecdotal observations that preceded the more formal research.

ANECDOTAL EVIDENCE OF EFFECTS OF URTIS ON HUMAN ERROR

Tye (1960) summarized a number of case histories that suggested a link between influenza and accidents (both at work and outside the workplace). These case histories were supported by road accident statistics from the 1950s which showed an increase in accidents in years when there was an increased prevalence of influenza (see Smith, 1992a). Tye's report also cites cases where performance was impaired before the development of the symptoms and after they had gone. This is an important observation, in that a person with severe symptoms may go to bed and refrain from safety-critical activities, but this is unlikely to be the case prior to or after the illness. Grant (1972) also reports case histories of post-influenzal effects on the decision-making of highly skilled technicians. The primary features of these errors were that they were made by individuals who had been ill with influenza but no longer had the primary symptoms; the errors went unnoticed and the person rejected advisory comments from colleagues; and the errors could not be attributed to poor motivation or general lack of ability. Overall, these results suggested that further research was required on the effects of such illnesses on performance efficiency.

EXPERIMENTAL STUDIES OF INFECTION

Warm & Alluisi (1967) concluded that "data concerning the effects of infection on human performance are essentially non-existent." They were referring to infections that do not cause structural damage to the brain (e.g. HIV infection) and they went on to examine the impact of severe infections on the performance of individuals and groups (Alluisi, Thurmond & Coates, 1971; Alluisi et al., 1973; Thurmond et al., 1971). In one study (Alluisi, Thurmond & Coates, 1971) those who became ill showed an average drop in performance of about 25 percent and after recovery they were still 15 per cent below the level of the control group. There was some evidence that working memory tasks showed a greater decrement that sustained attention tasks.

The illnesses studied in the above experiments were very severe and in most occupations or in education they would lead to absenteeism. In the case of more minor illnesses, such as the common cold, it is likely that individuals may continue to work or study. Heazlett & Whaley (1976) examined the effects of having a cold on the performance of 13- year-old pupils. The results showed that having a cold had no effect on reading comprehension but did reduce performance on tasks involving auditory or visual perception. These selective effects of minor illnesses clearly required replication and extension. This was done in a series of studies carried out at the MRC Common Cold Unit, Salisbury, in the late 1980s. The main findings from these studies are described in the next section.

EFFECTS OF EXPERIMENTALLY INDUCED COLDS AND INFLUENZA

The routine of the MRC Common Cold Unit is described in detail elsewhere (e.g. Smith, 1992a). The main features can be briefly summarized as follows. Volunteers aged 18–50 years came to the unit for a 10-day stay, during which they agreed to receive an infecting virus inoculation (all procedures of the Common Cold Unit were approved by the Harrow District Ethical Committee and carried out with the informed consent of the volunteers). On the first day of the trial, volunteers were given a medical examination and then put into quarantine (in apartments of 1–3 people) and were isolated from outside contact (apart from the staff of the unit) for the rest of the trial. Baseline measures of mood and performance were collected during this period. Volunteers were given a virus challenge (or saline) on day 4 and about one-third of the volunteers developed symptoms between 24 and 96 hours later (depending on the virus). Another third developed subclinical infections (as indicated by virological assays) and the remainder could not be shown to have the virus present (the "uninfected" group). The testing was repeated at a time when those with illnesses were symptomatic. The symptoms were rated by the unit's clinician and objective measures of illness recorded (e.g. nasal secretion weights; sublingual temperature).

EFFECTS OF EXPERIMENTALLY INDUCED INFLUENZA ILLNESSES

The initial studies of the behavioural effects of experimentally induced URTIs have been described in a number of publications and are summarized in Smith (1992a). If one considers influenza first, the initial study showed that those with an influenza B illness had slower reactions when they did not know exactly when or where the stimulus was going to occur (Smith, Tyrrell, Coyle & Willman, 1987). Other tasks showed no significant effect of influenza although this may have reflected the small sample size. Smith, Tyrrell, Al-Nakib, Coyle et al. (1988) found that those with influenza B illnesses and subclinical infections were impaired on a visual search task with a high memory load. This impairment was present in the incubation period prior to the development of symptoms. Again, no impairment was seen in tasks involving other functions (e.g. a pegboard task, a logical reasoning task and a semantic processing task). Two further studies examined the effects of influenza A illnesses. In the first study,[1] influenza led to impaired performance when the person did not know which of two possible locations a target was going to be presented in, but had no effect when the location was known. The final study (Smith, 1992b examined the effects of influenza on resistance to distraction (as measured by the Stroop Colour–Word task) and showed that those with influenza were more easily distracted by irrelevant stimuli. In summary, these studies of experimentally induced influenza demonstrated selective effects of the illnesses, with tasks involving unknown target locations, distraction or variable timing of the stimuli showing greatest impairment. Smith and colleagues (1992)[2] found those with influenza illnesses reported a general increase in negative affect, not just an increase in fatigue.

EFFECTS OF EXPERIMENTALLY INDUCED COLDS

Early studies of experimentally induced colds also showed selective impairments associated with the illness. However, the tasks which were impaired differed from those which were sensitive to the effects of influenza. For example, Smith, Tyrrell, Coyle & Willman (1987) showed that the performance of a tracking task was worse in the colds group, whereas detection tasks were not impaired by the cold. Similarly, Smith, Tyrrell, Al-Nakib et al. (1988) found that the colds group had slower performance on a pegboard task (involving transferring pegs from a full solitaire set to an empty one as quickly as possible), but not on a search task, a logical reasoning task or a semantic processing task. Smith, Tyrrell, Al-Nakib, Coyle, Donovan, Higgins & Willman (1987) confirmed that tasks involving hands–eye co-ordination and psychomotor speed (e.g. the five-choice serial response task) were impaired in both those with colds and those with subclinical infections. Smith and his colleagues (1989)[3] were able to investigate after effects of having a cold (volunteers were tested after the cold symptoms had gone). The results confirmed the effects of Alluisi, Thurmond & Coates (1971) in that the impairments continued into convalescence. Mood

---

1   See Smith, A.P., Tyrrell, D.A.J., Al-Nakib, W., Barrow, G.I., Higgins, P.G., Leekam, S. & Trickett, S. (1989).
2   Smith, A.P., Tyrrell, D.A.J., Barrow, G.I., Higgins, P.G., Willman, J.S., Bull, S., Coyle, K.B. & Trickett, S. (1992).
3   Smith, A.P., Tyrrell, D.A.J., Al-Nakib, W., Barrow, G.I., Higgins, P.G., Leekam, S. & Trickett, S. (1989).

changes were examined in these studies,[4] and they showed that fatigue increased with more severe illnesses and that this persisted after the primary symptoms of the illness had gone. In summary, these initial studies suggested that people with colds report greater fatigue and show psychomotor slowing. These effects may be present even when the person is asymptomatic. The performance impairments are selective and little evidence was found (Smith, Tyrrell, Coyle et al., 1990) for effects of having a cold on aspects of memory (episodic memory; working memory; and semantic memory). The psychomotor slowing, reduced alertness and insensitivity of other tasks appeared to replicate when different cold producing viruses were used. In contrast, effects on visual perception (contrast sensitivity and visual discomfort) appeared to be virus specific.[5] Overall, both the effects of influenza and the common cold appeared to be reliable but further research was required to determine whether effects could be observed with naturally occurring illnesses. These studies are described in the next section.

EFFECTS OF NATURALLY OCCURRING INFLUENZA ILLNESSES

The illnesses induced at the MRC Common Cold Unit were very mild and one must ask whether different effects might occur with more severe episodes. It was also the case that sample sizes were often small and it is possible that more extensive impairments may become apparent in studies with more power. Studying naturally occurring URTIs can be difficult because of the uncertainty of the infecting agent. In order to address this issue studies were conducted in the 1990s to examine the behavioural effects of naturally occurring colds and influenza. In some of the studies the nature of the infecting agent was identified using virological assays.

EFFECTS OF NATURALLY OCCURRING INFLUENZA B ILLNESSES

Smith, Thomas, Brockman et al. (1993) examined the effects of influenza B illnesses on mood and performance. The performance battery included both tests that had been shown to be sensitive to experimentally induced URTIs and those which had not. In the first study 92 volunteers were initially tested when healthy (the baseline measurement). Those who developed an illness were retested when ill and again a month later. Nasal swabs and blood samples were taken to identify infecting agents. Those who remained healthy were retested two and three months after recruitment. The results showed that volunteers with influenza B infections had a 38 percent increase in their reaction times to targets occurring at uncertain times. Similarly, their speed of response in a cognitive vigilance task and accuracy of performing a categoric search task were impaired. Those with the influenza illnesses had no impairment on memory tasks or reaction times in focused attention tasks. These findings confirm those observed in the studies of

---

4   See Smith, A.P., Tyrrell, D.A.J., Barrow, G.I., Higgins, P.G., Willman, J.S., Bull, S., Coyle, K.B. & Trickett, S. (1992).
5   Smith, A.P., Tyrrell, D.A.J., Barrow, G.I., Higgins, P.G., Bull, S., Trickett, S. & Wilkins, A.J. (1992).

experimentally induced influenza. A second study examined the effects of influenza on a 10 minute version of the variable fore period simple reaction time task. Those with influenza B illnesses showed a 19 percent increase in their reaction times compared to the healthy controls.

EFFECTS OF NATURALLY OCCURRING COLDS

Our first study of naturally occurring colds examined the effects and after effects on mood and performance (Hall & Smith, 1996a). This study confirmed that having a cold was associated with a more negative mood and psychomotor impairment but had little effect on other cognitive functions. These results were confirmed by Smith, Rich et al. (1999) in a study which also demonstrated that electrophysiological measures (speed of eye movements) are sensitive to the fatigue associated with the common cold. Smith, Thomas, Kent et al. (1998) used virological assays to study the effects of the common cold on mood and performance. Results from two studies showed that having a cold was associated with reduced alertness and psychomotor slowing. This was true for both illnesses where the infecting agent was identified and clinical illnesses where no virus was detected.

Overall, these initial studies of naturally occurring illnesses confirmed findings from the research on experimentally induced URTIs. Studies carried out by other research groups have also supported these findings (Matthews et al., 2001; Bucks et al., 2008) and extended them by showing that having a cold reduces task engagement (Matthews et al., 2001). In addition, Bucks et al. (2008) found that older adults with a cold showed much greater effects on memory processing speed than younger adults. This finding fits with research showing that infection or inflammation may have an exaggerated impact on the aged or diseased brain (Perry, Cunningham & Holmes, 2007). More recent studies looking at other issues (e.g. combined effects of URTIs and other factors; effects of different medication on malaise) have also replicated these effects. Smith (2002) reports findings from an investigation involving nearly 200 participants who completed a wide range of tasks. Results from this study confirmed that those with URTIs reported more negative affect (reduced alertness and hedonic tone) and had slower reaction times on simple and choice reaction time tasks. Those who were ill took longer to encode new information and this may underlie some of the effects reported in the literature. Speed of verbal reasoning and semantic processing were also slower in the URTI group which supports the findings of Bucks et al. (2008). Symptom scores were correlated with the mood changes but not the performance changes. These results have been replicated by Smith (2002) in a large-scale analysis of nearly 500 volunteers (188 of whom developed URTIs). These results also showed that the behavioural effects of the URTIs could not be accounted for by the psychosocial characteristics of those who developed an illness. The results of both of this last two studies showed that the most significant effects of having an URTI were those identified in the earlier small-scale studies (i.e. reduced alertness and slower simple and choice reaction). Other smaller effects were also significant but many of these were no longer significant after controlling for multiple testing. Before considering these studies and those aimed at identifying underlying mechanisms it is important to consider effects which have not been replicated and how the topic has been extended.

AFTER EFFECTS OF NATURALLY OCCURRING COLDS

Hall & Smith (1996a) found some evidence of impairments after the symptoms of the cold had gone. They interpreted the earlier findings looking at the after effects of experimentally induced colds as being due to poorer learning in the symptomatic phase which then had an effect on performance at a subsequent occasion after the symptoms had gone. Smith, Thomas & Whitney (2000b) investigated performance after cold symptoms had gone (but with no testing when the person was symptomatic). They found little evidence of impairments when the person was symptom free. The one exception was the variable fore-period simple reaction time task where impairments were still apparent immediately after the symptoms had finished and a week later. Given the number of analyses conducted, this could have been a chance result. However, Hall & Smith (1996a) also found that this task was sensitive to the after effects of having had a cold. This short-lived post-viral fatigue should be distinguished from more severe conditions that are found after influenza-like illnesses (see below).

The next section briefly considers other URTIs and summarizes the evidence that they induce fatigue both during and after the symptomatic phase.

ALLERGIC RHINITIS AND ITS TREATMENT

Allergic rhinitis can produce sedation and impairment of cognitive functioning (Bender, 2005; Marshall & Colon, 1993; Marshall, O'Hara & Steinberg, 2000; Wilken, Berkovitz & Kane, 2002). These impairments in performance do not appear to be correlated with the symptom severity. It can also impair restorative sleep, resulting in increased fatigue and decreased daytime performance. This has implications for education and for productivity and safety at work. Treatment with second-generation (non-sedating) antihistamines appears to prevent or reverse such effects.

INFECTIOUS MONONUCLEOSIS

Many upper respiratory tract illnesses have a strong resemblance to influenza but can have longer-lasting effects. Infectious mononucleosis (glandular fever) can lead to post-viral fatigue which can persist for several months. Hall & Smith (1996b) investigated the effects of both acute and chronic infectious mononucleosis on cognitive function. The acute effects were identical to influenza, with performance of variable fore-period simple reaction time and cognitive vigilance tasks being impaired. Those participants who reported chronic problems following the illness showed a different profile of impairments. The psychomotor slowing and impaired sustained attention present during the acute illness were no longer observed. However, working memory (e.g. logical reasoning) and episodic memory tasks (e.g. free recall) were impaired in the chronic phase. These selective effects possibly reflect the different profile of immunological changes occurring over time.

The next section examines the extent to which having an illness such as the common cold not only has direct effects on performance but also makes the person more sensitive to the effects of other factors that increase stress and fatigue.

COMBINED EFFECTS OF URTIS AND NOISE, PROLONGED WORK AND ALCOHOL

It has frequently been demonstrated that fatigued individuals are more sensitive to other factors which induce fatigue or stress. Research has examined this by comparing the effects of noise, prolonged work and alcohol in those with colds and in healthy individuals. Smith, Thomas, Brockman et al. (1993) found that individuals with a cold were more sensitive to noise than healthy individuals. For example, noise led to a slight improvement in simple reaction time in the healthy group, whereas reaction times were 17 percent slower in the colds/noise group than the colds/quiet group. Smith, Thomas & Whitney (2000a) examined the effects of upper respiratory tract illness on mood and performance over the working day. The effects of having a cold (reduced alertness and slower simple reaction time) became greater over the course of the day.

Smith, Whitney et al. (1995) examined the effects of a low dose of alcohol (1.5 ml of vodka per kg body weight) on the mood and performance of healthy volunteers and those with URTIs. The alcohol had no significant effects on the healthy volunteers. In contrast, those with URTIs reported greater negative affect after alcohol than placebo. The combination of being ill and alcohol also led to a large increase (42 percent) in reaction time in a choice reaction time task where the location of the stimuli varied. Overall, the results of these studies show that URTIs not only have a direct effect on mood and performance but can also make the person more sensitive to the influence of other factors. This means that safe levels based on studies of healthy volunteers may not be appropriate for those with URTIs.

The next section considers findings from studies of simulations of real-life activities and surveys of human error.

EFFECTS OF THE COMMON COLD ON SIMULATED DRIVING

There have been a large number of studies on the effects of fatigue on skills needed for competent driving and on simulated driving and research has examined effects of URTIs in these paradigms. In the first of these studies Smith (2006) found that those with a cold hit the kerb more frequently and responded more slowly to unexpected events than healthy volunteers. Other research (Smith & Jamson, submitted) has investigated whether those with colds have a reduced ability to detect potential collisions. This study used the OMEDA task (Object Movement Estimation under Divided Attention, Read, Ward & Parkes, 2000) in which two targets move toward the centre point of the screen, emerging at different times and different speeds. The targets reach the centre at:

1. the same time;
2. almost the same time; or
3. noticeably different times.

Volunteers have to detect times when the two targets will collide at the centre. A secondary task involves presentation of five patterns, one in the centre and the others in the four corners of the screen. The volunteer has to press a button if the centre pattern matches any in the four corners. Those with a cold detected fewer collisions, made more false alarms and more secondary task errors. The volunteers in this study were young adults (mean age: 20 years) and those with a cold performed at the same level as healthy 65 year olds.

The above studies examined skills necessary for competent driving performance rather than using a sophisticated simulator. Smith & Jamson (submitted) investigated the effects of upper respiratory tract illness on performance in the University of Leeds Advanced Driving Simulator. There was no effect of being ill on speed, lateral control, gap acceptance or overtaking behaviour. However, those with a cold were more likely to spend a larger proportion of time at a shorter headway (especially in the safety critical area of less than two seconds). Those who were ill were also less able to respond quickly to unexpected events (e.g. they were more likely to collide with pedestrians and violated traffic stop signals twice as often as when they were healthy).

These results suggest that having an URTI may increase the risk of driving accidents. This provides support for Tye's (1960) conclusion based on case studies. Given that having a minor illness also makes one more susceptible to other risk factors known to impair driving ability (e.g. alcohol; fatigue due to prolonged work) one might expect even greater effects of minor illnesses on driving safety. Further research is required to address this last issue. The next section reviews evidence that having an upper respiratory tract illness may influence productivity and safety at work.

IMPACT OF URTIS ON PRODUCTIVITY AND SAFETY

Bramley, Lerner & Sarnes, (2002) conclude that in the USA the economic cost of lost productivity due to the common cold approaches $25 billion, of which $16.6 billion is attribute to productivity loss, $8 billion to absenteeism, and $230 million to caregiver absenteeism. Similar research has demonstrated a significant effect of allergic rhinitis on productivity in school and at work (Stull et al., 2007; Nathan, 2007; Stull et al., 2009). In contrast, there is little evidence about the effects of URTIs on accidents at work (see Smith, Harvey et al., 1994) and further research is needed on this topic.

The next section considers mechanisms that underlie the behavioral effects of URTIs.

UNDERLYING MECHANISMS

CNS Effects of Cytokines

Unlike many other viruses that affect the brain and behaviour upper respiratory tract viruses are not thought to enter the CNS. The behavioural effects do not reflect the symptoms of the illness and a more likely mechanism is that they are due to effects of immunological changes on the brain. Indeed, Smith, Tyrrell, Coyle & Higgins (1988, 1991) showed that an injection of alpha-interferon mimicked effects seen in influenza. Further research is now required to assess the relative contribution of cytokines and other inflammatory mediators on cognition and mood.

Changes in Afferent Stimulation

Smith and his colleagues[6] found that sucking zinc gluconate lozenges and using a nasal spray containing nedocromil sodium removed the cold-induced performance impairment. It is likely that both forms of medication have multiple modes of action. For example, nedocromil sodium suppresses the release of mediators such as histamine, leukotriene C_4 and prostaglandin D_2. One way in which the two compounds could produce similar effects is through changes in afferent stimulation. URTIs influence the trigeminal nerve and compounds which increase afferent stimulation may produce changes in the brain stem. Indeed, the presence of menthol and similar compounds in cold medication is because they produce symptomatic relief by stimulating the trigeminal nerve. Recent research has shown that sucking a peppermint (Smith, in preparation, a) or chewing menthol gum (Smith & Boden, in press) increases alertness in those with a cold but does not remove the performance decrements.

Sleep Disturbance

It has been suggested that the effects of URTIs on performance and mood may reflect an effect of the illness on sleep. Drake et al. (2000) examined the effects of experimentally induced rhinovirus infections on polysomnographically recorded sleep. They found that in symptomatic individuals total sleep time decreased by 23 minutes and sleep efficiency decreased by 5 percent. Psychomotor performance was also impaired in those with a cold although there was no increase in daytime sleepiness in that group. In addition, there were no significant correlations between changes in sleep parameters and measures of fatigue.

Neurotransmitters

While it has been difficult to identify the viral or immunological pathways that lead to malaise, it has been easier to show which CNS changes may be involved. The starting point for this research was a study which showed that caffeine removed many of the impairments induced by URTIs (Smith, Thomas, Perry & Whitney, 1997). At the same time studies were in progress to examine whether changes in central noradrenaline could explain lapses of attention (Smith & Nutt, 1996) and the effects of caffeine in fatigued individuals (Smith, Brice, Nash et al., 2003). This led to a study examining whether increases in the turnover of central noradrenaline (produced by the drug idazoxan) could remove the impairments seen in those with a cold (Smith, Sturgess et al., 1999). The results confirmed that changes in central noradrenaline occur when a person has an URTI and that compounds that increase noradrenaline may reverse such effects. It is, of course, quite likely that changes in other neurotransmitter systems are also involved in the malaise induced by URTIs.

---

6 Smith, A.P., Tyrrell, D.A.J., Al-Nakib, W., Barrow, G.I., Higgins, P.G. & Wenham, R. (1991).

CONCLUSIONS

This review has shown that URTIs are associated with a general malaise that includes an increase in fatigue. This is associated with changes in cognitive function that resemble those seen in other fatigue states. When a person has an URTI they are also more sensitive to the negative effects of other factors (e.g. prolonged work, alcohol). The mechanisms underlying such effects can be considered at many different levels and while our knowledge of these is incomplete, evidence for cytokine changes, reduction in the turnover of central noradrenaline, trigeminal effects and indirect effects of sleep disturbance have been demonstrated. The fatigue induced by these illnesses has many implications for real-life performance and safety. Indeed, simulated driving studies suggest that URTIs may lead to impaired driving. It is also desirable to develop medications that not only treat the local symptoms but remove the fatigue produced by the illness. The addition of caffeine to many over-the-counter products may be one method of doing this. Further research is required to increase our knowledge of the behavioural effects of URTIs so that we gain a better understanding of the underlying mechanisms and the implications for policy and practice.

REFERENCES

Alluisi, E.A., Beisel, W.R., Bartelloni, P.J. & Coates, G.D. (1973). Behavioral effects of tularensis and sandfly fever in man. *Journal of Infectious Disease*, 128, 710–17.

Alluisi, E.A., Thurmond, J.B. & Coates, G.D. (1971). Behavioral effects of infectious disease *Pasteurella tularensis* in man. *Perceptual and Motor Skills*, 32, 647–88.

Bender, B.G. (2005). Cognitive effects of allergic rhinitis and its treatment. *Immunol Allergy Clin N Am.* 25, 301–12.

Bramley, T.J., Lerner, D. & Sarnes, M. (2002). Productivity losses related to the common cold. *Journal of Occupational and Environmental Medicine*, 44, 822–9.

Bucks, R.S., Gidron, Y., Harris, P., Teeling, J., Wesnes, K.A., Perry, V.H. (2008). Selective effects of upper respiratory tract infection on cognition, mood and emotion processing: A prospective study. *Brain, Behavior and Immunity*, 22, 399–407.

Drake, C.L., Roehrs, T.A., Royer, H., Koshorek, G., Turner, R.B. & Roth, T. (2000). Effects of an experimentally-induced rhinovirus cold on sleep, performance and daytime alertness. *Physiology and Behavior*, 71, 75–81.

Grant, J. (1972). Post-influenzal judgement deflection among scientific personnel. *Asian Journal of Medicine*, 8, 535–9.

Hall, S.R. & Smith, A.P. (1996a). An investigation of the effects and after-effects of naturally-occurring upper respiratory tract illnesses on mood and performance. *Physiology and Behavior*, 59, 569–577.

Hall, S.R. & Smith, A.P. (1996b). Behavioural effects of infectious mononucleosis. *Neuropsychobiology*, 33, 202–9.

Heazlett, M. & Whaley, R.F. (1976). The common cold: Its effects on perceptual ability and reading comprehension among pupils of a seventh grade class. *Journal of School Health*, 46, 145–7.

Marshall, P.S. & Colon, E.A. (1993). Effects of allergy season on mood and cognitive function. *Ann Allergy*, 71, 251–8.

Marshall, P.S., O'Hara, C. & Steinberg, P. (2000). Effects of seasonal allergic rhinitis on selected cognitive abilities. *Ann Allergy Asthma Immunol*, 84, 403–10.

Matthews, G., Warm, J.S., Dember, W.N., Mizoguchi, H. & Smith, A.P. (2001). The common cold impairs visual attention, psychomotor performance and task engagement. Proceedings of the

Human Factors and Ergonomics Society 45th Annual Meeting. Santa Monica, CA: Human Factors and Ergonomics Society, 1377–81.

Nathan, R.A. (2007). The burden of allergic rhinitis. *Allergy Asthma Proc.* 28, 3–9.

Perry, V.H., Cunningham, C. & Holmes, C. (2007). Systemic infections and inflammation affect chronic neurodegeneration. *Nature Reviews Immunology*, 7, 161–7.

Read, N., Ward, N. & Parkes, A. (2000). The role of dynamic tests in assessing the fitness to drive of healthy and cognitively impaired elderly. *J. Traffic Med.* 28, 34S–35S.

Smith, A.P. (1992a). Colds, influenza and performance, in A.P. Smith & D.M. Jones (eds), *Handbook of Human Performance*, Vol. 2: *Health and Performance*. London: Academic Press, 197–218.

Smith, A.P. (1992b). Effects of influenza and the common cold on the Stroop color–word test. *Perceptual and Motor Skills*, 74, 668–70.

Smith, A.P. (2002). The psychology of the common cold: An integrated approach. Final report on ESRC project R02250188. Swindon, UK.

Smith, A.P. (2006). Effects of the common cold on simulated driving, in P.D. Bust (ed.), *Contemporary Ergonomics 2006*, 621–4.

Smith, A.P. (in preparation, a). Effects of peppermint on malaise produced by the common cold.

Smith, A.P. & Boden, C. (in press). Effects of chewing menthol gum on the alertness of healthy volunteers and those with an upper respiratory tract illness.

Smith, A.P. & Nutt, D.J. (1996). Noradrenaline and attention lapses. *Nature*, 380, 291.

Smith, A.P., Brice, C.F., Nash, J., Rich, N. & Nutt, D.J. 2003. Caffeine and central noradrenaline: Effects on mood and cognitive performance. *Journal of Psychopharmacology*, 17, 283–92.

Smith, A.P., Harvey, I., Richmond, P., Peters, T.J., Thomas, M. & Brockman, P. (1994). Upper respiratory tract illnesses and accidents. *Occupational Medicine*, 44, 141–4.

Smith, A.P., Rich, N., Sturgess, W., Brice, C., Collison, C., Bailey, J., Wilson, S. & Nutt, D. (1999). Effects of the common cold on subjective alertness, simple and choice reaction time and eye movements. *Journal of Psychophysiology*, 13, 145–51.

Smith, A.P., Sturgess, W., Rich, R., Brice, C., Collison, C., Bailey, J., Wilson, S. & Nutt, D.J. (1999). Effects of idazoxan on reaction times, eye movements and mood of healthy volunteers and subjects with upper respiratory tract illnesses. *Journal of Psychopharmacology*, 13, 148–51.

Smith, A.P., Thomas, M. & Brockman, P. 1993. Noise, respiratory virus infections and performance. *Proceedings of 6th International Congress on Noise as a Public Health Problem*, Actes Inrets, 34(2), 311–14.

Smith, A.P., Thomas, M. & Whitney, H. (2000a). Effects of upper respiratory tract illnesses on mood and performance over the working day. *Ergonomics*, 43, 752–63.

Smith, A.P., Thomas, M. & Whitney, H. (2000b). After-effects of the common cold on mood and performance. *Ergonomics*, 43, 1342–9.

Smith, A.P., Thomas, M., Brockman, P., Kent, J. & Nicholson, K.G. (1993). Effect of influenza B virus infection on human performance. *British Medical Journal*, 306, 760–61.

Smith, A.P., Thomas, M., Kent, J. & Nicholson, K. (1998). Effects of the common cold on mood and performance. *Psychoneuroendocrinology*, 23, 733–9.

Smith, A.P., Thomas, M., Perry, K. & Whitney, H. (1997). Caffeine and the common cold. *Journal of Psychopharmacology*, 11(4), 319–24.

Smith, A.P., Tyrrell, D.A.J., Al-Nakib, W., Barrow, G.I., Higgins, P.G. & Wenham, R. (1991). The effects of zinc gluconate and nedocromil sodium on performance deficits produced by the common cold. *Journal of Psychopharmacology*, 5, 251–4.

Smith, A.P., Tyrrell, D.A.J., Al-Nakib, W., Barrow, G.I., Higgins, P.G., Leekam, S. & Trickett, S. (1989). Effects and after-effects of the common cold and influenza on human performance. *Neuropsychobiology*, 21, 90–93.

Smith, A.P., Tyrrell, D.A.J., Al-Nakib, W., Coyle, K.B., Donovan, C.B., Higgins, P.G. & Willman, J.S. (1987). Effects of experimentally-induced virus infections and illnesses on psychomotor performance. *Neuropsychobiology*, 18, 144–8.

Smith, A.P., Tyrrell, D.A.J., Al-Nakib, W., Coyle, K.B., Donovan, C.B., Higgins, P.G. & Willman, J.S. (1988). The effects of experimentally-induced respiratory virus infections on performance. *Psychological Medicine*. 18, 65–71.

Smith, A.P., Tyrrell, D.A.J., Barrow, G.I., Coyle, K.B., Higgins, P.G., Trickett, S. & Willman, J.S. (1990). The effects of experimentally induced colds on aspects of memory. *Perceptual and Motor Skills*, 71, 1207–15.

Smith, A.P., Tyrrell, D.A.J., Barrow, G.I., Higgins, P.G., Bull, S., Trickett, S. & Wilkins, A.J. (1992a). The common cold, pattern sensitivity and contrast sensitivity. *Psychological Medicine*, 22, 487–94.

Smith, A.P., Tyrrell, D.A.J., Barrow, G.I., Higgins, P.G., Willman, J.S., Bull, S., Coyle, K.B. & Trickett, S. (1992). Mood and experimentally-induced respiratory virus infections and illnesses. *Psychology and Health*, 6, 205–12.

Smith, A.P., Tyrrell, D.A.J., Coyle, K.B. & Higgins, P.G. (1988). Effects of interferon alpha on performance in man: A preliminary report. *Psychopharmacology*, 96, 414–16.

Smith, A.P., Tyrrell, D.A.J., Coyle, K.B. & Higgins, P.G. (1991). Effects and after-effects of interferon alpha on human performance, mood and physiological function. *Journal of Psychopharmacology*, 5, 243–50.

Smith, A.P., Tyrrell, D.A.J., Coyle, K. & Willman, J.S. (1987). Selective effects of minor illnesses on human performance. *British Journal of Psychology*, 78, 183–8.

Smith, A.P., Tyrrell, D.A.J., Coyle, K., Higgins, P.G. & Willman, J.S. (1988). Diurnal variation in the symptoms of colds and influenza. *Chronobiology International*, 5, 411–16.

Smith, A.P., Tyrrell, D.A.J., Coyle, K.B., Higgins, P.G. & Willman, J.S. (1990). Individual differences in susceptibility to infection and illness following respiratory virus challenge. *Psychology and Health*. 4, 201–211.

Smith, A.P., Whitney, H., Thomas, M., Brockman, P. & Perry, K. (1995). A comparison of the acute effects of a low dose of alcohol on mood and performance of healthy volunteers and subjects with upper respiratory tract illnesses. *Journal of Psychopharmacology*. 9, 225–30.

Stull, D.E., Roberts, L., Frank, L. & Heithoff, K. (2007). Relationship of nasal congestion with sleep, mood and productivity. *Curr Med Res Opin*, 23, 811–19.

Stull, D.E., Schaefer, M., Crespi, S. & Sandor, D.W. (2009). Relative strength of relationships of nasal congestion and ocular symptoms with sleep, mood and productivity. *Curr Med Res Opin*, 25, 1785–92.

Thurmond, J.E., Alluisi, E.A. & Coates, G.D. (1971). An extended study of the behavioral effects of respiratory *Pasteurella tularensis* in man. *Perceptual and Motor Skills*, 33, 439–54.

Tye, J. (1960). The invisible factor: An inquiry into the relationship between influenza and accidents. London: British Safety Council.

Warm, J.S. & Alluisi, E.A. (1967). Behavioral reactions to infections: Review of the psychological literature. *Perceptual and Motor Skills*, 24, 755–83.

Wilken, J.A., Berkowitz, R. & Kane, R. (2002). Decrements in vigilance and cognitive functioning associated with ragweed-induced allergic rhinitis. *Annals of Allergy. Asthma and Immunology*, 89, 372–80.

PART VII

Applied Contexts for Operator Fatigue

21
Long Work Hours, Fatigue, Safety, and Health

Roger R. Rosa

The share of workers working long hours has increased over the last three decades (Rones, Iig & Gardner, 1997; Jacobs & Gerson, 2004; Kuhn & Lozano, 2008) leading to concerns about risk if working hours become excessive (Caruso et al., 2006). The U.S. Census Bureau estimated that at least 38.5 million people reported working more than the standard 40-hour week and about 24.9 million, or at 17.7 percent of full-time workers, regularly worked at least 49 hours per week in 2007 (see Table 21.1 adapted from the U.S. Census Bureau Statistical Abstract, 2009).

Extended work duration and overtime hours appear to be distributed across all industry sectors. From an industry survey, Dawson Heightmann & Kerin (2004) reported an average overtime frequency of 12.6 percent, with the highest frequencies in the utilities, processing, and service industries, and about 10 percent or more observed in manufacturing, healthcare, and transportation. In agriculture, 42 percent of those surveyed reported working at least 49 hours per week and 23 percent reported working at least 60 hours per week (U.S. Census Bureau, Statistical Abstract, 2009).

Possible risks cited in the literature on long work hours include job-related injuries, health declines and related increases in healthcare costs, lost productivity, and employee turnover (Caruso et al., 2006). Despite these risks, U.S. workers generally are satisfied with their weekly hours with respect to earned income. When asked by the U.S. Census Bureau in 2001 if they preferred more or less hours per week, 60–70 percent of workers

Table 21.1 Persons at work by hours worked, 2007

	Millions	Percentage
Total	140	(100)
35 hours and over	108	(76.9)
41 hours and over	38.5	(27.5)
41 to 48 hours	13.6	(9.7)
49 to 59 hours	14.5	(10.3)
60 hours and over	10.4	(7.4)

Source: U.S. Census Bureau, Statistical Abstract. Available online at: http://www.census.gov/prod/2008pubs/09statab/labor.pdf

preferred to keep their schedules, 20–25 percent preferred more hours, and fewer than 10 percent preferred fewer hours (Golden & Gebreselassie, 2007). These results were consistent for schedules of up to 40, 50, 60, or more hours per week and appear to be relatively unchanged since the 1980s (Shank, 1986, cited by Golden & Gebreselassie, 2007). Worker demographics and occupations modulate these results somewhat. Women with young children, older workers, and salaried versus hourly workers, for example, were more willing to sacrifice some income for fewer working hours (Golden & Gebreselassie, 2007).

WORKER CHARACTERISTICS AND WELL-BEING

Characteristics of those working more than 40 hours per week were examined by researchers at the National Institute for Occupational Safety and Health (NIOSH) in a "quality of work life" module that was added to the General Social Survey in 2002 (Grosch et al., 2006). Topics covered in the module included work arrangements (e.g., permanent employee vs. independent contractor), type of employment (e.g., public vs. private sector), job tasks (e.g., bending, lifting), psychosocial working conditions (e.g., workload, social support), and several measures of worker health and well-being (e.g., being injured at work, job stress). Demographically, men reported working more than 40 hours per week more frequently than women and that difference widened with greater working hours (e.g., 63 percent vs. 37 percent in the 49- to 69-hour category). Younger workers and workers with less than 10 years of job tenure reported 41-plus hours per week more frequently than older or more experienced workers. That difference diminished at 70 or more weekly hours where the proportion of older and tenured workers increased. Organizationally, salaried workers, those in establishments with fewer than 10 employees, self-employed, independent, contingent, and on-call workers, and those working mandatory overtime, were proportionally higher among those working 48 hours or more. Workers who had split, irregular, or rotating shifts also had higher percentages working more than 48 hours per week. Mixed results were observed for worker well-being. Longer hours were associated with higher odds ratios for the job interfering with family life, feeling "used up" at the end of the day, and feelings of job stress, but also associated with reports of higher levels of participation in decision-making, opportunities to develop special abilities, and job satisfaction. Reported general health status was similar to, or better than, the 35 to 40-hour working group except in the 70 or more hours group, where reported health status was worse (Grosch et al., 2006).

POSITIVE AND NEGATIVE OUTCOMES

Both employees and employers may realize benefits from long hours of work. For the employee, extra wages or bonuses are obvious incentives while increased job satisfaction, recognition, or other intangible returns also may play a role (Grosch et al., 2006). Employers may encourage or require extra hours to meet production or service demands, to compensate for a shortage of skilled labor, or minimize the cost of new hires or extra benefits (Caruso, 2006). At the community or societal level, extra hours often are expected in emergency situations to establish safety, restore utilities, or maintain continuity of care as the magnitude of the event may outpace the number of skilled workers available to meet response demands within standard work hours. It is during such acute situations that society at large may recognize the potential risks to workers as they may become more fatigued than usual while performing emergency response and recovery operations. The millions of workers logging long hours on a regular basis, however, suggest that concerns about fatigue and associated risks extend beyond emergencies. Research supports this notion.

An overview of performance loss and injury and illness outcomes associated with overtime and extended work shifts (i.e., longer than eight hours) was reported by NIOSH (2004a). Fifty-two research reports published within a decade of the NIOSH report showed trends of poorer performance, safety, or health outcomes, or null effects. Methodologies, sampling schemes, work schedule definitions, and target outcomes varied among the studies and interacting factors were examined infrequently. In general, however, the observed outcomes point to caution in the utilization of extended duration schedules. Standardized psycho-physiological and cognitive measures presumed to reflect fatigue, for example, deteriorated through the 9th to the 12th hours of long shifts and the effects appeared to be more acute when long shifts were combined with more than 40 hours per week. Since the NIOSH (2004a) overview, several additional studies of extended work duration have been reported. This chapter highlights a sample of recent studies emphasizing injury and illness outcomes associated with working more than the standard 35 to 40-hour week practiced by the majority of U.S. workers.

SAFETY COMPROMISES

Studies using carefully selected national samples of workers have observed elevated injury risks similar to the studies reviewed in the NIOSH (2004a) report. Dembe, Erickson, Delbos, and Banks, (2005) analyzed injury outcomes in the U.S. Bureau of Labor Statistics National Longitudinal Survey of Youth.[1] That survey followed a cohort who were 14–22 years of age in 1979 using annual or biennial interviews on various aspects of work life. An injury outcome was any injury reported to be work-related by the respondent. In a regression analysis that adjusted for covariates such as age, gender, geographic region, high-risk versus low-risk occupation, and high-risk versus low-risk industry, Dembe et al. (2005) observed elevated injury hazard rates among respondents reporting any self-defined overtime work (61 percent higher), working more than 12 hours per day (37 percent higher), and working at least 60 hours per week (23 percent higher). Those rates were not a simple function of greater exposure time to job-related hazards, as exposure time was adjusted in the regression analysis along with participation in high-risk occupations or industries.

Lombardi et al. (2010) analyzed injury experience associated with weekly work hours reported in the U.S. National Health Interview Survey (NHIS) over the years 2004–2008. The NHIS is conducted yearly by the National Center for Health Statistics on a nationally representative sample of households. Injury in the survey included only those experiences requiring medical attention. Using multiple regression techniques and a 31- to 40-hour week as a referent, Lombardi et al., reported significantly increased odds of injury among respondents working 41–50, 51–60, and more than 60 hours per week. Covariates adjusted in the analysis included age, sex, race, occupation, type of pay, reported sleep duration,

---

1 National Longitudinal Survey of Youth (NLSY), see http://www.bls.gov/nls

and body mass index. Additional analysis of sleep duration suggested that short sleep duration also contributed to the risk of injury. The authors concluded that fatigue, both from hours of work and from sleep loss, was a major contributing factor to injury risk in those data.

Other research has examined work hours within industry sectors having higher than average injury rates such as transportation, construction, and manufacturing, and in sectors such as healthcare, where injury rates have increased in recent years (NIOSH, 2004b). Dong (2005) examined the injury experience of construction workers in the 1979 cohort of the NLSY, noting that they frequently started work earlier and worked more hours each day than average, but also worked fewer weeks per year given the seasonal nature of many construction jobs. After adjusting for covariates, several indices of extended duration work were associated with injury risk. The highest injury odds ratios were associated with working more than 50 hours per week (1.98), receiving overtime pay (1.64), overtime in the previous week (1.54), more than one job (1.51), starting work before 07:00 hours (1.28), shift work (1.21), usually working more than 10 hours per day (1.18), and ending work after 20:00 hours (1.18). Dong (2005) stated that the variable and unpredictable nature of construction work, with respect to knowing when the next job would be available, increased the pressure to maximize working hours during times of employment.

Vegso et al. (2007) used a "case-crossover" approach to examine the contribution of work hours to injury at a major multi-plant manufacturing corporation. Injury days (cases) were compared with non-injury days (controls) of the same worker with respect to length of work shift and hours worked in the previous week. Results indicated that weekly work hours were greater in the injury week. Working more than 64 hours per week was associated with an 88 percent higher probability of injury compared to working a 40-hour week.

In the transportation sector, Jovanis et al. (2005) analyzed work schedule contributions to crash risk among interstate commercial drivers, reporting highest crash risks at the 11th hour of driving and following multiple consecutive days of driving. Outside of the transportation sector, similar reports of motor vehicle crash risk after long hours of work have been reported by other researchers. Barger et al. (2006), for example, report higher odds of a motor vehicle crash (2.3) or near miss (5.9) among medical interns working extended shifts.

Dembe, Delbos & Erickson (2009) revisited the NLSY to evaluate working hours in the healthcare sector. Compared to conventional work weeks of about 40 hours, injury odds ratios were highest for those reporting overtime (2.11) or more than 60 hours per week (2.02). Trinkoff, Geiger-Brown & Lipscomb (2007) reported that nurses working 13 or more hours per day, mandatory overtime, and performing work during expected days off, rest breaks, or while ill, were factors increasing risk of needlestick injuries. Among nurses performing home care, Gershon et al. (2009) cited mandatory overtime (odds ratio=2.44) as second only to exposure to violence (odds ratio=3.47) for risk of percutaneous injuries. Ayas et al. (2006) reported higher incidence of percutaneous injuries among medical interns after working more than 24 hours compared to strictly day work. Olds & Clarke (2010) associated working over 40 hours per week, or taking voluntary overtime, with a variety of adverse events such as needlestick and other work-related injuries, patient injury after falling, nosocomial infections, and medication errors.

HEALTH COMPROMISES

Consistently working overtime or extended schedules over periods of months or years leads to questions about long-term effects on health. Van der Hulst (2003) addressed those questions in a systematic review that selected only peer-reviewed studies with clear definitions of extended hours, carefully selected participant samples, and systematic assessment methods. The most consistent results were observed for cardiovascular and diabetes disease outcomes, and for disability retirement, subjectively reported physical health, and subjective fatigue. There also was evidence of immunologic compromises and behavioral changes, such as sleep loss or lack of exercise, that may contribute to poor health.

More recent research is consistent with the notion that long hours of work can affect health. Trinkoff et al. (2006) reported that nurses working 13 or more hours per day, mandatory overtime, and performing work during expected days off, rest breaks, or while ill, were factors that increased the risk of musculoskeletal disorders (e.g., persistent neck, shoulder, or back pain). Risk of spontaneous abortion during the first trimester was highest among nurses working more than 40 hours per week (Whelan et al., 2007). Violanti et al. (2009) observed more frequent cardiovascular risk factors (elevated waist circumference and triglycerides, low HDL cholesterol, hypertension, and glucose intolerance) among police working overtime, especially when combined with night shifts. Three to four hours of overtime per day were linked with 1.6 times higher risk of coronary heart disease among British public sector workers in the Whitehall II study (Virtanen et al. 2010). Similar results in that cohort were observed for myocardial infarction and angina. Adjusting for several

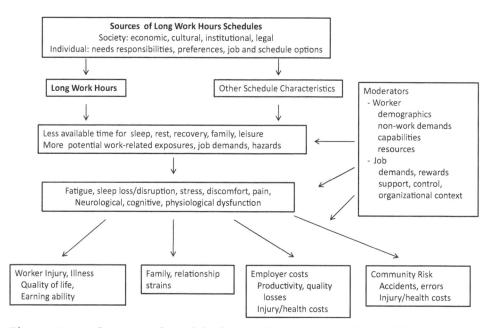

Figure 21.1   Conceptual model of potential negative effects of long work hours

known coronary risk factors (e.g., smoking, cholesterol, triglycerides) had little effect on the overtime risk estimates. A strength of the Virtanen et al. (2010) study was the ability to follow the participants' progression toward disease longitudinally for an average of 11 years. Other data from the Catalonia Health Survey linked working more than 50 hours per week with indicators of poorer health and well-being such as lower mental health status, self-reported hypertension, job dissatisfaction, smoking, no leisure time physical activity, and frequently sleeping less than six hours per night (Artazcoz et al., 2009).

In summary, recent studies are generally consistent with earlier studies included in the NIOSH (2004a) report suggesting increased safety and health risks associated with extended duration work schedules. Several definitions of long hours were examined in recent studies, such as frequent daily or weekly overtime, working more than 40 hours per week, shifts longer than eight hours, early starts or late finishes, or extra hours combined with other demands such as night or rotating shift work. While each of those factors may play a role, recent reports frequently emphasized extended weekly hours as a consistent base predictor of health or safety outcomes. Working 50 to 60 or more hours per week appears to be associated more consistently with negative outcomes. As is outlined in a conceptual model by Caruso et al. (2006; see Figure 21.1), reduced time for rest and recovery, reduced time for social and family interaction, and increased exposure to job demands and workplace hazards may play a role in those outcomes. Recent results, however, have been reported in a variety of industry sectors, occupations, and work settings. Some of those studies adjusted statistically for high-risk industries, high-risk occupations, or other exposures to emphasize the common effect of work hours. Consequently, cross-cutting elements associated with a broad array of industries and occupations need to be explored for their common contribution. Fatigue and associated reductions in opportunities for rest and recovery is one set of cross-cutting elements that has received considerable attention with respect to long hours of work (Caruso et al., 2006).

Reports of fatigue in the working population are fairly frequent irrespective of work hours. One recent national survey stated about 38 percent of respondents reported fatigue (defined as low levels of energy, poor sleep, feeling fatigued) within the previous two weeks of interview (Ricci et al., 2007). Among that 38 percent, about two-thirds reported health-related absenteeism or "presenteeism" (i.e., working at reduced capacity because of illness) which the authors estimated to cost U.S. employers $136.4 billion annually. With respect to working hours, lack of rest and recovery, and especially insufficient quantity and poor quality sleep, have been cited as key elements contributing to fatigue and potential health or safety risks. Indeed, a recent in-depth review of fatigue and safety pointed to insufficient quantity and quality of sleep and, relatedly, extended time awake, as prepotent factors contributing to performance decrements and safety risks (Williamson et al., 2011).

WORK HOURS, SLEEP, AND RECOVERY

The contribution of long work hours to restricted or poor quality sleep, and fatigue and sleepiness during waking hours, has been demonstrated. Among recent studies, Dahlgren, Kecklund & Åkerstedt (2006) compared a routine 40-hour week to a 60-hour week undertaken voluntarily by white-collar workers. With just a week of overtime, average sleep times (verified by actigraphs – automated measures of activity) were shorter by about 20 minutes and feelings of sleepiness and exhaustion were elevated by the end of the week. The 2008 Sleep in America Poll (National Sleep Foundation, 2008) surveyed sleep and work habits among a selected sample of adults. Reported sleep of those respondents decreased in a dose-response fashion with increasing hours worked. Analysis of data from the Whitehall II cohort by Virtanen et al. (2009) indicated that working more than 55 hours per week was associated with statistically adjusted higher odds of shortened sleeping hours, sleep disturbances, difficulty falling asleep, and waking unrefreshed. Nakashima et al. (2011) reported a dose-response association between increasing monthly overtime and shortened sleep, poorer sleep efficiency, and impaired daytime function in a survey of Japanese white-collar workers. Son et al. (2008) observed increasing odds of severe sleepiness associated with increasing overtime, especially when combined with night shift, among Korean manufacturing employees. Decreasing the extended work hours of medical interns, on the other hand, increased total sleep as demonstrated by Lockley et al. (2004). Reducing weekly schedules from 80 to 65 hours in that study increased average weekly sleep by 5.8 hours.

SHORT SLEEP, SLEEP DISTURBANCE, AND INJURY

It follows that restricted, disrupted, or disturbed sleep, either directly or in combination with work schedule demands, may contribute to injury risk. Reports of getting insufficient sleep were associated with injury among small or medium-sized enterprises in Japan, but sleeping fewer than six hours was not associated with injury risk (Nakata et al., 2005). Choi et al. (2006), on the other hand, reported that sleeping less than 7.5 hours per night increased the risk of injury by 61 percent in a sample of rural Iowa residents. Among commercial drivers, sleep prior to "critical incidents" (i.e., crashes, near-crashes, or sudden avoidance maneuvers) was more than one hour less than the usual amount of sleep reported by those drivers (Hanowski et al., 2007). Sleep disturbances measured by the Jenkins Sleep Problems Scale predicted the work-related injuries of Finnish public sector workers (Salminen et al., 2010). Similar results were reported in a national survey of Canadian workers (Kling, McLeod & Koehoorn, 2010). National estimates in the United States were provided by Lombardi et al. (2010), who evaluated various quantities of self-reported habitual sleep along with work hours in the NHIS. With adjustments for weekly work hours and other covariates, the odds of injury increased progressively with decreasing hours of sleep. Less than five hours of sleep per day were associated with an injury odds ratio of 2.65. Partially consistent results were reported by Valent et al. (2010) in a case-crossover study of roadway crash injuries requiring emergency room treatment. Injuries in that study were associated more consistently with being awake more than 16 hours prior to the crash than to total sleep before the crash. Most recently, the Centers for Disease Control and Prevention (CDC, 2011) observed the highest frequencies of nodding off or falling asleep while driving among respondents reporting less than seven hours of sleep in the Behavioral Risk Factor Surveillance System.

SLEEP, WORK HOURS, AND INJURY

The combination of working hours and sleep also has received recent attention. Stutts et al. (2003) reported that roadway crashes determined to be related to sleep were more frequent among drivers working multiple jobs, night shifts, or other unusual work schedules. Those drivers also obtained fewer hours of sleep, poorer quality sleep, or inadequate quantities of sleep and reported more daytime sleepiness, late-night driving, and drowsy driving. Among Japanese small or medium-sized businesses, individuals working long hours or reporting insufficient or poor-quality sleep had higher injury rates than a reference group working 6–8 hours per day or reporting good sleep. The highest rates were reported by those working long hours combined with reports of inadequate or poor sleep (Nakata, 2011a).

In summary, evidence suggests that extended work hours, insufficient sleep, or poor-quality sleep can contribute independently to injury risk. Combining those factors appears to exacerbate the potential effect on injury rates. Other work schedule demands, such as night or rotating shift work, irregular and unpredictable scheduling, or availability of rest breaks, may modulate the opportunities for sleep either independently or in combination with long hours (see Caruso & Rosa, 2006, and Rosa, 2001, for reviews). To model some of those effects, Folkard & Lombardi (2004, 2006) combined data from several injury studies to demonstrate increased risk with successive hours on shift, consecutive shifts, or night versus day shift. Injury risk during night shifts longer than eight hours was especially apparent. Availability of rest breaks, however, may help decrease injury risk (Folkard & Lombardi, 2004, 2006)

SLEEP AND HEALTH RISKS

A variety of studies explored potential links between sleep duration and health, irrespective of working hours. Mortality has been associated with sleep duration in a number of studies in the past decade. Gallicchio & Kalesan (2010) recently reported a systematic review of 23 of those studies, noting a combined relative risk associated with fewer than seven hours sleep of 1.10 for all-cause mortality and 1.06 for cardiovascular-related mortality. A systematic review focusing on cardiovascular outcomes reported similar results for short sleep duration (less than 5 or 6 hours) and likelihood of developing or dying from coronary heart disease or stroke (Cappuccio et al., 2011). Similar cardiovascular mortality and morbidity results were reported in a Finnish population (Kronholm et al., 2011). Risk factors contributing to cardiovascular disease have been associated with short habitual sleep or sleep loss in studies of weight gain (Chaput et al., 2008; Lyytikäinen et al., 2010; Watanabe et al., 2010), inflammation (Meier-Ewert et al., 2004; Irwin et al., 2006; van Leeuwen et al., 2009; Hayes, Babineau & Patel, 2011) and diabetes and impaired glucose tolerance (Chaput et al., 2007). Sleep complaints and sleep disturbance have been associated with metabolic syndrome (defined by the presence of three or more cardiovascular risk factors; Troxel et al., 2010). Evidence from the Whitehall II Cohort suggests that the combination of short sleep and disturbed sleep carries the greatest coronary heart disease risk (Chandola et al., 2010).

In summary, the relationship between short sleep and health risk has sharpened in focus in recent years with cardiovascular endpoints being frequent outcomes and metabolic and inflammatory responses being potential mediators. Combining short sleep with long work hours could intensify those mediating responses, not only by further reducing opportunities for rest but also by increasing exposure to stress from job factors or from reduced contact with family and friends (see Harma, 2006, for review). In that regard, Liu & Tanaka (2002) emphasized the combined effects of insufficient sleep and long hours on myocardial infarction among Japanese workers. More recently, Nakata (2011b) observed the poorest self-rated health among workers with the longest hours combined with short sleep duration.

CONCLUSION

Taken together, studies over the past 7–8 years add to accumulating evidence of acute risks to safety and long-term risks to health associated with long work hours. Fatigue, as characterized by reduced opportunities for rest and recovery and, especially, by limited opportunities for adequate sleep (an optimum state for rest and recovery) appears to be a major factor contributing to risk. From a population perspective, 50 to 60 work hours per week seems to be a consistent range for observing increases in adverse outcomes. With respect to sleep, habitually obtaining fewer than six hours of sleep, or having persistent complaints of disturbed, non-restorative sleep, may be reliable indicators of risk. More systematic reviews of those factors, with attention to magnitudes of effect across methodologically sound studies, would help validate these impressions. Factors that modulate risk to safety or health, as well as additional examination of family and social strains, risks to the community, and economic impacts, would round out the current narrative (see Caruso et al., 2006, for additional discussion). Ability to control work hours, for example, may have a positive effect on recovery and well-being (Takahashi et al., 2011) but more such studies need to be accomplished.

COUNTERMEASURES

Efforts to prevent excessive fatigue or mitigate its consequences are not new. Federal regulations limiting hours of service for commercial drivers and pilots, and federal law limiting hours of service for railroad personnel (49 U.S. Code Chapter 211 – Hours of Service) have been in place for decades and currently are undergoing revision in response to expanding scientific knowledge of fatigue, sleep, rest, and recovery (Federal Register, 2010a, 2010b, 2011). A growing number of states are passing laws limiting mandatory overtime for nurses (American Nurses Association, 2011). The Nuclear Regulatory Commission (2010) requires management of fatigue under its fitness for duty rules for nuclear power plant operations. For most of industry, however, the only potential government curbs against working extensive hours are requirements for additional compensation for working overtime in certain wage categories under the Fair Labor Standards Act (29 U.S. Code Chapter 8 – Fair Labor Standards). Thus, for the vast majority of U.S. workers, the only limits to working hours are access to work for pay and the will of the employer or worker. Even within regulatory limits, it is possible to work weekly hours associated with elevated injury risk. Commercial drivers, for example, may drive up to 77 hours in 7 days or 88 hours in 8 days and still comply with current federal regulations. A majority of drivers reported such extensive driving on a regular basis to meet demands or maximize economic returns (McCartt et al., 2008).

Given the broad limits of current laws or regulations, or their absence altogether, both organizations and individuals share responsibility for preventing excessive fatigue and its consequences. Human factors experts advise organizations to use a systems approach to fatigue risk management that includes, for example, establishment of a safety culture, safety management systems, rational approaches to work schedule design, adequate staffing and absentee planning, contingencies for scheduled and emergency overtime, engineering of job tasks to reduce risk, error and critical incident analysis, and training and awareness programs (Gander et al., 2011; Rosekind et al., 1996). An example of such an approach is a recent standard established for the refining and petrochemical industries (American National Standards Institute, 2010). Within the organization and within such programs, however, the individual must be responsible for actually getting rest or sleep when the appropriate opportunity arises. The challenge for all is to make those opportunities regularly available.

REFERENCES

American National Standards Institute (2010). Fatigue Risk Management Systems for Personnel in the Refining and Petrochemical Industries, ANSI/API Recommended Practice 755–2010.

American Nurses Association (2011). Position on mandatory overtime. Available at: http://www.nursingworld.org/MainMenuCategories/ANAPoliticalPower/State/StateLegislativeAgenda/MandatoryOvertime.aspx

Artazcoz, L., Cortès, I., Escribà-Agüir, V., Cascant, L. & Villegas, R. (2009). Understanding the relationship of long working hours with health status and health-related behaviours. *Journal of Epidemiology and Community Health*, 63(7), 521–7.

Ayas, N.T., Barger, L.K., Cade, B.E., Hashimoto, D.M., Rosner, B., Cronin, J.W., Speizer, F.E., et al. (2006). Extended work duration and the risk of self-reported percutaneous injuries in interns. *JAMA: The Journal of the American Medical Association*, 296(9), 1055–62.

Barger, L.K., Ayas, N.T., Cade, B.E., Cronin, J.W., Rosner, B., Speizer, F.E. & Czeisler, C.A. (2006). Impact of extended-duration shifts on medical errors, adverse events, and attentional failures. *PLoS Medicine*, 3(12), e487.

Cappuccio, F.P., Cooper, D., D'Elia, L., Strazzullo, P. & Miller, M.A. (2011). Sleep duration predicts cardiovascular outcomes: a systematic review and meta-analysis of prospective studies. *European Heart Journal*, 32(12), 1484–92.

Caruso, C.C. (2006). Possible broad impacts of long work hours. *Industrial Health*, 44(4), 531–6.

Caruso, C.C. & Rosa, R.R. (2006). Shiftwork and long work hours, in W.N. Rom (ed.), *Environmental and Occupational Medicine*, Fourth Edition. Philadelphia, PA: Lippincott-Raven, 1359–63.

Caruso, C.C., Bushnell, T., Eggerth, D., Heitmann, A., Kojola, B., Newman, K., Rosa, R.R., et al. (2006). Long working hours, safety, and health: toward a National Research Agenda. *American Journal of Industrial Medicine*, 49(11), 930–42.

CDC (2011). Unhealthy sleep-related behaviors – 12 States, 2009. *MMWR: Morbidity and Mortality Weekly Report*, 60(8), 233–8.

Chandola, T., Ferrie, J.E., Perski, A., Akbaraly, T. & Marmot, M.G. (2010). The effect of short sleep duration on coronary heart disease risk is greatest among those with sleep disturbance: A prospective study from the Whitehall II cohort. *Sleep*, 33(6), 739–44.

Chaput, J.-P., Després, J.-P., Bouchard, C. & Tremblay, A. (2007). Association of sleep duration with type 2 diabetes and impaired glucose tolerance. *Diabetologia*, 50(11), 2298–2304.

Chaput, J.-P., Després, J.-P., Bouchard, C. & Tremblay, A. (2008). The association between sleep duration and weight gain in adults: A 6-year prospective study from the Quebec Family Study. *Sleep*, 31(4), 517–23.

Choi, S.-W., Peek-Asa, C., Sprince, N.L., Rautiainen, R.H., Flamme, G.A., Whitten, P.S. & Zwerling, C. (2006). Sleep quantity and quality as a predictor of injuries in a rural population. *The American Journal of Emergency Medicine*, 24(2), 189–96.

Dahlgren, A., Kecklund, G. & Åkerstedt, T. (2006). Overtime work and its effects on sleep, sleepiness, cortisol and blood pressure in an experimental field study. *Scandinavian Journal of Work, Environment & Health*, 32(4), 318–27.

Dawson, T., Heitmann, A. & Kerin, A. (2004). Industry trends, costs, and management of long working hours. Proceedings of conference on long working hours, safety, and health: toward a national research agenda, Baltimore, MA, April 29–30.

Dembe, A.E., Delbos, R. & Erickson, J.B. (2009). Estimates of injury risks for healthcare personnel working night shifts and long hours. *Quality & Safety in Health Care*, 18(5), 336–40.

Dembe, A.E., Erickson, J.B., Delbos, R.G. & Banks, S.M. (2005). The impact of overtime and long work hours on occupational injuries and illnesses: New evidence from the United States. *Occupational and Environmental Medicine*, 62(9), 588–97.

Dong, X. (2005). Long work hours, work scheduling and work-related injuries among construction workers in the *United States. Scandinavian Journal of Work, Environment & Health*, 31(5), 329–35.

Federal Register (2010a). Department of Transportation, Federal Aviation Administration, 14 CFR Parts 117 and 121, Flightcrew Member Duty and Rest Requirements; Proposed Rule, Vol. 75, No. 177, 55851–55889.

Federal Register (2010b). Department of Transportation, Federal Motor Carrier Safety Administration, 49 CFR Parts 385, 386, 390, and 395, Hours of Service of Drivers; Proposed Rule, Vol. 75, No. 249, 87170–82198.

Federal Register (2011). Department of Transportation, Federal Railroad Administration, 49 CFR Part 228, Hours of Service of Railroad Employees; Substantive Regulations for Train Employees Providing Commuter and Intercity Rail Passenger Transportation; Conforming Amendments to Recordkeeping Requirements, Vol. 76, No. 55, 16200–16229.

Folkard, S. & Lombardi, D.A. (2004). Toward a "Risk Index" to assess work schedules. *Chronobiology International*, 21(6), 1063–72.

Folkard, S. & Lombardi, D.A. (2006). Modeling the impact of the components of long work hours on injuries and "accidents." *American Journal of Industrial Medicine*, 49(11), 953–63.

Gallicchio, L. & Kalesan, B. (2009). Sleep duration and mortality: A systematic review and meta-analysis. *Journal of Sleep Research*, 18(2), 148–58.

Gander, P., Hartley, L., Powell, D., Cabon, P., Hitchcock, E., Mills, A. & Popkin, S. (2011). Fatigue risk management: Organizational factors at the regulatory and industry/company level. *Accident; Analysis and Prevention*, 43(2), 573–90.

Gershon, R.R.M., Pearson, J.M., Sherman, M.F., Samar, S.M., Canton, A.N. & Stone, P.W. (2009). The prevalence and risk factors for percutaneous injuries in registered nurses in the home health care sector. *American Journal of Infection Control*, 37(7), 525–33.

Golden, L. & Gebreselassie, T. (2007). Overemployment mismatches: The preference for fewer work hours. *Monthly Labor Review*, April, 18–37.

Grosch, J.W., Caruso, C.C., Rosa, R.R. & Sauter, S.L. (2006). Long hours of work in the U.S.: Associations with demographic and organizational characteristics, psychosocial working conditions, and health. *American Journal of Industrial Medicine*, 49(11), 943–52.

Hanowski, R.J., Hickman, J., Fumero, M.C., Olson, R.L. & Dingus, T.A. (2007). The sleep of commercial vehicle drivers under the 2003 hours-of-service regulations. *Accident; Analysis and Prevention*, 39(6), 1140–45.

Härmä, M. (2006). Work hours in relation to work stress, recovery and health. *Scandinavian Journal of Work, Environment & Health*, 32(6), 502–14.

Hayes, A.L., Xu, F., Babineau, D. & Patel, S.R. (2011). Sleep duration and circulating adipokine levels. *Sleep*, 34(2), 147–52.

Irwin, M.R., Wang, M., Campomayor, C.O., Collado-Hidalgo, A. & Cole, S. (2006). Sleep deprivation and activation of morning levels of cellular and genomic markers of inflammation. *Archives of Internal Medicine*, 166(16), 1756–62.

Jacobs, J.A. & Gerson, K. (2004). *The Time Divide: Work, Family, and Gender Inequality*. Cambridge, MA: Harvard University Press.

Jovanis, P.P., Park, S.W., Chen, K.Y. & Gross, F. (2005). On the Relationship of Crash Risk and Driver Hours of Service. Proceedings of the 2005 International Truck and Bus Safety and Security Symposium, National Safety Council, Alexandria, VA, November 14–16.

Kling, R.N., McLeod, C.B. & Koehoorn, M. (2010). Sleep problems and workplace injuries in Canada. *Sleep*, 33(5), 611–18.

Kronholm, E., Laatikainen, T., Peltonen, M., Sippola, R. & Partonen, T. (2011). Self-reported sleep duration, all-cause mortality, cardiovascular mortality and morbidity in Finland. *Sleep Medicine*, 12(3), 215–21.

Kuhn P. & Lozano F. (2008). The expanding workweek? Understanding trends in long work hours among U.S. men, 1979–2002. *Journal of Labor Economics*, 26(2), 311–43

Liu, Y. & Tanaka, H. (2002). Overtime work, insufficient sleep, and risk of non-fatal acute myocardial infarction in Japanese men. *Occupational and Environmental Medicine*, 59(7), 447–51.

Lockley, S.W., Cronin, J.W., Evans, E.E., Cade, B.E., Lee, C.J., Landrigan, C.P., Rothschild, J.M., et al. (2004). Effect of reducing interns' weekly work hours on sleep and attentional failures. *The New England Journal of Medicine*, 351(18), 1829–37.

Lombardi, D.A., Folkard, S., Willetts, J.L. & Smith, G.S. (2010). Daily sleep, weekly working hours, and risk of work-related injury: US National Health Interview Survey (2004–2008). *Chronobiology International*, 27(5), 1013–30.

Lyytikäinen, P., Rahkonen, O., Lahelma, E. & Lallukka, T. (2010). Association of sleep duration with weight and weight gain: A prospective follow-up study. *Journal of Sleep Research*, 20(2), 298–302

McCartt, A.T., Hellinga, L.A. & Solomon, M.G. (2008). Work schedules of long-distance truck drivers before and after 2004 hours-of-service rule change. *Traffic Injury Prevention*, 9(3), 201–10.

Meier-Ewert, H.K., Ridker, P.M., Rifai, N., Reganm M.M., Price, N.J., Dinges, D.F., Mullington, J.M. (2004). Effect of sleep loss on C-reactive protein, an inflammatory marker of cardiovascular risk. *Journal of the American College of Cardiology*, 43(4), 678–83.

Nakashima, M., Morikawa, Y., Sakurai, M., Nakamura, K., Miura, K., Ishizaki, M., Kido, T., et al. (2011). Association between long working hours and sleep problems in white-collar workers. *Journal of Sleep Research*, 20(1, Pt. 1), 110–16.

Nakata, A. (2011a). Effects of long work hours and poor sleep characteristics on workplace injury among full-time male employees of small- and medium-scale businesses. *Journal of Sleep Research*, 20(4), 576–84.

Nakata, A. (2011b). Investigating the associations between work hours, sleep status, and self-reported health among full-time employees. *International Journal of Public Health*, March 8, 2011. Available at: http://www.springerlink.com/content/d356xw60gqx3776u/fulltext.pdf [accessed January 12, 2012].

Nakata, A., Ikeda, T., Takahashi, M., Haratani, T., Fujioka, Y., Fukui, S., Swanson, N.G., et al. (2005). Sleep-related risk of occupational injuries in Japanese small and medium-scale enterprises. *Industrial Health*, 43(1), 89–97.

National Institute for Occupational Safety and Health (NIOSH) (2004a). Overtime and extended work shifts: Recent findings on illnesses, injuries, and health behaviors. DHHS (NIOSH) Publication No. 2004–143, 1–36.

National Institute for Occupational Safety and Health (NIOSH) (2004b). *Worker Health Chartbook*, 2004. DHHS (NIOSH) Publication No. 2004–146.

National Sleep Foundation (2008). Sleep in America Poll: summary of findings. Available at: http://www.sleepfoundation.org/sites/default/files/2008%20POLL%20SOF.PDF

Nuclear Regulatory Commission (2010). Fitness for duty programs. 10 CFR, Part 26. Available at: http://www.nrc.gov/reading-rm/doc-collections/cfr/part026/full-text.html

Olds, D.M. & Clarke, S.P. (2010). The effect of work hours on adverse events and errors in health care. *Journal of Safety Research*, 41(2), 153–62. doi:10.1016/j.jsr.2010.02.002.

Ricci, J.A., Chee, E., Lorandeau, A.L. & Berger, J. (2007). Fatigue in the U.S. workforce: Prevalence and implications for lost productive work time. *Journal of Occupational and Environmental Medicine*, 49(1), 1–10.

Rones, P.L., Iig, R.E. & Gardner, J.M. (1997). Trends in hours of work since the mid-1970s. *Monthly Labor Review*, 120(4), 3–14.

Rosa, R.R. (2001). Examining work schedules for fatigue: It's not just hours of work, in P.A. Hancock and P.A. Desmond (eds), *Stress, Workload, and Fatigue*. Mahwah, NJ: Lawrence Earlbaum Associates, 513–28.

Rosekind, M.R., Gander, P.H., Gregory, K.B., Smith, R.M., Miller, D.L., Oyung, R., Webbon, L.L., et al. (1996). Managing fatigue in operational settings 2: An integrated approach. *Behavioral Medicine* (Washington, DC), 21(4), 166–70.

Salminen, S., Oksanen, T., Vahtera, J., Sallinen, M., Härmä, M., Salo, P., Virtanen, M., et al. (2010). Sleep disturbances as a predictor of occupational injuries among public sector workers. *Journal of Sleep Research*, 19(1, Pt 2), 207–13, doi:10.1111/j.1365-2869.2009.00780.x.

Shank, S. (1986). Preferred hours of work and corresponding earnings. *Monthly Labor Review*, 40–44.

Son, M., Kong, J.-O., Koh, S.-B., Kim, J. & Härmä, M. (2008). Effects of long working hours and the night shift on severe sleepiness among workers with 12-hour shift systems for 5 to 7 consecutive days in the automobile factories of Korea. *Journal of Sleep Research*, 17(4), 385–94.

Stutts, J.C., Wilkins, J.W., Scott Osberg, J. & Vaughn, B.V. (2003). Driver risk factors for sleep-related crashes. *Accident; Analysis and Prevention*, 35(3), 321–31.

Takahashi, M., Iwasaki, K., Sasaki, T., Kubo, T., Mori, I. & Otsuka, Y. (2011). Worktime control-dependent reductions in fatigue, sleep problems, and depression. *Applied Ergonomics*, 42(2), 244–50.

Trinkoff, A.M., Le, R., Geiger-Brown, J. & Lipscomb, J. (2007). Work schedule, needle use, and needlestick injuries among registered nurses. *Infection Control and Hospital Epidemiology: The Official Journal of the Society of Hospital Epidemiologists of America*, 28(2), 156–64.

Trinkoff, A.M., Le, R., Geiger-Brown, J., Lipscomb, J. & Lang, G. (2006). Longitudinal relationship of work hours, mandatory overtime, and on-call to musculoskeletal problems in nurses. *American Journal of Industrial Medicine*, 49(11), 964–71.

Troxel, W.M., Buysse, D.J., Matthews, K.A., Kip, K.E., Strollo, P.J., Hall, M., Drumheller, O., et al. (2010). Sleep symptoms predict the development of the metabolic syndrome. *Sleep*, 33(12), 1633–40.

U.S. Census Bureau (2009). *Statistical Abstract, 2009: Labor Force, Employment, and Earnings*. Available at: http://www.census.gov/prod/2008pubs/09statab/labor.pdf [accessed January 12, 2012].

van der Hulst, M. (2003). Long work hours and health. *Scandinavian Journal of Work, Environment & Health*, 29(3), 171–88.

van Leeuwen, W.M.A., Lehto, M., Karisola, P., Lindholm, H., Luukkonen, R., Sallinen, M., Härmä, M., et al. (2009). Sleep restriction increases the risk of developing cardiovascular diseases by augmenting proinflammatory responses through IL-17 and CRP. PloS One, 4(2), e4589.

Valent, F., Di Bartolomeo, S., Marchetti, R., Sbrojavacca, R. & Barbone, F. (2010). A case-crossover study of sleep and work hours and the risk of road traffic accidents. *Sleep*, 33(3), 349–54.

Vegso, S., Cantley, L., Slade, M., Taiwo, O., Sircar, K., Rabinowitz, P., Fiellin, M., et al. (2007).

Extended work hours and risk of acute occupational injury: A case-crossover study of workers in manufacturing. *American Journal of Industrial Medicine*, 50(8), 597–603.

Violanti, J.M., Burchfiel, C.M., Hartley, T.A., Mnatsakanova, A., Fekedulegn, D., Andrew, M.E., Charles, L.E., et al. (2009). Atypical work hours and metabolic syndrome among police officers. *Archives of Environmental & Occupational Health*, 64(3), 194–201.

Virtanen, M., Ferrie, J.E., Gimeno, D., Vahtera, J., Elovainio, M., Singh-Manoux, A., Marmot, M.G., et al. (2009). Long working hours and sleep disturbances: The Whitehall II prospective cohort study. *Sleep*, 32(6), 737–45.

Virtanen, M., Ferrie, J.E., Singh-Manoux, A., Shipley, M.J., Vahtera, J., Marmot, M.G. & Kivimäki, M. (2010). Overtime work and incident coronary heart disease: The Whitehall II prospective cohort study. *European Heart Journal*, 31(14), 1737–44.

Watanabe, M., Kikuchi, H., Tanaka, K. & Takahashi, M. (2010). Association of short sleep duration with weight gain and obesity at 1-year follow-up: A large-scale prospective study. *Sleep*, 33(2), 161–7.

Whelan, E.A., Lawson, C.C., Grajewski, B., Hibert, E.N., Spiegelman, D. & Rich-Edwards, J.W. (2007). Work schedule during pregnancy and spontaneous abortion. *Epidemiology*, 18(3), 350–55.

Williamson, A., Lombardi, D.A., Folkard, S., Stutts, J., Courtney, T.K. & Connor, J.L. (2011). The link between fatigue and safety. *Accident; Analysis and Prevention*, 43(2), 498–515.

22

Fatigue and Road Safety: Identifying Crash Involvement and Addressing the Problem within a Safe Systems Approach

R.F. Soames Job, Andrew Graham, Chika Sakashita and Julie Hatfield

INTRODUCTION

Endeavours to address fatigue as a factor in traffic crashes rest in part on an uncomfortable combination of disagreements and agreements. On the one hand there is extensive disagreement regarding the definition of fatigue, and there are significant differences in the criteria applied to crash characteristics to determine whether fatigue was a likely causal contributor. On the other hand, despite these disagreements, there is broad agreement that driver (including rider) fatigue is a major contributor to road trauma and that it must be addressed.

THE CONTRIBUTION OF FATIGUE TO ROAD TRAUMA

Our best available estimation of the role fatigue in crashes in the crash database of New South Wales (or NSW, the most populous state of Australia) reveals that this decade,

Table 22.1   Trend in number of fatal crashes to which behavioural factors contribute in NSW, over years

Behavioural Factor	2000	2001	2002	2003	2004	2005	2006	2007	2008	2009
Excessive Speed	235	226	256	209	195	190	197	140	152	213
Fatigue	122	78	110	75	84	95	89	87	61	81
Illegal Alcohol	107	99	130	102	84	83	112	91	79	80
Restraint Non Usage	92	89	83	72	65	57	57	47	46	64
Total	603	524	561	539	510	508	496	435	374	460

Note: More than one factor may be identified in a crash, and in some crashes no behavioural factors are identified.

fatigue has contributed to 14 percent to 20 percent of fatalities from year to year (see Table 22.1). This places fatigue third among the behavioural contributors to road-related fatalities in NSW, behind speed (32 percent to 46 percent of fatalities from year to year) and drinking–driving (16 percent to 23 percent, much lower than average for western jurisdictions, following Random Breath Testing, see Job, 1997). These estimates for fatigue involvement are consistent with a converging international estimate of fatigue involvement in around 20 percent of crashes (McLean, Davies & Thiele, 2003) or more for truck crashes (e.g., US National Transportation Board, 1995).

These estimates of fatigue involvement are generally larger than earlier estimates, which often ranged from 1 percent to 4 percent of crashes or fatalities (e.g., Knipling & Wang 1994). Earlier estimates of the role of fatigue may have been limited by the absence of prompted police reporting on this factor and the retrospective analysis of information in search of evidence of fatigue (see Radun & Radun, 2009). Long-term trends in fatigue in crashes many also be affected by police reports of crashes being "typically assigned to the cause of current interest" (Ogden & Moskowitz, 2004).

Nonetheless, a significant role of fatigue in road trauma causation is apparent. Furthermore, fatigue not only impairs driving alone, but also exacerbates the effects of alcohol (Horne, Rayner & Barrett, 2003; Banks et al., 2004), possibility on a variety of medical conditions (Bajaj, 2009; Matthews et al., 2000) and drowsing medications (Ceutel, 1995).

PUBLIC PROMOTION OF FATIGUE AS A FACTOR IN ROAD TRAUMA

Fatigue is well recognized as a major contributor in road safety circles (Connor, Norton & Ameratunga, 2002; Hakkanen & Summala, 2001; Hartley, 2004; Friswell & Williamson, 2008; Radun & Radun, 2009). Furthermore, whereas speeding is often underestimated as a contributor to road trauma, by the general public and motoring advocacy groups, fatigue is often overestimated as a factor.

Fatigue is often promoted to the public as a contributing factor in crashes. Numerous government road safety agencies and non-government agencies (e.g., the National Roads and Motorist Association, see NRMA, 2010) advertise and promote to the public the need to avoid driving when fatigued (e.g., the NSW Roads & Traffic Authority and the Transport Accident Commission (TAC), 2010), and the need for regular breaks from driving. For example, extensively in NSW television, radio and print media campaigns have promoted the need for rest stops, the risks of microsleeps and the impact of circadian rhythms in fatigue risk. The public is often exposed to such messages and they are often readily accepted, whereas speeding messages and advertising to the community are often rejected. For example after the speed limit on the Newell Highway (a major NSW highway of over 1,000km) was reduced from 110km/h to 100km/h, a number of local newspapers argued that this would cause more fatalities through increased fatigue (e.g., Anonymous, 2010; Gordon, 2010).

It is worthwhile to understand why this occurs and three reasons are apparent. First, drivers often do not see their own speeding as risky, especially if they feel in control (Walker et al., 2009), whereas fatigue is often only identified by drivers as present when they have a near-miss incident. Second, there appears to be no ulterior motive for governments and road safety experts to promote fatigue as a factor to the general public. This is because in most jurisdictions there is no specific offence in relation to fatigue, except for heavy vehicle drivers, or enforcement is essentially limited to post-crash analysis (Radun & Radun, 2009). Thus, there is no accusation of revenue-raising, a common catchcry against speed enforcement by journalists (e.g., Squires, 2010, Hildebrand, 2010) opposition politicians (e.g., see Maley, 2010) and in public letters to major daily newspapers which describe increased speed enforcement as "pure revenue raising" or similar (Christy, 2010; McGregor, 2010). It is noteworthy that despite this vocal, media-fed view, independent surveys show that speed enforcement is viewed more positively by the majority of motorists (Walker et al., 2009). Third, while drivers show an optimistic bias, including the belief in their superior abilities and future compared with those of their peers in relation to driving skills (Job, 1990; Fernandes, Hatfield & Job, 2009; Svenson, 1981) and in relation to driving while fatigued (Dalziel & Job, 1997), the nature of the bias appears to be different. The view that the solution to the speeding problem is increased driver training rather than better speed control is so commonly accepted as to be taken as self-evident. High-speed motor sports are often seen as evidence for this view, despite the large number of raceway crashes from which drivers are saved by excellent protection in their vehicles and forgiving race tracks with wide clear zones and effective barriers.

However, while drivers may have a variety of erroneous views about how to deal with fatigue while driving, car-handling skills do not appear to be part of the list of solutions (Vanlaar et al., 2008). Thus, the extensive scientific evidence that speed is a major factor in fatalities is often dismissed by the public on the basis that skill is the solution (e.g., Scott-Young, 2010), but this does not appear to occur for fatigue. (Recommendations for training for fatigue management in specific medical groups do not refer to car-handling skills: e.g., Bajaj et al., 2009).

For these reasons, we have the opportunity to address fatigue with greater community acceptance than attempts to address other behavioural contributors such as speeding. However, a number of core barriers remain in attempts to reduce fatigue and trauma. These include the following claims, each of which is considered below.

- A lack of consensus on the definition of fatigue;
- Empirical difficulties in identifying fatigue as a contributing causal agent in crashes;
- The constrained set of policies and measures typically considered to address fatigue explicitly.

A Lack of Consensus on the Definition of Fatigue

Definitions of fatigue vary widely in terms of a number of features, especially when applied to road safety. This literature has been extensively reviewed elsewhere (Job & Dalziel, 2001: Matthews et al., 2000), and is thus only briefly summarized here. The features on which preferred definitions vary include:

1. Causes of fatigue: the specific task – hours of driving (or specific task) versus any task, versus hours awake. (Desmond & Hancock, 2001, also suggested separation of active fatigue [due to task] and passive [due to monotony or boredom]).

2. The nature of fatigue: biological state (Job & Dalziel, 2001) versus changed performance.
3. Defining consequence of fatigue: the desire to stop performing the task (Brown, 1994) versus decrements of performance (caused in various ways: Desmond & Matthews, 1997).
4. Measurements of fatigue: by cause (hours on task or hours away) versus by subjective impact (self-report of fatigue state) versus by change in performance (task or eye movements).

The myriad options yield a large variety of formal and practical definitions of fatigue, which are reflected to varying extents in attempts to determine the contribution of fatigue to specific crashes.

Empirical Difficulties in Identifying Fatigue as a Contributing Causal Agent in Crashes

The essential problem with determining fatigue is the absence of any biological marker such as blood alcohol concentration (though this may be developed in future). In almost every jurisdiction a significant proportion of the road network (often the majority) is rural and typically fatal crashes are over-represented on the rural component of the network, including crashes for which fatigue is identified as a factor (e.g., NSW Centre for Road Safety, 2009; Main Roads Queensland, 2004). To understand the problem it is worthwhile to consider the nature of many crashes on this network (and, indeed, a number on more urban and metropolitan roads) through a hypothetical example. At 10.00 hours, police are called out to a crash to find a single crashed vehicle, with a single male occupant (deceased). The car has left the roadway on a slight left-hand curve, and rolled; speed unknown; no witnesses; no survivors; travel origin unknown; destination unknown; no other factors are apparent. Clearly, it is very difficult to determine the role, if any, of fatigue. The driver's address might give some hint as to the length of the journey prior to the crash, but no clear evidence, no information on recent rest stops, no information on work or other activities prior to driving, no information on how well or recently the victim slept. In an ideal world of unlimited resources, Police would perhaps investigate these factors. In reality, in most states and most countries this will not occur, though a coroner's investigation might reveal at least some of these factors in some countries. At the time of completing the crash report, the attending police officer must make a judgment, but in the absence of witnesses that the driver was tired or sleepy, fatigue is not likely to be identified as a causal contributor.

In a number of states in Australia and other countries, the holders of the crash records may set criteria for assigning fatigue as a factor in a crash. These also vary. Information from states of Australia exemplify the variation (with thanks to those states for the information). In New South Wales, fatigue is considered to have been involved as a contributing factor to a road crash if an involved motor vehicle controller is assessed as having been fatigued because:

1. The controller was described by the police as being asleep, drowsy or fatigued; or
2. The vehicle performed a manoeuvre which suggested loss of concentration of the controller due to fatigue, that is, the vehicle travelled onto the incorrect side of

a straight road and was involved in a head-on collision (and was not overtaking another vehicle and no other relevant factor was identified); or the vehicle ran off a straight road or off the road to the outside of a curve and the vehicle was not directly identified as travelling at excessive speed and there was no other relevant factor identified for the manoeuvre.

Thus, under the NSW coding practice, the above hypothetical crash would most likely be flagged as a fatigue crash. However, in the Northern Territory, unless the police identify fatigue as a factor, or it is identified as a factor by a coroner's enquiry, this would not be classified as a fatigue crash. Queensland has additional criteria. Fatigue is recorded as a factor if attending police identify it, or in the case of a single vehicle crash in a speed zone of 100km/h or greater, occurring during the typical fatigue times of 14.00–16.00 or 22.00 to 06.00 hours. Thus, because this crash occurred at 20.00 hours, it would not be classified as a fatigue crash.

Other countries and studies have similarly variable criteria for the assignment of fatigue as a casual factor in crashes. For example, in analysis of French crash data, Philip et al. (2001) ascribed a crash to fatigue if the driver could have avoided the crash but took no evasive action (and had low or zero blood alcohol). In the USA the NCSRD/NHTSA Expert Panel on Driver Fatigue and Sleepiness (1998) add other criteria based on evidence of increased fatigue crash likelihood with these features (e.g., single vehicle, vehicle leaves the road, single occupant, high-speed road), as well as a time of day criterion similar to that of Queensland. More restrictive criteria, such as good weather and clear visibility have also been applied (e.g., Horne & Reyner, 1995). In addition to the inconsistent criteria, the data collected are subjective (the opinion of the attending police), and even the collection of information such as hours awake or hours of driving is difficult and, even when collected is not precisely associated with fatigue. Finally, self-report by survivors can be motivated by self-protection, and thus can bias outcomes (see Howarth et al., 1989, for an example).

The Constrained Set of Policies and Countermeasures Typically Considered to Address Fatigue

A number of countermeasures for fatigue in crashes are commonly recommended (see also Watson, Chapter 17, this volume). These are reviewed below.

Education

Behaviour change in various forms dominates the suggested countermeasures for fatigue, and of these suggestions, education is the most popular. Education is the first recommended option from the NCSDR/NHTSA Expert Panel on Driver Fatigue and Sleepiness (1998) and education is a commonly applied countermeasure (e.g., Transport Accident Commission – TAC, 2010). Education is often recommended to remedy identified erroneous beliefs about fatigue (e.g., Vanlaar, et al., 2008) or identified crash patterns (e.g., Chang & Ju, 2008). Despite the regular recommendation of education (or training in management of fatigue), evaluations are generally absent or based only on changes in knowledge, self-reported behaviour, or belief in the program (see Gander et al., 2005) but not in terms of crash or trauma reductions.

Roadside reminders, removal of behavioural barriers and inducements
Another approach to changing road user behaviour in relation to fatigue is to encourage rest stops. There are three broad elements of roadside furniture available for this, in addition to education programs conducted via various media outlets: roadside signage, provisions of rest opportunities, and inducements to stop. Various elements may be employed although a comprehensive roadside fatigue countermeasure program will involve all the elements. NSW has such a comprehensive program, which is described briefly below:

1. Rest areas. Government-funded rest areas are supplied across the road network of NSW – and, indeed, all states of Australia, as well as in many other countries. In NSW there are over 1,400 rest areas of various forms depending on demand. These may vary from simple parking bays to separate heavy vehicle and light vehicle facilities, toilets, washrooms, and children's play equipment. Analyses of crashes and surveys of usage identify safety benefits of rest areas (King, 1989). Furthermore, they may be seen as an integral part of the infrastructure necessary to allow effective regulation and enforcement of heavy-vehicle driving: if we are to mandate rest stops, then the physical possibility of stopping is essential, though it may be argued that the provision of such opportunities is not the responsibility of government.
2. Permanent signage. NSW employs broad fatigue message signage reminding drivers of the value of rest, with signage reading: "Stop Revive Survive." In addition, rest areas are signposted well in advance and signs may include information about how far it is to subsequent rest opportunities, allowing drivers to plan rest stops for their journey.
3. Incentives to stop. In addition to toilets and children's play equipment at some rest areas, during peak holiday travel times additional incentives programs operate (with the help of local volunteers and sponsors organized across the state). These include free tea, coffee and snacks for all who stop at the designated rest areas. Both users and the volunteers groups who run it have strong faith in this program.

Warnings of leaving the roadway
Many countries are making increasing use of devices in the roadside which warn drivers that they are diverging from their lane. These are audio–tactile edge lines, or profile line marking, which include raised bumps of paint or other devices, or milled/grooved shallow transverse grooves in the road surface. These result in vibration and noise when a vehicle drives onto them, thus warning (and perhaps awakening) the driver. These are typically placed so that some road shoulder remains outside the edge line, allowing the alerted driver some space to correct the direction of the vehicle. They may also be placed on the centreline of two-way roads, ideally with a wide median to allow room for the driver to correct before reaching oncoming traffic. Evaluations show that these devices yield significant reductions in crashes and trauma (Job & Hatfield, 2002; Hatfield, Murphy, Job & Wei, 2009; Torbic, 2009), as do wider shoulders and median separations (Levett, Job & Tang 2009). Lane departure warnings are also being added to some vehicles to achieve an in principle similar driver learning function. Though these warning systems achieve only moderate acceptability to novice drivers (Young et al., 2003), they are likely to yield road safety gains (Bayly et al., 2007). The benefits of audio–tactile line marking may be enhanced if drivers are discouraged from continuing to drive while fatigued, relying on the lines to avoid a crash (Hatfield, Murphy & Job, 2008).

Although a number of the above measures count as engineering in terms of delivery (rest areas, rumble strips, signage), they are classified here as behavioural measures on the basis of their intended outcome – to change driver behaviour either in advance of serious risk or at the point of critical danger.

Regulation and enforcement
Another key element of behaviour change strategies is regulation and enforcement. This broadly occurs for three specific target groups, depending on the country or state:

1. Heavy vehicle and professional drivers. Many countries, including the states of Australia, limit the hours for which heavy vehicle drivers can drive or work. These regulations are generally based on the research evidence of the limits of hours before significant impairment occurs (Chang & Hwang, 1991; and see Chang & Ju, 2008: 1844) and may be enforced through checks of driving records (Compulsory "log books" or "work diaries") or company records. "Chain of responsibility" laws may also be used to extend the sanctions for driving time breaches to trucking companies, consignors, or other parties who cause drivers to exceed the regulated hours.
2. Novice drivers. Many countries and states have introduced various nighttime driving curfews as components of (graduated) licensing programs. A number of evaluations of such restrictions have identified material reductions in crash and casualty involvement of novice drivers with these programs (Foss, Feaganes & Rodgman, 2001; and for review see Williams & Preusser, 1997). While these successes are achieved by addressing many aspects of road safety risk (total exposure, darkness and drink-driving) they also address fatigue, which is known to increase as a crash risk factor from late night through early morning (NCSDR/NHTSA Expert Panel, 1998).
3. All drivers. Although well recognized as a causal factor in road trauma, fatigued driving is rarely explicitly covered in road law. A key reason for this is the absence of an ambiguous marker of fatigue equivalent to blood alcohol. In some jurisdictions fatigue is considered to be broadly covered by laws relating to "being in proper control of the vehicle," or similar wording. However, a small number of jurisdictions have explicit laws regarding driving while fatigued. In the USA state of New Jersey, Maggie's Law allows severe penalties for a driver who causes a fatal crash after being awake for more than 24 hours, and in Finland, the Road Traffic Act explicitly prohibits driving while tired (Radun & Radun, 2009).

ADVANTAGES AND LIMITATIONS OF ATTEMPTS TO REGULATE AND ENFORCE AGAINST DRIVING WHILE FATIGUED

Road safety history is littered with critical examples of the smaller behaviour change effects of education to improve voluntary compliance and the large safety benefits of regulation with effective enforcement. Examples include:

1. Seatbelts. Large-scale education campaigns produced small or non-existent increases in seatbelt-wearing. On the other hand, regulation making the wearing of a seatbelt

compulsory, along with police enforcement and education of the threat, caused dramatic increases (in NSW to well over 90 percent, Job, 1988.

2. Drink-driving. Australia (specifically the states of NSW and Victoria) was one of the first countries to introduce random breath-testing (RBT). RBT allows effective enforcement regardless of driving overconfidence, which existed even for driving after drinking (Job, 1990). Without the randomness of the check, drivers who believe they do not drive in a manner which attracts attention or warrants a breath test do not believe that they will be caught by conventional policing. RBT overcomes this perception and thus substantially increases the perceived risk of being caught. In NSW and Victoria, driving accounted for 40 to 50 percent of fatalities prior to RBT. With the introduction of RBT and extensive media campaigns, drink-driving fatalities halved (Job, 1990; Job, Prabakar & Lee, 1997).

3. Speeding. An extensive literature attests to the efficiency of effective enforcement of speeding through fixed cameras (Mountain, Hirst & Maher, 2004; Pilkington & Kinra, 2005) or mobile cameras (Christie et al., 2003); or point to point (Stephan, 2006).

4. Technology. Carsten and Tate's (2005) evaluation of the benefits of intelligent speed adaptation technology shows many times the crash reduction benefits for mandatory (limiting) application compared with voluntary (advisory).

Thus, despite psychological differences in approaches to these behavioural problems (Fernandes, Hatfield & Job, 2010), broad deterrence of driving while fatigued, along with effective enforcement and education, including warnings of the enforcement, could be seen to be the most powerful behaviour change technique available. The claimed value of enforcement is further supported by the evidence that driving while fatigued is often a risk-taking behaviour rather than inadvertent (Hatfield, Murphy, Kasparian & Job, 2007). However, in the case of fatigue, the barriers to effective enforcement are substantial:

1. No unambiguous reliable, agreed marker currently exists for measurement of fatigue. Although work is progressing for such a marker, it has been slow. Broadbent considered the feasibility of a test of fatigue in 1979. Recent research suggests potentially fruitful directions (Baulk, Briggs, Reid, van den Heuvel & Dawson, 2008; Liu, Hosking & Lenne, 2009). Until a marker exists and is accepted, enforcement of fatigue will remain problematic. This problem precipitates the ensuing problems. The lack of legal challenges to driving while fatigued charges in the Netherlands should not be taken as evidence that the charges are effective and unsurmountable. Rather, as Radun & Radun (2009) note, the penalties for driving while fatigued may be less than the alternative charges for the error, such as crossing to the wrong side of the road. Thus, for defendants there is little incentive to challenge the charge.

2. Time awake (or time driving) as markers are difficult to prove.

3. A set criterion for time away (or time driving without a break) may normalize dangerous levels below the criterion. The New Jersey fatigue law sets 24 hours without sleep as a criterion. It is necessary to have such a high level, which is clearly accepted as unassailable evidence for fatigue, in order to ensure that the creation of the offence is politically acceptable as well as supported by the judiciary who will determine penalties. The difficulty with setting such a high- level criterion lies in the message it sends for all the levels below. On road, many work on the legal status quo as supported by our road safety messages, "Illegal equals unsafe and legal equals

safe." Would any of us really want a criterion which sends the message that driving after 23 hours awake is safe?

THE SAFE SYSTEMS APPROACH TO FATIGUE

In summary, the traditional countermeasures to fatigue provide some benefits, but considerable fatigue-related risk and traffic crash trauma remains. The Safe System approach to road safety has been formally adopted in numerous countries. This approach accepts that human error will occur and argues that it must be accommodated within the transport system. Death and serious injury are not inevitable or acceptable consequences of mobility. Accommodation is achieved by limiting the forces to which the human body is exposed (in the event of error and a crash) to that which is tolerable without debilitating injury or death. While gains will, and should, continue to be made by improving on-road behaviour, including minimizing driving while fatigued, the safe systems approach dictates that the transport systems must accommodate impaired (including fatigued) drivers. The essential elements of the transport system to be manipulated in the safe systems approach are speed, the roads and roadsides, and vehicles. Each of these can be manipulated to accommodate crashes caused by fatigue. In fact fatigue crashes, in particular, may benefit more than most from safe-systems measures which reduce physical forces in the event of a crash, because fatigue crashes tend to be more severe (Symmons, 2007). Great force is to be explecated because in many fatigue crashes the driver has had a microsleep or is asleep, resulting in no braking or evasive action to mitigate crash impact. Each safe system element is considered briefly, in relation to fatigue.

Speed
The evidence that speeding contributes significantly to road trauma is extensive and undeniable (Elvik, Christensen & Amundsen, 2004; Kloeden et al., 1997; Kloeden, Ponte & McLean, 2001), as is the evidence that reduced speed limits benefit safety (Nilsson, 1991; Kloeden, Wooley & McLean, 2007). Effective speed enforcement saves lives. Even if a crash occurs for another reason, such as fatigue, the crash forces will be less severe if the fatigued driver is driving more slowly.

Vehicles
A thoroughgoing safe systems approach would entail that vehicle-related countermeasures do not rely on the driver, because this reliance reintroduces the risk of human error circumventing the safety system. Thus, under safe system principles, when fatigue is detected the vehicle would be disabled with safe warning. This would be a fatigue-specific countermeasure, in addition to the well documented safety benefits of technology such as electronic stability control, which helps the driver maintain control when suddenly alerted to leaving the road (for example, by profile line markings), and the occupant protection safety features increasingly being fitted to vehicles.

Roads and roadsides
The evidence for the safety benefits of flexible barriers, which reduce crash forces, is undeniable (Wegman, 2007). Similarly, the safety benefits of numerous engineering treatments (curve-straightening, improved delineation and wide shoulders) are the basis

of worldwide "blackspot" treatment programs (but see Elvik, 1997, and Job & Sakashita, 2009, for consideration of regression to the mean and category shift risks in evaluations). A more precise and thus more cost-effective use of these treatments has been demonstrated to arise from a program of multidisciplinary highway road safety reviews (de Roos, M.P., Job, R.F.S., Graham, A., Levett, S. 2008). These reviews have been conducted and engineering works arising from the works completed for two major highways in NSW. The works from these reviews have resulted in benefit cost ratios several times higher than the blackspot program in the same state. Comparing casualties for the year prior to the works with the year they were completed reveals powerful road safety benefits on each highway, with fatality reductions of 63 percent. These results are presented in Table 22.2. We conducted retrospective analyses to determine if these works (mainly median, shoulder, wire rope barriers, curve improvements, delineation improvements

Table 22.2   Reductions in fatalities and injuries comparing the calendar year before and after Multidisciplinary Highway Road Safety Reviews and works on the Pacific and Princes Highways in NSW

Severity	Pacific Highway	Princes Highway	Total Savings (%)
Fatalities			
Before	55	24	79
After	25	4	29
Savings	30	20	50 (63%)
Injuries			
Before	617	324	941
After	483	294	777
Savings	134	30	164 (17%)

Table 22.3   Reductions in fatalities and injuries associated with the Multidisciplinary Highway Road Safety Reviews in NSW, separated into fatigue-related and non-fatigue-related crashes (Pacific and Princes Highways combined)

Severity	Fatigue-related	Non-fatigue-related
Fatalities		
Before	9	70
After	8	21
Savings	1 (11%)	49 (70%)
Injuries		
Before	160	781
After	113	664
Savings	47 (29%)	117 (15%)
TOTAL casualty savings	**48 (28%)**	**166 (20%)**

and increased speed enforcement by fixed speed cameras) reduced severe fatigue-related crashes, in line with the safe systems approach. The results are presented in Table 22.3.

The data in Table 22.3 provide direct evidence for what we would expect: treatments which broadly address road safety by reducing the risk of off-road and head-on crashes through improved alignment, flexible wire rope barriers and speed enforcement do significantly reduce fatigue related serious crashes. In fact, across the two highways the reductions in fatigue-related casualty crashes are broadly similar to the reductions in non-fatigue related serious crashes. Clearly, the safe system approach to the road environment is effective for addressing fatigue-related road trauma.

CONCLUSIONS

Measures focussed on fatigue are strongly biased toward behaviour change solutions. While useful, most of these are not as powerful as safe systems countermeasures. However, in two instances in particular, critical potential improvements would yield significant road safety gains. First, with a reliable marker or test of fatigue, regulation and enforcement could be considerably more powerful tools and could overcome the risk of normalizing unacceptable levels of wakefulness or driving. Second, strengthening the consequence of an in-vehicle fatigue detection system (through company policy or technology) such that the driver must stop driving will increase the benefits of such detection systems.

While safe systems countermeasures may be seen as weaker because they are not focussed on fatigue, this is in fact their strength. More forgiving roads and roadsides, better crash-protected vehicles, and safer speed will mitigate the casualty consequences of fatigue as well as other crashes. The re-analysis presented here of the benefits of the safe system approach on highways in NSW confirms that fatigue-related casualties are substantially reduced.

REFERENCES

Anonymous (2010). Newell Highway speed limit here to stay. *West Wyalong Advocate*. P33, 27 January.

Bajaj, J.S., Hafeezullah, M., Zadvornova, Y., Martin, E. Schubert, C.M. & Gibson, D.P. (2009). The effect of fatigue on driving skills in patients with hepatic encephalopathy. *American Journal of Gastroenterology*, 104, 898–904.

Banks, S., Catcheside, P., Lack, L., Grumstein, R.R. & McEvoy, R.D. (2004). Low levels of alcohol impair driving simulator performance and reduce perception of crash risk in partially sleep deprived subject. *Sleep*, 27, 1063–7.

Baulk, S.D., Biggs, S.N., Reid, K.J., van den Heuvel, C.J. & Dawson, D. (2008). Chasing the silver bullet: Measuring driver fatigue using simple and complex tasks. *Accident Analysis and Prevention*, 40, 396–402.

Bayly, M., Fildes, B., Regan, M. & Young, K. (2007). Review of research effectiveness of intelligent transport systems. TRACE (Traffic Accident Causation in Europe) Project No. 027763.

Broadbent, D.E. (1979), Is a fatigue test now possible? *Ergonomics*, 22, 1277–90.

Brown, I.D. (1994). Driver fatigue. *Human Factors*, 36, 298–314.

Carsten, O.M.J. & Tate, F.N. (2005). Intelligent speed adaptation: Accident savings and cost–benefit analysis. *Accident Analysis and Prevention*, 37, 407–16.

Ceutel, C. (1995). Risk of traffic accident injury after a prescription for a benzodiazepine. *Annals of Epidemiology*, 5, 239–44.

Chang, H.L. & Hwang, H.H. (1991). The effect of prolonged driving on accident risk for highway truck operations. *Transportation Planning Journal*, 20, 295–312.

Chang, H.L & Ju, L.S. (2008). Effect of consecutive driving on accident risk: A comparison between passenger and freight train driving. *Accident Analysis and Prevention*, 40, 1844–49.

Christie, S.M., Lyons, R.A., Dunstan, F.D. & Jones, S.J. (2003). Are mobile speed cameras effective? A controlled before and after study. *Injury Prevention*, 9, 302–6.

Christy, W. (2010). Letter to *Daily Telegraph*. P22. 12 July.

Connor, J., Norton, R. & Ameratunga, S. (2002). Driver sleepiness and risk of serious injury to car occupants: Population based case control study. *British Medical Journal*, 325, 1125.

Dalziel, J.R. & Job, R.F.S. (1997). Motor vehicle accidents, fatigue and optimism bias in taxi drivers. *Accident Analysis and Prevention*, 29, 489–94.

de Roos, M.P., Job, R.F.S., Graham, A. & Levett, S. (2008). Strategic road safety successes from multi-disciplinary highway safety reviews. Proceedings of Safer Highways of the Future Conference, Brussels, Belgium, February, 2008.

Desmond, P.A. & Matthews, G. (1997). Implications of task-induced fatigue effects for in-vehicle countermeasures to driver fatigue. *Accident Analysis and Prevention*, 29, 513–23.

Desmond, P.A. & Hancock, P.A. (2001). Active and passive fatigue states, in P.A. Hancock & P.A. Desmond (eds), *Stress, Workload and Fatigue*. Mahwah, NJ: Lawrence Erlbaum Associates, 455–65.

Elvik, R. (1997). Evaluations of road accident blackspot treatment: A case of the iron law of evaluation studies? *Accident Analysis and Prevention*, 29, 191–9.

Elvik, R., Christensen, P. & Amundsen, A.H. (2004). Speed and road accidents: An evaluation of the power model. Report 740. Institute of Transport Economics, Oslo.

Fernandes, R., Hatfield, J. & Job, R.F.S. (2010). A systematic investigation of the differential predictors for speeding, drink-driving, driving while fatigued, and not wearing a seat belt, among young drivers. *Transportation Research*, Part F: *Traffic Psychology and Behaviour*, 13, 179–96.

Foss, R.D., Feaganes, J.R. & Rodman, E.A. (2001). Initial effects of graduated driver licensing on 19 year old driver crashes in North Carolina. *Jama*, 286, 1588–92.

Friswell, R. & Williamson, A. (2008). Exploratory study of fatigue in light and short haul transport drivers in NSW, Australia. *Accident Analysis & Prevention*, 40, 410–17.

Gander, P.H., Marshall, N.S., Bolger, W. & Girling, I. (2005). An evaluation of driver training as a fatigue countermeasure. *Transportation Research*, Part F, 8, 47–58.

Gordon, L. (2010). Keneally's "out of touch" speed limit claim sparks mass uproar. *Daily Liberal*, P9, 25 June.

Hakkanen, H., & Summala, H. (2001). Fatal traffic accidents among trailer truck drivers and accident causes as viewed by other truck drivers. *Accident Analysis & Prevention*, 33(2), 187–96.

Hartley, L. (ed.) (2004)*Managing Fatigue in Transportation*. Oxford: Pergamon.

Hatfield, J., Murphy, S. & Job, R.F.S. (2008). Beliefs and behaviours relevant to the road safety effects of profile lane-marking. *Accident Analysis and Prevention*, 40, 1872–9.

Hatfield, J., Murphy, S., Job, R.F.S. & Wei, D. (2009). The effectiveness of audio–tactile lane-marking in reducing various types of crash: A review of evidence, template for evaluation, and preliminary findings from Australia. *Accident Analysis and Prevention*, 41, 365–79.

Hatfield, J., Murphy, S., Kasparian, N. & Job, R.F.S. (2007). Risk perceptions, attitudes, and behaviours regarding driver fatigue in NSW youth: The development of an evidence-based driver fatigue educational intervention strategy. Online. Available at: http://www.irmrc.unsw.edu.au/documents/interventionstrategy%20report.pdf

Hildebrand, J. (2010). Blackspot cameras turned on. *Daily Telegraph*, 12 July, 3.

Horne, J.A. & Reyner, L.A. (1995). Sleep related vehicle accidents. *British Medical Journal*, 310, 565–7.

Horne, J.A. Reyner, L.A. & Barrett, P.R. (2003). Driving impairment due to sleepiness is exacerbated by low alcohol intake. *Occupational and Environment Medicine*, 60, 689–92.

Howarth, N.L., Heffernan, C.J. & Horne, E.J. (1989). Fatigue in truck accidents. Monash University Accident Research Centre Report. Clayton, Vic: MUARC.

Job, R.F.S. (1988). Effective and ineffective use of fear in health promotion campaigns. *American Journal of Public Health*, 78, 163–7.

Job, R.F.S. (1990). The application of learning theory to driver confidence: The effects of age and the impact of Random Breath Testing. *Accident Analysis and Prevention*, 22, 97–107.

Job, R.F.S. (2007). Strategic road safety successes from multi disciplinary highway safety reviews. Paper presented to Australasian Road Safety Research, Policing and Education Conference, 2007, Melbourne, Australia.

Job, R.F.S. & Dalziel, J. (2001). Refining fatigue as a condition of the organism and distinguishing it from habituation, adaptation and boredom, in P.A Hancock & P.A Desmond (eds), *Stress, Workload and Fatigue: Human Factors in Transportation*. Mahwah, NJ: Lawrence Erlbaum Associates, 466–75.

Job, R.F.S. & Hatfield, J. (2002). Review of the evidence for profile line markings as a road safety countermeasure: I. A revised cost-benefit analysis for rural roads. *Roadwise (Journal of the Australasian College of Road Safety)*, 13(4), 16–18.

Job, R.F.S., Prabhakar, T. & Lee, S.H.V. (1997). The long term benefits of random breath testing in NSW (Australia): Deterrence and social disapproval of drink-driving. In C. Mercier-Guyon (ed.), Proceedings of the 14th. International Conference on Alcohol, Drugs and Traffic Safety, 1997. France: CERMT, 841–8.

Job R.F.S. & Sakashita, C. (2009). The psychology of punishment: The category shift problem in road safety and how to address it. Paper presented to the 2009 Road Safety 2020: Smart Solutions, Sustainability, Vision, Perth, 5–6 November. Available at: http://www.acrs.org.au/acrsconferen ces/2009roadsafety2020perth.html

King, G.F. (1989). *Evaluation of Safety Roadside Rest Areas*. NCHRP Report 324. Washington, DC: Transportation Research Board.

Kloeden, C.N., McLean, A.J., Moore, V.M. & Ponte, G. (1997). *Travelling Speed and the Risk of Crash Involvement*. Vol. 1: *Findings*. NHMRC Road Accident Research Centre. Adelaide: University of Adelaide.

Kloeden, C.N., McLean, J., Ponte, G., (2001). Travelling speed and the risk of crash involvement on rural roads. NHMRC Road Accident Research Centre. Adelaide: University of Adelaide.

Kloeden, C.N., Wooley, J. & McLaen, J. (2007). A follow-up evaluation of the 50km/h default urban speed limit in South Australia. Proceedings of the Australasian road safety research, policing and education conference, Melbourne, 2007. Clayton, Vic: Vic Roads.

Knipling, L.R. & Wang, J.S. (1994). Research note: crashes and fatalities related to driver drowsiness/ fatigue. Washington, DC: U.S National Highway Traffic Safety Administration.

Levett, S.P., Job, R.F.S. & Tang, J. (2009). Centreline treatment countermeasures to address crossover crashes. Proceedings of the Australasian Road Safety Research, Policing and Education Conference. Sydney, NSW: Roads and Traffic Authority.

Liu, C.C., Hosking, S.G. & Lenne, M.G. (2009). Predicting driver drowsiness using vehicle measures: Recent insights and future challenges. *Journal of Safety Research* P40, P239–45.

Main Roads (Queensland) (2004). Driver fatigue guidelines for road based driver fatigue management in rural areas. Brisbane: Main Roads.

Maley, J. (2010). Road blitz to fill NSW coffers. *Sydney Morning Heard*, 12 July.

Matthews, G., Davies, D.R., Westerman, S.J. & Stammers, R.B. (2000). *Human Performance: Cognition, Stress and Individual Differences*. East Sussex, UK: Psychology Presentation.

McGregor, P. (2010). Try another tack. Letter to The *Daily Telegraph*, P22, 12 July.

McLean, A.W., Davies, D.R.T. & Thiele, K. (2003). The hazards and prevention of driving while sleepy. *Sleep Medicine Reviews*, 7, 507–21.

Mountain, L.J., Hirst, W.M. & Maher, M.J. (2004). Costing lives or saving lives: A detailed evaluation of the impact of speed cameras. *Traffic Engineering & Control*, 45(8), 280–87.

NCSDR/NHTSA, Expert Panel on Driver Fatigue and Sleepiness (1998). Drowsy driving and automobile crashes. DOT HS808707. Washington, DC NHTSA.

Nilsson, G. (1991). Speed limits, enforcement and other factors influencing speed, in M.J. Koornstra & J. Christensen (eds), *Enforcement and Rewarding: Strategies and Effects*. Proceedings of the International Road Safety Symposium in Copenhagen, Denmark, September 19–21, 1990. Leidschendam, SWOV Institute for Road Safety Research, Chapter 10.

NRMA (2010). NSW Drivers are burning the candle at both ends. Media release, 21 July.

NSW Centre for Road Safety (2009). Roads and traffic crashes in New South Wales: Statistical statement for the year ended 31 December 2008. North Sydney, NSW: RTA. Available at: www.rta.nsw.gov.au

Ogden, E.J. & Moskowitz, H. (2004). Effects of alcohol and other drugs on driver performance. *Traffic Injury Prevention*, 5, 185–98.

Philip, P., Vervialle, F., Le Breton, P., Taillard, J. & Horne, J.A. (2001). Fatigue, alcohol, and serious road crashes in France: Factorial study of national data. *British Medical Journal*, 322, 829–30.

Pilkington, P. & Kinra, S. (2005). Effectiveness of speed cameras in preventing road traffic collisions and related casualties: Systematic review. *BMJ*, 330, 331–4.

Radun, I. & Radun, J.E. (2009). Convicted of fatigued driving: Who, why and how? *Accident Analysis and Prevention*, 41, 869–75.

Scott-Young, B. (2010). This Easter bunny. *Daily Telegraph*, P22, 12 July.

Squires, R. (2010). Another fast buck. *Sunday Telegraph*, P9, 11 July.

Stephan, C. (2006). Section control – automatic speed enforcement in the Kaisermuhlen Tunnel (Vienna, A22 Motorway). Vienna: Austrian Road Safety Board.

Svenson, O. (1981). Are we all less risky and more skilled than our fellow drivers? *Acta Psychologia*, 4, 143–48.

Symmons, M. (2007). Fatigue as a crash factor: Applying the ATSB definition for a fatigue involvement crash in Victoria's crash data. Proceedings of the Road Safety Research, Policing and Education Conference, Melbourne. Melbourne: VicRoads.

Takayama, L. & Nass, C. (2008). Assessing the effectiveness of interactive media in improving drowsy driver safety, *Human Factors*, 50, 772–81.

Torbic, D.J. (2009). Guidance for the design and application of shoulder and centerline rumble strips. NCHRP Report 641. Washington, DC: Transportation Research Board.

Transport Accident Commission (TAC) (2008). Audience reaction to driver fatigue campaigns, Transport Accident Commission, Melbourne, Vic. Online. Available at: http://www.tac.vic.gov.au/jsp/content/NavigationController.do?areaID=13&tierID=3&navID=AE6832F77F0000010085AA408AA47571&navLink=null&pageID=396 (viewed 8 September 2008).

Transport Accident Commission (TAC) (2010). Reducing fatigue – a case study. Available at: www.tac.vic.gov.au

US National Transportation Safety Board (1995). Factors that affect fatigue in heavy truck accidents. Safety Study 95/01. Washington, DC: National Transportation Board.

Vanlaar, W., Simpson H., Mahew, D. & Robertson, R. (2008). Fatigued and drowsy driving: a survey of attitudes, opinions and behaviours. *Journal of Safety Research*, 39, 303–9.

Walker, E., Murdoch, C., Bryant, P., Barnes, B. & Jonson, B. (2009). Quantitative study of attitudes, motivations and beliefs related to speeding and speed enforcement. Proceedings of the National Road Safety Policing, Education and Enforcement Conference, Sydney, NSW: Roads and Traffic Authority.

Wegman, F. (2007). Improving traffic safety culture in the United States. Washington, DC: AAA Foundation for Traffic Safety.

Williams, A.F. & Preusser, D.F. (1997). Night driving restrictions for youthful drivers: A literature review and commentary. *Journal of Public Health Policy*, 18, 334–5.

Young, K.L., Regan, M.A., Mitsopoulos, E. & Hawanth, N. (2003). Acceptability of in vehicle intelligent transport systems to young novice drivers in NSW. MUARC Report 1999. Clayton, Victoria: MUARC.

23

Driver Fatigue and Safety: A Transactional Perspective

Catherine Neubauer, Gerald Matthews and Dyani J. Saxby

INTRODUCTION

Generally speaking, fatigue refers to a mental state characterized by psychological or physical tiredness, which may be accompanied by a variety of additional affective, cognitive and physiological responses. In the driving context, fatigue states can also be influenced by both psychological (e.g., attentional demands) and environmental aspects (e.g., time of day) of the task (Wijesuriya, Tran & Craig, 2007). Brown (1994) described fatigue as "a subjectively experienced disinclination to continue performing the task at hand." emphasizing its motivational aspect. Numerous approaches have been used to measure various aspects of driver fatigue, which include, but are not limited to, objective measures of performance decrement, physiological measures aimed to capture the driver's physical tiredness, and psychological, self-report measures of the driver's subjective feelings of fatigue (Wijesuriya et al., 2007). Fatigue may also coexist with stress derived from work, life and the demands of the driving task itself (Rowden et al., 2011).

Driver fatigue is a well-known contributor to potentially fatal car crashes (Lee, 2006). In recent years, The National Highway Traffic Safety Administration (NHTSA) reported that drowsiness/fatigue accounted for approximately 56,000 annual fatigue-related crashes, as cited by on-site police officers (NHTSA Expert Panel on Driver Fatigue and Sleepiness, 1997), which may account for up to 15 percent of fatal large-truck crashes (Pratt, 2003). According to Fletcher et al. (2005), fatigue is named as a causal factor in crashes if police describe the driver as being drowsy, asleep, and/or if the vehicular path line suggests loss of control by the driver. Williamson et al. (2011) reviewed studies that attempted to quantify the risks of sleepiness. The increase in risk (odds ratio) varied from roughly 2–3 times that of normal driving for minor sleep deprivation to 10 times normal or more for severe sleep loss. Risks associated with long drive durations and circadian rhythms appear to be of lesser magnitude.

Some experts suggest that the actual number of fatigue-related car crashes may be even higher than is typically reported, due to several factors (Wijesuriya et al., 2007). First, there is the persistent issue of an agreed upon definition and measurement of fatigue. Williamson et al. (2011) stress that fatigue has multiple components, with symptoms including mental, physical and muscular fatigue, which, they claim, are the most important aspects of fatigue within the driving context. In addition, fatigue may be determined by both physiological and psychological factors, which may elicit other potentially hazardous state changes such as stress (Matthews & Desmond, 1998; Saxby et al., 2007). Second, drivers may underreport their involvement in fatigue-related crashes due to legal or insurance issues (Wijesuriya et al., 2007). In this chapter, we review the psychology of the driver impairments that fatigue may produce. We will focus especially on cognitive psychological models that conceptualize driving as a task that requires active regulation of task demands. In developing the psychological theory, we will review evidence from field studies, discuss conceptual models that may capture some of the complexity of fatigue response, and outline a transactional model that sees fatigue as a relationship between the driver and the traffic environment. We will also describe the implications of the model for developing countermeasures, and finish with some general conclusions.

DRIVER FATIGUE AND ROAD SAFETY

The contribution of fatigue to fatal crashes, as well as the risks of driving while experiencing symptoms of fatigue, may indeed be comparable to the risk associated with alcohol intoxication (Fletcher et al., 2005). Professional drivers may be particularly susceptible to fatigue effects, since it is estimated that the risk for being involved in a fatigue-related accident is almost 20 times higher for long-haul truck drivers than for other motorists (Hitchcock & Matthews, 2005). Lal and Craig (2001) suggest that a number of factors put long-haul truck drivers at risk. First, irregular schedules that require professional drivers to continue driving after several consecutive days may lead to cumulatively greater feelings of not only subjective fatigue but also stress. Second, physical demands of the job such as cargo-handling or increased sleep debt may also increase physical fatigue, while decreasing physiological arousal, a well-known risk associated with fatigue-related decrements. In addition, professional truck drivers are required to be away from their families and friends for extended periods of time, which may also contribute to stress and fatigue (Driver Health and Wellness, 2011).

According to Feyer and Williamson (2001), a considerable number of professional drivers report that they have driven while experiencing increased symptoms of fatigue, which is associated with poorer vehicle control and reduced arousal. Moreover, Guppy and Guppy (2003) found driving while tired to be a significant factor in near-miss and accident encounters, as well as increased variation in driving and close calls (Morrow & Crum, 2004). It is clear that the majority of commercial vehicle drivers complete their task many times while tired, the effects of which are detrimental to road safety. In an interview study with long-distance truck drivers (N=593), Lee (2006) found that 47.1 percent of drivers reported having fallen asleep at least once in their career. To better explore these effects, Friswell and Williamson (2008) conducted a survey among commercial light and short haul truck drivers and found that, of those drivers who had been subject to feelings of fatigue, 71.1 percent of them believed that fatigue negatively affected their performance, as seen by a slowing of reaction time, decreased situation awareness, and poorer vehicle control. In addition, of those sampled by Friswell and Williamson (2008), 91.7 percent claimed that fatigue accounted for at least one potentially dangerous event, the most prevalent being near-miss incidents, nodding off at the wheel, running a red light, colliding with something or running off the road. There are various measures used which explore the relationship between the operational practice of commercial truck drivers and fatigue, which include subjective measures of fatigue response and objective measures of performance (Feyer & Williamson, 2001). According to Morrow and Crum (2004), subjective measures are superior to other approaches, suggesting that drivers can be aware of their fatigue states. Other subjective measures of fatigue have been distributed to truck drivers previously involved in crashes in order to obtain information on their sleep habits and duty history 72 hours prior to a crash (Gander et al., 2006). The results of the Gander et al. (2006) study suggest that duration of continued wakefulness, amount of sleep loss and time of day are common problems for those drivers involved in an accident, although Williamson et al. (2011) argue that self-report measures may be subject to increased response variability and recall bias. By contrast, Dingus et al. (2006) argue that empirical methods, which utilize test tracks and driving simulators, are not ideal as they may result in drivers modifying their behavior. Rather, they suggest using an incident-based, naturalistic data collection approach, which can be implemented by using in-vehicle technologies to secure data from commercial vehicle operation.

It appears that pre-trip levels of fatigue, which may stem from a lack of sleep the night before and physical work such as loading cargo, can potentially contribute to later fatigue (Feyer & Williamson, 2001; Friswell & Williamson, 2008). Feyer and Williamson (2001) found that drivers who did more work prior to driving, such as loading their own cargo, had significantly higher levels of fatigue throughout their trip. In addition, multiple studies have found a wide variety of factors that are potential contributors to an increase in fatigue and crash-related outcomes for commercial truck drivers. Morrow and Crum (2004) argue that work overload and schedule irregularity, which is incompatible with natural circadian rhythms, are leading contributors, while others suggest that fatigue-related problems may be linked to time of day and shift rotations (Guppy & Guppy, 2003; Lee, 2006). Furthermore, Friswell and Williamson (2008) claim that long driving hours and inadequate amounts of sleep before a trip were most commonly identified as fatigue contributors. Perhaps most importantly, it appears that the actual nature of commercial vehicle driving and working conditions are ill suited to meet the requirements of the natural human circadian cycle, which can contribute to reduced alertness in drivers (Fournier, Montreuil, & Brun, 2007).

It is essential that countermeasures be put in place in order to reduce the harmful effects of commercial vehicle driving and fatigue, as discussed by Williamson (Chapter 27, this volume). Current efforts may be limited by their primary focus of regulating driving hours only (Feyer & Williamson, 2001). By contrast, Guppy and Guppy (2003) argue that it would be most effective if schedules were created that supported break-taking and increased driver autonomy. Theoretical advances that differentiate qualitatively different forms of fatigue may guide more effective countermeasures (May & Baldwin, 2009).

CONCEPTUALIZING DRIVER FATIGUE

There is much evidence for the dangers of fatigue, but making progress in research requires a psychological conceptualization of fatigue. There are two interlinked issues here (Matthews, 2001; Matthews et al., 2011). The first is how to understand the impact on the driver's state of mind of the multiple external and personal factors that influence the fatigue state. There are certainly a number of potential causes of driver fatigue, each akin to a particular theoretical perspective. Some experts argue that cognitive psychological factors such as workload and attention are central to driver fatigue (Hancock & Verwey, 1997), while others stress physiological factors such as circadian rhythms, neurocognitive correlates of sleepiness (see Banks, Jackson & Van Dongen, Chapter 11, this volume), and shift work (Lal & Craig, 2001), although Williamson et al. (2011) claim that further research is needed to clarify these relationships. The second issue is how to determine which components of information-processing mediate the impact of fatigue states on driver performance and safety. Potentially, fatigue may produce a variety of changes in attention, psychomotor control and decision-making, which are difficult to differentiate in studies of real crashes. Driver simulator studies provide multiple performance metrics that may discriminate the effects of fatigue on different processing mechanisms (Baulk et al., 2008; Matthews, 2001; Matthews et al., 2011). The traditional approach in performance studies refers to arousal theory; under-arousal, induced from monotonous highway driving, for example, may result in loss of alertness, especially for professional drivers who may find themselves exposed to this particular type of environment for several hours (Thiffault & Bergeron, 2003). Williamson et al. (2011) highlight a term, "highway hypnosis," which is often used by professional drivers to describe a dangerous state of unconscious awareness that results from highly monotonous roadway environments. However, while the dangers of falling asleep are obvious, arousal theory may be inadequate to capture the full spectrum of changes in affect, motivation, and compensatory coping that are typical of the driver who is fatigued but wakeful (Hitchcock & Matthews, 2005).

In the "wakeful-but-tired" driver, performance impairments are sometimes attributed to depletion of some general attentional capacity or resource, resulting from prolonged high workload (Matthews, 2001; Matthews et al., 2011). Resource theories suggest that the driver has a limited supply or reserve of attentional processing capabilities and when task demands increase above this available quantity driver performance suffers, especially when the task is attentionally demanding (Desmond & Matthews, 1997). However, resource theory does not seem to apply straightforwardly to low workload driving scenarios, in which mental underload may lead to loss of situation awareness. For example, Young and Stanton (2007) cite several studies in which automated systems such as adaptive cruise control (ACC) reduced workload but also slowed driver response times. Such outcomes may be attributed to misconceived effort regulation: drivers attempt to adapt to changing workloads dynamically, but do not always allocate resources optimally (Hancock & Verwey, 1997). In a simulator study, Matthews and Desmond (2002) found that fatigue became more damaging to driver performance as workload decreased, contrary to prediction from resource theory. They argued that, in line with adaptive models of the stress and fatigue process (see Hancock, Desmond, & Matthews, Chapter 4, this volume), the fatigued driver might underestimate the need to apply effort when the task is easy (Desmond & Matthews, 1997). The attribution of performance deficit to a failure to regulate effort effectively is consistent with broader theoretical accounts of fatigue (Hancock, Desmond & Matthews, Chapter 4, this volume; Hockey, Chapter 3, this volume).

Active and Passive Fatigue

As has previously been mentioned, there may be multiple forms of fatigue, with differing implications for safety. Desmond and Hancock (2001) have proposed an account of fatigue that explains how different forms of driver task load may elicit different patterns of fatigue response, active and passive fatigue. Active fatigue may result from a need for prolonged effortful control of the vehicle (e.g., driving from New York to Boston in heavy traffic), while passive fatigue is typically triggered by low workload and monotonous driving tasks (e.g., driving across Kansas). While driving, active fatigue may ensue during conditions of overload, so that the driver has insufficient attentional capacity or resources to maintain effortful compensation (Hancock & Warm, 1989). Conversely, passive fatigue may then result during periods of underload, leading to withdrawal of effort from the task, as is evidenced by reduced small-magnitude steering movements in the Matthews and Desmond (2002) studies. May and Baldwin (2009) suggest that vehicle automation may be an effective countermeasure for active fatigue, whereas interactive technologies that increase the novelty and demand of the driving task may work better to counter passive fatigue.

In a simulator study, Saxby et al. (2007) differentiated the effects of active and passive fatigue. Active fatigue was induced through high workload wind gusts, which required the driver to actively steer and alternate acceleration changes, while passive fatigue was induced through required automation use and chronic under-stimulation. Their results indicated that drivers are primarily subject to symptoms of increased distress during active fatigue induction, but experience of the passive fatigue induction increased task disengagement and distractibility. Essentially, the findings of Saxby et al. (2007) confirm some of the advantages of a multidimensional approach to defining and assessing subjective fatigue states, as well as providing support for Desmond and Hancock's (2001) theory of active and passive fatigue, which suggests that driver responses can differ according to the nature of the fatigue induction.

A further study (Saxby et al., 2008) found that passive fatigue was associated with impaired alertness, but active fatigue was not. Drivers exposed to the monotony of full vehicle subsequently exhibited slowed braking and steering response times to a van unexpectedly pulling out in front of them. They were also more likely to crash into the van, compared with drivers in active fatigue and control (normal driving) conditions. Interestingly, passive fatigue had no effect on vehicle control as measured by standard deviation of lane position, demonstrating the need for multiple performance indices (Baulk et al., 2008). Use of a spectrum of both performance measures and self-report measures may help to identify the specific mechanisms for loss of safety.

A Transactional Perspective on Fatigue Effects

The crux of what we have argued so far is that there is no simple, direct link between fatigue in the wakeful driver and increased crash risk. The safety impacts of fatigue depend on the environmental factors that elicit fatigue, the nature of the fatigue state that develops, and the specific cognitive demands of the driving task. In addition, there are individual differences in fatigue vulnerability (Desmond & Matthews, 2009; Matthews, 2002). The complexities of driver fatigue may be understood in terms of the transactional model of stress and emotion (Lazarus, 1999; Szalma, Chapter 5, this volume), which attributes stress (broadly defined) as an interaction between person and environment factors. Cognitive processes, including appraisal of task demands, and choice of coping strategy, play a key role in mediating the interaction between external demands and stress response. In the case of fatigue, drivers may vary in whether they appraise long drives as potentially dangerous, and in the coping strategies they employ once they appraise themselves as experiencing fatigue. Maladaptive coping may contribute to loss of safety; recent studies of bus drivers have shown that strategies including emotion-focus (e.g., self-blame) and avoidance are related to performance impairment (Dorn et al., 2010; Machin & Hoare, 2008). Matthews (2002; Matthews & Desmond, 2001; Matthews et al., 2011) developed a transactional model of driver stress, based on Lazarus (1999), which suggests that different forms of driver stress, including anxiety, anger and fatigue, correspond to different patterns of appraisal and coping. These cognitive processes are shaped both by external factors, such as congestion or icy road surfaces, and by stable personality traits. The Driver Stress Inventory (DSI) (see Matthews et al., 1997) assesses five trait dimensions linked to different styles of driver cognition; Fatigue Proneness, Aggression, Dislike of Driving, Thrill Seeking, and Hazard Monitoring. The first three dispositions are characterized by feelings of tiredness, anger, and anxiety, respectively. Hazard Monitoring describes the driver's tendencies to search actively for dangerous events, whereas Thrill Seeking refers to enjoyment of danger. Fatigue-proneness is modestly, but significantly, associated with higher crash risk in bus drivers, as established from company records (Dorn et al., 2010), with various stress symptoms, and with a higher incidence of self-reported lapses and errors on the road (Rowden et al., 2011).

Appraisal and Coping Process in Driver Fatigue

A transactional account of fatigue requires a specification of how environmental factors that promote fatigue interact with personal factors to elicit appraisal and cognitive processes that generate the subjective and behavioral consequences of fatigue (Desmond & Matthews, 2009). Monotonous roadway situations are a major external influence on fatigue (Oron-Gilad & Hancock, 2005). Thiffault and Bergeron (2003) explored the effects of roadway type on driver fatigue and found that driver performance deteriorated and varied more during monotonous road environments, as compared to normal or high workload environments. The Fatigue Proneness scale of the DSI has been shown to predict state fatigue response in both simulator studies (Matthews & Desmond, 1998), and studies of truckers and nonprofessional drivers in real life (Desmond & Matthews, 2009). The latter study also found that fatigue proneness was associated with stress symptoms such as tension and worries about the driving task. Similarly, Stanton and Young (2005) argued that in very under-arousing or monotonous driving environments, drivers may find the lack of physical stimulation not only fatiguing but also stressful.

Thus, person and environment factors interact to affect driver fatigue (see Figure 23.1). For example, a late-night drive along an under-arousing, monotonous highway may be especially dangerous for the individual with a fatigue-prone personality. Fatigue is a function of the driver/environment relationship. In terms of the Desmond and Hancock (2001) model, monotony drives passive fatigue. Active fatigue may correspond to a different kind of driver/environment configuration, characterized by cognitive overload rather than underload. Figure 23.1 does not show the feedback processes that are an important element of the transactional model. For example, over the short term, the driver's awareness or metacognitions of their own fatigue symptoms may motivate coping efforts (adaptive or otherwise). Over longer time periods, the experience of fatigue might motivate use of public transport, buying a more comfortable vehicle or planning journeys to include more rest breaks.

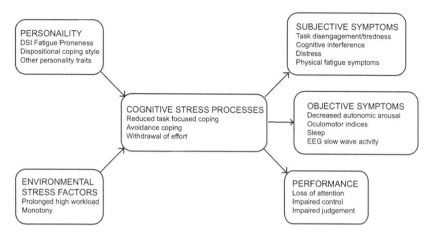

Figure 23.1   Fatigue as a cognitive process: an outline transactional model

Note: Feedback loops omitted.

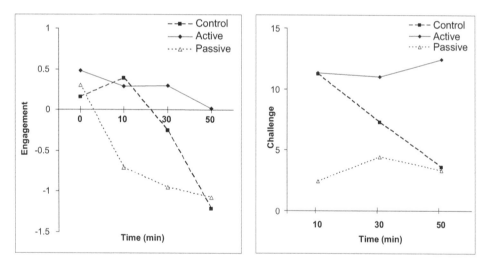

Figure 23.2   Subjective task engagement (left panel) and challenge appraisal (right panel) as a function of drive duration and fatigue manipulation in a study of simulated driving

The transactional model predicts that environmental and personal factors should interact to influence appraisal and coping. Figure 23.2 shows data from the Saxby et al. (2007) study previously described, which used three different driving durations. The left panel illustrates how subjective task engagement decreases most rapidly in a monotonous, passive fatigue condition (i.e., more rapid build-up of mental fatigue). The right panel shows temporal changes in challenge appraisal in each condition. The sharp decline in challenge seen in the passive fatigue condition corresponds to the loss of task engagement.

Matthews et al. (1997) found that DSI fatigue proneness was related to higher levels of avoidance coping, and also emotion-focused coping. Similar correlations were found in unpublished data from the Saxby et al. (2008) study that showed performance impairment in fatigue states. As discussed by Matthews, Desmond and Hitchcock (Chapter 9, this volume), evidence from other performance contexts also links mental fatigue (loss of task engagement) to a pattern of changes in cognitive processing, including lowered challenge appraisal and increased avoidance (as well as reduced task-focused coping). Additional associations between fatigue-proneness and stress and emotion-focused coping may be attributed to the driver's increasing awareness that fatigue constitutes a threat to personal safety (Desmond & Matthews, 2009).

Thus, consistent with more general analyses of fatigue (Hancock, Desmond & Matthews, Chapter 4, this volume; Hockey, Chapter 3, this volume), fatigue may elicit avoidance coping and withdrawal of effort from the task. Simulator data suggest that the patterns of coping characteristic of fatigue contribute to the loss of safety observed in real driving (Lee, 2006). As has previously been mentioned, Matthews and Desmond (2002) found that, following a fatigue induction, driver performance was impaired on straight rather than curved road sections (i.e., in underload conditions). The appraisal that there is little need for effort in following a straight road (with little traffic) may encourage avoidance coping and withdrawal of task-directed effort, leading to loss of alertness (Saxby et al., 2008).

Countermeasures to Driver Fatigue: Transactional Perspectives

We will discuss countermeasures for driver fatigue only briefly as they are reviewed in other chapters (Job, Graham, Sakashita & Hatfield, Chapter 22, this volume; Williamson, Chapter 27, this volume). Popular measures aimed to combat driver fatigue include socially accepted stimulants such as caffeinated beverages, alternating tasks, taking a rest break or nap, and psychophysiological monitoring for sleepiness or loss of alertness. We will focus on those issues that relate directly to the transactional perspective on driver fatigue. The transactional model implies that countermeasures may be directed toward environmental influences, driver characteristics, and the appraisal and coping processes that mediate fatigue responses. It is often difficult to change the physical environment to reduce monotony. However, it may be possible to design in-vehicle systems to alleviate boredom, by enhancing the driver's interest and sense of challenge. Oron-Gilad, Ronen and Shinar (2008) showed that an alertness-maintaining task (answering trivia questions) enhanced simulator performance in monotonous conditions. Another example relates to the monotony associated with driving the long road tunnels found in some European nations. Vashitz, Shinar and Blum (2008) used a simulation of tunnel driving to show that providing a highly informative display reduced boredom without substantially increasing distraction. Stanton and Young (2005) also highlight the importance of good design of in-vehicle systems that maintain driver situation awareness.

It may be more difficult to address stable personality characteristics associated with vulnerability to fatigue. Commercial organizations might usefully employ scales such as the DSI (Matthews et al., 1997) in selecting professional drivers who are resilient to fatigue. Interestingly, the professional sample in the Desmond and Matthews (2009) study obtained low fatigue proneness scores compared to norms for the DSI, suggesting that some formal or informal selection for fatigue vulnerability had already taken place. It may also be possible to train enhanced coping skills in vulnerable individuals. Machin (2003; Machin & Hoare, 2008) has advocated training adaptive coping with fatigue in bus drivers. Training is personalized by presenting drivers with specific work-related scenarios, and having them evaluate the likely outcomes. Training may be directed not only toward vehicle operation, but also toward strategies for recovering from fatigue. Driving simulators may also be used to generate scenarios that test the driver's coping skills (Dorn et al., 2010). As Machin and Hoare (2008) also state, it is important that person-focused approaches are complemented by changes in organizational practices to mitigate fatigue.

The transactional perspective may also be useful for understanding drivers' reactions to in-vehicle technologies. As Williamson (Chapter 27, this volume) discusses, fatigue warning devices, such as those that monitor eye movements, potentially provide useful information to drivers regarding their level of alertness. In transactional terms, such devices are intended to generate an appraisal of personal danger, which elicits a coping response of ceasing to drive. However, the fatigue state may impair drivers' abilities to assess their own feelings of fatigue during driving (Brown, 1997). Faced with time and job pressures, long-haul truck drivers may consciously choose to persist, unsafely, with their journey. The fatigued driver may not be able to gauge exactly how tired they really are until it is too late. Brown (1997) suggests that fatigued drivers may run the risk of continuing to drive until they are consciously aware of lane drifting or even by experiencing near collisions. To the extent that judgment is impaired in fatigue states (Brown, 1997), the driver may be inclined to ignore the warning (i.e., a form of avoidance coping). Again, drivers may need to be trained to use the warning as an aid to effective coping.

Another relevant technology is vehicle automation. With emerging technology such as Adaptive Intelligent Cruise Control and Automated Highway Systems, the role of the driver shifts from one of active monitor to passive observer. These advances are designed to help fatigued drivers by potentially relieving some of the task demand placed upon them but, in actuality, automation may elicit poorer performance and control by reducing task engagement, increasing driver drowsiness (Hancock & Verwey, 1997), and decreasing situation awareness (Stanton & Young, 2005). Automated systems may cause the driver to feel less in control, thereby reducing driver trust in the vehicle as well as decreasing situation awareness. Desmond, Hancock and Monette (1998) confirmed that prolonged automation use may decrease situation awareness. They suggest that fatigued drivers are more likely to under-mobilize effort when faced with low task demands, suggesting that partial automation, which requires the driver to remain somewhat involved in the driving experience, is superior to full automation. Consequently, the interaction between the fatigued driver and automated systems may significantly reduce drivers' perceptions and regulation of task demands during driving (Desmond, et al., 1998). Findings from our laboratory confirm that full automation has stronger effects on passive fatigue symptoms than partial automation (Funke et al., 2007; Saxby et al., 2007, 2008).

The studies just cited suggest that exposure to full automation produces passive fatigue. A complementary question is how the driver's awareness of fatigue influences decisions to cope with fatigue by initiating automation. No doubt it may be tempting to the exhausted driver to let the vehicle do the work. In a recent study (Neubauer et al., 2011), we explored the relationship between fatigue and stress states and voluntary use of automation. Drivers who were more fatigued at the start of a monotonous drive were more likely to choose to use automation. However, automation failed to reduce fatigue, and, indeed, tended to elevate stress, so that the strategy seemed counter-productive.

Consistent with the transactional model, drivers high in fatigue-proneness seemed more vulnerable to these effects. Thus, although vehicle automation is attractive as a method for reducing driver workload and increasing safety, it appears to interfere with the driver's ability to cope adaptively with fatigue, especially when the driver transitions from automated to normal driving (Saxby et al., 2007, 2008).

CONCLUSIONS

Ours is a culture where driving is essential, whereby time and job constraints continue to place more and more demands on today's driver. As these demands increase, which they surely will in the future, both professional and non-professional drivers will face a flux of fatigue-related issues. Fatigue effects are dangerous not only for the driver but also for others on the road. There are certain issues researchers need to be concerned with. First, researchers need to understand how to appropriately characterize the fatigued driver. Within the driving context, fatigue can be assessed by a range of physical, affective, and motivational symptoms, which can impair a range of performance tasks (Williamson, Feyer & Friswell, 1996; Hitchcock & Matthews, 2005). Fatigue is typically experienced most markedly by individuals such as professional long-distance truck drivers. As a result of inconsistent work schedules and time pressures, the professional driver may face an increase in sleep debt, which may raise their risk for experiencing a significant reduction in attentional capabilities or falling asleep at the wheel (Nguyen, Jauregui & Dinges, 2001).In addition, it appears that the driving environment in which commercial vehicle drivers are engaged can negatively affect driver fatigue. Stanton and Young (2005) argue that in very under-arousing or monotonous driving environments, drivers may find the lack of physical stimulation not only fatiguing but also stressful. Driver fatigue may also be determined by person factors, whereby the very fatigued driver may engage in a reduction of active, task-focused coping (Matthews, 2002; Desmond & Matthews, 2009). The transactional model conceptualizes driver fatigue as a function of an environment–person interaction, whereby prolonged monotony interacts with personal vulnerabilities to elicit cognitive processes, including avoidance coping, which may lead to down-regulation of effort, especially in underload conditions. Although under-arousal and overload of attentional capacity may play a role in fatigue effects, it appears that adequate effort matching is the primary key to understanding driver fatigue.

The transactional approach highlights the need to ensure that driver appraisals and cognitions maintain task engagement during drives of extended duration, an issue that may require both design solutions and driver training. It also appears that recent technological advances such as automated systems may accentuate driver fatigue. Such systems aim to remove the task load placed upon the driver, but in actuality their use may result in the driver under-mobilizing effort (Saxby et al., 2007). Such systems should be designed to keep the driver involved in the driving task without loss of situation awareness.

REFERENCES

Baulk, S.D., Biggs, S.N., Reid, K.J., van den Heuvel, C.J., & Dawson, D. (2008). Chasing the silver bullet: Measuring driver fatigue using simple and complex tasks. *Accident Analysis and Prevention*, 40, 396–402.

Brown, I.D. (1994). Driver fatigue. *Human Factors*, 17, 298–314.

Brown, I.D. (1997). Prospects for technological countermeasures against driver fatigue. *Accident Analysis and Prevention*, 29, 525–31.

Desmond, P.A. & Hancock, P.A. (2001). Active and passive fatigue states, in P.A. Hancock & P.A. Desmond (eds), *Stress, Workload, and Fatigue*. Mahwah, NJ: Lawrence Erlbaum, 455–65.

Desmond, P.A. & Matthews, G. (1997). Implications of task-induced fatigue effects for in-vehicle countermeasures to driver fatigue. *Accident Analysis and Prevention*, 29, 515–23.

Desmond, P.A. & Matthews, G. (2009). Individual differences in stress and fatigue in two field studies of driving. *Transportation Research. Part F, Traffic Psychology and Behaviour*, 12, 265–76.

Desmond, P.A., Hancock, P.A. & Monette, J.L. (1998). Fatigue and automation-induced impairments in simulated driving performance. *Transportation Research Record: Journal of the Transportation Research Board*, 1628, 8–14.

Dingus, T.A., Neale, V.L., Klauer, S.G., Peterson, A.D. & Carroll, R.J. (2006). The development of a naturalistic data collection system to perform critical incident analysis: An investigation of safety and fatigue issues in long-haul trucking. *Accident Analysis and Prevention*, 38, 1127–36.

Dorn, L., Stephen, L., af Wåhlberg, A. & Gandolfi, J. (2010). Development and validation of a self-report measure of bus driver behaviour. *Ergonomics*, 53, 1420–33.

Driver Health and Wellness (2011). Online. Available at: http://www.drivershealth.com/info/health/managing_stress.jsp?usertrack.filter_applied=true&NovaId=2935376876913518703 (retrieved January 22, 2011).

Feyer, A.M. & Williamson, A.M. (2001). Broadening our view of effective solutions to commercial driver fatigue, in P.A. Hancock & P.A. Desmond (eds), *Stress, Workload, and Fatigue*. Mahwah, NJ: Lawrence Erlbaum, 550–65.

Fletcher, A., McCulloch, K., Baulk, S.D. & Dawson, D. (2005). Countermeasures to driver fatigue: A review of public awareness campaigns and legal approaches. *Australian and New Zealand Journal of Public Health*, 29, 471–6.

Fournier, P.S., Montreuil, S. & Brun, J.P. (2007). Fatigue management by truck drivers in real life situations: Some suggestions to improve training. *Work: A Journal of Prevention, Assessment, and Rehabilitation*, 29, 213–24.

Friswell, R. & Williamson, A. (2008). Exploratory study of fatigue in light and short haul transport drivers in NSW, Australia. *Accident Analysis and Prevention*, 40, 410–17.

Funke, G.J., Matthews, G., Warm, J.S. & Emo, A. (2007). Vehicle automation: A remedy for driver stress? *Ergonomics*, 50, 1302–23.

Gander, P.H., Marshall, N.S., James, I. & Le Quesne, L. (2006). Investigating driver fatigue in truck crashes: Trial of a systematic methodology. *Transportation Research*, Part F, 9, 65–76.

Guppy, J.A. & Guppy, A. (2003). Truck driver fatigue risk assessment and management: A multinational survey. *Ergonomics*, 46, 763–79.

Hancock, P.A. & Verwey, W.B. (1997). Fatigue, workload and adaptive driver systems. *Accident Analysis and Prevention*, 29, 495–506.

Hancock, P.A. & Warm, J.S. (1989). A dynamic model of stress and sustained attention. *Human Factors*, 31, 519–37.

Hitchcock, E.M. & Matthews, G. (2005). Multidimensional assessment of fatigue: A review and recommendations. Proceedings of the International Conference on Fatigue Management in Transportation Operations, Seattle, WA, September.

Lal, S.K.L. & Craig, A. (2001). A critical review of the psychophysiology of driver fatigue. *Biological Psychology*, 55, 173–94.

Lazarus, R.S. (1999). *Stress and Emotion: A New Synthesis*. New York: Springer.

Lee, J.D. (2006). Driving safety, in R.S. Nickerson (ed.), *Reviews of Human Factors and Ergonomics*. Santa Monica, CA: Human Factors and Ergonomics Society, 172–218.

Machin, M.A. (2003). Evaluating a fatigue management training program for coach drivers, in L. Dorn (ed.), *Driver Behaviour and Training*. Aldershot, UK: Ashgate, 75–83.

Machin, M.A. & Hoare, N.P. (2008). The role of workload and driver coping styles in predicting bus drivers' need for recovery, positive and negative affect, and physical symptoms. *Anxiety, Stress & Coping: An International Journal*, 21, 359–75.

Matthews, G. (2001). A transactional model of driver stress, in P.A. Hancock & P.A. Desmond (eds), *Stress, Workload and Fatigue*. Mahwah, NJ: Lawrence Erlbaum, 133–63.

Matthews, G. (2002). Towards a transactional ergonomics for driver stress and fatigue. *Theoretical Issues in Ergonomics Science*, 3, 195–211.

Matthews, G. & Desmond, P.A. (1998). Personality and multiple dimensions of task-induced fatigue: A study of simulated driving. *Personality and Individual Differences*, 25, 443–58.

Matthews, G. & Desmond, P.A. (2001). Stress and driving performance: Implications for design and training, in P.A. Hancock & P.A Desmond (eds), *Stress, Workload, and Fatigue*. Mahwah, NJ: Lawrence Erlbaum, 211–31.

Matthews, G. & Desmond, P.A. (2002). Task-induced fatigue states and simulated driving performance. *The Quarterly Journal of Experimental Psychology*, 55A, 659–86.

Matthews, G., Desmond, P.A., Joyner, L.A. & Carcary, B. (1997). A comprehensive questionnaire measure of driver stress and affect, in E. Carbonell Vaya & J.A. Rothengatter (eds), *Traffic and Transport Psychology: Theory and Application*. Amsterdam: Pergamon, 317–24.

Matthews, G., Saxby, D.J., Funke, G.J., Emo, A.K. & Desmond, P.A. (2011). Driving in states of fatigue or stress, in D. Fisher, M. Rizzo, J. Caird & J. Lee (eds), *Handbook of Driving Simulation for Engineering, Medicine and Psychology*. Boca Raton, FL: CRC Press/Taylor and Francis, 29-1–29-10.

May, J.F. & Baldwin, C.L. (2009). Driver fatigue: The importance of identifying causal factors of fatigue when considering detection and countermeasure technologies. *Transportation Research, Part F, Traffic Psychology and Behaviour*, 12, 218–24.

Morrow, P.C. & Crum, M.R. (2004). Antecedents of fatigue, close calls, and crashes among commercial motor-vehicle drivers. *Journal of Safety Research*, 35, 59–69.

Neubauer, C., Langheim, L.K., Matthews, G. & Saxby, D.J. (2011). Fatigue and voluntary utilization of automation in simulated driving. *Human Factors*. DOI: 10.1177/0018720811423261.

Nguyen, L.T., Jauregui, B. & Dinges, D.F. (2001). Changing behaviors to prevent drowsy driving and promote traffic safety: Review of proven, promising, and unproven techniques. AAA Foundation for Traffic Safety, July. Available at: www.aaafts.org/text/research/drowsyfinalreport.htm (retrieved September 23, 2010).

NHTSA Expert Panel on Driver Fatigue and Sleepiness (1997). Drowsy driving and automobile crashes. Available at: http://www.nhtsa.gov/people/injury/drowsy_driving1/Drowsy.html (retrieved September 23, 2010).

Oron-Gilad, T. & Hancock, P.A. (2005). Road environment and driver fatigue, in *Proceedings of the Third Driving Assessment Symposium*, 318–24.

Oron-Gilad, T., Ronen, A. & Shinar, D. (2008). Alertness maintaining tasks (AMTs) while driving. *Accident Analysis and Prevention*, 40, 851–60.

Pratt, S.G. (2003). Work-related roadway crashes: Challenges and opportunities for prevention. Technical report: NIOSH Hazard Review.

Rowden, P., Matthews, G., Watson, B. & Briggs, H. (2011). The relative impact of occupational stress, life stress, and driving environment stress on driving outcomes. *Accident Analysis and Prevention*, 44, 1332–40.

Saxby, D.J., Matthews, G., Hitchcock, E.M. & Warm, J.S. (2007). Development of active and passive fatigue manipulations using a driving simulator. *Proceedings of the Human Factors and Ergonomics Society*, 51, 1237–41.

Saxby, D.J., Matthews, G., Hitchcock, E.M., Warm, J.S., Funke, G.J. & Gantzer, T. (2008). Effects of active and passive fatigue on performance using a driving simulator. *Proceedings of the Human Factors and Ergonomics Society*, 52, 1252–6.

Stanton, N. & Young, M.S. (2005). Driver behaviour with adaptive cruise control. *Ergonomics*, 48, 1294–1313.

Thiffault, P. & Bergeron, J. (2003). Monotony of road environment and driver fatigue: A simulator study. *Accident Analysis and Prevention*, 35, 381–91.

Vashitz, G., Shinar, D. & Blum, Y. (2008). In-vehicle information systems to improve traffic safety in road tunnels. *Transportation Research*, Part F: *Traffic Psychology and Behaviour*, 11, 61–74.

Wijesuriya, N., Tran, Y. & Craig, A. (2007). The psychophysiological determinants of fatigue. *International Journal of Psychophysiology*, 63, 77–86.

Williamson, A.M., Feyer, A.M. & Friswell, R. (1996). The impact of work practices on fatigue in long distance truck drivers. *Accident Analysis and Prevention*, 28, 709–19.

Williamson, A., Lombardi, D.A., Folkard, S., Stutts, J., Courtney, T.K. & Connor, J.L. (2011). The link between fatigue and safety. *Accident Analysis and Prevention*, 43, 498–515.

Young, M.S. & Stanton, N.A. (2007). What's skill got to do with it? Vehicle automation and driver mental workload. *Ergonomics*, 50, 1324–39.

24

Understanding and Managing Fatigue in Aviation

John A. Caldwell

INTRODUCTION

Recent events in the U.S. aviation sector have highlighted the problem of fatigue in flight operations while making it clear that the time has come for implementation of new fatigue-management strategies and policies. In 2004, Corporate Airlines flight 5966 crashed on approach to Kirksville, MO, because the fatigued pilots were on their sixth flight of the day after having been on duty for 14 hours. These veteran pilots ignored published procedures, failed to respond to ground proximity alerts, and crashed into trees after losing situational awareness. In February of 2008, the pilots of Go! Airlines flight 1002 fell asleep on the flight deck during the short trip from Honolulu to Hilo, HI (a trip of only 150 miles and 50 minutes of flight time) and overshot their destination by more than 30 miles. In October of 2009, a similar event occurred when the sleepy pilots of Northwest Airlines flight 188 remained unresponsive to air traffic control communications for almost 90 minutes and overflew their Minneapolis destination by 150 miles because they evidently had dozed off at the controls. In recent times, stories about fatigue and sleepiness on the flight deck have made the headlines in numerous major media outlets and garnered significant public attention. However, the pivotal event that made pilot fatigue a priority problem for airlines and government regulators was the 2009 crash of a Continental Connection flight near Buffalo, NY, in which 50 people were killed. During the crash investigation, the National Transportation Safety Board discovered that prior to the fateful flight one of the pilots had been awake all night and the other pilot reported for duty following a lengthy commute that terminated in a likely inadequate sleep period in the crew lounge of Newark Liberty airport. Based on this and other information, the NTSB ultimately concluded that "the pilots' performance was likely impaired because of fatigue," and this finding led the Federal Aviation Administration to charter an Aviation Rulemaking Committee to update the flight-duty limitations for commercial pilots. A primary charge to the ARC was to recommend modifications to flight-duty regulations, which would bring them more in line with our current scientific understanding of the causes and consequences of human fatigue. Fortunately, the new U.S. airline flight and duty regulations released in January of 2012 have in fact broadened the focus of flight-time limitations to include an emphasis on the sleep/wake factors that obviously – or fundamentally – are at the root of the fatigue problem; nevertheless, updated guidance on select fatigue countermeasures remains to be seen.

PHYSIOLOGICAL FATIGUE FACTORS

Years of scientific study indicate that fatigue is not a one-dimensional phenomenon limited to singular variables such as hours on duty, but rather the product of several factors related to physiological sleep needs and internal biological rhythms (Caldwell, Caldwell, & Schmidt, 2008; Banks, Jackson & Van Dongen, Chapter 11, this volume; Waterhouse, Chapter 16, this volume). These factors exert an interactive effect on alertness and performance, and both must be taken into account when considering the impact of crew scheduling on flight safety.

Sleep Needs

The average adult needs 7–9 hours of sleep every day in order to function most effectively (Balkin et al., 2008; Roth, 2004), and failing to fulfill this quota rapidly leads to cognitive impairments from which recovery is slower and more difficult than previously thought/ believed. Periods of moderate or prolonged sleep loss produce dangerous decrements in alertness and effectiveness, and recent evidence suggests that people who consistently sleep insufficiently suffer from an increased rate of on-duty performance deterioration (Cohen & Ernst, 2010). Such findings help explain Goode's report (2003) that flights longer than 10 continuous hours are at significantly greater risk for fatigue-related mishaps, as well as a conclusion from an NTSB study of major accidents in domestic air carriers from 1978 through 1990 that "Crews comprising captains and first officers whose time since awakening was above the median for their crew position made more errors overall, and significantly more procedural and tactical decision errors" (National Transportation Safety Board, 1994: 75).

Biological Rhythms

The fatiguing effects of duty periods that extend into the subjective night are in part related to the biological propensity toward sleepiness and inactivity at night and arousal and activity during the day. Dinges et al. (1990) found that vigilance lapses during long-haul flights were approximately five times greater during the night than during the day, and that vigilance deteriorations associated with time on-task were steeper during night flights than during day flights. Such results are congruent with the Lyman & Orlady (1981) finding that the majority of fatigue-related flight incidents in one sample of NASA's Aviation Safety Reports System occurred between midnight and 06.00 in the morning; the Klein et al. (1970) report that simulator flight performance at 04.00 in the morning was 75–100 percent below the level seen at 15.00 in the afternoon; and the de Mello et al. (2008) finding that the risk of flight crew operational errors was 50 percent higher in the pre-dawn hours than later in the morning daylight hours. Taken together, these reports highlight the importance of the natural 24-hour rhythm for alertness and performance and this, combined with the fact that human internal clocks are highly resistant to change

(Monk, 1994) helps explain fatigue-related problems associated with inconsistent duty times and rapid/frequent time-zone crossings.

The Combined Impact of the Two Primary Fatigue Factors

The combined impact of the sleep (homeostatic) process and the circadian process on both the timing/quality of sleep and the time course of waking vigilance (or, conversely, waking drowsiness) has been well established (Achermann, 2004). In personnel who work during the day and sleep during the night, the sleep and circadian processes work together to create a relatively constant level of alertness and performance during the day, and a reasonable sleep quality during the night. Unfortunately, personnel who are chronically sleep-restricted or who are required to remain awake for extended duty periods are often compromised due to the elevated sleep pressure (and increased drowsiness or cognitive instability) that results. This problem is further compounded by a requirement to operate during times when circadian-driven sleep pressure also is high. Whenever low homeostatically driven arousal and low circadian-driven arousal converge, alertness, performance, and safety are seriously compromised.

RECOMMENDATIONS FOR COUNTERING FATIGUE IN AVIATION

The conflict between operational scheduling demands and human physiological makeup is at the heart of fatigue-related problems in aviation. Humans simply were not designed to operate effectively on the ever-changing 24/7 schedules that often define today's flight operations. Because of this, it is unlikely that aircrew fatigue will ever be completely *eliminated*; however, it can be *effectively managed* with a well- planned, science-based, fatigue-management strategy.

Optimization of Crew Scheduling Procedures

Since scheduling factors are often cited as the number one contributor to pilot fatigue, the development and implementation of more "human-centered" work routines should be considered paramount for promoting on-the-job alertness. Fortunately, crew-scheduling practices in aviation have very recently begun to incorporate the advanced knowledge of fatigue, sleep, and circadian rhythms that has been gained over the past 20 years (see Tucker & Folkard, Chapter 28, this volume). The most recently-released flight-duty-period regulations from the Federal Aviation Administration have begun to highlight the importance of scheduling crews in a manner which recognizes: 1) that adequate daily sleep is essential for the maintenance of safe and effective on-duty functioning; 2) that circadian rhythms affect on-the-job alertness and performance as well as off-duty sleep quality and duration; and 3) that these physiological realities cannot be overcome by training, professionalism, motivation, or increased monetary incentives. Although it is unrealistic to expect airlines to operate on schedules that are perfectly suited to human physiological ideals, it seems reasonable for airlines to optimize scheduling practices through the use of tools now available for routine application. These tools can predict the impact of scheduling factors on performance and fatigue risk by tracking the sleep of personnel working the schedules and then processing these sleep data through a validated performance-readiness/fatigue-risk prediction model. The utility of this strategy lies in the facts that: 1) fatigue is known to stem primarily from insufficient and/or disrupted sleep in combination with circadian factors (Caldwell, Caldwell & Schmidt, 2008); 2) the fatigue impact of different sleep/wake schedules and different levels of sleep quality/quantity has been well established by over 100 years of scientific research (Balkin et al., 2008; Roth, 2004); 3) sleep and circadian information can be processed through computerized mathematical models which can effectively translate this information into predictions of alertness, performance, and fatigue risk (Van Dongen, 2004; Hursh et al., 2004; French & Neville, Chapter 29, this volume; Van Dongen & Belenky, Chapter 30, this volume); and4) wrist actigraphy is able to accurately and unobtrusively track individual sleep information and aspects of circadian rhythms to be used as inputs for fatigue modeling (Sadeh & Acebo, 2002). One example of an integrated scheduling optimization approach is the Fatigue Science program in which the Sleep, Activity, Fatigue, and Task Effectiveness (SAFTE) model is used to examine the fatigue impact of proposed multi-day crew schedules. This is further augmented by a program in which a combination of ReadiBand wrist-worn actigraphs and the SAFTE model are used to determine the fatiguing effects of schedules once crew members are actually working those schedules. The first strategy is to simply examine the duty schedules and *estimated* sleep information while the second strategy involves the actual *measurement* of sleep for input into the validated fatigue-prediction model.

Fatigue modeling
The SAFTE model (Hursh et al., 2004; Van Dongen & Belenky, Chapter 30, this volume) mathematically simulates the primary physiological processes that determine the level of fatigue (i.e., performance effectiveness) at any given point in time. The SAFTE model has been independently validated as the most accurate model predictor of sleep restriction on subjective fatigue ratings and objectively measured performance (Van Dongen, 2004). In addition, the SAFTE model has been shown to accurately predict the impact of sleep and scheduling factors on human factors accident risk (Hursh et al., 2006).

Wrist actigraphy
One disadvantage of simply examining proposed duty schedules is that the sleep expected to be gained by personnel must be estimated rather than actually measured. And, of course, the accuracy of these estimations directly impacts the accuracy of fatigue-risk calculations. However, direct, empirical measurements of sleep and sleep/wake timing can be obtained via wrist actigraphs (Morgenthaler et al., 2007) – an example being the Fatigue Science ReadiBand (see Figure 24.1). The accuracy of ReadiBand sleep/wake classifications have been verified in a study of 50 patients undergoing polysomnographic evaluation, with results indicating 92 percent accuracy in comparison to gold-standard polysomnography. Thus, it is possible to accurately measure bedtimes, wake-up times, and sleep times for each 24-hour period, so that once modeled, an accurate description of fatigue risk can be calculated.

Schedule optimization
Assessing proposed aircrew schedules with validated fatigue modeling and assessing potentially problematic schedules via a combination of wrist actigraphy and fatigue

Figure 24.1   Fatigue Science ReadiBand is one type of wrist-worn actigraph that can quantify sleep quality and quantity as well as sleep/wake timing in operational contexts

modeling would permit airlines to eliminate or modify the schedules most likely to severely compromise crew alertness and performance. Schedule pass/fail fatigue criteria can be established based on information obtained from laboratory studies and evaluations that other industries have conducted, or they could be established via comparisons against operational measures such as the exceedance data that already is routinely collected as part of Flight Operations Quality Assurance (FOQA) efforts. Another alternative would be comparing model outputs to the results of pilot self-reported operational fatigue events available from Aviation Safety Action Programs (ASAP). Such comparisons will lead to empirically established criteria by which proposed multi-day crew schedules can be assessed from a fatigue standpoint.

In-flight Counter-fatigue Strategies

The optimization of crew scheduling procedures via fatigue modeling, wrist actigraphy, and better use of operational exceedance data will no doubt enhance fatigue-management efforts in aviation operations; however, the fact remains that the non-standard work/sleep schedules and rapid/frequent time-zone crossings that are integral parts of airline operations will continue to challenge the adaptability of flight crews. Thus, schedule-optimization strategies must be accompanied by the implementation of effective in-flight fatigue countermeasures, as well as effective pre-flight interventions.

On-board sleep
Ensuring adequate on-board, out-of-cockpit bunk sleep is one of the most important in-flight countermeasures that can be implemented to address sleep loss and circadian

disruption during extended aviation operations (Flight Safety Foundation, 2005). In-flight bunk sleep periods sometimes can be difficult to properly schedule, since they need to be arranged in consideration of operational demands and yet still be of a duration that allows all crew members to receive ample rest periods to ensure the safety of the flight. When scheduling on-board sleep, it should be kept in mind that the body is programmed for two periods of sleepiness throughout the day, with the maximum sleep propensity occurring during the early morning hours (in the latter half of the habitual sleep episode) and a second period of increased sleep propensity occurring in mid-afternoon (Dinges, 1986; Van Dongen & Dinges, 2005). Crews can estimate the times during a flight at which there is an increased risk of inadvertent sleepiness, and consider these the best times (from a circadian standpoint) to schedule in-flight bunk sleep opportunities. If flight demands permit, utilizing these periods of increased physiological sleep propensity will help to increase both the quantity and quality of bunk sleep, subsequently reducing physiological sleepiness across the remainder of the flight (Carskadon, 1989; Dinges, 1984; Dijk & Franken, 2005). If it is not feasible to account for such circadian considerations, it may be difficult for the crew member to initiate or maintain on-board sleep, especially if the bunk-sleep timing places the sleep period within a few hours prior to the individual's normal habitual bedtime, since the circadian propensity for wakefulness is strong at that point (Lavie, 1986).

Cockpit naps

A strategy related to out-of-cockpit bunk sleep is the in-seat cockpit nap. From non-aviation studies, there is an abundance of evidence that a nap taken during long periods of otherwise continuous wakefulness is extremely beneficial for promoting subsequent alertness (Bonnet, 1990; 1991; Dinges, Whitehouse et al., 1988; Matsumoto & Harada, 1994; Rogers et al., 1989; Rosa, 1993; Vgontzas, et al., 2007; Webb, 1987). A direct examination of the effectiveness of a 40-minute cockpit nap opportunity, resulting in an average of 26 minutes asleep, revealed significant improvements in subsequent pilot physiological alertness and psychomotor performance in the aviation context (Rosekind et al., 1994). In the nap group, reaction times were faster and lapses (failures to respond) were fewer than in the non-nap group, particularly during the last 90 minutes of flight. In fact, from top of descent to landing, the naps virtually eliminated inadvertent and uncontrollable drowsiness episodes. Many international airlines now utilize cockpit napping on long flights, and cockpit napping is sometimes authorized for U.S. military flight operations as well. However, cockpit napping has not yet been approved for U.S commercial aviators, despite the fact that 86 percent of the general public who were surveyed in a National Sleep Foundation poll agreed that "an airline pilot who becomes drowsy while flying should be allowed to take a nap if another qualified pilot is awake and can take over during the nap" (National Sleep Foundation, 2002).

Controlled rest breaks

Tasks requiring sustained attention, such as monitoring aircraft systems and flight progress, can be especially challenging for already fatigued personnel (Dinges & Powell, 1988). This is in part why pilots often implement some type of work break strategy (chatting, standing up, walking around, etc.) to help sustain alertness during lengthy flights. Neri et al. (2002) found that simply offering a 10-minute hourly break during a 6-hour simulated night flight significantly reduced slow eye movements, theta-band EEG activity, unintended sleep episodes, and subjective sleepiness ratings. Although positive benefits were transient (15–20 minutes), they were noteworthy and particularly evident near the time of the circadian trough. All of the reasons why rest breaks are helpful are a matter for debate, but it may be that their effectiveness is partially attributable to physiological factors as well as to the temporary relief of mental boredom or physical discomfort. Caldwell, Prazinko & Caldwell (2003) found that simply assuming a standing posture, as opposed to remaining seated, reduced the amount of slow-wave EEG activity and enhanced performance on a 10-minute vigilance task during the later part of a 28-hour sleep-deprivation cycle. Thus, taken together with the results from Neri et al. (2002), it appears that periodic breaks involving nothing more than simply leaving the flight deck and conversing with other crew members during long-duration flights can help to sustain alertness in the cockpit.

Caffeine

For promoting alertness, caffeine is a valuable pharmacological fatigue countermeasure. Numerous studies have shown that caffeine increases vigilance and improves performance in sleep-deprived individuals, especially those who normally do not consume high doses (Nehlig, 1999). Caffeine in the form of coffee, tea, or soft drinks is widely used throughout the adult population, and it is both safe and effective in daily doses of up to 800 mg (Committee on Military Nutrition Research, 2001). Caffeine is most effective when it is used intermittently for the short-term elevation of cortical arousal, whereas chronic caffeine ingestion may lead to tolerance and various undesirable side effects such as elevated blood pressure, stomach problems, nervousness, irritability, dehydration, and insomnia. A caffeine dose of 100–200 mg noticeably affects the nervous system within 15 to 20 minutes after consumption, enhancing alertness for 4 to 5 hours. Personnel should routinely consume caffeine sparingly and save the arousal effect until they really need it. For instance, pilots who are flying primarily daytime schedules and who are getting sufficient sleep every night should reduce or eliminate caffeine consumption until they find themselves on more fatiguing schedules. This will ensure that caffeine can exert the most beneficial effects when these effects are truly needed.

Hydration and nutrition

It is commonly believed that the maintenance of adequate hydration will help to stave off cognitive and physical fatigue, but at least one recent study found that while 24 hours without water increased subjective tiredness, drowsiness, perceived effort, and concentration, cognitive–motor or neurophysiological functioning were not degraded (Szinnai et al., 2005). Nonetheless, aircrew personnel should strive to ensure adequate hydration throughout flights if for no other reason than to maintain a higher level of motivation and a more positive self-appraisal of performance capabilities. As for the role of nutrition, it is likely important as well. Two recent studies indicate that meals with high fat content reduce alertness and vigilance, and thus suggest it may be worthwhile to emphasize the ingestion of higher protein meals immediately prior to flights and during flights. In addition, only "light" meals should be consumed, since Lieberman (2003) has reported that while the precise macronutrient content of meals is apparently of little importance, large meals tend to be more soporific than smaller ones.

PRE-FLIGHT COUNTER-FATIGUE STRATEGIES

In the previous section, on-duty techniques for mitigating the impact of air crew fatigue were briefly reviewed. All of these (with the exception of hydration and nutrition) have been scientifically proven to at least temporarily counter the fatigue associated with arduous flight schedules. However, there are pre-flight strategies that are important as well. These strategies focus on maximizing the benefits of every available sleep opportunity while minimizing as much circadian disruption as possible before reporting for duty.

Education
Education about the importance of sleep and proper sleep hygiene, the dangers and signs of fatigue, and the physiological mechanisms underlying sleepiness on the flight deck is one of the keys to addressing fatigue in operational aviation contexts. Ultimately, the pilots themselves and those scheduling flight routes and duty timelines must be convinced that sleep and circadian rhythms are important and that a high quality of off-duty sleep is the best possible protection against on the job fatigue. Educational programs should focus primarily on conveying the absolute importance of prioritizing adequate consolidated off-duty sleep; utilizing "good sleep habits" to optimize sleep quantity and quality; and supplementing shortened or disrupted sleep periods with naps (Caldwell & Caldwell, 2003). In addition, crews should be taught that: 1) fatigue is a physiological problem that cannot be overcome by motivation, training, or willpower; 2) people cannot reliably judge their own level of fatigue-related impairment; 3) some people are far more affected by fatigue than others, and fatigue susceptibility cannot be reliably predicted; and 4) there is no one-size-fits-all "magic bullet" that offers a complete fatigue solution for every person in every situation.

Sleep hygiene
Obtaining the required quantity of sleep on a day-to-day basis obviously is important, but obtaining high quality sleep is beneficial as well. Sleep fragmentation is known to degrade memory, reaction time, vigilance, and mood (Bonnet & Arand, 2003). Mother Nature has equipped humans with all that is needed to obtain good, restorative sleep on a day-to-day basis, but people often thwart their own attempts to sleep well by developing counterproductive habits. A number of specific strategies are helpful for optimizing the quality of every sleep opportunity.

- When possible, wake up and go to bed at the same time every day to avoid circadian disruptions.
- Use the sleeping quarters only for sleep and not for work, television, or internet.
- If possible, establish a consistent and comforting bedtime routine (e.g., reading, taking a hot shower, and then going to bed).
- Perform aerobic exercise every day, but not within two hours of going to bed.
- Make sure the sleeping quarters are quiet, totally dark, and comfortable.
- Keep the sleep environment cool ($20°–22°C$ if you are covered).
- Move the alarm clock out of sight so you cannot be a clock-watcher.
- Avoid caffeine in drinks and other forms within 6 hours of bedtime.
- Do not use alcohol as a sleep aid (it may make you sleepy, but you will not sleep well).
- Avoid cigarettes or other sources of nicotine right before bedtime.

- Do not lie in bed awake if you do not fall asleep within 30 minutes – instead, leave the bedroom and do something relaxing and quiet until you are sleepy.

Sleep-promoting compounds
Although the use of "sleeping pills" is often discouraged within the aviation context, certain medications can be extremely helpful in minimizing the sleep loss (and subsequent fatigue) associated with schedules and/or conditions that are not conducive to obtaining the sufficient day-to-day sleep that is so crucial for on-duty alertness and performance. Clear examples are:

1. situations in which sleep is difficult because of circadian factors (as is the case of jet lag or shift lag associated with rapid time-zone changes, night duty, and/or rapidly changing work/rest schedules), and/or
2. situations in which the sleep environment is poor (as is the case when personnel must sleep in uncomfortable or noisy hotels).

Under such circumstances, sleep medications may be the only way to ensure adequate restful sleep prior to the next flight. Although complete reliance on medications for the control of sleep cycles is not recommended, it is clear that enhancing sleep with today's medications (which have short half-lives and minimal side effects compared to older medications) is far preferable to being sleep-deprived. Choosing the best hypnotic for each situation requires consideration of a variety of factors. For instance, from a strictly pharmacological standpoint, temazepam (Restoril®) or a similar compound with an 8-to-10 hour half-life, is useful for maintaining sleep for relatively long periods during the night and/or for optimizing the daytime sleep of night-working personnel (Caldwell et al., 2003; Rosenberg, 2006; Simons et al., 2006). However, it may take more than 12–18 hours post-dose for any residual effects of temazepam to fully dissipate, and for that reason, it would not be recommended for promoting sleep during short layover periods. However, some of the newer hypnotics such as zolpidem (Ambien®) and zaleplon (Sonata®) help with sleep maintenance without the extended half-life of temazepam and without the increased probability of persistent post-sleep grogginess (Caldwell & Caldwell, 1998; Dooley & Plosker, 2000; Whitmore, Fischer & Storm, 2004; Pandi-Perumal et al., 2006). Zaleplon has a half-life of only 1 hour, the original formulation of zolpidem has a half-life of only 2.5 hours, and extended-release zolpidem (which improves sleep maintenance beyond that of original zolpidem (Greenblatt et al., 2005)) has a half-life of 2.8 hours. In addition to these very short-acting hypnotics, eszopiclone (Lunesta®) may be an option for certain situations since it has a half-life of 5–6 hours with minimal residual drug effects after as little as 10 hours post dose (Leese et al., 2002). Ramelteon (Rozerem®) is a newer type of sleep medication that targets the melatonin receptors in the brain to regulate the sleep–wake cycle. In line with research reported by Lieberman, J.A. (2007), ramalteon may be another alternative for individuals who need help with falling asleep more quickly in situations counter to the circadian rhythm.

Melatonin
Melatonin is a hormone normally secreted by the pineal gland that has a central role in the regulation of circadian rhythms, and as such, it is often used as an aid for overcoming the jet lag associated with rapid time-zone changes. It appears that 0.5–5mg melatonin, taken close to the target destination bedtime (10pm to midnight), decreases the jet-lag associated

with flights crossing five or more time zones (Herxheimer & Petrie, 2002). Melatonin is more effective after eastward flights when the objective is to advance the bedtime than after westward flights when the objective is to delay the bedtime. However, the exact timing of the melatonin dose should be carefully considered since melatonin consumed at the wrong time (i.e., early in the day rather than at the destination bedtime), is liable to cause daytime sleepiness and a delay rather than a facilitation in the adaptation to local time. Most likely concerns over this dose-timing issue is what has led the U.S. Air Force to prohibit its pilots from using melatonin, despite the fact that the U.S. Federal Aviation Administration has not placed similar restrictions on civil aircrew personnel.

CONCLUSIONS

As the demand for air travel continues to escalate while economic constraints continue to challenge aircrew staffing practices, pilots and crews increasingly will face fatigue-related issues associated with the complex and often difficult schedules that characterize modern 24/7 flight operations. Extended duty periods, variable work/sleep schedules, night operations, and rapid/frequent time-zone transitions all pose significant challenges to basic human capabilities; however, these challenges can be mitigated through scientifically based fatigue-management practices. Scheduling practices based on up-to-date information about human sleep needs and circadian factors are essential, and the implementation of validated in-flight and pre-duty fatigue countermeasures will optimize the safety and well-being of crews and passengers. Fortunately, the most recently-released flight duty and rest requirements developed by the Federal Aviation Administration demonstrate progress towards better aligning current crew-scheduling practices with available scientific data; however, a greater focus on updated fatigue-countermeasures guidance remains to be seen.

REFERENCES

Achermann, P. (2004). The two-process model of sleep regulation revisited. *Aviation, Space, and Environmental Medicine*, 75(3, Suppl), A37–43.

Balkin, T.J., Rupp, T., Picchioni, D. & Wesensten, N.J. (2008). Sleep loss and sleepiness. *Chest*, 124(3), 653–60.

Bonnet, M.H. (1990). Dealing with shift work: Physical fitness, temperature, and napping. *Work Stress*, 4(3), 261–74.

Bonnet, M.H. (1991). The effect of varying prophylactic naps on performance, alertness and mood throughout a 52-hour continuous operation. *Sleep*, 14(4), 307–15.

Bonnet M.H. & Arand, D.L. (2003). Clinical effects of sleep fragmentation versus sleep deprivation. *Sleep Medicine Review*, 7(4), 293–5.

Caldwell, J.A. & Caldwell, J.L. (1998). Comparison of the effects of zolpidem-induced prophylactic naps to placebo naps and forced rest periods in prolonged work schedules. *Sleep*, 21, 79–90.

Caldwell, J.A. & Caldwell, J.L. (2003). *Fatigue in Aviation: A Guide to Staying Awake at the Stick*. Burlington, VT: Ashgate.

Caldwell, J.A., Caldwell, J.L. & Schmidt, R.M. (2008). Alertness management strategies for operational contexts, *Sleep Medicine Reviews*, 12(4), 257–73.

Caldwell, J.A., Prazinko, B.F. & Caldwell, J.L. (2003). Body posture affects electroencephalographic activity and psychomotor vigilance task performance in sleep deprived subjects. *Clinical Neurophysiology*, 114(1), 23–31.

Caldwell, J.L., Prazinko, B.F., Rowe, T., Norman, D., Hall, K.K. & Caldwell, J.A. (2003). Improving daytime sleep with temazepam as a countermeasure for shift lag. *Aviation, Space, Environmental Medicine*, 74, 153–63.

Carskadon, M.A. (1989). Ontogeny of human sleepiness as measured by sleep latency, in D.F. Dinges & R.J. Broughton (eds), *Sleep and Alertness: Chronobiological, Behavioral, and Medical Aspects of Napping*. New York: Raven Press, 53–84.

Cohen, P.A. & Ernst, E. (2010). Safety of herbal supplements: A guide for cardiologists. *Cardiovascular Therapeutics*, 28(4), 246–53.

Committee on Military Nutrition Research (2001). *Caffeine for the Sustainment of Mental Task Performance: Formulations for Military Operations*. Washington, DC: National Academy Press.

de Mello, M.T., Esteves, A.M., Pires, M.L., Santos, D.C., Bittencourt, L.R., Silva, R.S. & Tufik, S. (2008). Relationship between Brazilian airline pilot errors and time of day. *Brazilian Journal of Medical and Biological Research*, 41(12), 1129–31.

Dijk, D.J. & Franken, P. (2005). Interaction of sleep homeostasis and circadian rhythmicity: dependent or independent systems? In M.A. Kryger, T. Roth & W.C. Dement (eds), *Principles and Practice of Sleep Medicine*. Philadelphia, PA: Elsevier Saunders.

Dinges, D.F. (1984). The nature and timing of sleep. Transportation Study. Philadelphia, PA: College of Physicians Philadelphia, 6(3), 177–206.

Dinges, D.F. (1986). Differential effects of prior wakefulness and circadian phase on nap sleep. *Electroencephalography and Clinical Neurophysiology*, 64, 224–7.

Dinges, D.F., Graeber, R.C., Connell, L.J., Rosekind, M.R. & Powell, J.W. (1990). Fatigue-related reaction time performance in long-haul flight crews. *Sleep Research*, 19, 117.

Dinges, D.F. & Powell, J.W. (1988). Sleepiness is more than lapsing. *Sleep Research*, 17, 84.

Dinges, D.F., Whitehouse, W.G., Orne, E.C. & Orne, M.T. (1988). The benefits of a nap during prolonged work and wakefulness. *Work Stress*, 2(2), 139–53.

Dooley, M. & Plosker, G.L. (2000). Zaleplon: A review of its use in the treatment of insomnia. *Drugs*, 60, 413–45.

Flight Safety Foundation (2005). Lessons from the dawn of ultra-long-range flight. *Flight Safety Digest*, 24, 1–60.

Goode, J.H. (2003). Are pilots at risk of accidents due to fatigue? *Journal of Safety Research*, 34, 309–13.

Greenblatt, D.J., Zammit, G., Harmatz, J. & Legangneux, E. (2005). Zolpidem modified-release demonstrates sustained and greater pharmacodynamic effects from 3 to 6 hours postdose as compared with standard zolpidem in healthy adult subjects. *Sleep*, 28(Suppl), A245.

Herxheimer, A. & Petrie, K.J. (2002). Melatonin for the prevention and treatment of jet lag. *Cochrane Database of Systematic Reviews*, 2, No. CD001520. DOI: 0.1002/14651858.CD001520.

Hursh, S.R., Raslear, T.G., Kaye, A.S. & Fanzone, J.F. (2006). *Validation and Calibration of a Fatigue Assessment Tool for Railroad Work Schedules, Summary Report* (Technical report DOT/FRA/ORD-06/21). Washington, DC: U.S. Department of Transportation, Federal Railroad Administration, Office of Research and Development.

Hursh, S.R., Redmond, D.P., Johnson, M.L., Thorne, D.R., Belenky, G., Balkin, T.J., Storm, W.F., Miller, J.C. & Eddy, D. (2004). Fatigue models for applied research in warfighting. *Aviation, Space, and Environmental Medicine*, 75, 3(Suppl), A44–A53.

Klein, K.E., Bruner, H., Holtmann, H., Rehme, H., Stolze, J., Steinhoff, W.D., et al. (1970). Circadian rhythm of pilots' efficiency and effects of multiple time zone travel. *Aerospace Medicine*, 41(2), 125–32.

Lavie, P. (1986). Ultrashort sleep-waking schedule. III. "Gates" and "forbidden zones" for sleep. *Electroencephalography and Clinical Neurophysiology*, 63, 414–25.

Leese, P., Maier, G., Vaickus, L. & Akylbekova, E. (2002). Esopiclone: Pharmacokinetic and pharmacodynamic effects of a novel sedative hypnotic after daytime administration in healthy subjects. *Sleep*, 25(Suppl.), A45.

Lieberman, H. (2003). Nutrition, brain function, and cognitive performance. *Appetite*, 40, 245–54.

Lieberman, J.A. (2007). Update on the safety considerations in the management of insomnia with hypnotics: Incorporating modified-release formulations into primary care. *Primary Care Comp Journal of Clinical Psychiatry*, 9, 25–31.

Lyman, E.G. & Orlady, H.W. (1981). *Fatigue and Associated Performance Decrements in Air Transport Operations* (NASA Contractor Report No. 166167). Moffett Field, CA: NASA Ames Research Center.

Matsumoto, K. & Harada, M. (1994). The effect of night-time naps on recovery from fatigue following night work. *Ergonomics*, 37(5), 899–907.

Monk, T. (1994). Shiftwork, in M.H. Kryger, T. Roth & W.C. Dement (eds) *Principles and Practice of Sleep Medicine*. Philadelphia, PA: Saunders.

Morgenthaler, T., Alessi, C., Friedman, L., Owens, J., Kapur, V., Boehlecke, B., Brown, T., Chesson, A., Coleman, J., Lee-Chiong, T., Pancer, J. & Swick, T.J. (2007). Practice parameters for the use of actigraphy in the assessment of sleep and sleep disorders: An update for 2007. *Sleep*, 30(4), 519–29.

National Sleep Foundation (2002). *Sleep in America Poll*. Washington, DC: National Sleep Foundation.

National Transportation Safety Board (1994). A review of flightcrew-involved, major accidents of U.S. air carriers, 1978 through 1990 (NTSB Safety Study No. SS-94–01). Washington, DC: National Transportation Safety Board.

Nehlig, A. (1999). Are we dependent upon coffee and caffeine? A review on human and animal data. *Neuroscience Biobehavior Review*, 23, 563–76.

Neri, D.F., Oyung, R.L., Colletti, L.M., Mallis, M.M., Tam, P.Y. & Dinges, D.F. (2002). Controlled activity as a fatigue countermeasure on the flight deck. *Aviation, Space, and Environmental Medicine*, 73(7), 654–64.

Pandi-Perumal, S.R., Verster, J.C., Kayumov, L., Lowe, A.D., Santana, M.G., Pires, M.L.N., Tufik, S. & Mello, M.T. (2006). Sleep disorders, sleepiness and traffic safety: A public health menace. *Brazilian Journal of Medical and Biological Research*, 39, 863–71.

Rogers, A.S., Spencer, M.B., Stone, B.M. & Nicholson, A.N. (1989). The influence of a 1 h nap on performance overnight. *Ergonomics*, 32(10), 1193–205.

Rosa, R.R. (1993). Napping at home and alertness on the job in rotating shift workers. *Sleep*, 16(8), 727–35.

Rosenberg, R.P. (2006). Sleep maintenance insomnia: Strengths and weaknesses of current pharmacologic therapies. *Annals of Clinical Psychiatry*, 18, 49–56.

Rosekind, M.R., Graeber, R.C., Dinges, D.F., Connell, L.J., Rountree, M.S., Spinweber, C.L., et al. (1994). Crew factors in flight operations IX: Effects of planned cockpit rest on crew performance and alertness in long-haul operations. (Report No: DOT/FAA/92/24). Moffett Field, CA: NASA Ames Research Center.

Roth, T. (2004). Measuring treatment efficacy in insomnia. *Journal of Clinical Psychiatry*, 65(suppl 8), 8–12.

Sadeh, A. & Acebo, C. (2002). The role of actigraphy in sleep medicine. *Sleep Medicine Reviews*, 6(2), 113–24.

Simons, R., Koerhuis, C.L., Valk, P.J. & Van den Oord, M.H. (2006). Usefulness of temazepam and zaleplon to induce afternoon sleep. *Military Medicine*, 171, 998–1001.

Szinnai, G., Schachinger, H., Arnaud, M.J., Linder, L. & Keller, U. (2005). Effect of water deprivation on cognitive–motor performance in healthy men and women. *American Journal of Physiological Regulation Integrated Comparative Physiology*, 289, R275–R280.

Van Dongen, H.P. (2004). Comparison of mathematical model predictions to experimental data of fatigue and performance. *Aviation, Space, and Environmental Medicine* 75(3 Suppl), A15–A36.

Van Dongen, H.P. & Dinges, D.F. (2005). Sleep, circadian rhythms, and psychomotor vigilance. *Clinical Sports Medicine*, 24(2), 237–49.

Vgontzas, A.N., Pejovic, S., Zoumakis, E., Lin, H.M., Bixler, E.O., Basta, M., et al. (2007). Daytime napping after a night of sleep loss decreases sleepiness, improves performance, and causes beneficial changes in cortisol and interleukin-6 secretion. *American Journal Physiology and Endocrinology Metabolism*, 292, 253–61.

Webb, W. (1987). The proximal effects of two and four hour naps within extended performance without sleep. *Psychophysiology*, 24(4), 426–29.

Whitmore, J.N., Fischer, J.R. & Storm, W.F. (2004). Hypnotic efficacy of zaleplon for daytime sleep in rested individuals. *Sleep*, 27, 895–98.

25

Soldier Fatigue and Performance Effectiveness: Yesterday, Today and Tomorrow

Gerald P. Krueger

They said good night. (*General James*) Longstreet watched the old man (*General Robert E. Lee*) back to his tent. Then he mounted and rode alone back to his camp to begin the turning of the army, all the wagons and all the guns, down the narrow mountain road that led to Gettysburg. It was still a long dark hour till dawn. He sat alone on his horse in the night and he could feel the army asleep around him, all those young hearts beating in the dark. They would need their rest now. He sat alone to await the dawn, and let them sleep a little longer. (Michael Shaara, *The Killer Angels*, 1974).

INTRODUCTION

In his novel, *The Killer Angels*, about the four-day Civil War Battle at Gettysburg in July 1863, Michael Shaara repeatedly points to combatants' need for sleep. The story describes difficulties Confederate and Union troops had in obtaining sleep during lengthy forced marches on foot and horseback, and while moving horse-drawn cannons about the hills of Virginia, Maryland and Pennsylvania. He especially depicts frequent interruptions military leaders encounter receiving status reports, enacting plans, preparing or giving orders, all snatching away opportunities for rest and sleep. Shaara superbly highlights many facets of *soldier fatigue*. "Some things never change; or at least they seem to remain the same no matter what we do." Or do they? What is the impact of decades of very painstaking research and consulting by people in this handbook, and others, who have assisted military forces (soldiers, marines, sailors, airman, and coast guardsmen) in everyday efforts to cope with "soldier fatigue?"

To help military forces sustain combat operations some among us mathematically modeled how units should be staffed with adequate numbers of personnel, proposed preferred work–rest schedules and recommended unit sleep discipline policies and suggested myriad fatigue countermeasures, including carefully placed napping and the use of pharmaceuticals. After several decades of applications work, a remaining question is: "Has research on *fatigue* made things better for soldiers?" Another cliché suggests we should "review the bidding." This chapter attempts to reassess from whence we came (*yesterday*), where we are *today*, and to imagine where we are headed in the future (*tomorrow*) regarding *soldier fatigue*.

It is tempting to focus on the most important variables contributing to *soldier fatigue*: the *need for sleep* – the *quantity and quality of sleep* obtained, or missed; on sustained vigilance; long bouts of psychomotor task performance; on field leaders' operational contingency planning and decision-making, which may be affected by acute sleep loss and information overload. As this handbook attests, our science teased apart many contributing factors, and elucidated general principles of "operator fatigue." Thus, several fatigue effects on soldier performance can be highlighted:

1. Mental–cognitive fatigue is often observable sooner than physical fatigue.
2. Soldiers generally employ trade-off strategies to preserve accuracy over speed of performance.
3. Vigilance and attention suffer because of long work stints, especially those including lengthy, boring performance requirements.
4. Soldiers and their leaders have difficulty obtaining sleep during military operations.
5. With partial or full sleep deprivation during sustained operations, cognitive facets of operational planning become less sharp, and battlefield decision-making can be compromised.
6. In maintaining alertness, soldiers need 7–8 hours of sleep per 24-hour day; or else they accumulate a sleep debt, which leads to degraded performance.
7. Due to circadian physiology, alertness varies during the day and manifests in recognizable, predictable peaks and lulls in mood and performance.
8. Time-of-day fatigue effects can be particularly apparent during sleep deprivation, when homeostatic drives toward sleep interact to produce differential impacts on performance.
9. Judicious use of nap-taking, when soldiers safely can, is useful for augmenting sleep obtained or replacing sleep missed; short "power naps" help boost cognitive alertness, albeit like other fatigue countermeasures, for a short duration.

Soldier fatigue research often focuses primarily (and simply) on sleep deprivation and subsequent performance measurement. However, numerous additional variables associated with the *natural environmental* and many *man-made stressors* of the battlefield should be given due weighting since at least they affect a soldier's opportunity and ability to obtain restful, restorative sleep, and therefore impact soldier alertness, fatigue, performance and health.

Environmental Stressors

A variety of *environmental stressors* play significant roles in determining a soldier's level of alertness or the counterpart: "fatigue." Stressors involving the geographic environment and seasonal weather changes include: intense heat, high humidity, extreme cold, high terrestrial altitude, storms (wind-, sand-, dust-, rain-, and snowstorms) and even indigenous insects and other annoying critters affect a soldier's ability to obtain satisfactory sleep. Imagine trying to sleep while wearing bulky protective uniform clothing, lying on lumpy frozen ground in high altitude mountainous terrain, amidst

Figure 25.1 Soldiers resting and attempting to sleep in hastily dug fox holes in the desert

Source: Photo courtesy of U.S. Marine Corps photo gallery, Quantico, Virginia.

the countless movement of other nearby troops who are going about their duties on their own work shift, moving heavy equipment, communicating over radios, or operating weapons. Noise, turbulence and heat plague would-be sleepers. Resting in the open on a hot dry desert, or in a humid, tropical jungle presents different sleep challenges. Sleeping curled up inside cramped crew compartments of military vehicles is beyond uncomfortable. Locating a good place to obtain 4–7 hours of solid sleep in a combat zone is never easy.

Most *natural environment* features do not change much; but the geography where military forces focus their activity shifts with each battlefront. World Wars I and II presented battlefields of open plains, forests, and river valleys in Europe; whereas the South Pacific Islands in WWII, and then Vietnam and Panama, offered hot, humid, tropical jungles. Cold wintry battles were hard fought in Europe during World War II, in Korea, and in 1982 in the Falkland Islands. World War II armor forces also fought in North Africa's deserts; dozens of Middle East desert clashes have ensued since. Middle East fighting recently morphed into limited warfare in high mountains in Afghanistan (8–12,000 ft. above sea level). House-to-house, street-to-street search-and-clear operations that took place in bombed European cities during World War II reemerged in Bosnia, Iraq, and Afghanistan. Natural terrain and environmental features of battlefields are significant in development of *soldier fatigue* – if for no other reason than to affect soldiers' ability to obtain restful sleep. These factors are usually unaccounted for in laboratory sleep studies.

Man-made Environmental Stressors

Soldier performance and fatigue levels also are affected by *man-made environmental stressors* (e.g. noise, blast waves, heat, toxic fumes, acceleration, vibration, claustrophobic vehicle platforms) that emanate from weapons and matériel systems military personnel employ. Soldiers endure high physical loads, carrying weapons, ammunition, protective clothing, body armor, food, water, first aid kits, and other items (Marshall, 1950). The proliferation of battlefield computers, even body-worn computers, brings soldiers a constant supply of data/information presenting risks of cognitive overload. These man-made environmental stressors continually plague soldiers and have a direct impact on *soldier fatigue* (Krueger, 1991; 2008; 2010b).

Operational Considerations of Fatigue (Technology and Weaponry)

Often "operator fatigue" is used to refer to *soldier fatigue* because many safety and performance concerns revolve around soldiers operating equipment, driving armored vehicles, trucks, flying helicopters, or operating individual or crew-served weapon systems (tank guns, artillery, attack helicopters). Operating equipment entails psychomotor control and skill performance. Success criteria are expressed in terms of accuracy of tracking and aiming, and errors of omission (did not turn where one is supposed to) or commission (hit the wrong target). More cognitively laden operations include use of computerized consoles in command centers wherein vigilance and interpreting displays, procedural matters, carrying out machine-assisted communications and computer actions are important. With soldier fatigue we say cognitive performance (reaction time, attention to detail, analytical reasoning, judgment, decision-making) deteriorates before physical performance, (lifting, carrying, digging) even before psychomotor skill performance (e.g. tracking, driving, firing). Haslam (1985; Haslam & Abraham, 1987) demonstrated that tired riflemen can maintain tight shot groups on target, but due to cognitive fatigue are sometimes confused about which targets they should shoot at – a point accentuated during Colonel Gregory Belenky's 1991 interviews of U.S. Army combatants in Iraq. Cognitive fatigue concerns include those of a leader's situational awareness, and his/her ability to continually make sense of the battlefield, or to formulate battle planning, to determine the next course of action. Mental sharpness and the quality and timeliness of decision-making are of paramount concern to command and control personnel, for leaders at all levels of the military structure. Fatigue studies of decision-making usually infer that if a participant makes poorer decisions while sleep deprived in the laboratory, performance is not likely to improve in the field of battle. The issue is what to do about that.

Continuous and Sustained Military Operations

Continual advances in weaponry and technology, but also changes in battlefield doctrine, affect notions of *soldier fatigue*. During the U.S. involvement in Vietnam (1960s to mid-1970s) the advent of sophisticated night vision, reconnaissance, and electronic surveillance systems first permitted military forces to conduct night and day operations indistinguishably. In the 1980s, Western alliance forces in Europe, equipped with technologies of night-fighting systems, (e.g. portable radar, light-amplification goggles, infrared sensors, thermal weapon sights, etc.) which initially their adversaries did not have, envisioned battling in Continuous Operations–Sustained Operations (CONOPS–SUSOPS) a NATO military doctrine that espoused fighting non-stop, around-the-clock for days, weeks, or even months at a time (DeWulf, 1987). In CONOPS it was envisioned that sufficient numbers of military units, equipped with night vision systems, would take alternating turns fighting the battle; each would have sufficient breaks to regroup and to rest; but the fighting would continue around-the-clock.

During CONOPS, it was also anticipated that small teams of combatants would conduct *sustained operations* (SUSOPS) – working steadily for long periods without relief. Individual combatants would not be relieved on the job by others; they would not get much opportunity for sleep; and they would be expected to continue working until a mission objective was achieved. Such nonstop Sustained Operations involve sleep loss, produce stress and fatigue, lead to poor individual and military unit performance, accidents, battle-weary psychological stress casualties, and reduced mission effectiveness (Krueger, 1989, 1991). Military psychologists and medical scientists launched extensive research programs to examine a myriad of human performance and soldier-fatigue related issues associated with the NATO CONOPS-SUSOPS doctrine (Krueger, 1991, 2010b). Today, since our 24-hour society has adopted the 24/7/365 mantra, it may be useful to re-examine some human elements of the 30-year old doctrinal notions of CONOPS–SUSOPS, and update what research and experience have taught us.

Much in our world has changed since the break-up of the Soviet Union (circa 1989) and the terrorist attacks in New York City and Washington, D.C. in September 2001. No longer do Western forces anticipate employing large-sized units (e.g. tank or infantry divisions) to fight on a conventional battlefield for weeks at a time (CONOPS). Recent large-scale force attacks involving armored tank battles in Middle East deserts were of relatively short duration – weeks, occasionally months, (e.g. between Israel and Lebanon, 1982, 2006; U.S.-led coalition forces in Kuwait and Iraq in 1991; again in Iraq in 2003). They were not years of sustained battle like in the two World Wars. Departing from pitched-conventional battlefield scenarios envisioned for Europe, today's asymmetric warfare, involving insurgents and counterinsurgency, calls for smaller-sized units employing different tactics; but also involving slightly different forms of CONOPS and elements of SUSOPS.

Deployment of hundreds of robotic systems, surveillance systems and unmanned weapons permits greater fine-tuning to around-the-clock warfare. Some aerodynamic drone-like surveillance systems stay aloft over battlegrounds for days at a time, providing video stream images of adversary movements. Unmanned stand-off capabilities decrease battlefield exposure risks for combatants seated at control consoles in remote locations out of harm's way, who presumably adhere to well-designed work-shift schedules that permit adequate sleep — in a bed. Countless other engaged ground troops still carry on close-contact operations, accomplishing counterinsurgency, peacekeeping security, and nation-building missions (Krueger, 2010b). For deployed soldiers in this different form of CONOPS, the norm still is working work shifts of 12 hours or longer, sometimes 6–7 days per week, for months at a stretch. This is so even in the relatively low-intensity peacekeeping and nation-building activities in Iraq and Afghanistan. Ten years of comprehensive examinations of U.S. Navy shipboard operations illustrate the variety of work-rest schedules and the paucity of sleep opportunities for sailors on naval vessels repeatedly result in chronic sleep deprivation and fatigue (Miller, Matsangas & Kenney, 2012).

Operational Tempo has Changed

In World War II, troops were expected to remain deployed until the Nazis were ejected from Western Europe – few soldiers envisioned returning home any time soon. Pitched battles were often fought in pulses, permitting forces to rearm, refuel, rest, and to plan next encounters. *Battle fatigue* involved compromised soldier performance and abilities to sustain the fight after extended exposure to weeks, months, even a couple of years of constant bombardment by artillery and other enemy fires (Marlowe, 1986). In Vietnam, U.S. forces could anticipate returning home after 12-month overseas tours; some Vietnamese career military veterans engaged in warfare for 25 years. In the past decade U.S. troops (Marines and army soldiers) deployed overseas in Iraq and Afghanistan for eight months to a year, sometimes longer. They also experienced shortened "dwell times" back home between repeated redeployments back to the combat zone. Such extended time in combat, and repeating the pattern every couple of years (Operational Tempo) contributes to *battle fatigue* for these combatants (Castro & Adler, 2005).

SOLDIER FATIGUE: THE NEED FOR REST-SLEEP HAS NOT CHANGED

In the 1980s–1990s "soldier fatigue and its effects on military performance" were hot topics in military research communities in countries that adhered to NATO doctrine. The Continuous and Sustained Operations Subgroup, a part of the U.S. Department of Defense Human Factors Engineering Technical Advisory Group (DoD HFE TAG) had over 600 members from nine countries on its membership roster. Participants in this Subgroup published results of numerous laboratory and field studies on soldier fatigue (for a collection of them, see Englund & Krueger, 1985; Krueger & Englund, 1985; Krueger, Cardinales-Ortiz & Loveless, 1985; Krueger & Barnes, 1989; Cox & Krueger, 1989). Until recently, this HFE TAG Subgroup met twice annually to continue their collaborative quest to elucidate human dimensions of CONOPS–SUSOPS.

Military strategy, doctrine, tactics, weapons technology, and many man-made stressors changed significantly over the past 30 years; but soldiers have not changed much, at least not physiologically and cognitively. Thirty years of concentrated behavioral and medical research addressing military sustained performance do not seem to have altered, but rather confirm the aforementioned soldier-fatigue related statements as being universal, lasting principles of human dimensions in CONOPS–SUSOPS. Today's predictions of sustained individual soldier or team performance during training or actual combat remain virtually the same (see Krueger 1989, 1991, 2010b).

Soldiers still require 7–8 hours of sleep per 24-hr day to maintain adequate alertness on the job. If soldiers do not obtain 7–8 hours of sleep, they accumulate a *sleep debt* during their work week. As the debt accrues they will not perform as well. Tired soldiers are likely to exercise poor judgment, decision-making becomes less sharp; they may make more errors and mistakes and become involved in more accidents. Equipment operators, security guards and others whose work necessitates sustained vigilance still experience two daily lull periods wherein their performance is likely to be noticeably degraded due to changes in circadian rhythm physiology. Time-of-day performance degradation is more pronounced if soldiers are partially sleep-deprived (<5–6 hrs sleep per 24-hr day). Night fighters expected to sleep during the day (difficult in most combat zones) are apt to get only brief, scattered, fragmented sleep, and they often develop significant sleep debts; as do perpetual shift-workers repeatedly changing work–rest schedules. Sustained workload combines with fatigue, especially after one or more nights of complete sleep loss or longer periods of reduced or fragmented sleep, to degrade performance, productivity, safety, and mission effectiveness. Sleep loss interacts with workload, resulting in decreased vigilance, slower reaction time, perceptual and cognitive distortions, and changes in mood and affect, all of which vary according to circadian time of day (Krueger, 1991, 2010b).

SOLDIER FATIGUE: A MATTER OF PERSPECTIVE

Lightening Soldiers' Load

Military forces, in particular the U.S. Army and the U.S. Marine Corps, made numerous attempts to lighten the soldier's physical load – the uniform items of clothing he/she wears, and the items carried, including weapons, combat helmet, and body armor. While military operations and missions vary widely, it is not uncommon for U.S. Army Infantrymen and U.S. Marines to wear and carry upwards of 80–120 pound loads, to do this while maneuvering on foot over arduous terrain (e.g. rocky mountains at 10,000 feet above sea level in Afghanistan, or over wet jungle floors of the tropics), and to travel considerable distances – not unlike Shaara's Civil War forced marches. Personal load carriage threatens these soldiers with physical fatigue.

Attempts to lighten the soldier's load have modified uniforms soldiers wear, the things they carry, and load-carrying systems such as backpacks and shoulder harnesses. Such equipment development programs are conducted at the U.S. Army Program Executive Office (PEO) for Soldier Systems at Fort Belvoir, VA, at the Soldier Systems Center at Natick, MA, and at the Marine Corps Systems Command at Quantico, VA. While working on design of the electronic-computerized infantryman ensemble, (i.e. the Army's Land Warrior system) one experienced older sergeant was heard to say about load carriage: "In carrying 85–90 or more pounds of stuff – no matter whether it be weapons, ammunition, night vision systems, electronic gear, batteries, water, food, first aid kits, or bags of sand or feathers, soldiers/marines will almost always be carrying about 90 pounds. If design improvements lighten the soldier's load, no doubt commanders simply require soldiers to carry more ammunition and water." How do we change that?

Managing Cognitive Workload to Reduce Overload

Cognitive workload does not lend itself to simple "pounds carried" measurements. Battlefield computer systems, providing assistance to command and control personnel, can help one skim through significant amounts of data and information, and can aid in battlefield situation awareness and decision-making. Applications of technological devices are so pervasive that today even front-line individual soldiers/marines have access to volumes more information than ever before, and likely more than they can use. Automated systems can clutter up cognitive processing by overloading soldiers with too much unsorted data, or too much information, which can lead to confusion, slowed reactions, and questionable decision-making.

Many military psychological research projects have tackled system design and operational use issues to help equipment designers configure automated systems to maximize assistance to soldiers, without making things too complicated or complex. While implications of battlefield information systems for evolving or easing "soldier fatigue" are numerous, only a case-by-case, mission scenario treatise could do justice to the topic. Whole chapters could be written about such projects. One far-reaching, advanced technological approach fostered by the U.S. Defense Advanced Research Projects Agency (DARPA), and recently turned over to the individual services, is under the project name "Augmented Cognition." In greatly simplified terms, this forward-reaching research program aimed at "removing the burdens of technology" as it advances human physiological and psychological monitoring to enable "automatics" to offer augmented control and decision-making assistance to equipment operators, thereby unburdening them of some cognitive processing requirements during high workload (Schmorrow & Stanney, 2008).

On another count, recent military trends pushed battlefield decision-making to the lowest echelons of command (Alberts & Hayes, 2003), increasing pressures on young soldiers/marines "in the front lines" to make quick decisions with potentially immediate international ramifications. General Charles Krulak, a former U.S. Marine Corps Commandant, refers to such marines as "strategic corporals" (Krulak, 1999). Making independent life and death decisions, (whether to fire a weapon or not at a security checkpoint) during "nation-building" and "peacekeeping" missions, can be overbearing cognitively and can ultimately be very fatiguing on a daily basis.

EDUCATION ABOUT FATIGUE AND MILITARY DOCTRINE

"Soldier fatigue management" advocates that each of the U.S. military services should insert into military training and operational guidance a set of human dimension principles acknowledging the need of the body and the brain for quantity and quality sleep and rest, and an admonition to provide that sleep, or risk the consequences. These attempts have met with considerable success regarding aviator flight-hour restrictions and crew rest considerations. Decades ago, sleep discipline guidance for aviators was provided by flight surgeons within the NATO Advisory Group for Aerospace Medicine Research and Development: AGARD (Nicholson & Stone, 1982); by the U.S. Air Force School of Aerospace Medicine (see United States Air Force, 1994); at the United States Army Aeromedical Research Laboratory (1997); and for all military forces by research psychologists at the U.S. Naval Health Research Center (Naitoh, Englund & Ryman, 1986), and by the Navy Surgeon's office at the United States Navy Bureau of Medicine and Surgery (2000). Each organization promulgated sleep medicine guidance. Many of those guidance principles made their way into aviator flight-limit regulations. For U.S. Air Force and U.S. Navy aircrew members, more recent official guidance to avoid fatigue is reported in Caldwell et al. (2009).

For ground-fighting forces (e.g. soldiers and marines not in aviation) institutionalization of "education regarding soldier fatigue" has lagged behind. One U.S. Army argument (advocated by researchers at the Walter Reed Army Institute of Research: WRAIR) is that soldiers need to obtain sufficient sleep to meet one of the body's *logistic needs*, in a way similar to ensuring that units acquire adequate resupply of ammunition, fuel, food, water, etc. Commanders must proactively plan for allocation of adequate sleep, for themselves and their subordinates.

In a significant development, in 2009, the U.S. Army incorporated a seven-page chapter on *Sleep Deprivation* into its doctrinal Field Manual 6–22.5: *Combat and Operational Stress Control Manual for Leaders and Soldiers* (United States Army, 2009: Chapter 4). Army leaders are told the sleep guidance in the manual is supported by decades of research, much of it accomplished by scientists at WRAIR. The doctrinal guidance applies to all levels of military operations, including training and tactical environments. The chapter offers guidance to formulate *unit sleep plans*.

Amidst other coverage, the synopsis presented for sleep and performance principles includes:

Sleep is a biological need, critical for sustaining the mental abilities needed for success on the battlefield. For optimal performance and effectiveness, soldiers require 7 to 8 hours of good quality sleep every 24-hour period to sustain operational readiness. As daily total sleep time decreases below this optimum, the extent and rate of performance decline increase. ...

Soldiers who lose sleep will accumulate a *sleep debt* over time that will seriously impair their performance. The only way to *pay off* this debt is by obtaining the needed sleep. The demanding nature of military operations often creates situations where obtaining sleep may be difficult or even impossible for more than short periods. While essential for many aspects of operational success, sheer determination or willpower cannot offset the mounting effects of inadequate sleep. This concept is applicable for all levels of military operations including basic training and in all operational environments. ...

Individual and unit military effectiveness is dependent upon initiative, physical strength, endurance, and the ability to think clearly, accurately, and quickly. The longer a soldier goes without sleep, the more his/her thinking slows and becomes confused, and the more mistakes he or she will make. Lapses in attention occur and speed is sacrificed in an effort to maintain accuracy. Degradation in the performance of continuous work is more rapid than that of intermittent work.

Tasks such as requesting fire, integrating range cards, establishing positions, and coordinating squad tactics are more susceptible to sleep loss than well-practiced, routine physical tasks such as loading magazines and marching. Without sleep, soldiers can perform the simpler and/or clearer tasks (lifting, digging, and marching) longer than more complicated tasks requiring problem solving, decision-making, or sustained vigilance. For example, soldiers may be able to accurately aim their weapon, but not select the correct target. Leaders should look for erratic or unreliable task performance and declining planning ability and preventive maintenance not only in subordinates, but also in themselves as indicators of a lack of sleep. ...

In addition to declining military performance, leaders can expect changes in mood, motivation, and initiative as a result of inadequate sleep. Therefore, while there may be no outward signs of sleep deprivation, soldiers may still not be functioning optimally. ...

U.S. *Army Field Manual* 6–22.5 (March 2009), Chapter 4 (see United States Army, 2009)

That Chapter provides scheduling information for planning sleep routines during all activities (predeployment, deployment, precombat, combat, and post-combat), basic sleep environment information and other related factors (these are listed in tabular form in Table 25.1 and Table 25.2 as extracted from the Army manual).

Table 25.1   Basic Sleep scheduling factors (Table 4–1 extracted from *Army Field Manual* 6–22, 5 March 2009)

Factor	Effect
Timing of sleep period	Because of the body's natural rhythms (called "circadian rhythms"), the best quality and longest duration sleep is obtained during nighttime hours (2300–0700) These rhythms also make daytime sleep more difficult and less restorative, even in sleep-deprived soldiers. The ability to fall and stay asleep is impaired when bedtime is shifted earlier (such as from 2300 to 2100 hours). This is why eastward travel across time zones produces greater deficits in alertness and performance than westward travel.
Duration of sleep period	**IDEAL** sleep period equals 7 to 8 hours of continuous and uninterrupted nighttime sleep each and every night. **MINIMUM** sleep period – there is no minimum sleep period. Anything less than 7 to 8 hours per 24 hours will result in some level of performance degradation.
Napping	Although it is preferable to get all sleep over one sustained 7 to 8 hour period, sleep can be divided into two or more shorter periods to help the soldier obtain 7 to 8 hours per 24 hours. Example: 0100–0700 hours plus a nap from 1300–1500 hours. Good nap zones (when sleep onset and maintenance is easiest) occur in early morning, early afternoon, and nighttime hours. Poor nap zones (when sleep initiation and maintenance is difficult) occur in late morning and early evening hours when the body's rhythms most strongly promote alertness. Sleep and *rest* are not the same. While resting may briefly improve the way the soldier feels, it does not restore performance the way sleep does. There is no such thing as too *much* sleep – mental performance and alertness always benefit from sleep. Napping and sleeping when off duty are not signs of laziness or weakness. They are indicative of foresight, planning, and effective human resource management.
Prioritize sleep need by task	**TOP PRIORITY** is leaders making decisions critical to mission success and unit survival. Adequate sleep enhances both the speed and accuracy of decision making. **SECOND PRIORITY** is soldiers who have guard duty, who are required to perform tedious tasks such as monitoring equipment for extended periods, and those who judge and evaluate information. **THIRD PRIORITY** is soldiers performing duties involving only physical work.
Individual differences	Most soldiers need 7 to 8 hours of sleep every 24 hours to maintain optimal performance. Most leaders and soldiers underestimate their own total daily sleep need and fail to recognize the effects that chronic sleep loss has on their own performance.

Note: This table is not copyrighted but is available through U.S. Government auspices in the Public Domain.

Table 25.2   Basic sleep environment and related factors (Table 4–2 extracted from *Army Field Manual* 6–22, 5 March 2009)

Ambient noise	A quiet area away from intermittent noises/disruptions is **IDEAL** Soldiers can use earplugs to block intermittent noises. Continuous, monotonic noise (such as a fan or white noise) also can be helpful to mask other environmental noises.
Ambient light	A completely darkened room is **IDEAL** For soldiers trying to sleep during daytime hours, darken the sleep area to the extent possible. Sleep mask/eye patches should be used if sleep area cannot be darkened.
Ambient temperature	Even small deviations above or below comfort zone will disrupt sleep. Extra clothing/blankets should be used in cold environments. Fans in hot environments (fan can double as source of white noise to mask ambient noise) should be used.
Stimulants (caffeine, nicotine)	Caffeine or nicotine use within 4 to 5 hours of a sleep period will disrupt sleep and effectively reduce sleep duration. Soldiers may not be aware of these disruptive effects.
Prescription sleep-inducing agents (such as Ambien®, Lunesta®,and Restoril®)	Sleep inducers severely impair soldiers' ability to detect and respond to threats. Sleep inducers should not be taken in harsh (for example, excessively cold) and/or unprotected environments. Soldiers should have nonwork time of at least 8 hours after taking a prescribed sleep inducer.
Things that do not improve or increase sleep	Foods/diet – no particular type of diet or food improves sleep, but hunger and thirst may disrupt sleep. Alcohol induces drowsiness but actually makes sleep worse and reduces the duration of sleep. Sominex®, Nytol®, melatonin, and other over-the-counter sleep aids induce drowsiness but typically have little effect on sleep duration and are, therefore, of limited usefulness. Relaxation tapes, music, and so forth may help induce drowsiness but they do not improve sleep.

Note: This table is not copyrighted but is available through U.S. Government auspices in the Public Domain.

SOLDIER FATIGUE OF TODAY MORPHS INTO TOMORROW

Biomathematical Models of Alertness and Work–Rest Schedules

Neurobehavioral and biomathematical models of alertness and fatigue, predictive of job performance, can assist planners in determining optimum personnel staffing of military units (e.g. brigade, battalion, company, platoon). The operational usefulness of these models depends upon how well they are incorporated into work–rest and sleep time schedules to maximize safety by:

1. predicting times when performance is likely to be optimal;
2. identifying timeframes when recovery sleep will be most restorative; and
3. determining what impact work–rest schedules will have on overall neurobehavioral functioning (Mallis et al., 2004; Hursh et al., 2004; Caldwell et al., 2009; Hursh, 2010).

These model-based schedules, along with promulgated sleep discipline plans, make intuitive sense in personnel operations policies. Such models work best when schedules can be reliably carried out, e.g. for commercial airline scheduling. On a practical level, in military operations, adherence to such policies or plans often gets discarded during intense training, or during exigencies of actual combat. Nicely configured and documented policies often are compromised or ignored while maintaining a continuing high level of combat readiness during the relative quiescence in peacekeeping missions overseas.

In a more encouraging application, the U.S. Army Research Lab's Human Research and Engineering Directorate incorporated the Sleep, Activity, Fatigue and Task Effectiveness (SAFTE) model (Hursh, 2010) into the human engineering matériel system design Improved Performance Research Integration Tool: IMPRINT. *Fatigue modeling* helps assess crew workload to determine optimum crew size for new armor vehicles.

Sleep Loss Affects Soldier Decision-making and Judgment

The effects of sleep loss on alertness and vigilance are broadly understood; while more subtle, insidious impacts on complex mental functions of judgment and decision-making have only recently been researched. In battlefield problem-solving, when soldiers and their leaders have ample time for employing a systematic approach, to develop multiple courses of action, to compare evaluation weighting criteria to alternative solutions to problems, the results may be recognized as logical decisions (rational decision-making). Under conditions of severe time pressure during intense conflict when high stakes decisions must be made, leaders often employ gut feel, time-saving cognitive heuristic strategies that intuitively recognize situational patterns from past experiences – they quickly make needed decisions (recognition-primed naturalistic decision-making), see Kilgore (2010). Soldiers must possess sound judgment and the ability to utilize intuitive decision-making skills under fire, capacities that involve integration of emotion and cognition. Kilgore and his colleagues provide a neurobiological model describing how the major brain regions (systems) involved in judgment, and the more intuitive aspects of the decision-making process have a neurobiological basis and may be disrupted by inadequate sleep during periods of prolonged wakefulness – a repeated concern in discussions of soldier fatigue (see Killgore, Chapter 15, this volume).

Cogniceuticals: Hypnotics, Stimulants, and Nutritional Supplements

Discussion of soldier fatigue would not be complete without commentary on the prolific research and seeming increased use of chemical substances to provide assistance for soldiers to:

1. to stay alert and awake while performing satisfactorily during lengthy missions; and
2. obtain needed sleep even when the noisy battlefield or his/her own circadian physiology suggests it is not a particularly good time to fall asleep.

Due to inherent problems of issuing such medications to troops (i.e. potential for misuse or abuse, adverse consequences and side effects of some drugs) most sanctioned use of stimulants and hypnotics is limited to highly specialized missions (e.g. special operations work). Most applications are in controlled and safety-monitored operational settings (for use in military aviation, see Caldwell et al., 2009).

While not addressing numerous issues and research findings on use of psychoactive medications with soldiers in this chapter, suffice it to say: competent, trustworthy laboratory research by military medical scientists in the U.S. and elsewhere demonstrated efficacy of various drug use protocols for military performance situations. These studies and significant findings are covered elsewhere (see, for example, Caldwell et al., 2009). However, insufficient practical applications research in the field has been conducted to render definitive recommendations for widespread use of "cogniceuticals" (psychoactive medications).

Soldier use of stimulants to stay alert, especially the U.S. military's approved, but controlled use of amphetamines in operations (Caldwell et al., 2009) conjures up debate as to whether or not drug affected performance is *acceptable performance*. While studying cultural differences and public opinions about the use of pharmacologic countermeasures to fatigue, Russo (2007) pointed out that in military operations, some cognitive performance enhancement technologies are more controversial than others. Use of psychoactive pharmaceutical agents to alter behavior in healthy humans, specifically cogniceuticals that alter alertness, attention, and perception, present controversial *ethical challenges* among leaders in coalitions formed to support one another in international security scenarios. Russo points out that while use of such medications may be increasing, doctrine for their use remains limited. In today's distributed battlespace, national laws regarding psychoactive agents differ widely, even as soldiers of multiple nationalities fight shoulder to shoulder. The relatively innocuous supplement synthetic *melatonin*, used by some militaries as a "natural sleep-inducing hormone" is not even approved for allied forces in some countries. Assuming operationally equivalent efficacy, cogniceuticals selected should be those most broadly acceptable within the coalition framework. Russo proposed conditions for their employment (Russo, 2007).

Despite some controversy over stimulant use, it is accepted that soldiers will consume caffeine, mostly in beverages. Use of caffeine to improve cognitive functioning during sustained military operations has been extensively researched (National Research Council (NRC), 2001). Lieberman (2003) reported that caffeine improves vigilance in rested and sleep-deprived individuals, improves target detection speed without adversely affecting rifle-firing accuracy, improves decision response (choice reaction) time, improves learning and memory, and improves mood, including perceptions of fatigue and sleepiness. Benefits are seen with as little as 100–200 mg. caffeine (~1.4 to 2.8 mg/kg body weight). Although optimal caffeine doses vary widely among individuals, caffeine can improve cognition and mood in both habitual caffeine users and nonusers (Haskell et al., 2005). In addition to issuing caffeinated coffee in meal ration packets, the U.S. Army also issues caffeine-laden chewing gum to troops in the field. Research at Walter Reed demonstrated the efficacy of such caffeine use (~100 mg. caffeine per stick of gum) for temporarily stimulating alertness in soldiers. Caffeine in gum enters the blood stream sublingually ~7 minutes after chewing – generating a faster stimulating effect than caffeine in beverages which must traverse the digestive system and so take about 20 minutes to obtain the same effect (McLellan et al., 2003–2004).

Modafinil (ProVigil®, Alertec®), the relatively new eugregoric stimulating compound, offers boosts in alertness similar to those brought about by caffeine. Modafinil, uniquely, permits the user to "overrule" the stimulating effect to initiate a nap. While modafinil is less addictive than other stimulants, in the U.S. it is a prescription non-narcotic schedule

IV controlled substance, and therefore must be administered in off-label prescription use by the military. Wesensten et al. (2005) at the WRAIR, and other military medical researchers continue to explore potential fatigue countermeasures that modafinil and other stimulating compounds offer. Monitoring such forward-looking research on alternative chemical substances is recommended.

Research also is ongoing to explore the efficacy of providing a variety of nutritional supplements in the diets of soldiers in the hope that such nutrients will boost energy levels and sustain or enhance performance (e.g. programs at the U.S. Army Research Institute of Environmental Medicine). In the interim, commercially available products such as functional energy drinks (FEDs) like Red Bull® or vitamin-laced waters, and other advertised energy boosters are consumed by soldiers in the field, as they are readily available at "convenience stores" in the theater of operations. While such products often contain taurine, ginseng, guarana, et al. the predominant psychoactive substance is caffeine (Krueger, 2010a; Krueger, Leaman & Bergoffen, 2011). Consumption of FEDs and other supplements should be monitored. Excessive ingestion of large quantities of FEDs can invoke adverse physiological symptoms (e.g. high blood pressure), which can surreptitiously interact with other substances, medications, or existing medical conditions.

Susceptibility to Fatigue: Individual Differences

A recent line of research pursues establishing individual variability in susceptibility to sleep deprivation, or fatigue. Individuals respond differently to fatigue factors. Under the same circumstances some people become fatigued at different times, and to a different degree of severity. When expressing preferences for work shift choices some people perceive themselves to be "owls" or "larks"– presumably attributable to individual differences in their predominant circadian rhythm patterns. An important avenue for improving management of sleep deprivation in sustained operations may be to identify, and optimally utilize innate differences in the ability of healthy adults to resist the cognitive effects of sleep loss (Balkin et al., 2008; Dinges & Goel, 2010). Differences among individuals in the magnitude of cognitive and behavioral changes which occur with sleep loss have more to do with differences in the brain's responses to sleep deprivation than to more malleable psychological aspects of behavior such as motivation and training (Goel et al., 2009). The differences appear to be independent of basal sleep need. When sleep-deprived, some individuals are highly vulnerable to cognitive performance deficits, while others show remarkable levels of cognitive resistance to sleep loss. Identifying the approximate degree of vulnerability or resistance to sleep loss among individuals would permit a greater utilization of military personnel, by providing a way of determining those individuals who need countermeasures early and often, and those who can withstand longer periods with little to no sleep (Van Dongen et al., 2007; Dinges & Goel, 2010).

The National Research Council Committee on Opportunities in Neuroscience for Future Army Applications recommended pursuing this line of research, to search for selection criteria and signs of soldier resilience, when they wrote: "An important lesson from neuroscience is that the ability to sustain and improve performance can be increased by identifying differences in individual solders and using individual variability to gauge optimum performance baselines, responses to performance-degrading stressors, and responses to countermeasures to such stressors" (National Research Council, 2009: 4).

Boredom, Stress Hardiness and Resilience

In an understated form of soldier fatigue, chronic *boredom* and cognitive underload present another stressor for soldiers who often spend countless hours doing little more than nothing (on sentry duty, or engaged in sustained vigilance tasks). But then, in just a few anticipatory moments of terror, as all hell breaks loose during an attack, they experience additional transitional stresses necessitating a call to high-spirited action. Because battles are fought in pulses, this state is likely followed again by relative quiescence, and again later by moments of terror. The transitions from one state to another, involving negative physiological feedback loops in the body, contribute to additional fatiguing stresses of temporal desynchrony (Hancock & Krueger, 2010). Following the notion of individual differences, some soldiers who cope better with such boredom may be more stress hardy and resilient than others.

As Bartone (2006) points out, many combatants suffer physical and mental health decrements following exposure to the stresses of military operations (undoubtedly contributing to soldier fatigue); while many others show remarkable resilience, remaining healthy despite high stress levels, even when working in the same military unit. Bartone, Barry & Armstrong (2009) address ways that leaders of small military units can help foster stress hardiness in individual soldiers, and work toward building resilience in their units. Germain (2010) points out that sleep is a critical component of psychological resilience to stress. Hancock & Krueger (2010), Bartone, Barry & Armstrong (2009) and Germain (2010) each acknowledge that PTSD in some soldiers, while not attributable to a single traumatic event, can be represented as a chronic effect of prolonged stress over which a soldier has little to no control. Such maladaptive conditions can involve continuous threat of traumatic fatality and exposure to conditions and events for which an individual is unprepared. Military combat and its sequalae often expose individuals to events for which no normal human being can be fully prepared (Paulson & Krippner, 2007). Hoge et al. (2008), Pastel (2010) and Germain (2010) all describe the importance of chronic sleep deprivation and soldier fatigue in the development of susceptibility to PTSD in combat. Germain (2010) and Pastel (2010) highlight the importance of obtaining substantial amounts of sleep and regularized sleep patterns during recovery and treatment of PTSD as well as for Traumatic Brain Injury (TBI) once an affected soldier is withdrawn from the combat zone.

Neurocognitive Sustenance of Performance During High-tempo Operations

This *Handbook of Operator Fatigue* contains many valuable findings and principles for application to soldier fatigue issues. But *soldier fatigue* is only one of a diversity of human dimension concerns inherent in military operations that warrant attention from our science. Identification of other neurocognitive needs has energized neuroscientists and other behavioral and medical researchers, pressing them into action to lend their expertise. A recent series of annual conferences convened to address *Sustaining Performance Under Stress* (Kornguth & Matthews, 2009; Kornguth, Steinberg & Matthews, 2010), and a special NRC Committee commissioned to examine *Opportunities in Neuroscience for Future Army Applications* (National Research Council, 2009) provide particularly current state-of-the art coverage of neuroscience and neurocognitive approaches to elucidation and resolution of many health and performance issues facing military forces.

One recent encouraging line of pursuit addresses whether or not soldiers can "bank or store up" sleep to stave off fatigue later. In the first of a series of studies on such questions, Rupp et al. (2009) at WRAIR suggest the extent to which sleep restriction impairs alertness and performance, and the rate at which these impairments are subsequently reversed by recovery sleep, varies as a function of the amount of nightly sleep obtained prior to the sleep restriction period.

In assessment of a very different, but related issue, Frings (2011) inquires whether fatigued individuals operating in small teams of problem solvers experience more Einstellung (low levels of cognitive flexibility), and whether group membership exacerbates or ameliorates such effects. In line with a group monitoring hypothesis (group members being cognizant of their partners' level of alertness or fatigue) Frings' test participants showed increased Einstellung effects when fatigued, while the groups did not; and groups with a member who was relatively less fatigued outperformed other groups (akin to becoming the group's leader). This line of research could point out to military leaders of small groups how important it is for them to obtain their own needed sleep.

Revisiting 1990 Strategies: Coping with Fatigue to Sustain Soldier Performance

Our research findings of the past 25 years fill volumes, and notwithstanding the guidance in *Army Field Manual* 6.22.5, the needs of our soldiers engaged in CONOPS–SUSOPS are still very much reflective of what military leaders were advised in 1990 (Krueger, 1991).

- *Personnel staffing.* Military units must staff their organizations with sufficient personnel to "do the mission." There should be personnel redundancy to allow soldiers to work in shifts, especially in high cognitive workload organizational elements (i.e. communications, command and control personnel), transportation, support, and logistics elements. With pressures to downsize military units, this goal is difficult to achieve.
- *Modify tasks, divide workload, cross-train.* Tasks should be modified to minimize effects of sleep loss (e.g. design vigilance tasks to permit operator rest breaks). Cognitive workload should be reduced as much as possible and equitably distributed among members of teams so everyone gets chances to rest. Team members must be cross-trained to do each other's jobs; and should "overlearn" select tasks to be less subject to performance decrement during sustained efforts. With the press of "do more with less," or at least with fewer people, this goal too is difficult to achieve.
- *Soldier load, physical fitness, training.* Planners should lighten load-carrying requirements to permit soldiers to conserve physical strength. High levels of physical fitness should be attained before beginning operations. "Training as you plan to fight" should include exercise for load-carrying muscles in rehearsals. Our forces do this pretty well, but recall the sergeant's warning that commanders will continue to load up their soldiers.

- *Create "night fighter" teams.* Teams can be selected and trained to work the "night shift." Measures should be taken to allow night fighters to pre-adjust to night work rhythms; and efforts to ensure rest conditions should especially promote quality sleep during the day. Much progress has been made, but supervisors still seem too ready to interrupt soldiers' daytime sleep to "bring more bodies into the action."
- *Rest and sleep discipline planning.* Commanders should develop, practice, and adhere to unit rest and sleep management plans. Issues of rest and sleep should specifically be addressed during mission planning and their importance reinforced to troops in training. Naps should be encouraged when combatants or leaders determine it is safe to take them. Leaders, and command and control personnel, who do much cognitive work, are susceptible to sleep deprivation effects, and therefore should adhere to their own rest plans, setting the example for others. All combatants should practice common courtesy in not interrupting others attempting to sleep.
- *Recovery sleep between arduous work episodes and combat pulses.* Interludes between battles offer opportunity for recuperation. Those who engaged in SUSOPS should be permitted "recovery sleep" to restore alertness and recharge motivation. Recent three-day attacks with overwhelming force or 100-hour battles (Iraq 1991), where recovery sleep was quickly obtained at battle's end, have seemingly given way to continual non-stop 24-hr operations in CONOPS of peacekeeping and nation-building. One might anticipate these issues are more manageable; in some cases they are (e.g. in Iraq and Afghanistan); but in many cases, they have frustrated thousands of deployed military who wonder when their promised rest schedule will become reality.

(The above is adapted from Krueger in the *Handbook of Military Psychology*, 1991)

As with most other sets of guidance principles issued, the last cliché here should be: "the devil is in the details." While researchers continue to explore our science, leaders and soldiers in the field continue to seek their rest and sleep. As Confederate Generals Lee and Longstreet, and Union Colonel Joshua Chamberlain attest in *The Killer's Angels*, perhaps their only truly recuperative sleep was going to be obtained in the days and nights after the Battle of Gettysburg wound itself out. Some things never change!

REFERENCES

Alberts, D.S. & Hayes, R.E. (2003). *Power to the Edge: Command, Control in the Information Age*. Washington, DC: Command and Control Research Program, Center for Advanced Concepts and Technology.

Balkin, T.J., Rupp, T., Picchioni, D. & Wesensten, N.J. (2008). Sleep loss and sleepiness: Current issues. *Chest*, 134, 653–660.

Bartone, P.T. (2006). Resilience under military operational stress: Can leaders influence hardiness? *Military Psychology*, 18, S131–S148.

Bartone, P.T., Barry, C.R. & Armstong, R.E. (2009). To build resilience: leader influence on mental hardiness. *Defense Horizons* No. 69. Washington, DC: National Defense University Center for Technology and National Security Policy.

Caldwell, J.A., Mallis, M.M., Caldwell, J.L., Paul, M.A., Miller, J.C., & Neri, D.E. (2009). Fatigue countermeasures in aviation. *Aviation, Space, and Environmental Medicine*, 80(1), 29–59.

Castro, C.A. & Adler, A.B. (eds) (2005). Operations Tempo, Special Issue of *Military Psychology*, 17(3), 131–246.

Cox, T. & Krueger, G.P. (1989). Stress and sustained performance: Editorial for special issue of nine articles on sustained work, sleep loss and performance. *Work and Stress*, 3(1), 1–2.

DeWulf, G.A. (ed.) (1987). Continuous operations study (CONOPS) Final Report. (CACDA Technical Report No. ACN 073194). Fort Leavenworth, KS: U.S. Army Combined Arms Combat Development Activity, 1987. (Defense Technical Information Center No. AD: B111–424L).

Dinges, D.F. & Goel, N. (2010). Identification and prediction of substantial differential vulnerability to the neurobehavioral effects of sleep loss, in S. Kornguth, R. Steinberg & M.D. Matthews (eds), *Neurocognitive and Physiological Factors During High-tempo Operations*. Burlington, VT: Ashgate Publishing, Chapter 7, 93–103.

Englund, C.E. & Krueger, G.P. (1985). Methodological approaches to the study of sustained work–sustained operations: Introduction to a special section of six articles on this topic. *Behavior Research Methods, Instruments, and Computers*, 17(1), 3–5.

Frings, D. (2011). The effects of group monitoring on fatigue-related Einstellung during mathematical problem solving. *Journal of Experimental Psychology: Applied*, 17(4), 371–81.

Germain, A. (2010). Vires per Somnen: enhancing psychological strength through sleep. Grand Rounds Lecture given September 16th, 2010 at the National Intrepid Center of Excellence, National Naval Medical Center, Bethesda, MD, by Dr. Anne Germain, professor of Psychiatry, University of Pittsburgh Medical Center Sleep Medicine Institute.

Goel, N., Rao, H., Durmer, J.S. & Dinges, D.F. (2009). Neurocognitive consequences of sleep deprivation. *Seminars in Neurology*, 29(4), 320–39.

Hancock, P.A. & Krueger, G.P. (2010). Hours of boredom, moments of terror: Temporal desynchrony in military and security force operations. Defense Technology Paper No. 78. Washington, DC: National Defense University Center for Technology and National Security Policy.

Haskell, C.F., Kennedy, D.O., Wesnes, K.A. & Scholey, A.B. (2005). Cognitive and mood improvements of caffeine in habitual consumers and non-consumers of caffeine. *Psychopharmacology* 179(4), 813–25.

Haslam, D.R. (1985). Sustained operations and military performance. *Behavior Research Methods, Instruments, and Computers*, 17(1), 90–95.

Haslam, D.R. & Abraham, P. (1987). Sleep loss and military performance, in G. Belenky (ed.), *Contemporary Studies in Combat Psychiatry*. Westport, CT: Greenwood Press, Chapter 12, 167–84.

Hoge, C.W., McGurk, D., Thomas, J.L., Cox, A.L., Engel, C.C. & Castro, C.A. (2008). Mild traumatic brain injury in U.S. soldiers returning from Iraq. *New England Journal of Medicine*, 358(5), 453–63.

Hursh, S.R. (2010). Army research psychology: Moving from science to solutions, in P.T. Bartone, R.H. Pastel & M.A. Vaitkus (eds), *The 71F Advantage: Applying Army Research Psychology for Health and Performance Gains*. Washington, DC: Center for Technology and National Security Policy, National Defense University Press, Chapter 2, 47–57.

Hursh, S.R., Redmond, D.P., Johnson, M.L., Thorne, D.R., Belenky, G., Storm, W., et al. (2004). Fatigue models for applied research in warfighting. *Aviation, Space, and Environmental Medicine*, 75(3. Supp), A44–A53.

Killgore. W.D.S. (2010). Asleep at the trigger: Warfighter judgment and decisionmaking during prolonged wakefulness, in P.T. Bartone, R.H. Pastel & M.A. Vaitkus (eds), *The 71F Advantage: Applying Army Research Psychology for Health and Performance Gains*. Washington, DC: Center for Technology and National Security Policy, National Defense University Press, Chapter 3, 59–77.

Kornguth, S. & Matthews, M.D. (eds) (2009). Sustaining soldier high operations tempo performance: Proceedings of the sustaining performance under stress symposium. *Military Psychology*, 21(Supp), 1.

Kornguth, S. Steinberg, R. & Matthews, M.D. (eds) (2010). *Neurocognitive and Physiological Factors During High-tempo Operations.* Burlington, VT: Ashgate.

Krueger, G.P. (1989). Sustained work, fatigue, sleep loss and performance: A review of the issues. *Work and Stress,* 3(2), 129–41.

Krueger, G.P. (1991). Sustained military performance in continuous operations: combatant fatigue, rest and sleep needs, in R. Gal & A.D. Mangelsdorff (eds) *Handbook of Military Psychology.* Chichester, UK: Wiley, Chapter 14, 255–77.

Krueger, G.P. (2008). Contemporary and future battlefields: Soldier stresses and performance, in P.A. Hancock & J.L. Szalma (eds), *Performance Under Stress.* Aldershot, UK: Ashgate, Chapter 2, 19–44.

Krueger, G.P. (2010a). Psychoactive medications, stimulants, hypnotics, and nutritional aids: Effects on driving alertness and performance. *Washington Academy of Sciences Journal,* 96(3), 51–85.

Krueger, G.P. (2010b). Sustaining human performance during security operations in the new millennium, in P.T. Bartone, B.H. Johnsen, J. Eid, J.M. Violanti & J.C. Laberg (eds), *Enhancing Human Performance in Security Operations: International and Law Enforcement Perspectives.* Springfield, IL: Charles C. Thomas, Chapter 10, 205–28.

Krueger, G.P. & Barnes, S.M. (1989). Human performance in continuous/sustained operations and the demands of extended work/rest schedules: an annotated bibliography. Vol. II. USAARL Technical Report No. 89–8, June 1989. Fort Rucker, AL: U.S. Army Aeromedical Research Laboratory. (Defense Technical Information Center No. ADA 210–504).

Krueger, G.P. & Englund, C.E. (1985). Methodological approaches to the study of sustained work/ sustained operations: Introduction to second special section of nine articles. *Behavior Research Methods, Instruments, and Computers,* 17(6), 587–91.

Krueger, G.P., Cardenales-Ortiz, L. & Loveless, C.A. (1985). Human performance in continuous/ sustained operations and the demands of extended work/rest schedules: an annotated bibliography. Vol. I. WRAIR Technical Report No. BB-85-1. Washington, D.C.: Walter Reed Army Institute of Research, May 1985. (Defense Technical Information Center No. ADA 155–619).

Krueger, G.P., Leaman, H.M. & Bergoffen, G. (2011). Effects of psychoactive chemicals on commercial driver health and performance: Stimulants, hypnotics, nutritional and other supplements. Commercial Truck and Bus Safety Synthesis Program: Synthesis No. 19. Washington, DC: Transportation Research Board of the National Academies of Science. Available at: www.trb.org

Krulak, C.C. (1999). The strategic corporal: leadership in the three block war. *U.S. Marines Corps Gazette,* 83, 18–22.

Lieberman, H.R. (2003). Nutrition, brain function and cognitive performance. *Appetite,* 40(3), 245–54.

Mallis, M.M., Mejda., S., Nguyen, T.T. & Dinges, D.F. (2004). Summary of the key features of seven biomathematical models of human fatigue and performance. *Aviation, Space, and Environmental Medicine,* 2004, 75, 3(Supp), A4–A14.

Marlowe, D.H. (1986). The human dimension of battle and combat breakdown, in R. Gabriel (ed.), *Military Psychiatry.* Westport, CT: Greenwood Press, 7–24.

Marshall, S.L.A. (1950). *The Soldier's Load and the Mobility of a Nation.* Quantico, VA: The Marine Corps Association.

McLellan, T.M., Bell, D.G., Lieberman, H.R., & Kamimori, G.H. (2003–2004). The impact of caffeine on cognitive and physical performance and marksmanship during sustained operations. *Canadian Military Journal: Military Medicine,* 4(Winter), 47–54.

Miller, N.L., Matsangas, P. & Kenney, A. (2012). The role of sleep in the military: Implications for training and operational effectiveness, in J.H. Laurence & M.D. Matthews (eds), *The Oxford Handbook of Military Psychology.* Oxford: Oxford University Press, Chapter 20, 262–81.

Naitoh, P., Englund, C.E. & Ryman, D.H., (1986). Sleep management in sustained operations: user's guide. NHRC Technical Report No. 86–22. San Diego, CA: U.S. Naval Health Research Center.

National Research Council (NRC) (2001). Caffeine for the sustainment of mental task performance: Formulations for military operations. Washington, DC: The National Academies Press.

National Research Council (NRC) (2009). Committee on opportunities in neuroscience for future army applications. Washington, DC: The National Academies Press.

Nicholson, A.N. & Stone, B.M. (1982). Sleep and wakefulness handbook for flight medical officers. NATO AGARDOgraph No. 270(E). Neuilly-sur-Seine, France: North Atlantic Treaty Organization Advisory Group for Aerospace Research and Development. [Defense Technical Information Center ADA-115-076].

Pastel, R.H. (2010). Preparing security forces for extreme stress: The case of chemical, biological, radiological, or nuclear threat. The importance of sleep for security forces preparing and reacting to the extreme stress of chemical, biological, radiological and nuclear incidents, in P.T. Bartone, B.H. Johnsen, J. Eid, J.M. Violanti, & J.C. Laberg (eds.) *Enhancing Human Performance in Security Operations: International and Law Enforcement Perspectives*. Springfield, IL: Charles C. Thomas, Chapter 19, 375–97.

Paulson, D.S. & Krippner, S. (2007). Haunted by combat: Understanding PTSD in war veterans including women, reservists, and those coming back from Iraq. Westport, CT: Praeger Security International.

Rupp, T.L., Wesensten, N.J., Bliese, P.D. & Balkin, T.J. (2009). Banking sleep: Realization of benefits during subsequent sleep restriction and recovery. *Sleep*, 32(3), 311–21.

Russo, M.B. (2007). Recommendations for the ethical use of pharmacologic fatigue countermeasures in the U.S. military. *Aviation, Space, and Environmental Medicine*, 78(5.Section II), B119–B127.

Schmorrow, D.D. & Stanney, K.M. (eds.) (2008). *Augmented Cognition: A Practitioner's Guide*. Santa Monica, CA: Human Factors and Ergonomics Society.

Shaara, M. (1974). *The Killer Angels*. New York: Random House Publishing Group.

United States Air Force (1994). Warfighter endurance management during continuous flight and ground operations: An Air Force counter fatigue guide. Memorandum for Air Force Professionals. Washington, DC: Boling Air Force Base, Headquarters, U.S. Air Force.

United States Army (2009). *Combat and Operational Stress Control Manual for Leaders and Soldiers*. U.S. Army Field Manual No. 6–22.5. Washington, DC: Department of the U.S. Army.

United States Army Aeromedical Research Laboratory & U.S. Army Safety Center (1997). Leaders' guide to crew endurance. Fort Rucker, AL: USAARL & USASC.

United States Navy Bureau of Medicine and Surgery (2000). *Performance Maintenance during Continuous Flight Operations: A Guide for Flight Surgeons*. NAVMED P-6410, 1 Jan. Naval Strike and Air Warfare Center (NSAWC), Naval Fighter Weapons School (TOPGUN).

Van Dongen, H.P., Mott, C.G., Huang, J.K., Mollicone, D.J., McKenzie, F.D. & Dinges, D.F. (2007). Optimization of biomathematical model predictions for cognitive performance impairment in individuals: Accounting for unknown traits and uncertain states in homeostatic and circadian processes. *Sleep*, 30, 1129–43.

Wesensten, N.J., Killgore, W.D.S. & Balkin, T.J. (2005). Performance and alertness effects of caffeine, dextroamphetamine, and modafinil during sleep deprivation. *Journal of Sleep Research*, 14, 255–6.

Operational Countermeasures

26

Adaptive Automation for Mitigation of Hazardous States of Awareness

Chad L. Stephens, Mark W. Scerbo and Alan T. Pope

HUMAN ERROR IN AVIATION INCIDENTS AND ACCIDENTS

A major challenge for civil aviation safety organizations, such as the National Aeronautics and Space Administration (NASA) and the Federal Aviation Administration (FAA), is to improve the safety record of an industry with an already exceptionally high level of safety. There have been substantial improvements in aviation safety made during the past century of passenger flight due to the aerospace industry's success at developing increasingly advanced and robust technology (FAA, 1995). A closer look at the causes of aviation accidents reveals that human error accounts for 66 percent of air carrier incidents and accidents, 79 percent of commuter air transport accidents, and 88 percent of general aviation incidents and accidents (FAA, 1990; Shappell, et al., 2007; Wiegmann et al., 2005). The problem of human error plagues all operators in the National Airspace System (NAS), including air traffic controllers. Between 1996 and 2000, the number of controller errors increased 51 percent, with an average of 2.6 operational errors committed each day (Stefani, 2000). While commercial aviation accident rates have continued to decline, the proportion of human error-related incidents and accidents remains remarkably constant in air carrier, commuter/air taxi operations, and general aviation; consequently some have questioned whether the current accident rate is as good as it gets (Shappell et al., 2007). Further, this consistency is unacceptable when considering projections for increasing traffic volume (FAA, 2012), which will lead to more incidents and accidents unless a more complete understanding of operator error is achieved and significant remediation practiced.

The role of human error in incidents and accidents is explicable considering the responsibilities of aircrew during flight: constant awareness of the current state of the aircraft in three-dimensional space and as it changes over time and the environment (air traffic, weather information, and terrain). Also, if a fault occurs at any level of the system (an aircraft component fails), the aircrew must compensate for this failure whether or not functionality is restored. Maintaining safe operations is an ongoing activity toward which all operators in the NAS must constantly strive. The importance of human operators and their ability to flexibly respond to changing demands of tasks and hazards in complex human–machine aviation systems has been the focus of aviation psychology (Fitts, 1951). Researchers noted that it was overly simplistic, if not naïve, to write off the causes of accidents to operator error (Heinrich, Petersen & Roos, 1980). The term "human error" is deceptive, as it implies that the human was the sole cause of the incident or accident. Conventional accident analyses focus almost exclusively on the actions of workers at the front lines; for aviation, the pilots, co-pilots, flight engineers, air traffic controllers, and dispatch operators. Human error may be observed because a human was the last line of defense in a series of failed system elements. Insightful researchers of human error in aviation (Maurino et al., 1995) suggested a broader approach, which considers that accidents and incidents emerge from a confluence of system failures – including individual human performance.

The work by Reason (1990) established that accidents cannot be attributed to a single cause, or in most instances, even a single individual. The "Swiss Cheese" model (or the cumulative act effect model) of accident causation is used in the risk analysis and risk management of human–machine systems (Reason, 1990). Reason proposed in the Swiss Cheese model that most accidents can be traced to one or more of four levels of failure: organizational influences, unsafe supervision, preconditions for unsafe acts, and the unsafe acts themselves. These failures are individual weaknesses conceived of as holes in the slices of cheese. The holes in the cheese were used to represent the flaws in individual safety barriers, which are rarely failure-proof in real world situations. Thus, when all the holes (continually varying in size and position in all slices) are aligned, risk is maximized and danger is imminent. The system as a whole produces failures during these momentary alignments, permitting "a trajectory of accident opportunity," so that a hazard passes through all of the holes in all of the defenses, leading to a failure. Reason's model of accident causation asserts the necessary levels of assessment required for investigations and more complete understanding of incidents and accidents.

The Human Factors Analysis and Classification System (HFACS; Shappell & Wiegmann, 2001) applied Reason's theoretical approach and provided a taxonomy which an investigator could use to identify causes of accidents from the "top" (organizational level) to the "bottom" (operational level), all of which impact aircrew behavior. This taxonomy included classifications for three forms of human error: skill-, rule-, and knowledge-based (Rasmussen, 1982). This classification of errors has allowed researchers to focus on the causes of incidents and accidents with finer resolution and examine predictable and preventable unsafe acts resulting from suboptimal human conditions. The theoretical Swiss Cheese model and the HFACS analytic methods reveal issues operators encounter when interfacing with complex systems that researchers and system designers can systematically examine to improve operator performance and safety.

HAZARDOUS STATES OF AWARENESS IN AVIATION

Fatigue is one of a set of human conditions that have been termed "hazardous states of awareness" (HSAs; Pope & Bogart, 1992). HSAs typically come about when operators of human–machine systems perform prolonged routine, habitual activities. A taxonomy of

HSAs has been developed to further understand types of HSAs and encourage systematic study of countermeasures (Scerbo et al., 1999).

The focus of this volume, fatigue, represents one of these HSAs that is clearly associated with degraded performance. The problem of fatigue in complex operational settings, such as aviation (Caldwell, Chapter 24, this volume), has high face validity, that is to say the general public, general aviation and commercial pilots, and human factors researchers understand that a fatigued state interferes with performance. Pilot fatigue has been identified as a major contributing factor to aviation accidents (United States Government, House of Representatives, 1999). A report from the National Transportation Safety Board (NTSB) indicated that 21 percent of the reports in the Aviation Safety Reporting System (ASRS; see Reynard et al., 1986) were related to fatigue (NTSB, 1999). This issue has been on the NTSB "Most Wanted" list since its creation in September 1990; making fatigue one of the four problems that have remained since the original list of 18 problems was published.

In response to a 1980 Congressional request, NASA Ames Research Center created the Fatigue/Jet Lag Program, to empirically investigate the impact of fatigue, sleep, and circadian rhythms on performance in flight operations (NTSB, 1999). In 1991, the Fatigue/Jet Lag Program became the Fatigue Countermeasures Group (FCG) to underscore the emphasis on the development and evaluation of countermeasures to mitigate the adverse effects of fatigue-inducing factors and maximize flight crew performance and alertness (United States Government, House of Representatives, 1999). FCG research studies were conducted in a variety of controlled laboratory environments, full-mission flight simulations and aviation environments, and this research resulted in specific recommendations to the FAA.

The FCG research program yielded an abundance of information related to fatigue based on a range of measures (e.g., performance, physiological, and behavioral). The impact of physiological mechanisms related to sleep, sleep loss, and circadian rhythms was clearly demonstrated in long-haul and overnight cargo operations (Gander, Gregory, Connell et al., 1998; Gander, Gregory, Miller et al., 1998). Short-haul operations were investigated and similar factors contributing to fatigue were identified, as well as the effect of long duty days (Gander, Gregory, Graeber et al., 1998). FCG identified the key mitigation factors to address fatigue: education and training, hours of service, scheduling practices, and technological interventions/countermeasures (Gander, Rosekind & Gregory, 1998; Mallis et al., 2007; Rosekind et al., 1996).

Brown (1994) underscored the challenge of addressing the fatigue problem by describing its nature as "sufficiently insidious that [operators] are unaware of their impaired state and hence in no position to remedy it." This defining characteristic, a lack of subjective awareness of fatigue, relates directly to research into fatigue detection, prediction, and mitigation through countermeasures. Some countermeasures in aviation are technologies that would identify fatigue states, improve a pilot's understanding of their state of fatigue, and alter that state. Such awareness monitoring and modification supports the operator's safe performance of necessary activities in the operational environment.

The pilot is the last line of defense; therefore, efforts to decrease human error represent an area for continued improvement in aviation safety. Human error in aviation has been dealt with principally by automating flight tasks. Although effective to the degree that this automation is reliable, it relegates the pilot to the role of backup to or cross-checker of automation (Palmer et al., 1995; Schutte, 1999). This results in additional problems (e.g., complacency, inattention, etc.) noted by pilots and researchers (Billings, 1991). Such issues instigated "human-centered" automation concepts, which aim to improve pilot–automation interaction, and maintain an active role for the pilot. One such concept, *adaptive automation*, permits flexible allocation of automation functions based on consideration of operator state (Scerbo, 1996). Some uses of adaptive automation have shown benefits in performance, increased situational awareness, and decreased cognitive workload in human–machine system users (Prinzel et al., 2000). Adaptive automation is a nascent area of research with numerous challenges remaining; now is an exciting time to be involved with this area of work.

ADAPTIVE AUTOMATION BACKGROUND

The idea of developing technology that could complement a pilot's abilities has interested researchers for many years. Historically, automated systems in the cockpit were developed with the promise of improving flight performance. Over the years, however, research began to accumulate showing that interaction with automation did not always make the pilot's job easier. Weiner (1989) noted that at times, automation could be downright clumsy, actually increasing workload, particularly when the pilot was already overburdened. Other researchers argued that automated systems can leave a pilot with too little to do. In the short term, the pilot risks becoming a mere passive observer instead of an active participant and may be less able to detect critical signals or warning conditions (Parasuraman, Mouloua, Molloy & Hilburn, 1996). Over longer intervals, however, Wickens (1992) suggested that extensive use of automation may cause an operator's manual skills to deteriorate.

The idea that systems designed to reduce pilot workload might have the opposite effect led some researchers to investigate the methods by which automation was implemented. Sheridan & Verplank (1978) originally described ten levels of automation. The lowest level represents manual systems that include no form of automation. At the opposite end of the spectrum are fully automated systems that required no operator intervention. Between these two extremes automation varies in terms of authority and autonomy. More specifically, lower levels of automation might offer suggestions for the user to consider, whereas higher levels of automation might have the authority to select and execute a course of action and then inform the user.

In the 1980s, researchers and designers began to discuss the requirements for some of these intermediate levels of automation. For instance, Rouse, Geddes & Curry (1987–1988) described a model to invoke automation relying on information about the current state of the system, external events, and expected operator actions. Hancock & Chignell (1987) discussed an adaptive aiding model in which tasks would be allocated to the operator or system based upon both current and future levels of operator workload. The first large-scale demonstration of these adaptive aiding concepts was the Pilot's Associate program sponsored by the Defense Advanced Research Projects Agency (DARPA; see Hammer & Small, 1995). The overall objective of the program was to employ intelligent systems to provide pilots with the right information, in the right format, at the right time. In the 1990s, the associate concept was extended further in the U.S. Army's Rotorcraft Pilot's Associate program. The goal of this development effort was to build a network of intelligent systems that could detect and organize incoming data, assess the internal status of the aircraft, assess external information about the mission, and feed this information into a planning and decision-making system (Miller & Hannen, 1999).

In the early 1990s, Parasuraman, Bahri, Deaton, Morrison & Barnes (1992) described automation that was adaptive in nature. They argued that designing automation to be adaptive could provide an optimal means to couple the level of automation to the level of operator workload. Parasuraman and his colleagues described several methods for triggering the automation or switching among modes of automation. For example, automation could be triggered by the occurrence of critical events. On the other hand, an operator's current level of performance could be compared against models or real-time assessments of operator performance. Last, they suggested that automation could be triggered by psychophysiological measures (see below).

Scerbo (1994) and others (Malin & Schreckenghost, 1992) noted the similarity between requirements for effective adaptive automation and teamwork concepts (e.g., collaboration, backing up one another, communication, etc.). Scerbo used the human teamwork taxonomy of Fleishman & Zaccaro (1992) to describe comparable functions for adaptive automation in aviation. For example, the pilot and the aircraft would need time to become oriented with one another. Tasks needed to complete a mission would have to be allocated between the pilot and the aircraft according to appropriate rules for resource allocation. Systems monitoring and procedure maintenance functions would be as necessary for adaptive automation as they are for human teams. Finally, Scerbo noted that overall mission effectiveness would ultimately be determined by how well the pilot and aircraft operated as a team and, therefore, necessitated monitoring joint performance of the pilot and the aircraft.

Scerbo (1996) took a comprehensive look at the notion of adaptive automation and described the current state of the technology, as well as the potential impact on the user if the technology were fully realized. Scerbo suggested that operators would have to learn the capabilities and limitations of the system and the system would have to learn those of the operator. Further, operators would require longer training times to achieve optimal levels of performance. The operator and the system would each have unique tasks to perform, but would need to collaborate on others. Perhaps most important, Scerbo argued that an adaptive system should be capable of assuming some of the operator's responsibilities during periods of high workload or anticipating the operator's actions and adjusting its behavior accordingly.

Interest in adaptive automation continued to grow. However, it soon became clear that there was no unanimity among researchers regarding terminology. For example, developers in the computer industry were describing adaptive interfaces that functioned more like coaching systems (Wilensky, Arens & Chin, 1984). Parasuraman, Sheridan & Wickens (2000) offered a framework for designers when considering different types of automation. Their first dimension is based on the different levels of automation described by Sheridan & Verplank (1978) and their second dimension describes system analogs of human information processing stages: information acquisition, information analysis, decision selection, and action implementation. Scerbo (2001) also offered a two-dimensional taxonomy of adaptive technology. The first dimension addressed the underlying source of flexibility in the system (i.e., whether the information displayed or the functions themselves were flexible). The second dimension addressed how the changes were invoked. In *adaptable* systems, changes in the display or in the functions would be initiated by the user. By contrast, in *adaptive* systems either the user or the system could initiate changes. Scerbo argued that adaptable and adaptive technology could be distinguished by authority. Specifically, the user maintains control in adaptable systems but *shares* control in adaptive systems.

Scerbo's (2001) definition of adaptive automation requires that control be shared between the operator and the system. However, there has been debate as to who should have control over changes among modes of operation. There are those who argue that operators should always have authority over the system because ultimately they are responsible for the system and are likely to be more efficient at managing the system's resources and their own (Billings & Woods, 1994; Malin & Schreckenghost, 1992). However, others have argued that the need for an operator to maintain control over changes in the automation is not always warranted. As Weiner (1989) noted, there may be times when the operator is not the best judge of when automation is needed, i.e., when changes in automation are needed precisely at the moment the operator is too busy to make those changes. Inagaki, Takae & Moray (1999) have demonstrated that the best piloting decisions concerning whether to abort a take-off are made when the pilot and the automation share control. In fact, a prototype adaptive automation system tested on the F-16D for ground collision avoidance could handle the aircraft quicker than any human pilot (Scerbo, 2007; Scott, 1999). Pilots who tested the system and were given the opportunity to override the automation did not do so, instead opting to maintain shared control with the aircraft.

PSYCHOPHYSIOLOGICALLY ADAPTIVE SYSTEMS (PAS)

Adaptive Automation as a Special Case of a Closed-Loop System

Researchers at NASA Langley Research Center established the first program to assess pilot mental state, specifically task engagement, via psychophysiological measures: electroencephalogram, event-related potentials, and heart-rate variability (EEG, ERP, and HRV) in the early 1980s (Pope & Bowles, 1982). The early work done by this group and research sponsored by this group transitioned basic psychophysiological research techniques to more applied/operational settings (Comstock, Harris & Pope, 1988; Cunningham & Freeman, 1994; Zacharias, Caglayan & Pope, 1985). A major impetus for this research effort came through an analysis of narratives in the ASRS, which identified suboptimal awareness states: complacency, boredom, diminished alertness, compromised vigilance, and lapsing attention, in addition to fatigue. This analysis led Pope & Bogart (1992) to identify hazardous states of awareness and investigate specific psychophysiological markers of operator engagement. This approach presented a promising technique for identifying diminished functional state. Pope & Bogart (1991) extended this work by focusing on modification of attention in individuals diagnosed with attention deficit disorder. The adaptive system was programmed to recognize each individual's characteristic EEG profile, using a technique similar to that used by Pope & Bogart (1992).

Science fiction is an influential source and has been identified as contributing to real-world technology, war, and politics. A recent publication noted that past and present scientists and military leaders contribute creative ideas to this realm of literature which have served as the basis for research and development of technology in the real world (Singer, 2010). Looking ahead to the future and considering the possibilities of blending human and machine potential, U.S. Air Force researchers described a prospective aviation technology – a symbionic cockpit capable of considering the mental state of the pilot to optimize mission performance (Reising & Moss, 1986). Inspired by these researchers and a description of a Physiological Control and Monitoring System (PCMS) from science fiction of two decades earlier (Poyer, 1969), and informed by work in the fields of psychophysiology and biofeedback, a closed-loop biocybernetic system was developed and experimental results from this system demonstrated the operation of a psychophysiologically adaptive automation system (Pope, Bogart & Bartolome, 1995). The biocybernetic system was based upon a closed-loop concept that involved adjusting or modulating (cybernetic, for governing) a person's task environment based upon a comparison of that person's psychophysiological responses (bio-) with a training or performance criterion.

This effort to construct a *biocybernetic* system benefited from a fortuitous confluence of developments and available capabilities. A shift in emphasis within the aerospace human factors community from the problems of high workload to concerns about underload issues from increasing automation, articulated by Hancock and colleagues (Hancock & Chignell, 1987; Hancock & Warm, 1989) and others (Gaillard, 1993; Hockey, 1986), resulted in a focus of the NASA Langley Research Center (LaRC) mental state research onto issues of complacency and boredom. The Multiple-Attribute Task (MAT) Battery, developed at LaRC (Comstock & Arnegard, 1992) provided a set of incorporated laboratory tasks resembling activities performed by pilots, including compensatory tracking, system monitoring, resource management, and communication tasks. The participants in experiments could be asked to perform the tasks individually or simultaneously and the tracking task could operate in automatic or manual mode while the other tasks were designed to operate in manual mode. The MAT Battery was not designed to be a flight simulator; the great utility of this application was (and continues to be) the possibility of testing human (i.e., non-pilot) behavior in a multitasking context that corresponds to aviation tasks.

These varied tasks map onto a comprehensive theoretical taxonomy previously described by Rasmussen (1982); the tracking task is a skill-based behavior, the system monitoring task is a rule-based behavior, the resource management task is a knowledge-based behavior (Weaver et al., 2003) and by extension the communications task can be categorized as a rule-based task. The application of the MAT Battery to aviation research has been questioned recently (Haarmann, Boucsein & Schaefer, 2009). While the MAT Battery cannot be used to validate specific hypotheses which would require a domain-specific simulation, its generic environment allows for validating foundational hypotheses that can then be addressed in more detail. One also should consider the mapping of the MAT Battery tasks on to a well established taxonomy of types of human errors (Rasmussen, 1983; Reason, 1990) to identify the potential for systematic investigation of human behavior.

A number of improvisations and expedients were made in order to implement the biocybernetic system prototype. Pope and Bogart (1992) introduced a conceptual model of HSA-inducing individual and situational factors and a methodology based on discriminant analysis of electrocortical frequency band power to distinguish between hazardous and optimal states. However, the discriminant analysis methodology did not lend itself to the real-time tracking of operator state necessary for timely adaptive response. The psychophysiological literature (Davidson, 1988; Lubar, 1991; Offenloch & Zahner, 1990; Streitberg et al., 1987) described shifts in EEG characteristics as fluctuating states of attention, which suggested that a ratio measure derived from the EEG might serve as a real-time responsive index of attentional state. This index was constructed to have higher values for subject states corresponding to greater degrees of mental engagement, that is, greater demands for operator involvement (Pope & Bogart, 1996). A sustained

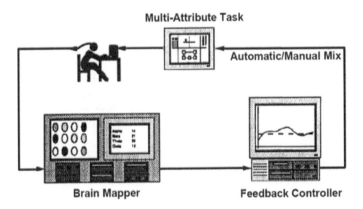

Figure 26.1   Biocybernetic closed-loop system used to monitor operator
               state and adjust level of automation of the Multi-Attribute Task
               Battery

reversal in slope of the trend of such an index over a time window, essentially the time
derivative of the index time series, was used as the criterion for indicating state change
prompting an adaptive system response. Pope, Bogart & Bartolome (1995) produced
two findings with direct applications to adaptive automation: 1) identification of a "high
performance" engagement index; and 2) a method by which to test future indices; these
applications were extended by further research conducted at Old Dominion University
(ODU).

Prinzel, Scerbo, Freeman & Mikulka (1997) conducted an experiment at ODU to
investigate the performance effects of using the system developed by Pope, Bogart &
Bartolome (1995) in an adaptive context (see Figure 26.1). The MAT Battery was the
primary task used for the study, which allowed for replication and extension of the work
by Pope, Bogart & Bartolome (1995). In their study, the MAT Battery tracking task was
programmed to shift between automatic and manual modes based on the state of the
operator; all other tasks required no input from the operator. Prinzel and his colleagues
(1997) evaluated the system under both negative and positive feedback conditions. The
investigators expected that under negative feedback conditions the system would result
in better tracking performance because it would switch to automatic mode when the
engagement index was rising and switch to manual mode when the engagement was
falling. The opposite pattern of task changes was expected in the positive feedback
condition. Thus, poorer tracking performance was expected under positive feedback
conditions because the system would demand manual operation during intervals when
the engagement index was rising and change to automatic mode when the engagement
index was falling. They recorded from four EEG sites (Cz, Pz, P3, and P4) and examined
system performance with three engagement indices (1/alpha, 10*beta/alpha, and 20*beta/
alpha+theta).

Prinzel and his colleagues (1997) examined task allocations between modes and values
of the engagement indices, as well as tracking performance. Their results showed the
system made more switches between modes under negative feedback conditions and that
the difference between positive and negative feedback was most pronounced with the

20*beta/alpha+theta engagement index. Also, under negative feedback conditions, the value of the 20*beta/alpha+theta engagement index was consistent with expectations: it was higher under automatic mode (when the index was rising) and lower under manual mode (when the index was falling). The opposite pattern was observed under positive feedback. Further, when looking only at periods of manual tracking, performance was better under negative feedback conditions across all engagement indices.

The findings obtained by Prinzel et al. (1997) are important in two regards. First, they showed that the brain-based biocybernetic system functioned as expected under both positive and negative feedback conditions and, second, the tracking data represented the first empirical demonstration that a brain-based adaptive system can facilitate operator performance. A similar pattern of results was obtained in a follow-up study by Freeman, Mikulka, Scerbo, Prinzel & Clouatre (2000). Again, more task switches and better tracking performance were observed under negative feedback conditions.

The results of numerous studies have demonstrated the potential for the brain-based system developed by Pope and his colleagues (1995) to be used as an adaptive interface (for review see Scerbo, Freeman & Mikulka, 2003). Additionally, studies using the same or a highly similar biocybernetic system have incorporated ERP and HRV to assess the state of the operator and drive the adaptive system (Parasuraman, Mouloua & Hilburn, 1999; Prinzel, Freeman, Scerbo, Mikulka & Pope, 2003; Prinzel, Parasuraman et al., 2003). While the previous research was thorough, it was not exhaustive in nature and there are still many methodological and theoretical stones left unturned.

FUTURE DIRECTIONS

Four applications of the closed-loop system design were proposed by Pope & Bogart (1996) and Pope, Bogart & Bartolome (1995): 1) psychophysiologically adaptive automation; 2) validation of candidate psychophysiological indices of operator state; 3) evaluating an interactive system design by determining optimal human/system task allocation "mixes;" and 4) a psychophysiological self-regulation training system based on the adaptive task concept. Of these, only the psychophysiologically adaptive automated system application has been researched to any extent, primarily by investigators at ODU, as chronicled above (cf. Scerbo, Freeman & Mikulka, 2000). Each of the four applications is discussed below.

Psychophysiologically Adaptive Automation: A Means for Maintaining Effective Operator State

Adaptive automation refers to systems that can modify their method of operation in response to dynamic situational requirements (cf. Bryne & Parasuraman, 1996; Scerbo, 1996). This allows for the restructuring of the task environment with regard to: a) what is automated; b) when it is automated; and c) how it is automated (Rouse, 1988; Scerbo, 1996). Adaptive automation is implemented to counteract the negative effects that often accompany static automation: rapid operator fatigue, impaired decision-making, degradation of manual skills, a loss of situational awareness, and a reduced ability to detect system failures (Parasuraman, Molloy & Singh, 1993; Scerbo, 1996). Central to the development of adaptive automation is when and how shifts in automation levels are to be made; and whether performance measures, operator workload, external demands or mental state should "trigger" these shifts. As is noted above, some investigators have suggested incorporating the use of psychophysiological measures such as heart rate variability, skin conductance, EEG, and ERPs into an adaptive system (Bryne & Parasuraman, 1996; Parasuraman, 1990; Wickens, 1990). A strength of psychophysiological measures, especially in the detection of task underload, is that they offer the possibility of continuous, real-time assessment of an individual's state.

Hettinger, Branco, Encarnacao & Bonato (2003) examined and expounded upon the capability of the biocybernetic system described by Pope, Bogart & Bartolome (1995). The concept of "performance equilibrium" was underscored by Hettinger et al. as the goal of an adaptive interface such that the human–machine system is maintained within the boundaries of a desired envelope with respect to some psychological variable(s). Hettinger et al. describe a neuroadaptive interface as a system of computer-based displays and controls, which are driven by specified cognitive and/or emotional states of the user. This adaptation can alter the information presented to the user, as well as modulate the control the user can exert on the system. Hettinger et al. clearly distinguish between brain–computer interfaces primarily intended to support the *control* of computer-based systems using neural signals and neuroadaptive interfaces which subtly change display and control characteristics to modify user state and ostensibly improve performance. This latter function may be thought of as psychophysiologically based *modulation* of the system that the operator is consciously controlling by indirect means. This difference between modulation and control distinguishes the biocybernetic paradigm in the LaRC and ODU work from other brain–computer interface (BCI) work.

Validation of Candidate Psychophysiological Indices of Operator State

Developing a biocybernetic system requires answering the question: How are psychophysiological parameter changes to be interpreted meaningfully to drive adaptation? Empirical research and development are required to examine the couplings between neuroadaptive interface concepts and nervous system activity in real world situations. Investigations are necessary to provide practical information about how such systems should be designed to allow for further specification and efficiency of such systems. An indirect benefit of this type of work includes useful knowledge about nervous system activity, which has not been obtained through other methods. Such a method of acquiring useful knowledge about the nervous system activity in real world situations based upon an "examination of neuroadaptive interface concepts" was demonstrated by Pope, Bogart & Bartolome (1995). The article included a method to evaluate the relative usefulness of candidate EEG indices for reflecting mental engagement in a task.

As previously described, Pope, Bogart & Bartolome (1995) introduced and described a psychophysiologically adaptive automation concept which was later explored in research conducted by investigators at ODU in collaboration with LaRC (Scerbo, Freeman & Mikulka, 2000), although they primarily presented a demonstration of the closed-loop methodology as a method for testing and validating psychophysiological indices of engagement for use in such systems. This validation method, further demonstrated by Freeman, Mikulka, Prinzel & Scerbo (1999), remains one of the largely unexplored contributions by Pope et al. (1995). The method offers an additional option for exploring functional relationships between task demand and physiological measures. The concept implies the proposition that the closed-loop paradigm that represents the adaptive configuration in which physiological indices are to have a steering role can also serve as a prior validation test bed for the indices themselves.

Selecting the psychophysiological signals for the task-modulating index is critical when implementing psychophysiologically based adaptive automation. Pope, Bogart & Bartolome (1995) demonstrated a method to evaluate the relative usefulness of candidate EEG indices for reflecting mental engagement in a task. In the initial biocybernetic system, candidate engagement indices were gleaned from the EEG literature on attention and vigilance (Davidson, 1988; Lubar, 1991; Offenloch & Zahner, 1990; Streitberg et al., 1987). Recent clinical research has employed a similar ratio construction as a diagnostic indicator in attention deficit hyperactivity disorder (Monastra et al., 2005). In other recent work computational approaches for identifying candidate cognitive indices have been implemented. These other methods do not involve feedback of psychophysiological parameters to modify the task, and thus are considered open-loop. Recent work by Russell et al. (2005) defined issues to be considered in selecting pattern classification algorithms for real-time operator state estimation. Russell & Wilson (2005) demonstrate the power of feature selection or saliency analysis methods for enhancing the practical utility of operator state estimation from classification of psychophysiological data collected in an open-loop task.

These algorithmic methods represent a valid and necessary form of investigation as they sometimes produce results that conflict with findings reported in prior literature. This affords novel information about the nervous system in unique laboratory and real-world settings. One example of contrary findings can be seen in a study by Russell & Wilson (2005) in which both the alpha and theta EEG bands were selected with the least frequency by all feature selection methods used in this study. The cognitive psychology literature typically associates these variables with changes in cognitive load; this finding requires further examination by some other means. The closed-loop index evaluation method offers one approach to validating competing candidate index definitions from the literature and from algorithmic methods, in the context of the closed-loop paradigm, the paradigm in which they will be employed in a PAS.

The problem of determining the relative usefulness of a cognitive index can be seen as one of determining the relative strengths of the functional relationships between the candidate indices and task characteristics (e.g., manual versus automatic modes) in a closed-loop configuration. This powerful method is an adaptation of a procedure first described by Mulholland (1977) in a biofeedback context. The closed-loop method is exceptional in that it enables an index of engagement to be identified, which is maximally sensitive to changes in task demand through the process of feedback. This methodology permits researchers to hypothesize about which features of brain signals best modify an automated system to suit human attentional capability and efficiently test the hypotheses in an experimentally controlled operational-like context. This adaptive method is essentially a feedback control process achieved by systematic adjustment of task demand for operator participation optimally resulting in stable task engagement.

If there is a functional relationship between an index and task mode (and, consequently, task demand), the index exhibits stable short cycle oscillations under negative feedback and longer and more variable periods of oscillation under positive feedback. The strength of the relationship is reflected in the degree of contrast between the behaviors of the index under the two feedback contingencies. The lack of a relationship is indicated by a finding of no difference between the behaviors of the index under the two feedback contingencies.

A strong functional relationship between a candidate index and task-operating mode (e.g., manual versus automatic) is reflected in stable operation under negative feedback and unstable operation under positive feedback. Candidate indices are judged on the basis of their relative strength in producing these expected feedback control system phenomena. That is, a better index choice causes the index to oscillate more regularly and stably under negative feedback than an inferior choice, and a better index increases the degree of contrast between the behaviors of the index under the two feedback contingencies (Kramer & Weber, 2000).

The method may also be characterized as a series of discrete replications of brief open-loop experiments in which task demand variations represent the "independent" variable, the effects of which on the "dependent" variable of EEG response are observed. The EEG response is, of course, programmed, in turn, to modulate the changes in task demand, which initiate each brief experiment. This paradigm is different from typical experimental methods of operator workload, in that the researcher bounds the independent variable within a range of possibilities, in the case of the LaRC and ODU work it was level of automation. But the initiation of the events is tied to the state of the experimental subject such that each subject experiences a distinctive chain of small experiments linked together by the decision rules implemented by the experimenter prior to the start of the overall experiment. This predefined control is akin to Scerbo's (2005: Section 3.1) description of complementary convergent and divergent tests implemented by the closed-loop evaluation method. Such an approach, which includes appropriate controls, is one way to exhibit that the system performs according to expectations under a variety of test conditions.

Engagement indices

Pope, Bogart & Bartolome (1995) examined the efficacy of several candidate engagement indices and suggested that there are still more indices that warrant study. They compared beta/alpha, beta/alpha+theta, and two versions of alpha/alpha recorded from different EEG sites. They found that the beta/alpha+theta index produced the greatest difference in task mode switches between positive and negative feedback conditions. In a follow-up study, Prinzel, Scerbo, Freeman & Mikulka (1997) also examined different indices (1/alpha, 10*beta/alpha, and 20*beta/alpha+theta) and found the 20*beta/alpha+theta index to be the best discriminator between positive and negative feedback conditions for task mode changes. Freeman, Mikulka et al. (1999) reported similar results and they also demonstrated that tracking performance was better with the beta/alpha+theta and beta/alpha indices under negative feedback. To date, no systematic investigation has been undertaken to determine the best performing engagement index on which to base mitigation intervention.

Individual differences

A commonality that this paradigm has with other paradigms is the necessity to consider the impact of individual differences. Since the biocybernetic system was based on an individual's electrocortical activity, it is expected that variations will occur in power spectral densities of brain wave frequencies. Pope & Bogart (1992) found it necessary to take individual differences into account when identifying the features of the EEG to be used as state discriminators. They found that there was minimal similarity of discriminant functions among individuals. The multiple state discrimination function set of each individual was sufficiently unique to discourage attempts to derive general sets that would be successful over a set of individuals. That is to say, the electrode sites and frequency bands that discriminated states for one individual were unlikely to be the most discriminating for another individual. Therefore, in the application of the state identification procedure, each individual must be "profiled" following a standard procedure that parallels the initial training/test session prior to the experimental session.

Pope & Bogart (1996) described such a profiling procedure to account for individual differences. The procedure would involve a quantitative EEG (QEEG; from at least 19 electrode sites) while a subject performs the MAT Battery, as it is systematically stepped through its range of automation levels. Considering that there are four tasks which can be operated in manual or automated mode, there would be 16 possible combinations of these task states. At each automation level, a spatial map of EEG activity, portraying the scalp distribution of activity within that band, would be generated for each of the alpha, beta, and theta frequency bands. Brain maps would be generated each corresponding to a level of task automation and its associated level of operator engagement. The EEG data represented in the entire set of brain maps would then be used to derive discriminant functions for each state experienced by the operator. This proposal to investigate the effect of various manual/automatic task mode mixes has yet to be fully realized in an empirical investigation. Furthermore, an exhaustive systematic investigation to determine the optimal electrode sites and frequency bands on which to base mitigation intervention has yet to be conducted, leaving room for improvement upon previous investigations involving a limited number of automation levels and electrode sites.

Slope vs. absolute

In addition to individual differences, the temporal trajectory of operator state changes requires consideration. This is necessary because temporal as well as magnitude changes in psychophysiological signals reflect operator state changes, which warrant mitigation. The biocybernetic system (Pope et al., 1995) was designed to mitigate task disengagement due to automation by triggering changes in task mode based on the fluctuations of an engagement index constructed as a ratio of EEG band powers. The adaptive automation configuration (negative feedback) of this system switched task mode to (or maintained at) manual when the engagement index was decreasing (negative slope) over several successive measurements, and to (or maintained at) automatic when the index was increasing.

However, subsequent work by Freeman, Mikulka, Scerbo & Hadley (1998) identified a means by which to improve the algorithm used for switching between modes. In the Pope et al. (1995) configuration, changes between modes were triggered by increases or decreases in the slope of successive values of the engagement index. Freeman et al. argued that under this arrangement, changes between modes could occur even when the absolute value of the engagement index indicated that no change was required. Consequently, they modified the algorithm that triggered changes between modes. Specifically, they adopted a criterion value for the engagement index and triggered task mode changes based upon values of the engagement index relative to the criterion. They found that the values of the engagement index were similar to those obtained by Prinzel et al. (1997). Under negative feedback conditions, the engagement index was higher in automatic than manual mode and lower in automatic than manual mode under positive feedback. However, Freeman and his colleagues did not observe better tracking performance under negative feedback and reasoned that the absence of a performance benefit may have been due to selecting an *arbitrary* criterion value.

In a subsequent study, Freeman et al. (1999) calculated the engagement index for each participant over a 5-min baseline interval prior to the experiment. The mean value of the engagement index obtained over this interval for each individual was used as the criterion to trigger task mode changes. The results showed that the engagement index was higher in automatic than manual mode under negative feedback and lower in automatic than manual mode under positive feedback. Moreover, tracking performance was significantly better under negative feedback conditions. Thus, when the criterion value was derived for each individual, the brain-based system performed as expected and facilitated tracking performance. Identification of an appropriate baseline against which to compare real-time operator states remains open to debate. A systematic comparison of the original slope method described by Pope et al. (1995) and the absolute method employed in later studies should reveal defining qualities of both of these methods in terms of strengths and weaknesses.

Window size
The "mitigation triggers" employed by Pope et al. (1995) in their biocybernetic system incorporated the dimension of time by using a moving window of successive EEG data points in computing both index slope and level. This moving window is another important variable that could impact the operation of the biocybernetic system. Pope et al. (1995) used a 40-s window in their initial study. Hadley, Mikulka, Freeman, & Scerbo (1997) reasoned that a smaller window might improve system's sensitivity and compared the original 40-s window to a smaller 4-s window. They found the same interaction between positive and negative feedback and automatic and manual modes that had been obtained in previous studies. Further, they found that the 4-s window exacerbated the feedback by mode interaction and also resulted in better tracking performance than the 40-s window. Freeman et al. (1999) also looked at 40-s and 4-s windows and found that the shorter window resulted in more task mode switches. In addition, the number of task mode switches made with the 4-s window was moderated by engagement index with the beta/alpha+theta and beta/alpha producing more switches than the 1/alpha index. However, these investigators did not observe a tracking performance benefit with the shorter window. To date, no systematic investigation has been undertaken to investigate the optimal windows of psychophysiological time series on which to base mitigation intervention.

A promising mechanism for triggering mitigations has been proposed which is based upon Statistical Process Control (SPC) rules for determining when psychophysiological data have departed from their nominal state sufficiently to warrant mitigation (Tollar, 2005). "The SPC rules have been formulated to detect when it is appropriate to intervene in a process. Such a point is an intuitive time to trigger a mitigation" (Tollar, 2005). The intent of this control mechanism is similar to that of the biocybernetic system, essentially "treat the operator's cognitive state as a process that needs to be kept in control" (Tollar, 2005: 418). The SPC approach explicitly treats operator state as a process unfolding over time and suggests that predicting from state temporal trajectories may render psychophysiological signal tracking more useful in determining events that are sufficiently significant to warrant intervention. SPC rules provide a structure to permit much-needed systematic investigation of temporal and magnitude parameter variations in the definition of changes that are significant. Intuitively, a dynamic method such as SPC should yield better results than a window with a defined range but this is a question for empirical investigation.

Workload

One potential advantage of an adaptive brain-based system would be the opportunity to moderate operator workload. Prinzel, Freeman, Scerbo, Mikulka & Pope (2000) conducted an experiment to examine the impact of the brain-based system developed by Pope et al. (1995) on workload. Once again, they used the MAT Battery and had participants perform under low and high workload conditions. Under low workload, they performed only the tracking task in manual mode and simply monitored the display in automatic mode. Under high workload in the manual mode, participants performed three tasks simultaneously: tracking, resource management, and system monitoring tasks. In the high workload automatic mode, the tracking task was allocated to the system, but the participants still performed the resource management and system monitoring tasks. Prinzel et al. also asked their participants to report subjective levels of workload using the NASA-Task Load Index (TLX; Hart & Staveland, 1988). This instrument provides an overall index of subjective mental workload as well as the relative contributions of six subscales: mental, physical, and temporal task demands; and effort, frustration, and perceived performance. The investigators recorded EEG from the Cz, Pz, P3, and P4 sites and used only the 20*beta/alpha+theta index. In addition, they ran a control group that performed the same three tasks for the same duration, but without the brain-based system.

Once again, the researchers found that the engagement index was higher in automatic than manual mode under negative feedback and lower in automatic than manual mode under positive feedback. They also found that the system made more task mode switches under negative feedback and further, more switches were made under the high workload condition. Tracking performance was better under negative feedback and also in the low workload condition. Participants rated the workload higher in the high as compared to low workload condition. Collectively, the results showed that increases in workload compromised tracking performance and elevated subjective reports of mental workload. Moreover, the system facilitated tracking performance under negative feedback and made more task mode switches under the high workload condition, suggesting it was sensitive to the increased task demands. Most important, the participants using the brain-based system had better tracking scores and lower workload ratings than those in the control group, highlighting the benefits of a real-time adaptive brain-based system.

Evaluating an Interactive System Design

The psychophysiologically adaptive system used at LaRC and ODU affords the capability of determining optimal human/system task allocation mixes. Hettinger et al. (2003) highlighted the method described by Pope, Bogart & Bartolome (1995) of providing "practical" information about how such systems should be designed based upon an examination of neuroadaptive interface concepts. The article presented a closed-loop method that was developed to evaluate a human/automation interface design based on its capacity to promote mental engagement. When applied for this purpose, the adaptive task method essentially implements a feedback control process whereby stable task engagement, reflected in stable short cycle oscillation of the engagement index, is eventually achieved after systematic adjustment of task demand for operator participation. The combination of automated and manual tasks, the task mix, presented to the operator in this stable oscillation condition may be considered optimal by this particular criterion of mental engagement derived from brain electrical activity. "The ultimate objective of this effort is the development of new methodologies to assess the extent to which automated flight deck systems maintain pilot engagement. This closed-loop method is intended to provide a dynamic, interactive method of adjusting a system design to optimize the operator's engagement" (Pope, Bogart & Bartolome 1995: 194). The method has yet to be applied for this purpose. The method was subsequently adapted at ODU for a slightly different purpose – the investigation of the human performance effects of psychophysiologically adaptive automation (Scerbo, Freeman & Mikulka, 2003).

Development of a Psychophysiological Self-regulation Training System Based on the Adaptive Task Concept

What are the implications for the design and application of a psychophysiologically adaptive system (PAS)? Evidence is accumulating that people can voluntarily control psychophysiological responses previously considered involuntary, even brain neuro-psychophysiological responses (deCharms et al., 2005; Vernon et al., 2003). One must consider that operators can self-regulate the psychophysiological signals that are monitored to reflect their state in a PAS. The ability to self-regulate psychophysiological state emerges and develops over time with exposure to reinforcement contingencies that encourage that ability. Prinzel, Pope, & Freeman (2002) demonstrated that participants given feedback on the accuracy of their estimates of engagement levels, across six levels of an EEG-based engagement index, were able to achieve a 70 percent level of correct identification of engagement level. Further, while interacting with an adaptive automation system, participants in a self-regulation condition were better able to maintain their task engagement level within a narrower range of task modes, thereby reducing the need for task mode changes. This resulted in increased task performance as well as a decrease in self-reported workload.

Following interactions with a PAS, operators could potentially learn psychophysiological self-regulation ability; this potential emphasizes the changeable relationship between psychophysiological indices and operator state. This possibility at the very least requires that factors affecting the fluidity of that relationship be investigated and modeled and, furthermore, recommends that self-regulation ability be utilized to optimize overall human–machine system effectiveness. Psychophysiologically adaptive systems will need to be designed to respond appropriately not only to transient changes and spontaneous drifts in operator state due to developing conditions such as fatigue but also to conditioning of psychophysiological changes as a result of operators' extended exposure to information feedback about psychophysiological state.

One of the original intentions for the closed-loop biocybernetic method was as an assessment procedure to determine the requirements for operator involvement that promote effective operator awareness states (Pope, Bogart & Bartolome, 1995). When applied for this purpose, the adaptive task method essentially implements a feedback control process whereby stable task engagement, reflected in stable short cycle oscillation of the engagement index, is eventually achieved after systematic adjustment of task demand for operator participation. As noted earlier, the mix of automated and manual tasks presented to the operator in a self-regulation condition may be considered optimal by this particular criterion of mental engagement derived from brain electrical activity (Prinzel, Pope & Freeman, 2002). This closed-loop testing setup was used to determine the level of automation that best kept test participants engaged in the experimental task. The assessment procedure may function as a training procedure in which the subject is reinforced for producing the EEG pattern that reflects optimal engagement by sharing more of the work between the operator and the automated system. With practice, a subject may learn how to deliberately control automation allocation to the level at which he or she prefers to work. This observation led to consideration of the adaptive task concept for an attention training application (Pope & Bogart, 1996); that is, it could be used as a brainwave biofeedback training system. It differs from conventional brainwave biofeedback systems in that the feedback and reward are not explicit on a display, but implicit in the subject's control of the task via his or her brainwaves. If the flight simulator is replaced with a video game, the resulting videogame biofeedback system becomes an entertaining way to deliver brainwave biofeedback training.

This type of psychophysiological training technology is described in NASA patented and licensed psychophysiological self-regulation training technologies (Pope & Prinzel, 2005). This technology uses psychophysiological signals (e.g., an engagement index) to continuously modulate parameters (e.g., character speed and mobility) of a game in real time while the game itself is operated by other means (e.g., a game controller). Such modification increases the challenge of the game and essentially opens up another modality of control for the player to utilize. The purpose of this system is to improve adherence to a psychophysiological self-regulation training regimen by delivering the training through engaging and motivating entertainment technologies (Palsson, Harris & Pope, 2002; Pope & Bogart, 1994).

The task modulation concept embodied in the videogame biofeedback technology may also be adapted for use in task simulators. Biofeedback training can help reduce the occurrence of HSAs by teaching pilots to maintain the necessary psychophysiological conditions for good cognitive and psychomotor performance under the circumstances which are most likely to produce inattention or dysfunctional stress. The training simulator embodiment of the closed-loop modulation concept is Instrument Functionality Feedback (IFF) Training, a concept for training pilots in maintaining the psychophysiological equilibrium suited for optimal cognitive and motor performance under emergency events in an airplane cockpit (Palsson & Pope, 1999). It is a training concept for reducing pilot error during demanding or unexpected events in the cockpit by teaching pilots self-regulation of excessive autonomic nervous system (ANS) reactivity during simulated flight tasks. The IFF training method (a) adapts biofeedback methodology to train psychophysiological balance during simulated operation of an airplane; and (b) uses graded impairment of control over the flight task to encourage the pilot to gain mastery over his/her autonomic functions. In IFF training, pilots learn to minimize their autonomic deviation from baseline values while operating a flight simulator. This can be achieved by measuring skin conductance and hand temperature and using deviations from their baseline values to impair the functionality of the aircraft controls. Trainees also receive auditory and visual cues about their autonomic deviation, and are instructed to keep these within pre-set limits to retain full control of the aircraft. IFF makes control of the simulator depend on more than the cognitive and physical skills of the trainee, by additionally making controllability inversely proportional to a physiological difference value that is computed as the absolute value of the difference between a predefined or selectable target physiological state and the current measured physiological state of the trainee. In effect, the fixed controllability that obtains in the absence of any modulation of the functionality of the simulator controls – the most controllability – is initially reduced by the modulation of various control functions to a less favorable controllability; and it is the biofeedback-moderated self-regulatory skill of the trainee that determines the degree to which the controllability is restored to its fixed maximum value for performing the flight training task. This action of the trainee's physiological state on the simulator controllability implements the principle of negative reinforcement, defined as the reinforcement of a response (the target physiological state) by removal or reduction of an undesirable condition or penalty (disruption of controllability) as a consequence of the response. IFF differs in this way from other biofeedback training methods that provide feedback display of trainee physiological parameters but that do not affect the functionality of the trainee's task environment (Cowings, 1997).

This IFF training method teaches trainees to incorporate autonomic or brainwave psychophysiological self-regulation into simulator training without the need for conscious attention to such regulation. The method works through a two-step mechanism of instrumental learning and classical conditioning. Through the trainee's effort and inherent motivation to master the flight task, and through repeated associations in accordance with established psychological learning principles, the psychophysiological control would be expected to transfer to the actual flight situation.

In order to study the effects of learned voluntary self-regulation on the functioning of psychophysiologically adaptive systems, it will be important to compare operator performance with an adaptive system before and after self-regulation training (which could include autonomic or brain activity or both.) Prinzel, Pope & Freeman (2002) noted that the neurofeedback provided during training may afford operators better management of their cognitive resources and thereby regulate their engagement state, allowing them to respond optimally to a change in automation mode. The results reported by Prinzel et al. support research designed to explore the potential of psychophysiological self-regulation training to enhance cognitive resource management skills of pilots and increase the benefits of adaptive automation.

APPLICATION OF NEURAL NETWORKS TO AA

One proposal for cognitive augmentation through adaptive aiding for air traffic control employs the method of inverse modeling using artificial neural networks to update the psychophysiologically adaptive automation paradigm introduced earlier (Pope, Bogart & Bartolome, 1995). The role that human error plays in aviation accidents is a consistent problem. The problem plagues all operators in the National Airspace System, particularly air traffic controllers. Research has shown the etiology of the problem to be an increase in both the amount and complexity of traffic without adequate automation support for controllers. Merely increasing automation, however, is not the answer as many errors are committed during periods of low workload. Instead, new approaches to automation design are required that can better meet the needs of controllers. One such approach would be to develop a neural network solution, based on an information-processing model of air traffic controllers, which can perform real-time adaptive modulation of human/automation task mode based upon cognitive state mapping.

As has been noted previously contemporary cognitive theories posit four stages of information processing that characterize controllers' management of air traffic. The four stages of processing carried out by human operators have a counterpart in system functions that can be automated. This solution, a traditional approach to automation design, fixes the level of automation during system development, and subsequently limits the responsiveness of the system to changing task demands. However, levels of automation need not be fixed at the system design stage. Instead, the level and even the type of automation could be designed to vary depending on situational demands during operational use, an approach that could be termed "adaptive" or "context-dependent" automation.

As an example, psychophysiologically adaptive automation applied to an air traffic control (ATC) system could be designed to provide decision (e.g., traffic conflict resolutions) and automated aids (e.g., automated handoffs and datalink clearances) to controllers when the monitoring technology identifies HSAs. Such a system would best support controller information processing because the system would be responsive to changes in the ATC environment, as well as to how the controller is managing these changes. To be effective, however, the inputs to the state monitoring and identification technology require sophisticated diagnostic mapping of cognitive states. This capability is becoming available with psychophysiological technologies such as real-time electroencephalographic (EEG) analysis and functional near infrared spectroscopy (fNIRS).

EEG (Craig & Tran, Chapter 12, this volume) and fNIRS (Warm, Tripp, Matthews & Helton, Chapter 13, this volume) imaging of the brain cortex provides both temporal and spatial information about changes in the brain during cognitive processing. These technologies detect variations in activated brain areas and provide a quantitative map of where the changes in cognitive demand occur. The map holds the promise of identifying the location and, consequently, the type of cognitive processing, essential for successful adaptive automation design.

An advanced adaptive automation method would adapt the method of inverse modeling using neural networks for the purpose of cognitive augmentation of a controller operator performing an ATC task. The first step of the procedure would involve inverse modeling of the operator under selected cognitive task demands. Brain imaging output data would be used to train a neural network to classify the brain responses exhibited by participants performing components of an ATC simulation task. Each task component would be designed to place a different type and degree of cognitive processing demand on the ATC operators, corresponding to the stages of information processing. In this way, the neural network can be trained to identify the type and degree of cognitive loading on the operator, based upon the operator's measured brain response.

Multi-layer Perceptron (MLP) neural networks would be trained with back-propagation to interpret the cortical activity associated with each of the selected cognitive processing states, thereby objectively indexing these states within human operators following an inverse modeling process (Figure 26.2, adapted from Principe, Euliano & Lefebvre, 1999). The neural network thus trained would subsequently be embedded in a real-time closed-loop paradigm to determine when and what types of excessive cognitive demands are being imposed on the operator. The closed-loop system would then adaptively aid the operator to moderate that type of cognitive demand. A further stage of this approach would create an even better cognitive augmentation system by implementing adaptive inverse model control to adjust model parameters in real time.

This approach represents an attempt to aid cognitive function based on modeling human performance and adaptively modulating the task environment to optimize controller information processing capabilities. Such an optimally adaptive human–machine system could reduce the potential for human error currently witnessed with "clumsy" automation by improving the capabilities of controllers to manage air traffic.

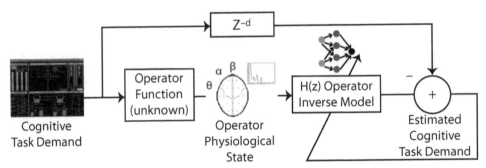

Figure 26.2   Inverse modeling of the operator under selected cognitive task demands based on neural network processing of operator psychophysiological signals

THE NATURALISTIC FLIGHT DECK

The Naturalistic Flight Deck (NFD; Schutte et al., 2007) is the implementation of a "clean-slate" design of a consistent and complete flight deck, which can serve as a platform for employing psychophysiologically adaptive automation to synergistically combine an innovative flight control design with self-regulation training. Modern automation design does not take pilot engagement into consideration; consequently, pilot state shifts between the extremes of boredom (when the automation is handling everything) and stress (when the automation returns control to the human, or fails). The NFD, built on the concept of complementary automation or "complemation" (as contrasted with "automation"), prescribes that the technology in the aircraft be applied with the goal of complementing the human (accentuating the pilot's strengths and compensating for weaknesses) rather than usurping the pilot's role. The NFD is one response to the question: "If automation fails, what is the back-up plan in terms of people, procedures, and automation?" (JPDO, 2007). This issue is of great concern to a number of governmental agencies involved in aviation safety; it is also currently prominent in considerations for developing the NextGen (Next Generation Air Transportation System). One proposal for the NFD integrates state monitoring and mitigation to augment an innovative flight control design that is itself already designed to enhance pilot engagement – the H-mode. In automated environments, operator engagement may be maintained by a controller design such as the H-mode (Flemisch et al., 2003), a haptic control system that keeps the pilot engaged in anticipation of possible automation degradation. The H-mode based design, in which the pilot initiates significant flight behaviors (e.g., turns, takeoffs, altitude changes) at or near the time of execution (i.e., no lengthy preprogrammed route executions) while automation handles inner-loop control and fosters pilot engagement.

Furthermore, the H-mode design could be augmented in a graduated fashion by employing psychophysiologically adaptive automation supplemented by self-regulation training. With the H-mode interface, "like a rider of a horse, the pilot can direct the vehicle through a combination of more automated, 'loose rein', behaviors in which the vehicle has a relatively high degree of autonomy, and less automated, 'tight rein', behaviors in which the pilot more directly controls the flight path of the vehicle" (Schutte et al., 2007: 2). "Loose rein" H-mode operation would likely suffice to maintain pilot engagement most of the time. However, when pilot state monitoring determines that the pilot is inattentive or workload challenged, cues to invoke attention management skill or stress management skill, respectively, would be provided. At the pilot's discretion, or in cases of extreme workload on the one hand or disengagement on the other, the H-mode functionality would assume a more automated role or the operator would be tasked to engage in planning and review, respectively. In this way, effective pilot engagement would be achieved by employing both a psychophysiologically adaptive capability and prior self-regulation training. The combination of these concepts has yet to be tested empirically but they hold much potential for improving aviation safety.

CONCLUSIONS

The applications of psychophysiologically based adaptive automation systems in aviation hold great promise. As designed and experimentally implemented these systems have demonstrated the capability of improving operator performance in multitask, human–machine systems. The potential for assessing operators in aviation, rail and highway transportation for the purposes of mitigating HSAs and increasing safety is very real. The future directions described in this chapter point out some important areas yet to be fully examined empirically. The NFD described in this chapter represents one experimental operator/vehicle system concept currently in use to investigate these issues. Also of note is the release of an updated Multi-Attribute Task Battery (MATB-II), which will allow researchers in aviation and other areas interested in multitasking situations to explore the effects of highly controlled and graduated levels of workload and automation mixes on operator state. Until these and other key factors are fully assessed, the use of biocybernetic or neuroadaptive technology will remain limited to the laboratory. The promise of these types of systems underscores the need to continue investing in their development and evaluation as a viable technology for mitigating HSAs and improving safety.

REFERENCES

Billings, C.E. (1991). Human-centered aircraft automation: A concept and guidelines. Moffett Field, CA: NASA Ames Research Center.

Billings, C.E. & Woods, D.D. (1994). Concerns about adaptive automation in aviation systems, in M. Mouloua & R. Parasuraman (eds), *Human Performance in Automated Systems: Current Research and Trends*. Hillsdale, NJ: Erlbaum, 264–9.

Brown, I.D. (1994). Driver fatigue. *Human Factors*, 36, 298–314.

Bryne, E.A. & Parasuraman, R.J. (1996). Psychophysiology and adaptive automation. *Biological Psychology*, 42, 249–68.

Comstock, J.R. & Arnegard, R.J. (1992). The multi-attribute task battery for human operator workload and strategic behavior research. (No. 104174). Hampton, VA: National Aeronautics and Space Administration.

Comstock, J.R., Harris, R.I. & Pope, A.T. (1988). Physiological Assessment of Task Underload. (No. N89-19846). Hampton, VA: National Aeronautics and Space Administration.

Cowings, P. (1997). USA Patent No. 5,694,939. U.S. Patent and Trademark Office.

Cunningham, S. & Freeman, F. (1994). The electrocortical correlates of fluctuating states of attention during vigilance tasks. (No. 95N14864). Hampton, VA: NASA Langley Research Center.

Davidson, R.J. (1988). EEG measure of cerebral asymmetry: Conceptual and methodological issues. *International Journal of Neuroscience*, 39, 71–89.

deCharms, R., Maeda, F., Glover, G., Ludlow, D., Pauly, J., Soneji, D. et al. (2005). Control over brain activation and pain learned by using real-time functional MRI. *Proceedings of the National Academy of Sciences*, 102, 18626–31.

FAA (1990). The national plan for civilian aviation human factors: An initiative for research and application.

FAA (1995). A revision of the national plan for civilian aviation human factors.

FAA (2012). FAA Aerospace Forecast Fiscal Years 2012–2032.

Fitts, P.M. (1951). Human engineering for an effective air navigation and traffic control system. Washington, DC: National Research Council.

Fleishman, E. & Zaccaro, S. (1992). Toward a taxonomy of team performance functions, in R. Swezey & E. Salas (eds), *Teams: Their Training and Performance*. Norwood, NJ: Ablex Publishing, 31–56.

Flemisch, F., Adams, C., Conway, S., Goodrich, K., Palmer, M. & Schutte, P. (2003). The H-Metaphor as a guideline for vehicle automation and interaction (No. TM-2003-212672). Hampton, VA: NASA Langley Research Center.

Freeman, F., Mikulka, P., Prinzel, L. & Scerbo, M. (1999). Evaluation of an adaptive automation system using three EEG indices with a visual tracking system. *Biological Psychology*, 50, 61–76.

Freeman, F., Mikulka, P., Scerbo, M. & Hadley, G. (1998). A comparison of two methods for use in adaptive automation. *Perceptual and Motor Skills*, 86, 1185–86.

Freeman, F., Mikulka, P., Scerbo, M., Prinzel, L. & Clouatre, K. (2000). Evaluation of a psychophysiologically controlled adaptive automation system, using performance on a tracking task. *Applied Psychophysiology and Biofeedback*, 25, 103–15.

Gaillard, A. (1993). Comparing the concepts of mental load and stress. *Ergonomics*, 36, 991–1005.

Gander, P.H., Gregory, K.B., Connell, L.J., Graeber, R.C., Miller, D.L. & Rosekind, M.R. (1998). Flight Crew Fatigue IV: Overnight Cargo Operations. *Aviation, Space, and Environmental Medicine*, 69, B26–B36.

Gander, P.H., Gregory, K.B., Graeber, R.C., Connell, L.J., Miller, D.L. & Rosekind, M.R. (1998). Flight Crew Fatigue II: Short-Haul Fixed-Wing Air Transport Operations. *Aviation, Space, and Environmental Medicine*, 69, B8–B15.

Gander, P.H., Gregory, K.B., Miller, D.L., Graeber, R.C., Connell, L.J. & Rosekind, M.R. (1998). Flight Crew Fatigue V: Long-Haul Air Transport Operations. *Aviation, Space, and Environmental Medicine*, 69, B37–B48.

Gander, P.H., Rosekind, M.R. & Gregory, K.B. (1998). Flight crew fatigue VI: A synthesis. *Aviation, Space, and Environmental Medicine*, 69, 849–60.

Haarmann, A., Boucsein, W. & Schaefer, F. (2009). Combining electrodermal responses and cardiovascular measures for probing adaptive automation during simulated flight. *Applied Ergonomics*, 40, 1026–40.

Hadley, G., Mikulka, P., Freeman, F. & Scerbo, M. (1997). An examination of system sensitivity in a biocybernetic adaptive system. Paper presented at the Proceedings of the 9th Symposium on Aviation Psychology.

Hammer, J. & Small, R. (1995). An intelligent interface in an associate system, in W. Rouse (ed.), *Human/Technology Interaction in Complex Systems*. Vol. 7. Greenwich, CT: JAI Press, 1–44.

Hancock, P. & Chignell, M. (1987). Adaptive control in human–machine systems, in P. Hancock (ed.), *Human Factors Psychology*. North Holland: Elsevier Science, 305–45.

Hancock, P. & Warm, J. (1989). A dynamic model of stress and sustained attention. *Human Factors*, 31, 519–37.

Hart, S.G. & Staveland, L.E. (1988). Development of NASA-TLX (task load index): Results of empirical and theoretical research, in P.A. Hancock & N. Meshkati (eds), *Human Mental Workload*. Amsterdam: North-Holland, 139–83.

Heinrich, H., Petersen, D. & Roos, N. (1980). *Industrial Accidents Prevention: A Safety Managment Approach*. Fifth edition. New York: McGraw Hill.

Hettinger, L., Branco, P., Encarnacao, L.M. & Bonato, P. (2003). Neuroadaptive technologies: Applying neuroergonomics to the design of advanced interfaces. *Theoretical Issues in Ergonomics Science*, 4, 220–37.

Inagaki, T., Takae, Y. & Moray, N. (1999). Automation and human interface for takeoff safety. Paper presented at the Proceedings of the 10th International Symposium on Aviation Psychology, Columbus, OH.

JPDO (2007). Actions needed to reduce risks with the next generation air transportation system. (No. AV-2007-031). Washington, DC: Federal Aviation Administration.

Kramer, A. & Weber, T. (2000). Applications of psychophysiology to human factors, in J. Cacioppo, L. Tassinary & G. Berntson (eds), *Handbook of Psychophysiology*. New York, NY: Cambridge University Press, 794–814.

Lubar, J.F. (1991). Discourse on the development of EEG diagnostics and biofeedback for attention deficit/hyperactivity disorders. *Biofeedback and Self-Regulation*, 16, 201–25.

Malin, J. & Schreckenghost, D. (1992). Making intelligent team players: Overview for designers. (No. 104751). Houston, TX: NASA Johnson Space Center.

Mallis, M., Neri, D., Rosekind, M., Gander, P., Caldwell, J. & Graeber, C. (2007). Understanding and counteracting fatigue in flight crews. NASA Tech Briefs, 31, 24.

Maurino, D., Reason, J., Johnston, N. & Lee, R. (1995). *Beyond Aviation: Human Factors*. Aldershot, UK: Ashgate Publishing.

Miller, C. & Hannen, M. (1999). The rotorcraft pilot's associate: design and evaluation of an intelligent user interface for cockpit information management. *Knowledge-Based Systems*, 12, 443–56.

Monastra, V., Lynn, S., Linden, M., Lubar, J.F., John, G. & LaVaque, T. (2005). Electroencephalographic biofeedback in the treatment of attention deficit/ hyperactivity disorder. *Applied Psychophysiology and Biofeedback*, 30, 95–114.

Mulholland, T. (1977). Biofeedback as scientific method, in G. Schwartz & J. Beatty (eds), *Biofeedback: Theory and Research*. New York: Academic Press, 9–28.

NTSB (1999). Evaluation of U.S. Department of Transportation efforts in the 1990s to address operator fatigue. Washington, DC: National Transportation Safety Board.

Offenloch, K. & Zahner, G. (1990). Computer aided physiological assessment of the functional state of pilots during simulated flight. Paper presented at the NATO Advisory Group for Aerospace Research and Development, September 1–9.

Palmer, M., Rogers, W., Hayes, P., Latorella, K. & Abbott, T. (1995). A crew-centered flight deck design philosophy for high-speed civil transport (HSCT) Aircraft. Hampton: NASA Langley Research Center.

Palsson, O., Harris, R.L. & Pope, A.T. (2002). Washington, D.C. Patent No. 6450820. U.S. Patent and Trademark Office.

Palsson, O. & Pope, A.T. (1999). Stress counterresponse training of pilots via instrumental functionality feedback. Paper presented at the Association for Applied Psychophysiology and Biofeedback, Vancouver, Canada.

Parasuraman, R. (1990). Event-related brain potentials and human factors research, in J. Rohrbaugh, R. Parasuraman & R. Johnson (eds), *Event-related Brain Potentials: Basic Issues and Applications*. New York, NY: Oxford University Press, 279–300.

Parasuraman, R., Bahri, T., Deaton, J., Morrison, J. & Barnes, M. (1992). Theory and design of adaptive automation in aviation systems. (No. NAWCADWAR-92033-60). Warminster, PA: Naval Air Warfare Center.

Parasuraman, R., Molloy, R. & Singh, I. (1993). Performance consequences of automation-induced "complacency." *International Journal of Aviation Psychology*, 3, 1–23.

Parasuraman, R., Mouloua, M. & Hilburn, B. (1999). Adaptive aiding and adaptive task allocation enhance human–machine interaction, in M. Scerbo & M. Mouloua (eds), *Automation Technology and Human Performance: Current Research and Trends*. Mahwah, NJ: Erlbaum, 119–23.

Parasuraman, R., Mouloua, M., Molloy, R. & Hilburn, B. (1996). Monitoring of automated systems, in R. Parasuraman & M. Mouloua (eds), *Automation and Human Performance: Theory and Applications*. Mahwah, NJ: Erlbaum, 91–115.

Parasuraman, R., Sheridan, T. & Wickens, C. (2000). A model of types and levels of human interaction with automation. *IEEE Transactions on Systems, Man, and Cybernetics – Part A: Systems and Humans*, 30, 286–97.

Pope, A. & Bogart, E. (1991). Extended attention span training. Paper presented at the Technology Transfer Conference and Exposition, San Jose, December 3–5.

Pope, A. & Bogart, E. (1992). Identification of hazardous awareness states in monitoring environments. (No. 921136): SAE 1992 Transactions: *Journal of Aerospace*.

Pope, A. & Bogart, E. (1994). Washington, D.C. Patent No. 5377100. U.S. Patent and Trademark Office.

Pope, A. & Bogart, E. (1996). Extended attention span training system: Video game neurotherapy for attention deficit disorder. *Child Study Journal*, 26, 39–50.

Pope, A. & Bowles, R. (1982). A program for assessing pilot mental state in flight simulators. Paper presented at the American Institute of Aeronautics and Astronautics.

Pope, A. & Prinzel, L. (2005). Recreation embedded state tuning for optimal readiness and effectiveness. Paper presented at the 11th International Conference on Human–Computer Interaction, Las Vegas, NV.

Pope, A., Bogart, E. & Bartolome, D. (1995). Biocybernetic system validates index of operator engagement in automated task. *Biological Psychology*, 40, 187–95.

Poyer, J. (1969). *North Cape*. Garden City: Doubleday & Company.

Principe, J.C., Euliano, N.R. & Lefebvre, W.C. (1999). *Neural and Adaptive Systems: Fundamentals through Simulations* (with CD-ROM). New York, NY: John Wiley & Sons, Inc.

Prinzel, L., Freeman, F., Scerbo, M., Mikulka, P. & Pope, A. (2000). A closed-loop system for examining psychophysiological measures for adaptive task allocation. *International Journal of Aviation Psychology*, 10, 393–410.

Prinzel, L., Freeman, F., Scerbo, M., Mikulka, P. & Pope, A.T. (2003). The effects of a psychophysiological system for adaptive automation on performance, workload, and event-related potential P300 component. *Human Factors*, 45, 601–13.

Prinzel, L., Parasuraman, R., Freeman, F., Scerbo, M., Mikulka, P. & Pope, A. (2003). Three experiments examining the use of electroencephalogram, event-related potentials, and heart rate variability for real-time human-centered adaptive automation design. (No. TP-2003-212442). Hampton, VA: NASA Langley Research Center.

Prinzel, L., Pope, A. & Freeman, F. (2002). Physiological self-regulation and adaptive automation. *International Journal of Aviation Psychology*, 12, 181–98.

Prinzel, L., Scerbo, M., Freeman, F. & Mikulka, P. (1997). Behavioral and physiological correlates of a bio-cybernetic, closed-loop system for adaptive automation, in M. Mouloua & J. Koonce (eds), *Human–Automation Interaction: Research and Practice*. Mahwah, NJ: Lawrence Erlbaum, 66–75.

Rasmussen, J. (1982). Human errors: A taxonomy for describing human malfunctions in industrial installations. *Journal of Occupational Accidents*, 4, 311–33.

Rasmussen, J. (1983). Skills, rules, and knowledge: Signals, signs, and symbols. *IEEE Transactions on Systems, Man, and Cybernetics*, 13, 257–66.

Reason, J. (1990). *Human Error*. New York: Cambridge University Press.

Reising, M. & Moss, R. (1986). 2010: The symbionic cockpit. *IEEE Aerospace and Electronic Systems*, 1, 24–7.

Reynard, W.D., Billings, C.E., Cheaney, E.S. & Hardy, R. (1986). *The Development of the NASA Aviation Safety Reporting System*. Moffett Field, CA.

Rosekind, M.R., Gander, P.H., Gregory, K.B., Smith, R.M., Miller, D.L., Oyung, R. et al. (1996). Managing fatigue in operational settings 2: An integrated approach. *Behavioral Medicine*, 21, 166–70.

Rouse, W. (1988). Adaptive aiding for human/computer control. *Human Factors*, 30, 431–43.

Rouse, W., Geddes, N. & Curry, R. (1987–1988). An architecture for intelligent interfaces: Outline of an approach to supporting operators of complex systems. *Human-Computer Interaction*, 3, 87–122.

Russell, C. & Wilson, G. (2005). Feature saliency analysis for operator state estimation. Paper presented at the 11th International Conference on Human–Computer Interaction, Las Vegas, NV.

Russell, C., Wilson, G., Rizki, M.T.W. & Gustafson, S. (2005). Comparing classifiers for real time estimation of cognitive workload. Paper presented at the 11th International Conference on Human–Computer Interaction, Las Vegas, NV.

Scerbo, M. (1994). Implementing adaptive automation in aviation: The pilot–cockpit team. *Human Performance in Automated Systems: Current Research and Trends*. Hillsdale, NJ: Lawrence Erlbaum Associates, 249–55.

Scerbo, M. (1996). Theoretical perspectives on adaptive automation, in R.J. Parasuraman & M. Mouloua (eds), *Automation and Human Performance: Theory and Applications*. Mahwah, NJ: Erlbaum Associates, 37–64.

Scerbo, M. (2001). Adaptive automation, in W. Karwowski (ed.), *International Encyclopedia of Ergonomics and Human Factors*. London: Taylor and Francis.

Scerbo, M. (2005). Biocybernetic systems: Information processing challenges that lie ahead, in D. Schmorrow (ed.), *Foundations of Augmented Cognition*. Mahwah, NJ: Lawrence Erlbaum Associates, 93–100.

Scerbo, M. (2007). Adaptive automation, in R. Parasuraman & M. Rizzo (eds), *Neuroergonomics: The Brain at Work*. Oxford: Oxford University Press, 239–52.

Scerbo, M., Freeman, F. & Mikulka, P. (2000). A biocybernetic system for adaptive automation, in R. Backs & W. Boucsein (eds), *Engineering Psychophysiology*. Mahwah, NJ: Lawrence Erlbaum Associates, 241–54.

Scerbo, M., Freeman, F. & Mikulka, P. (2003). A brain-based system for adaptive automation. *Theoretical Issues in Ergonomics Science*, 4, 200–219.

Scerbo, M., Freeman, F., Mikulka, P., Schoenfeld, V., Eischeid, T., Krahl, K. et al. (1999). Hazardous states of awareness: What are they and how do we measure them? *Aviation Psychology*, 2, 854–60.

Schutte, P.C. (1999). Complementation: An alternative to automation. *Journal of Information Technology Impact*, 1, 113–18.

Schutte, P.C., Goodrich, K., Cox, D., Jackson, B., Palmer, M., Pope, A.T. et al. (2007). The naturalistic flight deck system: An integrated system concept for improved single-pilot operations. Hampton, VA: NASA Langley Research Center.

Scott, W. (1999). Automatic GCAS: "You can't fly any lower." *Aviation Week and Space Technology*, 150, 76–9.

Shappell, S. & Wiegmann, D. (2001). Applying reason: The human factors analysis and classification system (HFACS). *Human Factors and Aerospace Safety*, 1, 59–86.

Shappell, S., Detwiler, C., Holcomb, K., Hackworth, C., Boquet, A. & Wiegmann, D. (2007). Human error and commercial aviation accidents: An analysis using the human factors and classification system. *Human Factors*, 49, 227–42.

Sheridan, T. & Verplank, W. (1978). Human and computer control of undersea teleoperators. Cambridge, MA: MIT Man–Machine Systems Laboratory.

Singer, P. (2010). *Wired for War: The Robotics Revolution and Conflicts in the Twenty-first Century*. London: Penguin.

Stefani, A. (2000). Actions to reduce operational errors and deviations have not been effective. Washington, DC: Office of the Inspector General.

Streitberg, B., Röhmel, J., Hermann, W.M. & Kubicki, S. (1987). COMSTAT rule for vigilance classification based on spontaneous EEG activity. *Neuropsychobiology*, 17, 105–17.

Tollar, J. (2005). Statistical process control as a triggering mechanism for augmented cognition, in D. Schmorrow (ed.), *Foundations of Augmented Cognition*. Mahwah, NJ: Lawrence Erlbaum Associates, 414–20.

United States Government House of Representatives (1999). First Sess. 1–308. Pilot fatigue.

Vernon, D., Egner, T., Cooper, N., Compton, T., Neilands, C., Sheri, A. et al. (2003). The effects of training distinct neurofeedback protocols on aspects of cognitive performance. *International Journal of Psychophysiology*, 47, 75–86.

Weaver, J., Bradley, K., Burke, K., Helmick, J., Greenwood-Ericksen, A. & Hancock, P. (2003). An experimental investigation of skill, rule, and knowledge-based performance under noise conditions. Paper presented at the Human Factors and Ergonomics Society 47th Annual Meeting, Santa Monica, CA.

Weiner, E.L. (1989). Human factors of advanced technology ("glass cockpit") transport aircraft. (No. 117528). Moffett Field, CA: NASA Ames Research Center.

Wickens, C. (1990). Applications of event-related potentials research to problems in human factors, in J. Rohrbaugh & R. Parasuraman (eds), *Event-related Brain Potentials: Basic Issues and Applications*. New York, NY: Oxford University Press, 301–9.

Wickens, C. (1992). *Engineering Psychology and Human Performance*. Second edition. New York, NY: Harper Collins.

Wiegmann, D., Faaborg, T., Boquet, A., Detwiler, C., Halcomb, K. & Shappell, S. (2005) Human error and general aviation accidents: A comprehensive, fine-grained analysis using HFACS (Report Number DOT/FAA/AM-05/24). Washington, DC: Office of Aerospace Medicine.

Wilensky, R., Arens, Y. & Chin, D. (1984). Talking to Unix in English: An overview of UC. *Communications of the ACM*, 27, 574–93.

Zacharias, G., Caglayan, A. & Pope, A.T. (1985). Study to identify visual evoked response parameters sensitive to pilot mental state. (No. 85K10877). Hampton, VA: NASA Langley Research Center.

27

Countermeasures for Driver Fatigue

Ann Williamson

INTRODUCTION

It is well known that fatigue affects our ability to perform. Fatigue is a serious problem in transportation. Controllers of all types of vehicles, including trains, planes, ships, cars and trucks, are at greater risk of crashing when fatigued. Statistics on road crashes notoriously underestimate fatigue involvement (Horne & Reyner, 1999), but those available suggest that between 10 percent and 20 percent of crashes are sleep- or fatigue-related (Horne & Reyner, 1995; Maycock, 1997; Stutts et al., 1999). Furthermore, fatigue-related crashes are more likely to be fatal (Bunn et al., 2005).

Fatigue is a particular problem while driving due to the demands for continuous attention while at the wheel and because a temporary distraction can have potentially disastrous consequences (Brown, 2001). Fatigue produces a gradual withdrawal of attention from road and traffic demands, which is involuntary and almost impossible to resist (Brown, 1994). The driving task itself may increase the likelihood of a driver experiencing fatigue, especially when driving involves prolonged exposure to monotonous conditions such as on country roads (Matthews & Desmond, 2002; Neubauer, Matthews & Saxby, Chapter 23, this volume; Thiffault & Bergeron, 2003). Driving is also vulnerable to the effects of fatigue from other sources such as sleep deprivation or circadian influences (Brown, 1994).

Fatigue affects driving performance in a range of ways, including poorer steering control, speed tracking, visual search and attentional selectivity (McDonald, 1984). These can have adverse effects on performance, which in turn affect quality and safety of performance (Williamson, Lombardi et al., 2011). Driving or controlling a vehicle is particularly vulnerable to these effects. First, the driving activity is often unstimulating due to the terrain, or the repetitiveness of the route. Second, driving often requires long periods at the controls. Third, driving is often performed regardless of the person's status with respect to sleep: often after a long period awake due to work and/or social activities, often despite little or poor- quality sleep the night before and often during periods when the body clock or circadian rhythm is expecting sleep rather than being awake and in charge of a vehicle. While there is considerable evidence about the conditions likely to produce fatigue and increase the likelihood of falling asleep at the wheel, the causes are multifactorial, the relationships between the factors are not well understood and individual variability is high. Clearly, managing fatigue is a serious concern for road safety.

Road safety countermeasures for fatigue have in most countries relied on increasing awareness or education (see Job, Graham, Sakashita & Hatfield, Chapter 22, this volume). Unlike other road safety problems, there are no clear exposure limits and fatigue management approaches take the form of guidance rather than prescribing specific actions through regulation. Fatigue management strategies on the road involve suggestions of driving limits and advice to people to take regular breaks (e.g., every two hours or when they feel tired) (Fletcher et al., 2005). Even for professional long distance truck drivers, who have regulatory limits on working, including driving hours, the limits are extremes and drivers are advised that they should respond to personal experiences of fatigue in scheduling working and driving periods. A major assumption inherent in this advisory approach is that drivers have access to information about their levels of fatigue and drowsiness and are able to make the decision to stop and rest before their performance is sufficiently adversely affected that their risk of crashing becomes too high. There is considerable debate about the validity of this assumption.

The research on this issue suggests that people can detect decreasing alertness and increasing fatigue and sleepiness. Many studies have shown the expected decreases in alertness and increases in self-rated fatigue and sleepiness when sleep-deprived (Dinges et al., 1997), when required to work at vulnerable times in the circadian rhythm (Monk, 1991), or for prolonged periods without a break (Rosa & Colligan, 1988). It is not clear when performance effects begin for a fatigued person and whether fatigued people have the capacity to detect the effect of these changes in state of performance. These are critical questions for safety. It is not enough simply to be aware of changes in alertness or feelings of fatigue and sleepiness. Drivers need to be able to detect and, importantly, to respond to changes which have an impact on their capacity to drive safely.

Unfortunately, the evidence on the relationship between changes in alertness and sleepiness-related states and performance effects is equivocal. There is some evidence that increasing self-reported sleepiness is related to poorer performance in driving tasks. For example, Reyner & Horne (1998b) showed in a driving simulator that increasing subjective sleepiness was significantly associated with an increase in the number of safety-related incidents. Horne & Baulk (2004) also found that subjective sleepiness, EEG-recorded sleepiness and lane deviations in a driving simulator were highly correlated. In contrast, some studies have shown that self-rated alertness or fatigue is significantly correlated with self-rated performance, but that the correlation of changes in these attributes with changes in actual performance using a range of laboratory measures is only moderate at best (Dorrian et al., 2000; Dorrian et al., 2003). In addition, some on-road studies found no association between self-assessed fatigue and a number of non-driving performance measures (Williamson, Feyer, Friswell & Finlay-Brown, 2000) or a set of driving-related performance measures (Belz, Robinson & Casali, 2004). Further research is needed to clarify when performance effects begin to occur and become noticeable for a fatigued person.

A number of studies have highlighted the differentiation between detecting fatigue and sleepiness and detecting when these experiences might lead to falling asleep and potentially to crashing. Horne & Reyner (1999) found that drivers underestimated the probability of falling asleep when sleepy and seemed to underestimate their likelihood of crashing. There is also recent evidence that even partially sleep-deprived people who are sitting quietly in a darkened room doing a task requiring them to predict how close they are to falling asleep have limited ability to detect when they are going to fall asleep

(Kaplan, Itoi & Dement, 2008). In fact it seems that people may not be able to tell when they are in the early stages of sleep. There is evidence that people overestimate the time they take to fall asleep and they can be in the early stages of sleep without being aware of it (Baker, Maloney & Driver, 1999). Furthermore, it seems that people can be in the early stages of sleep and still be responding to external stimuli. For example, Ogilvie, Wilkinson & Allison (1989) showed that some individuals were able to continue responding to a simple reaction time task despite EEG recordings showing that they were in Stage 1 sleep. Of direct relevance to this chapter, an on-road study of EEG-measured sleep in US truck drivers by Mitler et al., (1997) showed that some drivers had episodes of Stage 1 sleep for many seconds while they were driving. These drivers did not have sleep disorders and were free to stop and rest or nap at any time.

It seems that drivers can detect that they are increasingly becoming fatigued or drowsy, but may be less able to respond to these sensations at the appropriate time by discontinuing driving. The reasons for this apparent failure are not clear. It could be due to poorer ability to differentiate the effects of fatigue as levels increase or to failure to judge when fatigue levels are having adverse effects on performance. It could also be due to risky decision-making or to other motivational factors that might keep the person driving, such as the need to finish the trip either because of internal state reasons (they want the unpleasant fatigue experience to be over quickly) or because there are external factors operating like deadlines or, for professional drivers, monetary compensation (Fletcher et al., 2005). Further research is needed to clarify these possibilities.

Unfortunately, road safety authorities have tended to ignore the fact that simply telling drivers to stop when they feel tired is unlikely to be effective. This means that other countermeasures are needed to reduce the problem of driver fatigue while driving. There is a range of other possible countermeasures available to reduce fatigue while driving; some are already employed by drivers to attempt to reduce the effects of fatigue. The

Table 27.1 **Types of fatigue countermeasures available to drivers for preventing fatigue before driving and for reducing fatigue effects while driving**

Fatigue Countermeasures	
Prevention of fatigue	**Reducing effects of fatigue**
Sleep-related	*Arousal-related*

• Commencing driving when rested
  – Healthy sleep practices
  – Effective work–rest schedules including Fatigue Risk Management Systems
	• Monitoring and detection of adverse effects of fatigue while driving
	  – Physiological monitoring
	  – Performance monitoring

• Avoiding driving during circadian low points (especially when already tired)	• Journey-related measures
	  – breaks from driving
	  – time of day

• Avoidance of alcohol and other depressants before driving	• Driver-related measures
	  – change of activity (including exercise, meals)
	  – napping
	  – stimulant drug use

	• Road-related measures
	  – rumble strips
	  – places to stop

following section describes the main countermeasures in use and those that could be used for fatigue management in road safety. These countermeasures have been separated into those that attempt to prevent drivers commencing the trip in a fatigued state and those designed to reduce fatigue that occurs while driving. Table 27.1 provides an overview of the types of countermeasures reviewed in this chapter.

COUNTERMEASURES TO PREVENT DRIVING WHEN FATIGUED

Many drivers are fatigued before they even get into a vehicle to start driving. Long or irregular hours of work can lead to high levels of fatigue during the commute home. Short rest breaks between shifts that do not allow time for restorative sleep can lead to fatigued drivers during the commuting journey to work. These sorts of problems are especially characteristic of much occupational driving such as long and short haul road transport (Friswell & Williamson, 2008) and taxi driving (Dalziel & Job, 1997). Poor balance between work and rest leading to inadequate and poor quality sleep are at the heart of problems of driver fatigue early in a journey, so countermeasures to prevent early driver fatigue mainly relate to sleep and choosing the best times to drive.

Countermeasures around Sleep

The most obvious countermeasure to fatigue is sleep. Most of the countermeasures recommended to prevent drivers being fatigued at the commencement of driving aim to ensure that rest and sleep are obtained in appropriate amounts and at the best times to be effective. As is discussed above, the road safety approach has traditionally focused on education and advice to encourage drivers to obtain sufficient sleep before they drive and to avoid driving during circadian lows such as midnight to dawn. For occupational driving, countermeasures around sleep are typically based on regulation. Whether simply guidance or regulation, the scientific basis for these countermeasures has been shaky, with little evidence for work/driving and rest limits. This has been improving over recent years with the implementation of less prescriptive regulatory approaches in some countries in the form of Fatigue Risk Management Systems (FRMS), which are based on available scientific evidence (e.g., Dawson et al., 1997). Jones, Dorrian & Rajaratnam (2005) reviewed the advantages and disadvantages of prescriptive and more flexible regulatory approaches to fatigue management and concluded that both approaches have weaknesses; thy suggested that a hybrid approach may be more successful. Hopefully the better use of scientific evidence on fatigue management will flow over to community road safety as well.

Overall, it is crucial to control pre-existing fatigue due to sleep factors before driving, since the driving task itself produces fatigue independently of sleep loss. Sleep lack just makes this worse. Effective countermeasures to prevent driver fatigue at the start of even short journeys are essential. Despite recent increases in research and policy activity in this area, however, further research and analysis is needed to formulate and evaluate new countermeasures.

Countermeasures around the Time of Day

Crash statistics show that fatigue-related crashes are more likely during the midnight to dawn period, which corresponds to a dip in the circadian rhythm (Horne & Reyner, 1995; Pack et al., 1995). There is also evidence of performance decrements in driving simulator performance during this period (Lenné, Triggs & Redman, 1997). This effect is related to sleep countermeasures, as the circadian trough period corresponds to the period of greatest sleep need, with highest feelings of fatigue and lowest driver alertness. This effect is naturally made worse if this period coincides with a long period awake, for example, driving home after midnight after being awake since early morning. A simple countermeasure to reduce fatigue while driving is to choose the best time of day to drive. Avoiding driving during the circadian trough, especially after a long period awake, is an excellent fatigue countermeasure.

COUNTERMEASURES TO REDUCE THE EFFECTS OF FATIGUE WHILE DRIVING

A range of alternative approaches has been suggested as countermeasures for the fatigue that occurs during driving. For the purposes of this review, these have been grouped into fatigue monitoring and detection, and driver-related and road-related countermeasures. The evidence for the effectiveness of each group will be discussed in turn.

Monitoring and Detection of Adverse Effects of Fatigue while Driving

As is discussed earlier, it is questionable whether drivers are able to self-monitor their fatigue and alertness levels sufficiently to maintain safe driving performance. A possible countermeasure for fatigue while driving is therefore to provide drivers with support to detect increasing fatigue, either in themselves or when their driving is becoming adversely affected.

A range of physiological and physical devices is becoming available for the monitoring of driver fatigue. These devices are based on the fact that fatigue produces a number of observable physiological changes, including yawning, eye closure rate, eyelid movement and slowing of the heart beat and rate, which potentially can be used to detect the onset of sleep. Devices are also available for monitoring driving performance data, including steering angle, speed, headway, lane tracking and response to secondary signals.

The advantage of these monitoring methodologies is that they provide additional information to drivers, so encouraging them to initiate appropriate countermeasures to reduce fatigue. In addition, driver performance monitoring devices provide information relevant to the task at hand and so contribute to the maintenance of safe performance. The disadvantage of all of these monitoring devices is that they do not actually act to reduce fatigue, rather, they just tell drivers that they should take action to avoid a period of very unsafe driving. Unfortunately, it may not be possible or safe for drivers to stop when this information is provided due to lack of stopping places along the route, for the type of vehicle being driven, or the time of day.

An increasing number of devices are under development or commercially available to monitor driver fatigue. The devices function by providing a warning to the driver via an auditory alarm or light when the physiological function being monitored reaches a preset level. To be useful as a fatigue countermeasure, it is essential that these devices can detect changes early enough in the development of fatigue to avoid adverse effects on driving performance. An excellent review of in-vehicle sleepiness detection devices (Wright et al., 2007) highlighted that many of the devices have problems of reproducibility, sensitivity to fatigue, timeliness of detection, and difficulties of implementation in the real world due to intrusiveness and driver acceptance.

Furthermore, it is argued that there is no commonly accepted method for detecting fatigue in an operational context (Yang et al., 2009). Liu, Hosking & Lenné (2009) reviewed the literature on driver drowsiness detection countermeasures and suggested that multiple measures with multiple criteria to distinguish drowsy from non-drowsy states may be needed to account for changes over time and individual differences. Such devices do exist, but currently they show many of the same problems as single channel devices.

In summary, although drivers know when they are becoming increasingly tired, their capacity to detect when they are too tired to drive is probably not reliable. Consequently, giving drivers better information about their current capacity to drive safely is useful, but it needs to be timely. Furthermore, this approach is clearly not a sufficient countermeasure, and other methods are needed to act to reduce fatigue. Over the next decade it is likely that we will see the fruits of further research and development work to produce effective devices for on-road monitoring of driver fatigue.

Driver-related Countermeasures

Countermeasures that focus on helping the driver to manage or overcome fatigue range from those that tackle the effects of long periods of exposure to the monotony of driving, mostly by changing to a different activity to those that attempt to attack fundamental arousal levels – such as the use of stimulants.

Changes to different activity
One of the problems of managing fatigue while driving is that since driving is a continuing activity that requires moment-by-moment attention, there is a limit to the range of countermeasures available to drivers. The countermeasure problem becomes one of overcoming the unstimulating and monotonous nature of driving, especially over long periods. Recently it was realized that tasks like this produce passive fatigue and have adverse effects on performance because they require the driver to maintain sufficient levels of alertness when there is little stimulation from the task itself or the environment in which it is located (Matthews & Desmond, 2002). Drivers are restricted in the possibilities of changing activities while driving.

From surveys of professional long-distance truck drivers (Williamson, Feyer, Coumarelos & Jenkins 1992; Williamson, Feyer, Friswell & Finlay-Brown, 2001), it is clear that drivers use strategies that involve additional activities while driving to assist in maintaining alertness. The most commonly used fatigue management strategies reported by more than three-quarters of truck driver participants were listening to music or the radio, sleep, and caffeine use, although only the latter two were reported to be most helpful for reducing fatigue. A driving simulator study confirmed the limited usefulness of passive in-vehicle strategies (Reyner & Horne, 1998a), as exposure to cold air and listening to the radio were to have marginal benefit for safe driving, at best.

There is recent evidence that there are benefits to making driving less monotonous. Simply driving on more demanding roads (curves rather than straight roads) reduced fatigue and its affects in a simulator (Matthews & Desmond, 2002), and even making the terrain more interesting improved fatigue, although to a small degree (Thiffault & Bergeron, 2003). Promising results on the effectiveness of a knowledge-based "trivia" task also suggest that the passive fatigue in driving can be overcome by engaging the driver in cognitive activities (Oron-Gilad, Ronen & Shinar, 2008; Gershon et al., 2009). In these studies not all secondary activities were effective, indicating that not all cognitive tasks reduce fatigue effects and the effects need to be tested on-road, but this approach has promise.

Automation of the driving task
There is increasing interest in developing effective methods for automation of aspects of the driving task, including adaptive cruise control, headway monitors and automatic route-finding. The purpose of automating these aspects of driving is to reduce the driver stress and workload which, in turn, should have the effect of reducing driver fatigue over a long trip (May & Baldwin, 2009). There is good evidence that strategic automation of driving tasks can benefit driver performance by reducing workload (e.g. Stanton et al., 2001; Young & Stanton, 2004).

Automation is not without its problems, however (Parasuraman & Riley, 1997) and automation of some aspects of the driving task may reduce mental workload, but change its nature to emphasize monitoring activities rather than active engagement (Neubauer, Matthews & Saxby, Chapter 23, this volume). It is not entirely clear what are the effects over time on driving performance when automation results in lower mental workload and passive task engagement. Certainly the results of studies showing the benefits of enhancing cognitive engagement during long-duration driving suggest that reducing task demand through automation would adversely affect driving performance. A study by Desmond, Hancock & Monette (1998) provides some support for this contention as it showed that drivers handled a driving emergency better if they had been driving manually than if the driving task had been automated. Few of the studies of the effectiveness of automation studied performance over long periods, so it remains to be seen whether automation and fatigue interact and under what conditions.

Breaks from driving
Taking a break from driving is an obvious countermeasure for task-related driver fatigue, and is currently the main countermeasure advocated by road safety authorities. Rest breaks from driving have been shown to reduce accident risk for professional drivers (Dalziel & Job, 1997; Hamad, Jaradat & Easa, 1998). Similarly, in industrial settings, injury risk has also been shown to increase linearly between successive breaks (Tucker, Folkard & MacDonald, 2003). In practice, however, this countermeasure may not be easy to achieve. As is discussed above, drivers may not be aware that they need to stop, they are often not motivated to stop driving as they have not reached their destination and in some circumstances there are no suitable places to stop. Due to these practical problems, it is important to ensure that drivers are using breaks as effectively as possible to manage fatigue in terms of the timing and content of breaks.

a) Frequency and timing of breaks:

Tucker (2003) reviewed the evidence on the effectiveness of rest breaks on accidents, performance and fatigue and highlighted the paucity of research on this issue, a situation that has hardly changed since this review. Nevertheless, Tucker concluded that rest breaks are beneficial for performance and self-rated fatigue, but evidence on timing is not strong. Some research suggests that breaks should be timed to coincide with the individual's onset of fatigue or at least the commencement of gross errors (e.g. Stave, 1977). Other studies have shown benefits of preplanned, regular breaks (Tucker, Folkard & MacDonald, 2003).

None of this research involved driving, so there is currently very little guidance on the best approaches to using breaks as a driver fatigue countermeasure. In the absence of good evidence, probably the best current advice for driving is as concluded by Tucker, Folkard & MacDonald (2003): rest breaks should be scheduled and taken at regular intervals, but drivers should also be encouraged to take additional breaks in response to the onset of feelings of fatigue. Just how often regular breaks should be planned is not clear. There is evidence that short, frequent breaks are more useful (e.g. a few minutes every hour, Kopardekar & Mital, 1994) than waiting for longer periods. Certainly, where driving is monotonous, frequent breaks that change activity are likely to have the effect of raising alertness. More research is badly needed to clarify the best approaches to using rest breaks while driving.

b) Content of breaks:

Exercise. Physical activity or exercise is often suggested as a countermeasure for driver fatigue. Long distance truck drivers in Australia commonly report walking around the vehicle and kicking the tyres as a way of managing fatigue, as well as checking tyre wear (Williamson, Feyer, Friswell & Finlay-Brown, 2001). There is some evidence to support this approach. Bonnet & Arand (1999) showed that the ability to maintain wakefulness over a 31-hour period was enhanced by standing and doing knee bends, compared to just standing, sitting or speaking. Sallinen et al. (2008) compared the effect of rest breaks in sleep-deprived (restricted to 2 hours' sleep a night) and non-sleep-deprived people and found that 10-minute rest pauses involving light upper limb exercise every 2 hours improved performance and subjective sleepiness, regardless of the amount of prior sleep, but only for about 15 minutes. Similarly, Horne & Reyner (1995) found enhanced alertness after 10-minute breaks for light, moderate or heavy exercise, but as only heavy exercise produced lasting effects, the authors argued this is not a practical countermeasure. This would certainly be true for driving. Further, recent evidence suggests that exercise produces a dissociation of effects on sleepiness and performance such that we feel less sleepy, but performance is not improved (Matsumoto et al., 2002; LeDuc, Caldwell & Ruyak, 2009). This finding is particularly important for driving, as currently, driver subjective fatigue is endorsed as the indicator of countermeasure use. If exercise artificially enhances alertness in the face of decreasing performance capacity, this countermeasure should be discouraged.

Naps. There is good evidence for the benefits of napping for reducing fatigue. The effectiveness of naps, not surprisingly, depends on current sleep status. A meta-analysis on the efficacy of naps as a fatigue countermeasure (Driskell & Mullen, 2005) concluded that naps benefit performance by at least maintaining baseline performance levels and sometimes resulting in improvement. The length of performance benefit was directly proportional to the length of the nap (e.g., a 15-min nap produced 2 hours of performance benefit, whereas a 4-hour nap produced 10 hours of benefit); and the benefit decreased as the post-nap interval increased, regardless of nap length. The evidence suggests that naps need to be longer with increasing sleep pressure, such as longer time awake or short prior sleep duration, in order to obtain the same performance benefits (Caldwell, Caldwell & Schmidt, 2008). Naitoh (1992) reviewed the evidence on duration of naps and concluded that 4 minutes was the shortest duration to maintain performance and that naps longer than 20 minutes had the disadvantage of progressing into full sleep, from which it is harder to wake. This is relevant to driving, as drivers who pull over for a nap will want to return to driving as soon as possible after they wake. Any residual effects on arousal during the waking process will have adverse effects on safe driving performance.

The timing of naps is very important. Naps can be prophylactic; a nap taken before a long period of work has been shown to benefit performance (Schweitzer, Muehlback & Walsh, 1992; Bonnet, 1991); this is likely to apply to driving as well. A study of long haul truck drivers (Macchi et al., 2002) showed that a 3-hour nap taken before night-time driving in a simulator improved sleepiness and fatigue and produced faster reaction speeds and more consistent performance on psychomotor tasks. Also important is when the nap is taken with respect to the circadian rhythm, as this determines the ease of falling asleep. Naps taken closer to the circadian peak will be harder to initiate and will not last as long (Gillberg, 1984), but those taken at the circadian trough are harder to awake from and can have adverse effects of "sleep inertia" on performance as a result (Dinges et al., 1985).

Use of stimulants. Stimulant use is one of the means of dealing with increasing fatigue while driving. The most widely used and well-documented stimulants are caffeine, amphetamines, and modafinil. The latter two are prescription drugs and so not normally available to drivers. Caffeine, on the other hand, is readily available and is a very common means of overcoming fatigue effects. There is good evidence (Smith, 2002; Bonnet, Balkin & Dinges, 2005) of the effectiveness of caffeine for increasing alertness and reducing fatigue, especially in low arousal situations such as driving. Also important for driving is the fact that caffeine improves performance for tasks requiring sustained attention, especially when alertness is already diminished such as after time without sleep. A study of the effectiveness of caffeine for managing fatigue while driving after restricted or no overnight sleep showed that around 200mg caffeine (equivalent to two to three cups) reduced safety-related incidents in a driving simulator (Reyner & Horne, 2000). Caffeine in the form of an energy drink produced similar beneficial effects on driving (Reyner & Horne, 2002). Further, caffeine in combination with a short nap (<15 minute duration) has been shown to be even more effective than either countermeasure alone in terms of benefits for driving performance and the duration of those performance benefits (Reyner & Horne, 1997; Sagaspe et al., 2007).

It should be noted that while there is general agreement that caffeine aids performance when sleepy, there is some debate about the overall benefits of caffeine use, especially due to its affects on subsequent sleep (Roehrs & Roth, 2008). Nevertheless, there is good evidence that caffeine is an effective short-term countermeasure for fatigue while driving.

Road-related Countermeasures

Rumble strips

This countermeasure operates to increase drivers' awareness that their driving performance has become impaired. Also called Profile Lane Marking, this countermeasure involves a narrow series of small raised bumps, either milled into the road surface or raised slightly above it, located either just on the edge of the road or in the centre line. The purpose of rumble strips is to create vibrations or noise when the wheel of the vehicle hits the strip, so letting drivers know they are leaving their lane and also increasing driver arousal. Some studies have suggested that rumble strips reduced accidents by between 15 percent and 58 percent (Mahoney, Porter & Donnell, 2003; Gardner & Davies, 2006). A simulator study of the effectiveness of rumble strips showed that driver alertness increased as measured by eye closure duration, EEG activity and lane deviation, although the effect was short, lasting only around 5 minutes (Anund et al., 2008). Criticisms of rumble strips suggest that they may produce unintended consequences, by causing crashes as drivers swerve out-of-lane when attempting to correct their course. An on-road study comparing roads with and without rumble strips and before and after their introduction showed that rumble strips were effective in reducing crashes and most effective if installed on both edge and centre line (Hatfield et al., 2009). This is a countermeasure of promise for fatigue detection, although like the in-vehicle detection measures, it does not act on the source of fatigue or reduce it.

Places to stop

A number of the countermeasures discussed have the effect of advising drivers of the need to stop (e.g., monitoring devices, rumble strips) and in each case the effectiveness of the particular countermeasure depends on the availability of places to stop. There is little guidance from previous research on this countermeasure. Reyner et al. (2010) looked at the effectiveness of motorway service centres for reducing fatigue-related crashes. In this study the crash incidence of 16km stretches of road pre- and post-motorway service sites on two sections of UK motorways were compared over a 2–3 year period. The results showed significantly lower fatigue-related crashes post-service area, indicating that the service areas may have played a part in reducing fatigue-related crash risk. Interestingly, the least effect on crashes was seen in the circadian trough period of 02:00 to 06:00, suggesting that drivers are either loath to stop at that time or that breaks may be less effective during that period. No information was available on whether crashed drivers used or failed to use the opportunity to stop, so further research is needed to confirm the effectiveness of this countermeasure.

CONCLUSIONS

This review of driver fatigue countermeasures has highlighted some promising approaches. Clearly, ensuring that drivers are not tired and sleepy before they commence driving is a prime countermeasure. So, too, is avoiding driving during the circadian low points, especially during the 02:00–06:00hrs period and especially when already sleep-deprived. These approaches are really talking about ideal circumstances. It may not be realistic to expect that drivers will always be in peak alertness when they get into their vehicle, although if attempting a long drive, especially for work, then occupation health and safety responsibilities should require this as a minimum. In the real world then, a major focus for fatigue countermeasures is helping drivers manage fatigue while they are driving. Current evidence indicates that the most successful countermeasure for this purpose is stopping driving for a nap. Also effective as a fatigue countermeasure is simply taking a break from driving, however this is likely to become less effective with increasing levels of fatigue. Unfortunately there is surprisingly little good guidance on the best timing and patterning of driving rest breaks, or on what activities the break should contain, with the exception of naps. Current advice is based largely on historical precedence and intuition and more research is badly needed.

For a range of reasons, discussed in this chapter, rest breaks from driving may not be feasible in the real world, or always acceptable to drivers. Consequently, approaches that help the driver overcome fatigue while driving are also needed. None of these approaches will be effective over the longer term, but some are likely to help a driver push on for long enough to reach a suitable place to stop and take a break. Based on the current evidence, the most effective countermeasures for fatigue while driving include stimulant use, in particular caffeine-containing drinks, for which there is good evidence of effectiveness. Evidence on the benefits of using suitable additional tasks that make the driving task less monotonous and increase cognitive engagement while driving is also very promising. Probably in the long run most benefit will be gained from mixtures of countermeasures, like the current evidence of the effects of a nap and caffeine on fatigue, however there has been too little research on such combinations on which to base any conclusions.

Finally, the approaches that focus on monitoring and detecting fatigue in either drivers or their driving performance are also showing some promise, although these approaches need to be implemented with awareness that they do not act to reduce fatigue. They simply alert drivers to the need to stop driving as soon as possible, so leading to the same issues as highlighted above around taking breaks.

REFERENCES

Anund, A., Kecklund, G., Vadeby, A., Hjälmdahl, M. & Åkerstedt, T. (2008). The alerting effect of hitting a rumble strip: A simulator study with sleepy drivers. *Accident Analysis and Prevention*, 40, 1970–76.

Baker, F.C., Maloney, S. & Driver, H.S. (1999). A comparison of selective estimates of sleep with objective polysomnographic data in healthy men and women. *Journal of Psychosomatic Research*, 47, 335–41.

Belz, S.M., Robinson, G.S. & Casali, J.G. (2004). Temporal separation and self-rating of alertness as indicators of driver fatigue in commercial motor vehicle operators. *Human Factors*, 46, 154–69.

Bonnet, M.H. (1991). The effect of varying prophylactic naps on performance, alertness and mood throughout a 52-h continuous operation. *Sleep*, 14, 307–15.

Bonnet, M.H. & Arand, D.L. (1999). Level of arousal and the ability to maintain wakefulness. *Journal of Sleep Research*, 8, 247–54.

Bonnet, M.H., Balkin, T.J. & Dinges, D.F. (2005). The use of stimulants to modify performance during sleep loss: A review by the sleep deprivation and stimulant task force of the American Academy of Sleep Medicine. *Sleep*, 28, 1163–87.

Brown, I. (1994). Driver fatigue. *Human Factors*, 36, 298–314.

Brown, I. (2001). Coping with driver fatigue: Is the long journey nearly over? In P.A. Hancock and P.A. Desmond (eds), *Stress, Workload and Fatigue*. Mahwah, NJ: Lawrence Erlbaum, 596–606.

Bunn, T.L., Slavova, S., Struttmann, T.W. & Browning, S.R. (2005). Sleepiness/fatigue and distraction/ inattention as factors for fatal versus nonfatal commercial motor vehicle driver injuries. *Accident Analysis & Prevention*, 37, 862–9.

Caldwell, J.A., Caldwell, J.L. & Schmidt, R.M. (2008). Alertness management strategies for operational contexts. *Sleep Medicine Reviews*, 12, 257–73.

Dababneh, A.J., Swanson, N. & Shell, R.L. (2001). Impact of added rest breaks on the productivity and well being of workers. *Ergonomics*, 44, 164–74.

Dalziel, J.R. & Job, R.F.S. (1997). Motor vehicle accidents, fatigue and optimism bias in taxi drivers. *Accident Analysis and Prevention*, 29, 489–94.

Dawson, D. & McCulloch, K. (2005). Managing fatigue: It's about sleep. *Sleep Medicine Reviews*, 9, 365–80.

Desmond, P.A., Hancock, P.A. & Monette, J.L. (1998). Fatigue and automation-induced impairments in simulated driving performance. *Transportation Research Record*, 1628, 8–14.

Dinges, D.F., Orne, M.T. & Orne, E.C. (1985). Assessing performance upon abrupt awakening from naps during quasi-continuous operations. *Behavior Research Methods Instrumentation and Computers*, 17, 37–45.

Dinges, D.F., Pack, F., Williams, K., Gillen, K.A., Powell, J.W., Ott, G.E., Aptowicz, C. & Pack, A.I. (1997). Cumulative sleepiness, mood disturbance, and psychomotor vigilance performance decrements during a week of sleep restricted to 4–5 hours per night. *Sleep*, 20, 267–77.

Dorrian, J., Lamond, N. & Dawson, D. (2000). The ability to self-monitor performance when fatigued. *Journal of Sleep Research*, 9, 137–44.

Dorrian, J., Lamond, N., Holmes, A.L., Burgess, H.J., Roach, G.D., Fletcher, A. & Dawson, D. (2003). The ability to self-monitor performance during a week of simulated night shifts. *Sleep*, 26, 871–7.

Driskell, J.E. & Mullen, B. (2005). The efficacy of naps as a fatigue countermeasure: A meta-analytic integration. *Human Factors*, 47, 360–77.

Fletcher, A., McCulloch, K., Baulk, S.D. & Dawson, D. (2005). Countermeasures to driver fatigue: A review of public awareness campaigns and legal approaches. *Australian & New Zealand Journal of Public Health*, 29, 471–6.

Friswell, R. & Williamson, A. (2008). Exploratory study of fatigue in light and short haul transport drivers in NSW, Australia. *Accident Analysis and Prevention*, 40, 410–17.

Garder, P. & Davies, M. (2006). Safety effect of continuous rumble strips on rural interstates in Maine. *Transportation Research Record*, 1953, 156–62.

Gershon, P., Ronen, A., Oron-Gilad, T. & Shinar, D. (2009). The effects of an interactive cognitive task (ICT) in suppressing fatigue symptoms in driving. *Transportation Research* Part F, 12, 21–8.

Gillberg, M. (1984). The effects of two alternative timings of a one hour nap on early morning performance. *Biological Psychology*, 19, 45–54.

Hamad, M.M., Jaradat, A.S. & Easa, S.M. (1998). Analysis of commercial mini-bus accidents. *Accident Analysis and Prevention*, 30, 555–67.

Hatfield, J., Murphy, S., Job, R.F.S. & Du, W. (2009). The effectiveness of audio-tactile lane-marking in reducing various types of crash: A review of evidence, template for evaluation, and preliminary findings from Australia. *Accident Analysis and Prevention*, 41, 365–79.

Horne, J.A. & Baulk, S.D. (2004). Awareness of sleepiness when driving. *Psychophysiology*, 41, 161–5.

Horne, J.A. & Reyner, L.A. (1995). Driver sleepiness. *Journal of Sleep Research*, 4, 23–9.

Horne, J.A. & Reyner, L.A. (1999). Vehicle accidents related to sleep: a review. *Occupational and Environmental Medicine*, 56, 289–94.

Jones, C.B., Dorrian, J. & Rajaratnam, S.M.W. (2005). Fatigue and the criminal law. *Industrial Health*, 43, 63–70.

Kaplan, K.A., Itoi, A. & Dement, W.C. (2008). Awareness of sleepiness and ability to predict sleep onset: Can drivers avoid falling asleep at the wheel? *Sleep Medicine*, 9, 71–9.

Kopardekar, P. & Mital, A. (1994). The effect of different work–rest schedules on fatigue and performance of a simulated directory assistance operator's task. *Ergonomics*, 37, 1697–707.

LeDuc, P.A., Caldwell, J.A. & Ruyak, P.S. (2000). The effects of exercise as a countermeasure for fatigue in sleep-deprived aviators. *Military Psychology*, 12, 249–66.

Lenné, M.G., Triggs, T.J. & Redman, J.R. (1997). Time of day variations in driving performance. *Accident Analysis and Prevention*, 29, 431–7.

Liu, C.C., Hosking, S.G. & Lenné, M.G. (2009). Predicting driver drowsiness using vehicle measures: Recent insights and future challenges. *Journal of Safety Research*, 40, 239–45.

Macchi, M.M., Boulos, Z., Ranney, T., Simmons, L. & Campbell, S.C. (2002). Effects of an afternoon nap on night-time alertness and performance in long-haul drivers. *Accident Analysis and Prevention* 34, 825–34.

Mahoney, K.M., Porter, R.J. & Donnell, E.T. (2003). Evaluation of centerline rumble strips on lateral vehicle placement and speed on two-lane highways. (No. FHWA-PA-2002-034-97-04 (111)). Harrisburg, PA: Pennsylvania Department of Transportation.

Matthews, G. & Desmond P.A. (2002). Task-induced fatigue states and simulated driving performance. *Quarterly Journal of Experimental Psychology, Section A – Human Experimental Psychology*, 55, 659–86.

Matsumoto, Y., Mishima, K., Satoh, K., Shimizu, T. & Hishikawa, Y. (2002). Physical activity increases the dissociation between subjective sleepiness and objective performance levels during extended wakefulness in humans. *Neuroscience Letters*, 326, 133–6.

May, J.F. & Baldwin, C.L. (2009). Driver fatigue: The importance of identifying causal factors of fatigue when considering detection and countermeasure technologies. *Transportation Research, Part F*, 12, 218–24.

Maycock, G. (1997). Sleepiness and driving: The experience of heavy goods vehicle drivers in the UK. *Journal of Sleep Research*, 5, 229–37.

McDonald, N. (1984). *Fatigue, Safety and the Truck Driver*. London: Taylor & Francis.

Mitler, M., Miller, J., Lipsitz, J., Walsh, J. & Wylie, C.D. (1997). The sleep of long-haul truck drivers. *The New England Journal of Medicine*, 337, 755–61.

Monk, T.H. (1991). Circadian aspects of subjective sleepiness: A behavioural messenger? In T.H. Monk, *Sleep, Sleepiness and Performance*. Chichester, UK: John Wiley, 39–63.

Naitoh, P. (1992). Minimal sleep to maintain performance: The search for the sleep quantum in sustained operations, in C. Stampi (ed.), *Why we Nap*. Boston, MA: Birkhauser, 198–219.

Ogilvie, R.D., Wilkinson, R.T. & Allison, S. (1989). The detection of sleep onset: Behavioural, physiological, and subjective convergence. *Sleep*, 12, 458–74.

Oron-Gilad, T., Ronen, A. & Shinar, D. (2008). Alertness maintaining tasks (AMTs) while driving. *Accident Analysis and Prevention*, 40, 851–60.

Pack, A.I., Pack, A.M., Rodgman, E., Cucchiara, A., Dingers, D.F. & Schwab, C.W. (1995). Characteristics of crashes attributed to the driver having fallen asleep. *Accident Analysis and Prevention*, 27, 769–75.

Parasuraman, R. & Riley, V. (1997). Humans and automation: Use, misuse, disuse, abuse. *Human Factors*, 39, 230–53.

Reyner, L.A. & Horne, J.A. (1997). Suppression of sleepiness in drivers: Combination of caffeine with a short nap. *Psychophysiology*, 34, 721–25.

Reyner, L.A. & Horne, J.A. (1998a). Evaluation "in-car" countermeasures to sleepiness: Cold air and radio. *Sleep*, 21, 46–50.

Reyner, L.A. & Horne, J.A. (1998b). Falling asleep whilst driving: Are drivers aware of prior sleepiness? *International Journal of Legal Medicine*, 111, 120–23.

Reyner, L.A. & Horne, J.A. (2000). Early morning driver sleepiness: Effectiveness of 200 mg caffeine. *Psychophysiology*, 37, 254–56.

Reyner, L.A. & Horne, J.A. (2002). Efficacy of a "functional energy drink" in counteracting driver sleepiness. *Physiology & Behavior*, 75, 331–5.

Reyner, L.A., Horne, J.A. & Flatley, D. (2010). Effectiveness of UK motorway services areas in reducing sleep-related and other collisions. *Accident Analysis and Prevention*, 42, 1416–18.

Roehrs, T. & Roth, T. (2008). Caffeine: Sleep and daytime sleepiness. *Sleep Medicine Reviews*, 12, 153–62.

Rosa, R.R. & Colligan, M.J. (1988). Long workdays versus restdays: Assessing fatigue and alertness with a portable performance battery. *Human Factors*, 30, 305–17.

Sagaspe, P., Taillard, J., Chaumet, G., Moore, N., Bioulac, B. & Philip, P. (2007). Aging and nocturnal driving: better with coffee or a nap? A randomized study. *Sleep*, 30, 1808–13.

Sallinen, M., Holm, A., Hiltunen, J., Hirvonen, K., Harma, M., Koskelo, J., Letonsaari, M., Luukkonen, R., Virkkala, J. & Muller, K. (2008). Recovery of cognitive performance from sleep debt: Do a short rest pause and a single recovery night help? *Chronobiology International*, 25, 279–96.

Schweitzer, P.K., Muehlback, M.J. & Walsh, J.K. (1992). Countermeasures for night work performance deficits: The effect of napping or caffeine on continuous performance at night. *Work Stress*, 6, 355–65.

Smith, A. (2002). Effects of caffeine on human behavior. *Food and Chemical Toxicology*, 40, 1243–55.

Stanton, N.A., Young, M.S., Walker, G.H., Turner, H. & Randle, S. (2001). Automating the driver's control tasks. *International Journal of Cognitive Ergonomics*, 5, 221–36.

Stave, A.M. (1977). The effects of cockpit environment on long term pilot performance. *Human Factors*, 19, 503–14.

Stutts, J.C., Wilkins, J.W. & Vaughn, B.V. (1999). Why do people have drowsy driving crashes? Input from drivers who just did (No. 202/638-5944). Washington, DC: AAA Foundation for Traffic Safety.

Thiffault, P. & Bergeron, J. (2003). Monotony of road environment and driver fatigue: A simulator study. *Accident Analysis and Prevention*, 35, 381–91.

Tucker, P. (2003). The impact of rest breaks upon accident risk, fatigue and performance: A review. *Work & Stress*, 17, 123–37.

Tucker, P., Folkard, S. & MacDonald, I. (2003). Rest breaks and accident risk. *The Lancet*, 361, 680.

Williamson, A.M., Feyer, A.M., Coumarelos, C. & Jenkins, T. (1992). Strategies to combat fatigue in the long distance road transport industry. Stage 1: The industry perspective. *Federal Office of Road Safety Report*, CR 108.

Williamson, A.M., Feyer, A.M., Friswell, R. & Finlay-Brown, S. (2000). Demonstration project for fatigue management programs in the road transport industry – summary of findings. *Federal Australian Transportation Safety Bureau report*, CR192, 1–20.

Williamson, A.M., Feyer, A.M., Friswell, R. & Finlay-Brown, S. (2001). Driver fatigue: A survey of long distance heavy vehicle drivers in Australia. *Australian Transportation Safety Bureau report*, CR198.

Williamson, A., Lombardi, D.A., Folkard, S., Stutts, J., Courtney, T.K. & Connor, J.L. (2011). The link between fatigue and safety. *Accident Analysis and Prevention*, 43, 498–515.

Wright, N.A., Stone, B.M., Horberry, T.J. & Reed, N. (2007). A review of in-vehicle sleepiness detection devices. *Transport Research Laboratories Report* PPR157, UK.

Yang, J.H., Mao, Z.H., Tijerina, L., Pilutti, T., Coughlin, J.F. & Feron, E. (2009). Detection of driver fatigue caused by sleep deprivation. *IEEE Transactions of Systems, Man, and Cybernetics – Part A: Systems and Humans*, 39, 694–705.

Young, M.S. & Stanton, N.A. (2004). Taking the load off: Investigation of how adaptive cruise control affects mental workload. *Ergonomics*, 47, 1014–35.

28
Work Scheduling

Philip Tucker and Simon Folkard

INTRODUCTION

Individuals who regularly work on abnormal work schedules (i.e. shiftwork of some type) are more prone to fatigue than typical day workers. This is due in large part to restricted opportunities for rest, recovery and sleep, which may impact on their performance at work and the likelihood of them making a mistake, possibly resulting in an accident. Prolonged exposure to excessive fatigue and sleep deprivation may also impact on the individual's physical and psychological well-being.

THEORETICAL BACKGROUND

Many of the fatigue-related problems that shift workers encounter stem from their disrupted biological rhythms. These rhythms have evolved in response to the periodic changes in the environment, such as the day–night cycle. They have become internalized such that the body adjusts many of our physiological and psychophysiological processes and these regular cyclical changes are known as circadian rhythms (from the Latin "about a day"). They are jointly controlled by an internal, or "endogenous", body clock and by external, or "exogenous", factors in the environment such as awareness of clock time, meal timings, social activity, etc. (see Waterhouse, Chapter 16, this volume).

From a fatigue perspective, one of the most important biological rhythms is the sleep–wake cycle. Borbély (1982) suggested that this rhythm is driven by two processes, the endogenous "body clock" (the circadian component) and the build-up of sleep need with time awake (the homeostatic component). [Later theories (Folkard et al. 1999) added a third homeostatic process which took account of the sleepiness that is experienced in the period immediately after waking (the wake-up component).] The endogenous body clock is only weakly influenced by exogenous factors and this inherent stability of the circadian component can pose problems if a mismatch arises between the internal timing system and external time cues. The simplest example of this occurs when people fly across time zones. Their sleep is disrupted for some days after arriving at their destination, until their body clocks have adjusted to the new time zone. The strongest exogenous influence on the timing of the body clock is the presence or absence of bright light (i.e. the day–night cycle). Thus exposure to daylight in the new time zone helps travellers to adjust the timing of their body clock, but when shift workers change from working day shifts to working night shifts, most environmental time cues remain constant and hence discourage adjustment of the circadian system.

SHIFT SYSTEM DESIGN

Shift systems vary widely along several dimensions, including whether the shifts rotate or not, the direction and speed of the shift rotation, the number of consecutive shifts, the length of shifts, the start and end times of each shift, and the number and placement of days off. Hence there are an almost infinite number of different shift systems in operation and none of them is anything like perfect! However, we can classify systems according to their features and examine the impact of these features by comparing within and between groups of individuals working on different systems.

Rotating versus Permanent Shifts

A key issue is whether workers should regularly "rotate" between different shifts (e.g. between day and night shifts) or whether they should always work the same shift ("fixed" or "permanent" shifts). The fundamental question is whether permanent night workers can adjust the timing of their body clock such that they can easily sleep during the day and remain alert throughout the night. A recent review of the available evidence concluded "that less than one in four permanent night workers evidence sufficiently "substantial" adjustment to derive any benefit from it" (Folkard, 2008: 215). Circadian adaptation to a nocturnal routine by rotating workers on the night shift probably occurs very slowly if at all. Indeed epidemiological studies of accident and injury risk indicate that risk increases over at least four successive night shifts; and at a markedly higher rate than is observed over successive day shifts (Folkard & Tucker, 2003). A key factor working against adaptation to a nocturnal routine is that workers tend to revert to a diurnal routine on their days off, counteracting the process of circadian adaptation and resulting in their having to start adapting all over again following even just a few days off.

In summary, the available evidence indicates that fixed or permanent night shifts should be avoided in most circumstances. A possible exception is situations where a nocturnal routine may be maintained on rest days (e.g. remote work sites where workers have limited exposure to daylight; Bjorvatn, et al., 2006).

Speed of Rotation

Speed of rotation refers to the number of shifts of one type (e.g. night shifts) that are worked before the worker either changes to another type of shift (e.g. day shifts) or has a day off. Most research findings tend to favour very rapidly rotating shift systems (i.e. ones that involve working only one to three consecutive shifts of the same type) over more slowly rotating ones (Sallinen & Kecklund, 2010). It has been argued that workers on a more slowly rotating shift system will rarely achieve sufficient adjustment to derive any benefit from it. Instead, they are likely to experience a substantial degree of circadian disruption, such that their circadian rhythms remain in a state that is neither fully diurnal nor nocturnal, and hence suffer disturbed sleep between shifts. Permanent night shifts can be viewed as a special case of a slowly rotating shift system (i.e. where the speed of rotation is effectively zero). Thus, it seems that shift systems should seek to minimize the number of night shifts that are worked consecutively (i.e. no more than three consecutive night shifts, but preferably fewer).

Direction of Rotation

In rotating shift systems, changing from one type of shift to another entails altering the timing of the sleep and most other aspects of daily routine (e.g. meal times, free-time activities, etc.). An individual who changes from working morning shifts (e.g. working from 06:00 to 14:00) to afternoon shifts (e.g. working from 14:00 to 22:00) will probably go to bed and wake later in the day following the change, while an individual who changes from afternoon shifts to morning shifts will probably go to bed and wake up earlier following the change. The majority of people tend to find it easier to delay sleep onset and waking than to advance the timing of their sleep and this is thought to reflect on the natural tendency of the body clock to cycle with a period of slightly longer than 24 hours (Aschoff & Wever, 1962).

Thus from a circadian perspective, "forward-rotating" shift systems that involve delaying the timing of one's rhythms have been argued to be preferable to "backward-rotating" ones that involve advancing the timing of circadian rhythms. However, the evidence is equivocal with regard to the effects on sleep and fatigue (Sallinen & Kecklund, 2010). Thus, although two recent Finnish studies reported improvements in sleepiness following a change from a system rotating slowly backward to a very rapidly forward-rotating system (Harma et al., 2006; Viitasalo et al, 2008), it is unclear whether it was the speed or the direction of rotation that was primarily responsible for the improvements. Moreover, systems rotating very rapidly backward are more likely to feature "quick returns" (short intervals between the end of one shift and the start of the next – see below) which may be a more important influence on fatigue than the direction of rotation per se (Tucker, et al. 2000). In short, it seems that very rapidly forward-rotating shifts systems are to be preferred to slowly backward-rotating ones, particularly when the latter involve quick returns.

Shift Start and End Times

Several studies have shown that people working on early-starting morning shifts (e.g. starting before 07:00) tend to have shorter sleeps the night before the shift and are sleepier during the shift (Sallinen & Kecklund, 2010). This appears to be a result of workers failing to compensate for an early start by going to bed sufficiently early the night before (Folkard & Barton, 1993). This is thought to reflect, at least in part, on the influence of the circadian rhythm in sleep propensity, which reaches its lowest point, "the forbidden zone" (Lavie, 1986), in the early evening (at around 20:00–22:00) before rising steadily to peak in the early hours of the morning (04:00–06:00). Hence workers may find it impossible to initiate sleep sufficiently early in the evening to compensate for an early start the next day.

At first sight, it might seem appropriate to delay the start of morning shifts so as to reduce fatigue on that shift. However, if the morning shift workers are replacing a team of night shift workers, a substantial delay in the changeover time may cause sleep problems for the night shift workers. Just as morning shift workers have difficulty initiating sleep in the early evening, so too night workers experience greater sleep problems when going to bed in the relatively "late" morning, after returning home from the night shift (Rosa et al., 1996; Tucker, Smith, Macdonald & Folkard, 1998b). Sleep propensity falls rapidly from its peak at about 04:00–06:00 until about 13:00, implying that the later a night worker goes to bed following a night shift, the more difficulty the worker may have initiating sleep. The worker may also have difficulty remaining asleep for long enough to achieve adequate recovery.

In the light of the trade-off between the sleep needs of those on the morning and night shifts, a shift change over time of 07:00 has been proposed as an appropriate compromise (Åkerstedt, 2003). However, even then workers on both shifts may experience impoverished sleep, particularly if they have long commuting journeys. In situations where the night shift hands over directly to the morning shift it is recommended that early-starting morning shifts should be limited to three in a row in order to minimize the accumulation of fatigue. If the morning shift does not take over directly from a night shift, then early morning starts should be avoided completely.

So far, the focus of our discussion has been on the way in which shift system design can be used to mitigate the effects of circadian disruption on fatigue, particularly when night work is involved. However fatigue can also be caused by non-circadian aspects of schedule design such as excessively long work hours, particularly when there is insufficient opportunity for rest and recovery during and between work shifts (see Rosa, Chapter 21, this volume).

Shift Length

Long shifts (e.g. 12 hours) are a means of compressing the working week into fewer longer shifts. They tend to be popular with workers, who appreciate the extended periods of time off and the reduced number of commuting journeys, but they require work effort to be sustained over a longer period without substantial rest. This could, in theory, result in fatigue accumulating to unsafe levels toward the end of the shift.

Comparisons of the effects of 8-hour and 12-hour shift systems on sleep and sleepiness have produced mixed findings (Tucker, 2006). Our research found that measures of sleep and sleepiness outcomes tended to vary between favouring either 8-hour or 12-hour shift systems, depending on the time of day, but with few substantial differences being observed at any point (Tucker, Barton & Folkard, 1996; Tucker, Smith, Macdonald & Folkard, 1998a; Tucker et al., 1998b). This suggests that, overall, 12-hour shifts are no more problematic than 8-hour shifts and several other studies have reported either neutral or beneficial effects of 12-hour shifts on sleep and sleepiness (Duchon, Keran & Smith, 1994; Lowden et al., 1998; Mitchell & Williamson, 2000; Williamson, Gower & Clarke, 1994). However, sleepiness may be higher at certain times of day, particularly toward the end of the night shift (Rosa, 1991; Rosa & Bonnet, 1993; Rosa & Colligan, 1989), and 12 hour shifts may be associated with poorer sleep in highly demanding work environments (Iskra-Golec et al., 1996). In many studies shift length has been confounded with other factors, often in a manner that favoured 12-hour systems. For example, among the above studies, 12-hour systems were compared with 8-hour systems that variously featured quick returns (Lowden et al., 1998), earlier changeover times (Rosa, 1991; Rosa & Bonnet, 1993; Rosa & Colligan, 1989), or irregular shift patterns (Williamson et al., 1994).

An early review in this area concluded that there was no clear evidence of an overall adverse effect of extended work shifts on performance and safety (Smith, Folkard, Tucker & Macdonald, 1998). However, a subsequent meta-analysis of accident data collated from four previously published studies identified a substantial increase in risk in the last 3 hours of a 12- hour shift, after correcting for exposure (Folkard & Tucker, 2003). Risk in the twelfth hour on shift was more than double the average hourly risk during the first 8 hours. Longer shifts (particularly those $\geq$ 12.5 hours) have also been linked to an increased risk of drowsy driving and involvement in driving accidents/near–misses on the journey home from work (Scott et al., 2007).

In summary, there are inconsistent findings regarding the impact of extended shifts on fatigue, performance and safety. This may reflect differences in the nature of the occupations or job-tasks being undertaken by the subjects of the various studies (Smith et al., 1998). For example, unstimulating work environments, monotonous tasks and the requirement to sustain attention all increase the likelihood of performance decrement over prolonged periods, although this may not be due to the development of fatigue per se (Williamson et al., 2011). Long (i.e.12 hour) shifts may cause increased sleepiness in situations with a high work load, inadequate staff resources, insufficient rest breaks or extended commuting time (Rosa, 1995). Thus it is difficult to specify a universally applicable recommendation for maximum shift duration. Nevertheless, on the balance of available evidence, and especially in light of the observed increase in accident risk with extended time on shift, it is recommended that shifts should not be scheduled to last more than 12 hours.

Weekly Work Hours

Long weekly work hours result not only in a greater need for recovery, due to the prolonged exposure to work demands, but also in a reduced opportunity for achieving that recovery. They are thus associated with shorter and more disturbed sleeps (Virtanen et al., 2009). This is likely to reflect not just the restricted time available for sleeping but also that for relaxing during leisure time, since being unable to fully relax and unwind can lead to increased sleep disturbance (Viens et al., 2003). This in turn is likely to result in increased fatigue the next day (Meijman & Mulder, 1998). Not surprisingly, long weekly work hours in combination with night work results in particularly high levels of sleepiness at work (Son et al., 2008; Tucker et al., 2010).

Dembe and colleagues (2005) identified a strong positive association between weekly work hours and risk of occupational injuries and illness. Working at least 60 hours per week was associated with a 23 percent higher risk (as compared to working < 60 hours), after adjusting for age, gender, occupation, industry and region, suggesting that the increased risk was not simply due to more hazardous occupations.

Historically, long work shifts have been a common feature in healthcare settings and recent intervention studies have demonstrated reductions in medical errors following reductions in both shift length and the overall number of hours worked per week. In a US study a system involving 24-hour on-call shifts and weekly hours of >80 hours was replaced with one in which shifts were never longer than 16 hours and weekly hours were reduced to approximately 65 hours. The new schedule resulted in a 30 percent reduction in medical errors (Landrigan et al., 2004) and a 50 percent reduction in the risk of the doctors being involved in motor vehicle crashes (Barger et al., 2005). In the UK, a study of doctors in which shift length was reduced from 12.5 hours to 9–11 hours and the total number of weekly hours was reduced from 56 to 48, also produced a significant (33 percent) reduction in medical errors (Cappuccio et al., 2009).

The question of what constitutes a safe maximum for weekly work hours is a vexed issue. Among full-time employees (i.e. ≥ 40 hours per week), both sleep problems and risk increase approximately linearly with the number of hours worked per week (Dembe et al., 2005; Virtanen et al., 2009), suggesting that workers should not be scheduled to work more than about 48 hours in any single week.

Distribution of Rest Days

The main key to minimizing the accumulation of excessive fatigue is the provision of adequate opportunities for rest and recovery, both during shifts (i.e. see "intra-shift rest breaks," below), between successive shifts (see "quick returns," below) and between blocks of shifts (i.e. rest days). Weekends provide an important opportunity for rest and recovery from the demands of working (Fritz & Sonnentag, 2005), but there is relatively little evidence regarding the optimum distribution of rest days within rotating shift schedules. One study found that a 12-hour shift system involving 2 days on, 2 days off, resulted in slightly higher alertness than one involving 4 days on, 4 days off (Tucker, Smith Macdonald & Folkard, 1999). Another study (Totterdell et al. 1995) showed that alertness and performance were impaired on the first three days back at work following a single rest day as compared to a span of two or three rest days. Likewise, Åkerstedt et al. (2000) after reviewing the literature, concluded that a single day of rest is never sufficient, that two usually is and that three or four are needed after periods of severely disturbed circadian rhythmicity (e.g. after working several night shifts). More generally, it seems probable that it is the ratio of work days to rest days that is important in enabling full recovery, and that under normal circumstances it may be appropriate "to limit spans of successive work days to not more than six and to require a minimum of two successive rest days" (Spencer et al., 2006: 40).

Quick Returns

Quick returns, i.e. short intervals between the end on one shift and the start of the next, often occur in backward-rotating shift systems (see above) and are also common when the overall weekly work hours are high (see above). They restrict the opportunity for sleeping and other non-work activities between shifts and are associated with (i) shorter sleeps (Axelsson, 2004; Kurumatani et al., 1994) and (ii) increased fatigue on the subsequent shift (Tucker, Brown et al., 2010; Tucker, Smith, Macdonald & Folkard, 2000). The impact of quick returns on fatigue is likely to be exacerbated by long commuting times and other factors such as family responsibilities (Kogi, 1982), but there is a lack of clear evidence of their effect on injury risk (see Macdonald, Smith, Lowe & Folkard, 1997; and Spencer, Robertson & Folkard, 2006).

It should be noted that quick returns are a means of compressing the working week, giving longer periods of rest between spans of work days (e.g. Barton, Folkard, Smith & Poole, 1994). For this reason, quick returns and the shift systems that feature them (e.g. some backward-rotating shift systems) are often popular with the workforce, but given their impact on sleep and recovery they are best avoided. Indeed, it is noteworthy that the European Working Time Directive requires that workers be allowed a minimum rest interval of 11 hours between successive duty periods.

Intra-shift Rest Breaks

A very few studies have examined the impact of rest breaks during a shift on injury or accident risk (Tucker, Folkard & Macdonald, 2003; Tucker, Lombardi et al., 2006). Nevertheless, they agree in showing that risk is reduced in the first half hour following a rest break, and that this effect is similar across all three shifts. The trends over subsequent half hours varied and this may reflect on the extent to which the work was either self-paced or machine-paced. Thus it seems that the beneficial effects of rest breaks may be relatively short-lived in at least some work environments. Relatively few studies have provided evidence regarding the optimum timing and duration of rest breaks and most of these have focussed on outcomes such as performance, physiological indices of strain, or subjective indices of fatigue and comfort. Their results suggest that frequent short breaks are beneficial and that fatigue management is improved when the timing of rest is at the discretion of the individual, although this is clearly not feasible in many situations. There is conflicting evidence regarding the optimum duration of rest breaks and it seems likely that both the optimum scheduling of rest breaks and their likely beneficial effects will be affected by the nature of the work (Tucker, 2003). Thus, while there is a clear need for more research on the optimum scheduling of rest breaks, the limited available evidence suggests that schedules should aim to incorporate frequent (e.g. 2 hourly) short (e.g. 15-minute) breaks, rather than fewer longer ones.

MODELLING THE EFFECTS OF SHIFT SYSTEM FEATURES IN COMBINATION

As has been noted previously, the fact that shift systems can vary along so many dimensions means that there are vast numbers of potential shift systems. Some of their features are almost inextricably linked, such that a change in one dimension almost inevitably means a change in another (e.g. changing from a backward- to a forward-rotating shift system typically eliminates quick returns). There are also potential conflicts between some of the recommendations, e.g. the need for morning shifts to start relatively late is countermanded by the need for night shifts to finish early. Further, the features may interact with each other, such that the impact of one feature on fatigue (e.g. shift length) depends critically on the value of another feature (e.g. the frequency of rest breaks; Folkard & Lombardi, 2006).

This complexity is a major problem for those wishing to design better shift systems. Further, organizational factors often result in potential shift systems having to be evaluated in the absence of any real data. In an attempt to address these difficulties, researchers have developed various mathematical models to predict fatigue levels on any given shift system. These models have typically been based on fatigue data from a broad range of sources including studies of the sort cited above (see chapters by Van Dongen & Belenky and by French & Neville, Chapter 29 in this volume). They allow the user to specify a shift system design in terms of a range of parameters (i.e. features of shift systems) and to obtain estimates of the relative level fatigue. However, a fundamental problem with these models is that variations in fatigue and the risk of injuries and accidents show systematic differences (Folkard & Åkerstedt, 2004).

In the light of this, an alternative model based on systematic trends in the risk of injuries and accidents is incorporated in the UK's Health and Safety Executive (HSE) Fatigue and Risk Indices (Spencer, Robertson & Folkard, 2006). Both these indices are constructed from three separate components, namely: "(i) a cumulative component based on the pattern of work leading up to any given shift, (ii) a duty timing component concerned with the effect of start time, shift length and the time of day throughout the shift, and (iii) a job type/breaks component which relates to the activity being undertaken and the provision of breaks during the shift." (Folkard, Robertson & Spencer, 2007: 177). The inclusion of the third component of the model reflects the importance of taking situational factors into account when seeking to optimize shift system design. The user of the model is able to input information regarding the typical workload involved in the job, the degree of continuous attention required in the job and the typical commuting time. However, like most models in this area they fail to take account of known individual differences associated with factors such as age (Harma, 1996), gender (Estryn-Behar et al., 1990) and personality (Kerkhof, 1985).

CONCLUSIONS

Working shifts, particularly if it involves night-working, is almost bound to result in workers experiencing greater fatigue than their day-working counterparts, at least on some occasions. However, the judicious design of shift schedules in the light of research findings can minimize these problems. Our understanding of the effects of shift work continues to improve, despite the fact that very few studies have ever managed a "pure" comparison of a single feature of shift systems. Nevertheless more research is needed, particularly on how fatigue is likely to be influenced by the shift system in combination with the type of work being undertaken and other aspects of the work environment, the nature of the organization, etc. For example, many of the above recommendations may be less applicable in "extreme" work environments, e.g. offshore oil rigs, where the possibilities for achieving full circadian adaptation to a nocturnal routine are greater (Bjorvatn et al., 2006). In short, the above recommendations are based on generalizations and their application needs to take into account the broader context in which the shift system is to be implemented, as do the outputs from the various models that have been developed.

Notwithstanding these qualifications, some basic principles of shift system design can be elucidated. Shift work is "unnatural" and its impact on fatigue and other negative outcomes can impair performance, safety, health and well-being. There can be no such thing as a "good" shift system, but better shift systems are those which minimize the build-up of fatigue, maximize the dissipation of fatigue through rest, and minimize sleep and circadian disruption. Finally, it is critical that organizations monitor the impact of any changes they make to work schedules on fatigue, risk, performance, sleep and health.

REFERENCES

Åkerstedt, T. (2003). Shift work and disturbed sleep/wakefulness. *Occupational Medicine*, 53, 89–94.

Åkerstedt, T., Kecklund, G., Gillberg, M., Lowden, A. & Axelsson, J. (2000). Sleepiness and days of recovery. *Transportation Research* Part F, 3, 251–61.

Aschoff, J. & Wever, R. (1962). Spontanperiodik des Menschen bei Ausschlus aller Zeitgeber. *Die Naturwissenschaften*, 49, 337–42.

Axelsson, J., Åkerstedt, T., Kecklund, G. & Lowden, A. (2004). Tolerance to shift work – how does it relate to sleep and wakefullness? *International Archives of Occupational and Environmental Health*, 77, 121–9.

Barger, L.K., Cade, B.E., Ayas, N.T., Cronin, J.W., Rosner, B., Speizer, F.E., et al. (2005). Extended work shifts and the risk of motor vehicle crashes among interns. *New England Journal of Medicine*, 352, 125–34.

Barton, J., Folkard, S., Smith, L. & Poole, C.J.M. (1994). Effects on health of a change from a delaying to an advancing shift system. *Occupational and Environmental Medicine*, 51, 749–55.

Bjorvatn, B., Stangenes, K., Oyane, N., Forberg, K., Lowden, A., Holsten, F., et al. (2006). Subjective and objective measures of adaptation and readaptation to night work on an oil rig in the North Sea. *Sleep*, 29, 821–9.

Borbély, A.A. (1982). A two-process model of sleep regulation. *Human Neurobiology*, 1, 195–204.

Cappuccio, F.P., Bakewell, A., Taggart, F.M., Ward, G., Ji, C., Sullivan, J.P., et al. (2009). Implementing a 48 h EWTD-compliant rota for junior doctors in the UK does not compromise patient safety: Assessor-blind pilot comparison. *Quarterly Journal of Medicine*, 102, 271–82.

Dembe, A.E., Erickson, J.B., Delbos, R.G. & Banks, S.M. (2005). The impact of overtime and long work hours on occupational injuries and illnesses: New evidence from the United States. *Occupational and Environmental Medicine*, 62, 588–97.

Duchon, J.C., Keran, C.M. & Smith, T.J. (1994). Extended workdays in an underground mine: A work performance analysis. *Human Factors*, 36, 258–68.

Estryn-Behar, M., Kaminski, M., Peigne, E., Bonnet, N., Vaichere, E., Gozlan, C., et al. (1990). Stress at work and mental health status among female hospital workers. *British Journal of Industrial Medicine*, 47, 20–28.

Folkard, S. (2008). Do permanent night workers show circadian adjustment? A review based on the endogenous melatonin rhythm. *Chronobiology International*, 25, 215–24.

Folkard, S. & Åkerstedt, T. (2004). Trends in the risk of accidents and injuries and their implications for models of fatigue and performance. *Aviation, Space and Environmental Medicine*, 75, A161–A167.

Folkard, S. & Barton, J. (1993). Does the "forbidden zone" for sleep onset influence morning shift sleep duration? *Ergonomics*, 36, 85–91.

Folkard, S. & Lombardi, D. (2006). Modeling the impact of the components of long work hours on injuries and "accidents." *American Journal of Industrial Medicine*, 49, 953–63.

Folkard, S. & Tucker, P. (2003). Shift work, safety and productivity. *Occupational Medicine*, 53, 95–101.

Folkard, S., Åkerstedt, T., Macdonald, I., Tucker, P. & Spencer, M.B. (1999). Beyond the three-process model of alertness: Estimating phase, time on shift, and successive night effects. *Journal of Biological Rhythms*, 14, 577–87.

Folkard, S., Robertson, K.A. & Spencer, M.B. (2007). A Fatigue/Risk Index to assess work schedules. *Somnologie*, 11, 9.

Fritz, C. & Sonnentag, S. (2005). Recovery, health, and job performance: Effects of weekend experiences. *Journal of Occupational Health Psychology*, 10, 187–99.

Harma, M. (1996). Ageing, physical fitness and shiftwork tolerance. *Applied Ergonomics*, 27, 25–9.

Harma, M., Tarja, H., Irja, K., Mikael, S., Jussi, V., Anne, B., et al. (2006). A controlled intervention study on the effects of a very rapidly forward rotating shift system on sleep–wakefulness and well-being among young and elderly shift workers. *International Journal of Psychophysiology*, 59, 70–79.

Iskra-Golec, I., Folkard, S., Marek, T. & Noworol, C. (1996). Health, well-being and burnout of ICU nurses on 12- and 8-h shifts. *Work and Stress*, 3, 251–6.

Kerkhof, G.A. (1985). Inter-individual differences in the human circadian system: A review. *Biological Psychology*, 20, 30.

Kogi, K. (1982). Sleep problems in night and shift work. *Journal of Human Ergology*, 11, 217–31.

Kurumatani, N., Koda, S., Nakagiri, S., Hisashige, A., Sakai, K., Saito, Y., et al. (1994). The effects of frequently rotating shiftwork on sleep and the family life of hospital nurses. *Ergonomics*, 37, 995–1007.

Landrigan, C.P., Rothschild, J.M., Cronin, J.W., Kaushal, R., Burdick, E., Katz, J.T., et al. (2004). Effect of reducing interns' work hours on serious medical errors in intensive care units. *New England Journal of Medicine*, 351, 1838–48.

Lavie, P. (1986). Ultrashort sleep-waking cycle schedule, III. 'Gates' and 'forbidden zones' for sleep. *Electroencephology and Clinical Neurophysiology*, 63, 414–25.

Lowden, A., Kecklund, G., Axelsson, J. & Åkerstedt, T. (1998). Change from an 8-hour shift to a 12-hour shift, attitudes, sleep, sleepiness and performance. *Scandinavian Journal of Work, Environment and Health*, 24, 69–75.

Macdonald, I., Smith, L., Lowe, S.L. & Folkard, S. (1997). Effects on accidents of time into shift and of short breaks between shifts. *International Journal of Occupational and Environmental Health*, 3, S40–S45.

Meijman, T.F. & Mulder, G. (1998). Psychological aspects of workload, in P.J.D. Drenth, H. Thierry & C.J. De Wolff (eds), *Handbook of Work and Organizational Psychology*. Hove, UK: Psychology Press, 5–33.

Mitchell, R.J. & Williamson, A.M. (2000). Evaluation of an 8 hour versus a 12 hour shift roster on employees at a power station. *Applied Ergonomics*, 31, 83–93.

Rosa, R.R. (1991). Performance, alertness, and sleep after 3.5 years of 12-hour shifts: A follow-up study. *Work and Stress*, 5, 107–16.

Rosa, R.R. (1995). Extended workshifts and excessive fatigue. *Journal of Sleep Research*, 4, 51–6.

Rosa, R.R. & Bonnet, M.H. (1993). Performance and alertness on 8h and 12h rotating shifts at a natural gas utility. *Ergonomics*, 36, 1177–93.

Rosa, R.R. & Colligan, M.J. (1989). Extended workdays: Effects of 8-hour and 12-hour rotating shift schedules on performance, subjective alertness, sleep patterns, and psychosocial variables. *Work and Stress*, 3, 21–32.

Rosa, R.R., Harma, M., Pulli, K., Mulder, M. & Nasman, O. (1996). Rescheduling a three shift system at a steel rolling mill: Effects of a one hour delay of shift starting times on sleep and alertness in younger and older workers. *Occupational and Environmental Medicine*, 53, 677–85.

Sallinen, M. & Kecklund, G. (2010). Shift work, sleep and sleepiness – differences between shift schedules and systems. *Scandinavian Journal of Work, Environment and Health*, 36, 13.

Scott, L.D., Hwang, W.T., Rogers, A.E., Nysse, T., Dean, G.E. & Dinges, D.F. (2007). The relationship between nurse work schedules, sleep duration, and drowsy driving. *Sleep*, 30, 1801–7.

Smith, L., Folkard, S., Tucker, P. & Macdonald, I. (1998). Work shift duration: A review comparing eight hour and 12 hour shift systems. *Occupational and Environmental Medicine*, 55, 217–29.

Son, M., Kong, J.O., Koh, S.B., Kim, J. & Harma, M. (2008). Effects of long working hours and the night shift on severe sleepiness among workers with 12-hour shift systems for 5 to 7 consecutive days in the automobile factories of Korea. *Journal of Sleep Research*, 17, 385–94.

Spencer, M.B., Robertson, K.A. & Folkard, S. (2006). The development of a fatigue/risk index for shiftworkers (No. 446): Health and Safety Executive.

Totterdell, P., Spelten, E., Smith, L., Barton, J. & Folkard, S. (1995). Recovery from work shifts: How long does it take? *Journal of Applied Psychology*, 80, 43–57.

Tucker, P. (2003). The impact of rest breaks upon accident risk, fatigue and performance: A review. *Work and Stress*, 17, 123–37.

Tucker, P. (2006). Compressed working weeks. Geneva: International Labour Office.

Tucker, P., Barton, J. & Folkard, S. (1996). Comparison of eight and 12 hour shifts: Impacts on health, wellbeing, and alertness during the shift. *Occupational and Environmental Medicine*, 53, 767–72.

Tucker, P., Brown, M., Dahlgren, A., Davies, G., Ebden, P., Folkard, S., et al. (2010). The impact of junior doctors' working time arrangements on their fatigue and wellbeing. *Scandinavian Journal of Work Environment & Health*, 36, 458–65.

Tucker, P., Folkard, S. & Macdonald, I. (2003). Rest breaks and accident risk. *The Lancet*, 361, 680.

Tucker, P., Lombardi, D., Smith, L. & Folkard, S. (2006). The impact of rest breaks on temporal trends in injury risk. *Chronobiology International*, 23, 1423–34.

Tucker, P., Smith, L., Macdonald, I. & Folkard, S. (1998a). The impact of early and late shift changeovers on sleep, health, and well-being in 8- and 12-hour shift systems. *Journal of Occupational Health Psychology*, 3, 265–75.

Tucker, P., Smith, L., Macdonald, I. & Folkard, S. (1998b). Shift length as a determinant of retrospective on-shift alertness. *Scandinavian Journal of Work, Environment and Health*, 24, 49–54.

Tucker, P., Smith, L., Macdonald, I. & Folkard, S. (1999). Distribution of rest days in 12 hour shift systems: Impact on health, wellbeing, and on shift alertness. *Occupational and Environmental Medicine*, 56, 206–14.

Tucker, P., Smith, L., Macdonald, I. & Folkard, S. (2000). Effects of direction of rotation in continuous and discontinuous 8 hour shift systems. *Occupational and Environmental Medicine*, 57, 678–84.

Viens, M., De Koninck, J., Mercier, P., St-Onge, M. & Lorrain, D. (2003). Trait anxiety and sleep-onset insomnia: Evaluation of treatment using anxiety management training. *Journal of Psychosomatic Research*, 54, 31–7.

Viitasalo, K., Kuosma, E., Laitinen, J. & Harma, M. (2008). Effects of shift rotation and the flexibility of a shift system on daytime alertness and cardiovascular risk factors. *Scandinavian Journal of Work Environment & Health*, 34, 198–205.

Virtanen, M., Ferrie, J.E., Gimeno, D., Vahtera, J., Elovainio, M., Singh-Manoux, A., et al. (2009). Long working hours and sleep disturbances: The Whitehall II prospective cohort study. *Sleep*, 32, 737–45.

Williamson, A.M., Gower, C.G.I. & Clarke, B.C. (1994). Changing the hours of shiftwork: A comparison of 8- and 12-hour shift rosters in a group of computer operators. *Ergonomics*, 37, 287–98.

Williamson, A., Lombardi, D.A., Folkard, S., Stutts, J., Courtney, T.K. & Connor, J.L. (2011). The link between fatigue and safety. *Accident Analysis and Prevention*, 43, 498–515.

29

Avoiding the Impact of Fatigue on Human Effectiveness

Jonathan French and Kelly J. Neville

INTRODUCTION

This chapter will discuss the use of biomathematical modeling tools as a fatigue countermeasure and the issues involving model development, methodology and validation. The discussion will avoid some of the complex technical and statistical issues associated with biomathematical models of human effectiveness. While these issues are important to theory development and ultimately will improve models, there are too many unknowns in our understanding of human fatigue to allow contention to impede what can be a meaningful tool for managers and planners now. It is our perspective that these issues will have little import to the operator, manager or planner who could find them extraordinarily useful. It is our hope that this chapter will inform the potential fatigue model users for whom this book is designed of their limitations and to persuade them that models can be used to support, not to make scheduling decisions.

The most frequent stressor encountered in society and one with arguably the greatest toll on human effectiveness is fatigue stress induced by inadequate sleep. Prolonged sleep loss of several days can be fatal (Rechtschaffen, Gilliland & Bergmann, 1983) and cognitive impairment from fatigue is well known to occur within hours of an atypical sleep cycle, as many of the chapters in this book aptly demonstrate. Predicting when cognitive ability will be impaired by fatigue is the goal of several computer-modeling tools and these represent a relatively recent and promising development in approaches to dealing with the high cost and dangers of on-the-job fatigue.

The cost to society of on-the-job accidents, missed work or decline in productivity from fatigue is staggering. One has only to look at some of the most infamous examples of operator fatigue such as the Three Mile Island, Chernobyl or Bhopal tragedies, mentioned earlier in this book, to realize the immensity of the consequences from lapses of attention in our highly technological society. A tool that provides a realistic estimate of operator effectiveness during conditions that produce fatigue would certainly lower the costs and improve the safety associated with endeavors in which fatigue is an occupational hazard. In many ways, the best countermeasure for excessive operator fatigue and the associated tragedies that could result is to avoid them altogether through effective planning. This is the promise of biomathematical models of fatigue. They attempt to provide an expert, data-based opinion on when conditions will produce fatigue levels that are likely to be too great. There are several good fatigue prediction tools available, even at this early stage in the technology, that are well thought out by experts in the field of fatigue management. They purport to mitigate the risks of fatigue by accurately predicting the probability of operator impairment from prolonged shift schedules, or atypical duty cycles.First, we will briefly review the history of models of human performance before making recommendations for their future. Then we will discuss their current state, in terms of their usefulness, limitations and validity as fatigue countermeasures. Like most good fatigue countermeasures however, biomathematical models suffer from the inability to show success, since one cannot know how many times tragedy was averted by solid planning (Friedl et al., 2004). It is far simpler to demonstrate a reduction of fatigue effects on performance through more direct countermeasures such as the ergogenic (stimulant-like) compounds. Nonetheless, this chapter is for the manager who would like to use a biomathematical model for guidance, for decision support as a fatigue management planning tool, and who additionally wants to know their constraints.

A BRIEF HISTORY OF FATIGUE MODELS

The history of the effort to realistically model the conditions that lead to human fatigue impairment is not long and may have had its origin in aviation. Some of the earliest reported human fatigue models occurred in the behavioral literature over 35 years ago, when human fatigue equations were developed to provide a more systematic and realistic means to schedule rest for aircrew. How long crews were idle on the ground and how long a duty day lasted were important business decisions and are still a matter of considerable debate (Buley, 1970; Nicholson, 1972; Gerathewohl, 1974; Mohler, 1976). The pilot rest models attempted to account for all the variables that might go into a crew rest decision, such as the number of time zones crossed, the number of landings in a duty day, or the time of the cycle when crew rest begins. With the development of increasingly longer-range aircraft, pilot fatigue was becoming a rate limiting factor in the time that aircraft could be airborne. Indeed, most aircraft with aerial refueling capability can stay aloft for many days.

There was an approximately 20-year gap in the 1980–1990s in which new models that purported to estimate fatigue were absent in the scientific literature until the 2002 Seattle Fatigue and Performance Modeling Workshop (Neri, 2004). The number of models presented at this cornerstone meeting indicated that the modeling approach was not dormant but that models were simply under development. The gap likely resulted from the tedious calculations which had to be done by hand in the early airline models. Personal computers were not introduced until 1981, and were neither readily accessible nor capable of the extensive programming that modeling can require (Bellis, 2007) until some years later. The incentive to reinvigorate crew-scheduling models occurred around 1990, when faster computers and the programming and mathematical software tools became readily available. There also was an incentive provided by the late Dr. Fred Hegge in 1986, who called for objective means to attempt to predict fatigue based on mathematical models of empirical data (Redmond, Sing & Hegge, 1982). Fred Hegge saw the need for reliable and predictive models of fatigue that could be used to protect operators from being pushed beyond their limits; and managers would have a decision tool to help them in their scheduling. Hegge also sponsored and encouraged model development (Rowe et al., 1992; Whitmore et al., 1992).

Seven models were showcased at the 2002 Seattle conference (Mallis et al., 2004). These can be easily found by the interested reader and, indeed, many of them are commercially available through the Internet to operators, managers and schedulers along with expert developers to customize the applications. Table 29.1 summarizes the major contributors and characteristics of these models. Most of the models are derived from empirical data, collected by the developer and mathematically modeled to predict fatigue based on recent sleep history, time awake, work patterns, rest periods, circadian phase, performance capability and/or risk. The recency of biomathematical models to describe cognitive and psychomotor effects following some stressor leads to an important discussion of the validity of these models. We will discuss model validity at length in a separate section, following a discussion of some contemporary issues of modeling.

Table 29.1   Characteristics of current fatigue models present at the 2002 Fatigue & Performance Modeling Workshop (Mallis, Mejdal, Nguyen & Dinges, 2004)

Author(s)	Model Name	Model Inputs	Model Outputs
Achermann (2004)	Two-process Model of Sleep Regulation	Sleep/wake time, light levels	Circadian phase estimate. Subjective alertness ratings. Subjective fatigue level
Åkerstedt, Folkard, & Portin (2004)	Sleep/wake Predictor Model (Three-process Model)	Work hours	Estimates of sleep accumulated. Sleep latency, sleep start/end. Performance effectiveness. Reaction time. Vigilance performance. Subjective alertness ratings. Subjective fatigue level
Belyavin & Spencer (2004)	SAFE Model: System for Aircrew Fatigue Evaluation	Aviation variables (bunk availability, number of pilots working, number of sectors, pilot augmentation), location, work hours	Subjective alertness ratings
Hursh, Redmond, Johnson, Thorne, Belenky, Balkin, Storm, Miller & Eddy (2004)	SAFTE Model SAFTE: Sleep, Activity, Fatigue & Task Effectiveness	Sleep length, Sleep quality, Sleep/wake time	Performance effectiveness
Jewett & Kronauer (1999)	Interactive Neurobehav-ioral Model	Sleep/wake time. Time allotted for sleep. Light levels	Circadian phase estimate. Cognitive throughput. Subjective alertness ratings
Moore-Ede, Heitmann, Guttkuhn, et al. (2004)	CAS Model: Circadian Alertness Simulator	Work hours	Sleep start/end. Subjective alertness ratings. Subjective fatigue level
Roach, Fletcher & Dawson (2004)	FAID Model: Fatigue Audit InterDyne	Work hours	Estimate of sleep accumulated. Violations based on risk threshold levels. Subjective fatigue level

CONTEMPORARY ISSUES

In this section, we will discuss three issues currently under consideration by model developers, which, it is important for the manager to understand, limit the reliability of the models. The issues to which we now turn are: What is fatigue? How accurately can it be predicted by a model, a biomathematical computer algorithm? and: Where is the cutoff, the threshold beyond which fatigue levels are too risky?

First, though, we would like to address the concern that human behavior may be far too complex to be simulated. Although some would argue that human behavior cannot be predicted by mathematical models, useful models of human behavior in stressful situations have appeared in the literature as early as 1908 with the Yerkes-Dodson law. This well-known function predicts an inverted U-shaped effect of arousal (or stress) on human performance. More comprehensive models of human performance limitations as a result of stress were developed during World War II for gun turret operators (as cited by Pew, 2008). During the nuclear stalemates of the postwar years, models were developed that related radiation exposure levels at increasing distances from ground zero with degraded performance when tactical nuclear weapons were considered a battlefield option (Openshaw & Steadman, 1983). As a more recent example, realistic estimates of body temperature limitations have been modeled based on firefighter workload and insulation levels and these describe how to reduce the numbers of firefighters needed and how long each could stay on-task before becoming impaired by heat stress (French & Miller, 2005; Hancock & Warm, 1989; Radakovic et al., 2007). Often experienced managers are not available to make these decisions and a model that can accurately predict when stress would impair or harm operators would be useful.

What is Fatigue?

There are many chapters in this book that discuss what fatigue is in far greater detail than is necessary here. The reader should not conclude that fatigue is well understood. Model developers, inadvertently perhaps, have defined fatigue by the data they use to generate their outcome measure; for some, fatigue is a subjective state; for others, it is a change in a throughput measure on a cognitive task; for still others it is a change in psychomotor vigilance task (see Table 29.1). These operational approaches are reasonable approaches for the manager who wishes to use a model to help make decisions about fatigue defined operationally as time away from adequate rest. One can choose from models that predict psychomotor ability for at-risk skills that require this dexterity. Cognitive ability or subjective state are predicted by still other models. The manager needs to find a modeling tool that predicts an outcome measure most akin to their task of concern. For the purposes of choosing a model to guide operational readiness, fatigue can be considered to be a lowered state of arousal resulting from too many hours awake, compounded by time of day (circadian) factors, which can lead to an increase in the likelihood of task error (see Banks, Jackson & Van Dongen, Chapter 11, this volume; Waterhouse, Chapter 16, this volume). This allows the manager to consider fatigue operationally, in relation to on-the-job performance, when task performance is at risk. For tasks, such as hazardous materials work or combat, the choice of models is more critical than simple routine job performance and should be based on models derived from tasks and work conditions as close as possible to the task of interest.

What Does Model Accuracy Mean?

Some criticize fatigue model predictions as being far too inaccurate to be useful at present. Major contributions to error in biomathematical models of fatigue are likely related to the inaccuracies inherent in our understanding of exactly what fatigue is. Models may become more accurate when we know how best to assess the neurophysiological factors involved in fatigue. On the other hand, the complexity and specificity of many of these factors and their interactions may not be amenable to quantitative predictions; the accuracy some seek may never be completely attainable. We know even less about the mechanisms by which sleep restores brain function than we do about the physiology of fatigue. It is our contention that the current accuracy rate, however, is high enough to support decision makers with useful guidance.

There are many discussions in the fatigue modeling and simulation literature that recommend incorporating individual differences (Van Dongen et al., 2004; Van Dongen et al., 2007), or light and homeostatic mechanisms (Jewett & Kronauer, 1999) as a means to reduce the variability of these models. We believe that a large part of the error in the models arises more from how little we know about the sources of fatigue in the brain than from individual differences (although we know very little about the nature of individual differences as well). Later in the chapter we will discuss how difficult it is, perhaps impossible, to model individual behavior due to the complex interaction of variables such as motivation, setting, experience and so on and that perhaps the best we can hope to model is the average individual as long as our model is predictive of the task at hand. The inability to satisfactorily define the physiological conditions that give rise to the phenomenon of fatigue would certainly impact the ability to predict its occurrence. Attempts to account for the fundamental lack of understanding of fatigue by referencing the current two-process model of fatigue (a simple summation of hours awake and circadian hours of alertness) seems quite simplistic an explanation of the cognitive impairment from fatigue. Fortunately, we do not have to be concerned with these discussions. We believe that there will be a larger variability in fatigue models due to our limited understanding of the biology of fatigue itself than to attempt to account for specific constraints such as individual differences in genetics or light exposure or homeostatic sensitivity for example. In other words, models that attempt to predict the average operator should be adequate, since averages tend to account for characteristics common to all in a population than models of individuals. It seems far easier to predict the average response than to predict all the motivational, genetic and experiential factors that constitute the individual responseAccording to some, the accuracy rate for these models is quite low, in the range of 20–30 percent (Van Dongen, 2004). Others, however, report much higher predictive success, on the order of 70–80 percent (Fletcher & Dawson, 2001; Walters, Archer & Yow, 2000). Using a model of fatigue derived from Air Force pilot performance on one cognitive task, French, Morris & Hancock (2003) were able to predict performance on another task another quite successfully, as is shown in Figure 29.1 on the following page.

Seemingly poor accuracy rates found by some likely come from applying a model of fatigue based on a limited set of data and conditions to a scenario involving dramatically different types of work and work conditions. The accuracy of a model should not hinge on its ability to predict a value for fatigue precisely; rather, an accurate model should make predictions that statistically are likely to happen or will occur with a reasonable certitude (e.g. Campbell & Bolton, 2005). This inaccuracy should not prevent a manager from

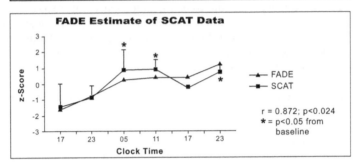

**Figure 29.1   Comparison of FADE prediction with the degraded performance
from the SCAT test (Mallis, Mejdal, Nguyen & Dinges, 2004)**

Note: The asterisks indicate where SCAT was significantly different from baseline scores approximately 21–23 hours at 05:00.

using models that could provide additional input for decisions about operator safety and quality of life. The situation is something like models predicting hurricanes; while based on imperfect knowledge of hurricane behavior, they are nonetheless useful in providing a warning that the probability of danger is higher than acceptable. We discuss the concept of model validity at length below.

How Much Fatigue is Too Much?

This operational question highlights the last and perhaps the most important of the issues facing the development of biomathematical models of fatigue; the issue of when is the risk too great. Consistent with our perspective of the model as countermeasure, we see the task of defining an operationally relevant level of fatigue, when behavior changes from well rested levels to just beginning to be impaired by fatigue levels. For example, this would be akin to the yellow caution zone in common use. The important question then is not so much what fatigue is or what accuracy levels exist between models, but rather: when does performance associated with inadequate rest begin to appear; where is the yellow zone for fatigue? We believe fatigue levels should be selected that limit duty time to these earliest negative behavioral changes (i.e. the yellow zone) rather than trying to discover a critical absolute limit of human alertness (i.e. the red danger zone).

The model, from our perspective, should not attempt to find the failure points, but instead should attempt to find a safe place to restrict duty time, well before the threshold of unacceptable risk is passed. It is far easier to identify when an individual is likely impaired than to attempt to determine when an operator's performance might have crossed a border from acceptable to unacceptable risk. Current models of fatigue leave it up to the manager or scheduler to decide if the risk is beyond an acceptable level. It is our experience, however, that managers often push the operators to their limits, that is, to their red light zone, if given a choice. They will accept a greater level of risk than the modelers intended. This means that the modeler should clearly define the safe limits of fatigue, the green zone. Modelers should additionally make use of other strategies that help decision makers balance short-term expenses against long-term benefits of fatigue management.In many sleep deprivation studies, prior to significant effects on cognitive tasks, researchers typically see lapses in wakefulness (responses that fall outside the distribution of most other responses) appear in performance measures in advance of overall statistically significant task impairments (Dorrien & Dinges, 2006; Neville, et al., 2000). These response lapses can be considered a foreshadowing that greater error risks are imminent and mark a point to identify the onset of operator performance difficulties. We propose that modelers and managers consider using this point, where lapses begin to occur in the performance data upon which their algorithms are based. In our opinion, the occurrence of lapses should be the beginning of a red zone.A dramatic means to determine operator limits was demonstrated in a study by Dawson (Dawson & Reid, 1997) in which performance on a response time task was impaired in subjects kept awake for 21 hours in a manner equivalent to the performance obtained from individuals whose blood alcohol content (BAC) was 0.1 percent, the legal intoxication standard in the US. As shown in Figure 29.2, this means that individuals kept awake for 21 hours will have the same performance levels as individuals who are legally impaired by alcohol. This suggests that a limit to operator performance could reasonably be set at the level of fatigue associated with 19–21 hours of wakefulness. It seems rational to expect that performance beyond that model result would indicate that the operator was seriously impaired, equivalent to one with a BAC of 0.1 percent. This does not imply that alcohol and sleep deprivation are the same physiologically or experientially. However, in terms of the performance output, the operator impairment, alcohol and fatigue can be equated, and after just 21 hours of sleep deprivation. The number that results from

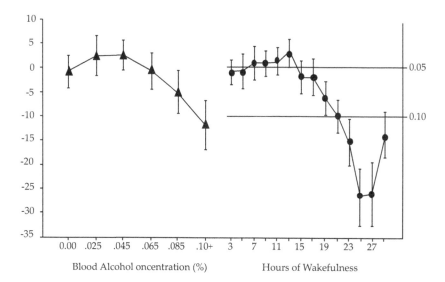

Figure 29.2   The relationship between blood alcohol level compared to sleep deprivation on response time measures

Note: The behavioral effects of a BAC of 0.1 percent is equivalent to 21 hours of sleep deprivation. These results suggest that fatigued individuals are performing about as well as someone legally intoxicated.

Source: Reprinted by permission of Macmillan Publishers Ltd from Dawson, D. & Reid, K. Fatigue, alcohol and performance impairment, *Nature*, 388: 235 © 1997

the model at this time point should be a "never exceed" limit. Dawson's work further suggests that 19 hours of wakefulness can produce performance impairment equivalent to 0.8 percent BAC. Again, this suggests that 19 hours is too impaired to be considered safe. Since most experts recommend 8 hours of continuous sleep, allowing for 16 hours of wakefulness in a 24 hour day, we believe the "red" zone on a model should be this point, somewhere between 16 and 19 hours of wakefulness, assuming an adequate rest period. Thus we have for any model a reasonable means to identify a limit, based on Dawson's alcohol equivalency data. The model result that is obtained at 16–18 hours of wakefulness mark can reasonably be used as a set point for when fatigue is too great for that model's prediction

Inherent in most models is some mechanism that diminishes the calculated fatigue or the likelihood of error with rest and this presents another issue; how do we know if the operator is truly resting as well as the model assumes? Most models will simply assume that any rest is of high quality. The manager needs to take this assumption into account. We recommend the manager err on the conservative side and treat the model's estimate as a best case estimate. This is part of a fatigue management system we will recommend at the end of this chapter to use actigraphy (wrist worn unobtrusive motion detectors during sleep) or some objective sleep measure to ensure operator rest. These devices cannot tell if the rest period is induced by drugs or alcohol or interrupted by sleep apnea or some other issue that would prevent adequate rest, but for the typical worker, for conservative estimates of the yellow zone, it seems adequate to us.

THE VALIDITY OF MODEL PREDICTIONS IN OPERATIONAL SETTINGS

The process of validating a model is described by Davis as:

> a process of determining (a) the manner in which and degree to which a model (and its data) is an accurate representation of the real world from the perspective of the intended uses of the model and (b) the subjective confidence that should be placed on this assessment. (1992: 6)

In their discussion of model validation, Glenn et al. (2005) similarly treat validity as relative, based on the appropriate use of a model given the accuracy of the model's predictions and the conditions of use. Sargent (1984) offers a somewhat more cut-and-dried definition, arguing that a model is valid only if its accuracy falls within an acceptable range, a range that "reflects the amount of accuracy required for the model's intended purpose" (37). These definitions suggest that validation is about adjusting a model's use to its level of accuracy. To take this definition to an extreme, an inaccurate model can be valid so long as its users understand that it is inaccurate and do not use it in situations that require accuracy. Zachary et al. (2003) note that validity has as much to do with the clarity of a model's boundaries and limitations as it has to do with the model's predictive accuracy. A model that specifies clear conditions of use will generally produce results that are more reliably accurate – and valid – than models that are ambiguous about their optimal conditions of use.

Are current biomathematical models of fatigue valid for use by real-world decision makers in operational settings? What are the boundaries on their valid use in operational

Table 29.2   Definitions and summaries of five types of validity relevant to the operational usefulness of biomathematical models of fatigue

Type of Validity	Definition	General Status of Models
Predictive Validity	The ability of a model to predict outcomes that do not differ from observed outcomes (e.g., Campbell & Bolton, 2005). Also called *operational validity* (Davis, 1992) and, more commonly, *concurrent validity*.	Current models effectively predict the effects of acute sleep loss and the circadian cycle on subjective fatigue and the performance of basic psychomotor tasks; models that attempt to account for a greater number of variables have been largely unsuccessful (Dinges, 2004; Van Dongen, 2004).
Event Validity	The ability of a model to make accurate predictions about the occurrence or frequency of a type of event such as an accident (Sargent, 1984).	Predictions based on sleep-opportunity histories and time of day have correlated with the risk of accidents attributed to human factors in the railroad industry (Dean et al., 2007). In general, however, there is a lack of validation and calibration work using accident data from operational settings (Folkard & Åkerstedt, 2004).
Face Validity	The perception of a model's accuracy and usefulness by domain experts and model users (e.g., Davis, 1992).	Evidence of model adoption and regular use in industry and the military is needed as support for model face validity. This evidence is currently lacking.
Conceptual Model Validity: Model Content	The extent to which a model accurately represents relevant theory and research. Similar to *construct validity* (Campbell & Bolton, 2005).	This form of conceptual model validity is not considered to be directly relevant to predicting fatigue or performance in operational settings.
Conceptual Model Validity: Model Structure	The extent to which a model's structure accurately reflects the nature of the relationships among variables and processes and does not impose an artificial structure (e.g., does not model concurrent processes as linear). Similar to Davis' (1992) *structural validity*.	A mathematical format has been effective and therefore seems valid for modeling the robust trends of time awake and circadian fluctuation. The format may be invalid for settings in which all variables and their interactions cannot be anticipated or in which a precise accurate prediction is sought.
Descriptive Validity	The ability of a model to explain relationships and phenomena (e.g., Davis, 1992).	Biomathematical fatigue models typically have descriptive counterparts that allow users to understand model functions and assumptions and that facilitate scrutiny by the scientific community.

settings? Are these boundaries clearly conveyed to the models' users? To answer these questions, we need to consider different types of model validity. The types of validity that are relevant to the use of fatigue models for operational decision-making are numerous. Five types that seem especially relevant are *predictive validity, event validity, face validity, conceptual model validity,* and *descriptive validity*. Definitions of these types of validity are given in the second column of Table 29.2. In the third column of Table 29.2, we summarize the general status of biomathematical fatigue model validity with respect to their use in operational settings. In the paragraphs that follow, each of the five types of validity is addressed in more depth.

Predictive Validity

A user of a biomathematical model who wants to know how much to trust the model's prediction of operator error rate, response time, or efficiency, for example, is interested in the predictive validity of the model. Predictive validity refers to the ability of a model to make accurate predictions (Campbell & Bolton, 2005; Davis, 1992; Zachary et al., 2003). In the case of fatigue models, those with high levels of predictive validity would accurately predict the occurrence, direction, and extent of change in measures of concern in a real-world work setting – accident risk, production rate, error rate, detection rate in monitoring tasks, health problems, and morale, for example. Validation efforts to date, however, focus almost exclusively on the ability of models to predict experimental performance data (e.g., reaction time and error measures) and fatigue self-assessments.

The predictive validity of six biomathematical models of fatigue was assessed in the 2002 Seattle workshop described previously in this chapter. No model or models stood out as more accurate in their predictions than other models (Van Dongen, 2004). Goodness of fit comparisons were made between each model's predictions and averaged experimental data.

Predictions of experimental data collected in three different sleep-loss scenarios were not much better than predictions made by a horizontal line representing no change across time. The models accounted for only 1–30 per cent more variance than the horizontal line. Model predictions were better in a fourth scenario, accounting for approximately two-thirds more variance than the horizontal line. This latter scenario was a continuous, i.e., acute, sleep-deprivation scenario, whereas the other three scenarios involved restricted sleep periods over multiple days, suggesting the models do not account well for performance when sleep deprivation is interrupted by sleep periods or naps. Dinges (2004) reviewed the six models that participated in the workshop plus four other biomathematical fatigue models and concluded that the models are most robust in their predictions of acute sleep deprivation effects on basic performance variables.Thus, current biomathematical fatigue models are limited in terms of the relevant variables they accurately model. Yet, as is noted above, a model with limited capabilities can be valid, so long as users know how to use it in ways consistent with its limitations. The user license agreement for the Fatigue Audit Inter Dyne (FAID) model (Roach, Fletcher & Dawson, 2004), for example, explains the purpose for which the model was designed and provides guidance on how to use the model (preferably in conjunction with a fatigue-risk management policy) and how much stock to place in the model (use individual's estimates of sleep when they differ from FAID estimates). The documentation does not indicate how well the FAID model predicts fatigue, nor does it describe conditions that are unaccounted for by the model but which may have an impact. In general, educating users about how to use current models appropriately and effectively is an area that deserves greater attention across the modeling board.

Event Validity

Event validity (Sargent, 1984) refers to the ability of the model to predict an event, in this case a negative event produced by fatigue. Predicted events could be aircraft accidents, single-vehicle traffic accidents, or system breakdowns; including those that were detected or prevented by operators. Information about the likelihood of disruptive events is especially useful to decision makers in operational settings. In contrast, information about likely changes in alertness and performance is useful in some operational settings, but generally is secondary to interest in increases in the risk of an accident or system failure that those changes may portend. Nonetheless, today's biomathematical models of fatigue by and large predict subjective fatigue and performance decrement in basic psychomotor tasks (e.g., Mallis et al., 2004). Folkard & Åkerstedt (2004) describe accident trends in industry that seem at odds with fatigue predictions, and accuse model developers of making an untested assumption that model predictions translate into real-world safety concerns.

Event prediction has been given little attention by fatigue model developers. Because significant negative events tend to occur infrequently, the validity of their prediction can be difficult to establish. There are other ways to assess validity, however, including strategies that rely on expert opinion, simulation-derived data, and general consistency with research and theory. Yet another possible reason for the lack of attention to event prediction may be the need to tailor models to specific domains so that the events and vulnerabilities of a given domain are taken into account by the model. It is also possible that a fatigue model is simply inadequate for predicting negative significant events, as these events tend to result from the co-occurrence of a number of unfortunate circumstances versus a change in a single variable such as fatigue.On the other hand, accident risk data in freight railroad operations has been favorably compared with effectiveness scores predicted by the Sleep, Activity, Fatigue, and Task Effectiveness (SAFTE) model (e.g., Dean et al., 2007). SAFTE effectiveness scores were calculated using 30-day histories of opportunities to sleep (i.e., time off from work) and time of day. These scores were calculated for workers involved in 500 accidents documented over a 2.5-year period. For each effectiveness value, accident risk was calculated for two types of accidents: accidents attributed to human factors and all other accidents as a group. Accident risk for each accident type was derived by dividing the proportion of accidents of that type by the total amount of time spent at work under a given set of effectiveness-score conditions across all workers over the 2.5-year period. The risk of accidents attributed to human factors was found to correlate significantly ($r=-0.93$) with effectiveness scores, indicating that accident risk grew linearly as a function of time of day (e.g., circadian troughs versus peaks) and sleep-opportunity history. (Time of day accounted for a significant portion of the variance in accident risk [$r=0.71$]).

Face Validity

According to Davis (1992), *face validity* is the extent to which users of a model perceive the model as useful and reasonable. "Perceived usefulness" is an interesting aspect of face validity and means that it is not enough for a model to be perceived as reasonably accurate; a model should be perceived as reasonably accurate in predictions that users find useful. A fatigue model can accurately predict that users' alertness and performance will be low at the time of their circadian trough – usually approximately 04:00 hours; however, if users find this prediction to be old news (and they soon would), the validity of the model is accordingly lowered.Determining how often modern biomathematical fatigue models are used in operational settings can provide insight into their face validity. Another strategy for assessing face validity is to query users about the usefulness of the models' prediction: what sorts of predictions and guidance do people in operational settings seek, and can current biomathematical models provide them, and at the needed level of accuracy? Searches of the fatigue literature produced no evidence of the adoption of biomathematical fatigue models by decision makers in industry or the military. The lack of evidence suggests there is scant use of the models in real-world work operations. Although there are other explanations of why biomathematical fatigue models are not yet embraced by their target users, the situation calls into question their face validity.

Conceptual Model Validity

Conceptual model validity can be considered a form of construct validity, a term that traditionally refers to the extent to which a measurement method or instrument measures what it is supposed to be measuring. Conceptual model validity has been described in terms of two main elements (e.g., Robinson, 2006; Rykiel, 1996; Sargent, 1984): a structural element (*Is the model's structure compatible with the nature of the relationships among the variables and processes being represented?*) and a content element (*Does the model accurately represent relevant theory and research?*). Researchers and theoreticians hope that models with high construct, or conceptual model, validity will predict better than other models, but there is no guarantee. Model users, on the other hand, are unlikely to care whether the model is true to underlying theory and research so long as it offers reliable predictions.

We bring conceptual model validity into this discussion because it may relate to fatigue models' poor record of accounting for variables other than time awake and circadian fluctuation. The lack of accuracy in these areas, according to Dinges (2004), will be overcome as more research is conducted and more data accumulates. We see another possibility, though. Specifically, the complexity of the relationships among the many relevant variables may be at odds with the structure of the biomathematical model. We question the structural element of the models' conceptual model. When modeling relationships that depend on complex, dynamic interactions among individuals, conditions, situations – variables for which precise values cannot be known in advance – a mathematical model might produce hopelessly inaccurate predictions. Similarly, the precise, stark quantitative output of a mathematical model is at odds with the complex reality of human performance responses to fatigue variables.

Descriptive Validity

Davis (1992) identifies descriptive validity as one of three critical types of model validity. (Structural and predictive validity are the other two). Davis describes it as the ability of a model to explain relationships and phenomena and to organize what is understood about variables, processes, and their relationships in a meaningful and useful way. Current biomathematical models were developed by researchers. Because understanding fatigue tends to be the goal of most of these researchers/model developers, their models tend to be tied to conceptual representations of hypothesized underlying variables and variable relationships. As a result, their descriptive validity tends to be high.

For example, Hursh's SAFTE model (e.g., Hursh et al., 2004) is tied to a conceptual model that specifies circadian oscillators, a sleep reservoir, and a sleep accumulation process moderated by sleep intensity and sleep fragmentation, which are in turn moderated by other variables in the model. As another example, Åkerstedt, Folkard, and Portins' (2004) *Three-Process Model of Alertness* is tied to a descriptive version of the model that synthesizes the research and theory underlying the three-modeled processes – circadian variation, time awake, and sleep inertia after awakening.

Overview of Validity in Operational Settings

Validation work to date suggests severe boundary conditions on the valid use of biomathematical fatigue models to predict even experimental data. The models are able to predict the relatively robust trends in fatigue and performance associated with time awake and circadian rhythmicity (e.g., van Dongen, 2004), but their strengths are currently limited to just these two variables. Some model developers believe more research into the biological mechanisms of fatigue will produce results that improve model accuracy; another possibility is that new modeling strategies are needed, for example, strategies based on complexity science.

Descriptive validity tends to be high for these models in that they tend to explain well, or include explanations of the relationships they predict. Going back to definitions of validity introduced at the outset of this section, biomathematical fatigue models may still be considered valid for use across a wide range of operational settings so long as their users are cognizant of their shortcomings and use the models' predictions accordingly. The development of user guidance to support the tempered use of current models should be given greater attention in order to benefit both model usability and validity. Likewise, developers should continue emphasizing the use of biomathematical fatigue models as decision aids, i.e., as one of many sources of information to be used by a decision maker (e.g., Hursh et al., 2004).

THE WAY AHEAD: IMPROVING MODEL VALIDITY

The above review of biomathematical fatigue model validity suggests a need to assess model validity in operational settings, using operational data, and with the involvement of domain experts. The frequent use of experimental data from laboratory and field studies to develop, evaluate, and calibrate biomathematical fatigue models makes it difficult to judge their validity for decision making in operational contexts. We are forced to make assumptions about validity in operational settings based on research conducted largely in controlled experimental settings. Some research suggests this may be acceptable (e.g., Dean et al., 2007); other studies demonstrate that the complexities of real-world operations, especially the additional variables that come into play, can render model predictions obsolete (e.g., Folkard & Åkerstedt, 2004). In either case, there is a clear need for model validation and calibration work to make use of more real-world data. Model validity may improve if models are tailored to specific domains in which they will be used. The models in the Seattle conference were best at predicting the effects of fatigue induced in a laboratory across a sleep deprivation period (Van Dongen, 2004). This is not surprising, given that most of the models were derived from data collected in similar settings. It follows that a model will make the best predictions about types of work and conditions that are most similar to those used to generate the data used to develop the model. For example, truck drivers in a driving simulator and trucking accident data would be the best to use in modeling the effects of truck drivers. This context-specificity may be critical for tasks in which injury or economic tragedy could result from fatigue; for example, from fatigue experienced by truckers hauling dangerous chemicals at night.Seemingly at odds with the limitations of biomathematical fatigue models, a wealth of relevant data and theory has accumulated for over more than a century. This rich body of work describes and documents symptoms of fatigue, biological underpinnings of fatigue, and ways in which fatigue impairs performance of many types, among other things. In light of the apparent inability of mathematical equations to predict complex relationships embedded in that rich body of work, we suggest that fatigue modelers consider descriptive modeling strategies and strategies rooted in complexity science.

In his 1999 paper, "Tossing algebraic flowers down the Great Divide," the late mathematician and computer scientist Joseph Goguen described his considerable efforts to represent the meaning of information – to capture what information means to the person interpreting it. He concluded that not all relationships can be represented quantitatively or mathematically. This 1999 paper is one of many written late in Goguen's career in which he defends the value and validity of qualitative data and models and decries the overuse of quantitative methods, which he viewed as context-devoid and generalizable to the point of being of limited value to any specific domain or situation.An example of a change in strategy may be seen in Australia's adoption of a fatigue management system for its airlines as a replacement for strict limits on crew rest and duty periods (McCulloch, Fletcher & Dawson, 2003). The fatigue management system, which can include the use of biomathematical fatigue models, may represent a paradigm shift toward a greater acceptance of descriptive models – of using knowledge in addition to numbers – to guide policies and decision-making. It may also represent an acknowledgment of boundaries on the valid use of biomathematical fatigue models in complex real-world settings.

Another recommendation for improving model validity involves the strength of the model inputs. People are notoriously poor at identifying their own fatigue (Dinges, 2004). An actigraph or wrist-worn accelerometer can measure the restlessness of sleep, as well as its onset and offset. It thereby provides an objective measure that can and should be used in this research to define fatigue or, rather, the sleep deprivation extent. It may be that an individual gets plenty of rest but is restless and the sleep is of poor quality. An actigraph could provide this sleep quality data for more accurate input values of recovery sleep duration or involuntary naps.

Actigraphs, however, suffer from stagnated development. The software to download them and analyze the data they collect is not well developed, or at least not in keeping with the importance of the tool. A simple infrared or radio transmitter download to a central computer was an exciting prospect for actigraphs 20 years ago but has not come to fruition. Actigraph functionality could be extended to collect additional forms of data and model inputs. They could record snoring, for example, or capture data that supports tracking the duration and intensity of apnea. Investigators at the Walter Reed Behavioral Biology Unit have proposed an actigraph that displays a gauge on the front of the sleep watch that maintained a dynamic estimate of the wearer's effectiveness (Balkin & McBride, 2005). Over time and with feedback, the actigraph "learns" how to judge the sleep quality of a particular wearer and the gauge enlarges as a measure of confidence in its estimate. This seems like a potentially valuable addition to the technology and worthy of evaluation. In general, actigraphs can and should be used to improve the capabilities and validity of biomathematical models of fatigue impairment. Finally, we reemphasize that current biomathematical models must be used as decision aids, not "fitness for duty tests," as Hursh et al. (2004) point out. The models provide a reasonable estimate of the composite performance ability of a population, but not of the performance of a specific individual. Planners must be aware that the outcomes of these models are ranges and "good guesses" and that they will need to use their expertise and take into account the characteristics of their personnel and work domain.At the same time, models might be used more if they offered more guidance to decision makers. For this reason, we argue for a very conservative "red zone" – a conservative precaution demarcation in the predicted data, indicating where performance may begin to be affected by fatigue. Additional guidance could take the form of flags in predicted data to mark levels of sleep deprivation associated with historic accidents and performance decrements in experimental data, for example. With this sort of descriptive guidance accompanied by descriptions of conditions on which it is based, the validity of existing models and their use in operational settings may be greatly benefited.

REFERENCES

Achermann, P. (2004). The two-process model of sleep regulations revisited. *Aviation, Space, and Environmental Medicine*, 75(3), A37–43.

Åckerstedt, T., Folkard, S. & Portin, C. (2004). Predictions from the three-process model of alertness. *Aviation, Space, and Environmental Medicine*, 75(3), A75–83.

Balkin, T.J. & McBride, S. (2005). Managing sleep and alertness to sustain performance in the operational environment, in *Strategies to Maintain Combat Readiness during Extended Deployments – A Human Systems Approach*. Meeting Proceedings RTO-MP-HFM-124, Paper 29. Neuilly-sur-Seine, France: RTO, 29-1–29-10).

Bellis, M. (2007). Inventors of the modern computer: The history of the IBM PC – International Business Machines. Online. The New York Times Company. Available at: http://inventors.about. com/library/weekly/aa031599.htm (retrieved March 28, 2007).

Belyavin, A.J. & Spencer, M.B. (2004). Modeling performance and alertness: The QinetiQ approach. *Aviation, Space, and Environmental Medicine*, 75(3), A93–103.

Buley, L.E. (1970). Experience with a physiologically-based formula for determining rest periods on long-distance air travel. *Aerospace Medicine*, 41(6), 680–83.

Campbell, G.E. & Bolton, A.E. (2005). HBR validation: Integrating lessons learned from multiple academic disciplines, applied communities and the AMBR project, in K.A. Gluck and R.W. Pew (eds), *Modeling Human Behavior with Integrated Cognitive Architectures: Comparison, Evaluation, and Validation*. Mahwah, NJ: Lawrence Erlbaum.

Davis, P.K. (1992). Generalizing concepts and methods of verification, validation, and accreditation (VV&A) for military simulations (Technical Report No. R-4249-ACQ). Santa Monica, CA: The RAND Corporation.

Dawson, D. & Reid, K. (1997). Fatigue, alcohol and performance impairment, *Nature*, 388, 235.

Dean, D.A. III, Fletcher, A., Hursh, S.R. & Klerman, E.B. (2007). Developing mathematical models of neurobehavioral performance for the "real world." *Journal of Biological Rhythms*, 22, 246–58.

Dinges, D.F. (2004). Critical research issues in development of biomathematical models of fatigue and performance. *Aviation, Space, and Environmental Medicine*, 75(3, Suppl), A181–91.

Dorrien, J. & Dinges, D. (2006). Sleep deprivation and its effects on cognitive performance, in T. Lee-Chiong (ed.), *Sleep: A Comprehensive Handbook*. Chichester: John Wiley, Chapter 19, 139–44.

Fletcher, A. & Dawson, D. (2001). A quantitative model of work-related fatigue: Empirical evaluations. *Ergonomics*, 44, 475–88.

Folkard, S. & Åkerstedt, T. (2004). Trends in the risk of accidents and injuries and their implications for models of fatigue and performance. *Aviation, Space, and Environmental Medicine*, 75(3, Suppl), A161–7.48.

French, J. & Miller, J. (2005). Estimating the effects of stress during operational conditions. Proceedings of the 2005 Winter Simulation Conference (ed.), M.E. Kuhl, N.M. Steiger, F.B. Armstrong & J.A. Joines.

French, J., Morris, C. & Hancock, P. (2003). Predicting fatigue degraded performance in models of human performance. Proceedings of the Human Factors and Ergonomics Society, Denver, CO.

Friedl, K.E., Mallis, M.M., Ahlers, S.T., Popkin, S.M. & Larkin, W. (2004). Research requirements for operational decision making using models of fatigue and performance. *Aviation, Space and Environmental Medicine*, March; 75(3 Suppl): A192–9.

Gerathewohl, S.J. (1974). Simple calculator for determining the physiological rest period after jet flights involving time zone shifts. *Aerospace Medicine*, 45(4), 449–50.

Glenn, F., Stokes, J., Neville, K. & Bracken, K. (2005). An investigation of human performance model validation. CHI Systems Technical Report 04009.050329. Fort Washington, PA: CHI Systems, Inc.

Gluck, K.A. & Pew, R.W. (eds) (2005). *Modeling Human Behavior with Integrated Cognitive Architectures: Comparison, Evaluation, and Validation*. Mahwah, NJ: Lawrence Erlbaum, V, 365–96.

Goguen, J. (1999). Tossing algebraic flowers down the Great Divide, in C.S. Calude (ed.), *People and Ideas in Theoretical Computer Science*. New York: Springer, 93–129.

Hancock, P.A. & Warm, J.S. (1989). A dynamic model of stress and sustained attention. *Human Factors*, 31(5), 519–37.

Hursh, S.R., Redmond, D.P., Johnson, M.L., Thorne, D.R., Belenky, G., Balkin, T.J., Storm, W.F., Miller, J.C., & Eddy, D.R. (2004). Fatigue models for applied research in warfighting. *Aviation, Space, and Environmental Medicine*, 75(3), A54–60.

Jewett, M.E. & Kronauer, R.E. (1999). Interactive mathematical models of subjective alertness and cognitive throughput in humans. *Journal of Biological Rhythms*, 14(6), 588–97.

Mallis, M.M., Mejdal, S., Nguyen, T.T. & Dinges, D. (2004). Summary of the key features of seven biomathematical models of human fatigue and performance. *Aviation, Space, and Environmental Medicine*, 75(3), A4–14.

McCulloch, K., Fletcher, A. & Dawson, D. (2003). Moving toward a non-prescriptive approach to fatigue management in Australian aviation: A field validation. Canberra, Australia: Civil Aviation Safety Authority.

Mohler, S.R. (1976). Physiological index as an aid in developing airline pilot scheduling patterns. *Aviation, Space, and Environmental Medicine*, 47(3), 238–47.

Neri, D.F. (2004). Preface: Fatigue and performance modeling workshop, June 13–14, 2002. *Aviation, Space, and Environmental Medicine*, 75(3), A1–3.

Neville, K., Takamoto, N., French, J., Hursh, S.R. & Schiflett, S.G. (2000). The sleepiness-induced lapsing and cognitive slowing (SILCS) model: Predicting fatigue effects on warfighter performance. Proceedings of the 44th Annual Meeting of the Human Factors and Ergonomics Society, Santa Monica, CA, 3-57–3-60.

Nicholson, A.N. (1972). Duty hours and sleep patterns in aircrew operating in world-wide routes. *Aerospace Medicine*, 43(2), 138–41.

Openshaw, S. & Steadman, P. (1983). Predicting the consequences of a nuclear attack on Britain: Model results and implications for public policy. Environment and Planning C: *Government and Policy*, Vol. 1, 205–28.

Pew, R.W. (2008). More than 50 years of history and accomplishments in human performance model development. *Human Factors*, 50(3), 489.

Radakovic, S.S., Maric, J., Surbatovic, M., Radjen, S. et al. (2007). Effects of acclimation on cognitive performance in soldiers during exertional heat stress, *Military Medicine* (Feb.), 172, 2.

Rechtschaffen, A., Gilliland, M.A., Bergmann, B.M. & Winter, J.B. (1983). Physiological correlates of prolonged sleep deprivation in rats. *Science,* 221(4606), 182–4.

Redmond, D.P., Sing, H.C., & Hegge, F.W. (1982). Biological time series analysis using complex demodulation, in F.M. Brown & R.C. Graeber (eds), *Rhythmic Aspects of Behavior*. Mahwah, NJ: Lawrence Erlbaum, 429–57.

Roach, G.D., Fletcher, A. & Dawson, D. (2004). A model to predict work-related fatigue based on hours of work. *Aviation, Space, and Environmental Medicine*, 75(3), A61–9.

Robinson, S. (2006). Issues in conceptual modelling for simulation: Setting a research agenda, in S. Robinson, S. Taylor, S. Brailsford, & J. Garnett (eds). Proceedings of the 2006 Operational Research Society Simulation Workshop. Birmingham, UK: Operational Research Society, 165–74.

Rowe, A.L., French, J., Neville, K.J. & Eddy, D.R. (1992). The prediction of cognitive performance degradation during sustained operations. Proceedings of the Human Factors Society, 111–15.

Rykiel, E.J. (1996). Testing ecological models: The meaning of validation. *Ecological Modeling*, 90, 229–44.

Sargent, R.G. (1984). Verification and validation of simulation models. Proceedings of the 1984 Winter Simulation Conference, Washington, DC, 121–30.

Van Dongen, H.P.A. (2004). Comparison of mathematical model predictions to experimental data of fatigue and performance. *Aviation, Space, and Environmental Medicine*, 75(3), A15–36.

Van Dongen, H.P.A., Baynard, M.D., Maislin, G. & Dinges, D.F. (2004). Systematic interindividual differences in neurobehavioral impairment from sleep loss: Evidence of trait-like differential vulnerability. *Sleep*, 27, 423–33.

Van Dongen H.P.A., Mott, C.G., Huang, J.K., Mollicone, D.J., McKenzie, F.D. & Dinges, D.F. (2007). Optimization of biomathematical model predictions for cognitive performance impairment in individuals: Accounting for unknown traits and uncertain states in homeostatic and circadian processes. *Sleep,* 30(9), 1129–43.

Walters, B., Archer, R. & Yow, A. (2000). Modeling the combined effects of performance shaping factors on performance: A comparison of two different techniques. *Human Factors and Ergonomics Society Annual Meeting Proceedings*, 44(17), 153–56(4).

Whitmore, J., French, J., Olenick, L. & Hall, J. (1992). Calculating crew rest intervals. Proceedings of the Aerospace Medicine Association Meeting, May 1992.

Yerkes, R.M. & Dodson, J.D. (1908). The relation of strength of stimulus to rapidity of habit-formation. *Journal of Comparative Neurology and Psychology*, 18, 459–82.

Zachary, W., Campbell, G., Laughery, R., Glenn, F., & Cannon-Bowers, J. (2003). The application of human modeling technology to the design, evaluation, and operation of complex systems, in E. Salas (ed.), *Advances in Human Performance and Cognitive Engineering Research*. Amsterdam: Elsevier Science, 201–50.

30
Model-based Fatigue Risk Management

Hans P.A. Van Dongen and Gregory Belenky

INTRODUCTION: ON THE NEUROBIOLOGY OF SLEEP AND FATIGUE

Fatigue in operational settings has many facets, as is discussed in various other contributions in this volume. One of the most pervasive manifestations of fatigue in today's 24/7 economy is in the form of human impairment resulting from extended work hours and night and shift work. Such work schedules lead to sleep insufficiency and temporal misalignment relative to the biological clock (i.e., circadian misalignment). These neurophysiological challenges promote fatigue and cognitive deficits (Åkerstedt, 1995) and lead to accidents (Folkard & Åkerstedt, 2004) and increased operational costs (Rosekind, 2005; Ricci et al., 2007). There is a well-established neurobiology underlying such operationally important aspects of work scheduling, and understanding the dynamics of how this neurobiology drives fatigue[1] is important for effectively managing fatigue risk. Figure 30.1 illustrates the two main neurobiological processes driving fatigue in the context of extended or shifted work schedules. They are a homeostatic process and a circadian process, both of which are fundamental to sleep/wake regulation (Borbély, 1982; Banks, Jackson & Van Dongen, Chapter 11, this volume).

The homeostatic process seeks to balance the amounts of wakefulness and sleep, and extended wakefulness therefore results in a progressive build-up of pressure for sleep. If this pressure for sleep is not met – i.e., if a person is sleep-deprived – fatigue results (Van Dongen & Dinges, 2005). Thus, fatigue is regulated by the homeostatic process as a function of sleep/wake history. The homeostatic process operates on short timescales (hours to days), as well as on longer timescales (days to weeks). As such, both acute

---

1 Whereas in operational settings the term "fatigue" is typically used, in basic research the term "sleepiness" is more often employed to describe the phenomenon discussed here. There is an ongoing scientific debate about this terminology and related phenomena (e.g., De Valck & Cluydts, 2003). However, there is no convincing evidence that fine distinctions between different subtypes of sleep-related fatigue/sleepiness are relevant vis-à-vis their temporal dynamics. For practical purposes, therefore, sleepiness and (sleep-related) fatigue are treated here as a unitary concept.

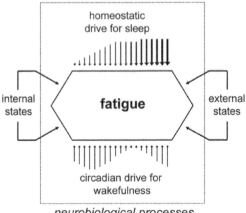

Figure 30.1   Simplified schematic of the two primary sleep-related
                  neurobiological processes underlying fatigue

Note: First, across time awake, there is accumulation of a homeostatic drive for sleep, which has a progressively fatiguing effect. The drive for sleep is dissipated during sleep periods. Second, over the 24 hours of the day, there is a waxing and waning of a circadian drive for wakefulness, which therefore opposes the homeostatic drive for sleep during the biological day but not during the biological night. The net level of fatigue resulting from these two processes is modulated by a variety of other factors, which are subsumed here under the broad categories of internal states (e.g., motivation, anxiety, illness) and external states (e.g., noise, light, time pressure). The effects of these internal and external states on fatigue is generally transient, whereas the neurobiological processes driving fatigue are enduring. As such, temporal variations in fatigue due to these neurobiological processes are predictable, and could therefore be manageable in the operational environment.

Source: Figure adapted from Van Dongen & Dinges, 2005, with permission.

total sleep deprivation and chronic sleep curtailment can lead to considerable levels of fatigue as evidenced by accumulating cognitive deficits (Van Dongen, Maislin, et al., 2003), although self-assessment of fatigue under conditions of chronic sleep curtailment may not be reliable – see Figure 30.2. Recuperation from prior sleep loss may take several days (Belenky et al., 2003; Banks, et al., 2010), even if recovery sleep duration is unrestricted.

The circadian process is a (near-)24-hour rhythm, produced by the endogenous biological clock (located in the suprachiasmatic nuclei of the hypothalamus) and synchronized to the day/night cycle primarily by light exposure (Wever, 1970; Waterhouse, Chapter 16, this volume). The circadian process promotes wakefulness during the day and withdraws the drive for wakefulness during the night, which results in fatigue at night even if enough sleep is obtained beforehand. Fatigue is, therefore, regulated by the circadian process as a function of time of day. Moreover, the circadian process makes it difficult to get sleep during the day, especially during the "wake maintenance zone" in the early evening (Dijk & Czeisler, 1994). Thus, the circadian process constrains the homeostatic process, causing sleep loss in individuals attempting to sleep during the day (Åkerstedt, 1998). The homeostatic process and the circadian process interact to determine the level of fatigue and the associated cognitive performance deficits (Van Dongen & Dinges, 2005). Indeed, fatigue is a function of time awake as well as time of day – see Figure 30.3.

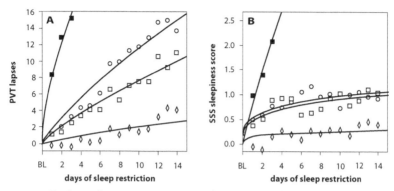

Figure 30.2   Fatigue in response to varying doses of sleep across days

Note: In a strictly controlled laboratory study, healthy young adults were restricted to 8 hours (diamonds; n=9), 6 hours (boxes; n=13) or 4 hours (circles; n=13) time in bed for 14 days (i.e., three different doses of chronic sleep restriction), or 0 hours (black boxes; n=13) for 3 days (i.e., 88 hours of total sleep deprivation). The graphs show group averages, relative to baseline (BL), for A) lapses of attention (defined as response times slower than 500 ms) on a psychomotor vigilance test (PVT; Dinges & Powell, 1985), and B) self-ratings of sleepiness on the Stanford Sleepiness Scale (SSS; Hoddes et al., 1973) measured every 2 hours and averaged within days. Upward corresponds to worse cognitive performance on the PVT and greater self-rated sleepiness (a correlate of subjective sleep-related fatigue) on the SSS. PVT performance degraded progressively across days as a function of sleep dose – the smaller the daily amount of sleep, the more rapid the performance degradation (the changes over days in the 8-hour conditions were not statistically significant). After 14 days with sleep restriction to 6 hours or 4 hours per day, performance deficits reached magnitudes equivalent to 2–3 days with no sleep at all. However, in the chronic sleep restriction conditions, self-rated sleepiness did not accurately track the objective performance deficits. On the contrary, in the 6-hour and 4-hour conditions, self-rated sleepiness leveled out after just a few days, and did not differentiate significantly between sleep doses. In other words, subjects exposed to chronic sleep restriction did not reliably assess their own level of fatigue (cf. Kaplan, Itoi & Dement, 2007).

Source: Figure adapted from Van Dongen, Maislin, Mullington & Dinges, 2003, with permission.

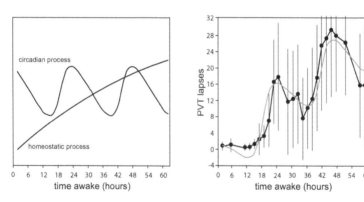

Figure 30.3   Dynamic influences on cognitive performance by the two primary sleep-related neurobiological processes underlying fatigue: The homeostatic process and the circadian process

Note: The left panel illustrates the gradual increase of the homeostatic drive for sleep during a 62-hour period of total sleep deprivation, and the rhythmic pattern of the circadian drive for wakefulness with a period of (approximately) 24 hours. The right panel shows the sum of these two processes as a model of cognitive performance impairment (gray curve), which matches well with actual measurements of performance (PVT lapses, see Figure 30.2, caption) measured during 62 hours of total sleep deprivation in a laboratory (black curve). The black dots in the right panel represent averages over 12 healthy young adults. The whiskers on each side represent one standard deviation over subjects, reflecting trait-like individual differences in cognitive performance responses to fatigue (Van Dongen, Baynard et al., 2004). These substantial individual differences are not addressed here (see Van Dongen, 2006; Van Dongen & Belenky, 2009).

Source: Figure adapted from Van Dongen & Belenky, 2009, with permission.

THE NATURE OF COGNITIVE PERFORMANCE IMPAIRMENT DUE TO FATIGUE

Of crucial importance in operational settings is the fact that fatigue from sleep loss and circadian misalignment is consistently associated with performance deficits across a wide range of cognitive domains (Van Dongen & Dinges, 2005; Jackson & Van Dongen, 2011). Although there are substantial individual differences in the extent to which workers are susceptible to cognitive performance impairment due to fatigue (Van Dongen, 2006), the temporal pattern of cognitive changes is highly replicable and predictable over hours and days (see below). However, self-assessment of fatigue tends to be unreliable both between individuals (Leproult et al., 2003; Van Dongen, Baynard et al., 2004) and within individuals over time (Van Dongen, Maislin et al., 2003; Axelsson et al., 2008). In part, this may be due to the stochastic and unpredictable nature of the performance deficits on a timescale of seconds and minutes (Van Dongen, Belenky & Krueger, 2011), which constitutes a fundamental property of fatigue-induced cognitive impairment (Williams, Lubin & Goodnow, 1959; Doran, Van Dongen & Dinges, 2001).

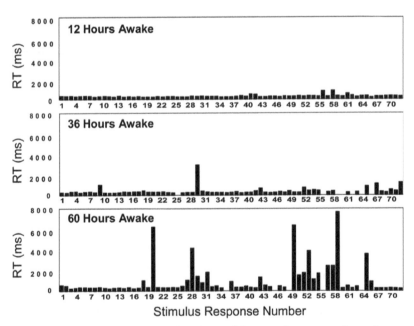

Figure 30.4   The stochastic nature of cognitive performance impairment due to sleep loss

Note: The panels show the raw response times (RT; on the ordinate) to individual stimuli (on the abscissa) on a 10-minute PVT for a single person undergoing total sleep deprivation under controlled laboratory circumstances. After 12 hours of wakefulness (with no fatigue due to sleep loss) performance was highly stable, with optimal RT, throughout the task duration. After 36 hours of continuous wakefulness, and particularly after 60 hours, fatigue from sleep loss degraded performance, which became a mixture of optimal RT, false starts (shown as gaps), and lapses of attention (increased RT) – especially as time-on-task progressed. This illustrates the stochastic nature of cognitive performance impairment due to fatigue from extended wakefulness as well as from extended time-on-task.

Source: Figure taken from Van Dongen & Hursh, 2010, as redrawn from Doran, Van Dongen & Dinges, 2001, with permission.

Figure 30.4 illustrates this phenomenon, showing that fatigue from sleep loss results in increased moment-to-moment variability of performance on a 10-minute sustained attention task (the psychomotor vigilance test or PVT; Dinges & Powell, 1985). Performance under baseline (non-fatigued) conditions tends to be stable, with response times on this task generally being optimal (ranging between approximately 200 and 250 ms) – see Figure 30.4 (top panel). However, after total sleep deprivation for one night (middle panel) and two nights (bottom panel), a mixture emerges of optimal response times, false starts (shown as gaps), and cognitive lapses (considerably slowed response times). Thus, cognitive impairment due to sleep loss does not constitute a gradual performance decline or a complete failure to perform, but rather takes the form of performance instability (Doran, Van Dongen & Dinges, 2001).

The graphs in Figure 30.4 also illustrate that performance when fatigued due to sleep deprivation is best in the beginning of each test bout, and deteriorates as time-on-task progresses. It is a result of fatigue from another source, namely the well-known time-on-

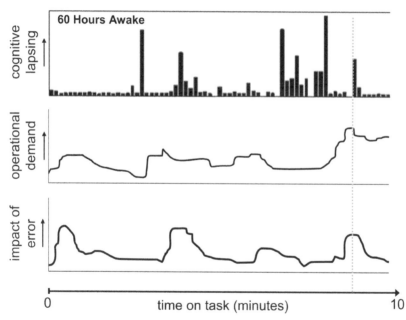

Figure 30.5   Heuristic for explaining how fatigue may contribute to accidents

Note: The top panel (copied from Figure 30.4) depicts stochastically occurring lapses of attention (longer bars indicting slowed response times) across a 10-minute span of task performance. These cognitive lapses were observed during a PVT taken at 60 hours of total sleep deprivation (see Figure 30.4), but for the sake of argument are assumed here to potentially occur regardless of the specifics of the task at hand. The middle panel shows a hypothetical pattern of changing operational demand across the duration of the task (e.g., owing to competing task requirements, time pressure, or environmental distractions). The bottom panel displays the hypothetical level of impact that human error would have over the course of the task (e.g., in terms of equipment damage or personal danger). In this view of how fatigue contributes to accident causation, cognitive lapses occurring when operational demand is high lead to human error, which in turn, if the consequences of error are significant, may result in an accident. Thus, for fatigue to lead to an accident, cognitive lapsing must line up in time with high operational demand and major impact of failure. Here, this is illustrated to occur at the dotted gray line. However, if the cognitive lapse coinciding with the dotted gray line took place just seconds earlier or later, or if the temporal profile of operational demand were different, or if failure were inconsequential, then no accident would occur. FRM approaches are conceptually concerned with improving safety by reducing the likelihood that the elements of fatigue risk coincide.

Source: Figure adapted from Van Dongen & Hursh, 2010, with permission.

task effect (Bills, 1931). This effect is especially evident in vigilance performance tasks (Davies & Parasuraman, 1982), and can be overcome by rest breaks (McCormack, 1958; Bergum & Lehr, 1962). Cognitive impairment due to fatigue from the time-on-task effect, like that from sleep loss or circadian misalignment, is characterized by performance instability (Bills, 1931; Van Dongen, Belenky & Krueger, 2011). Moreover, the effects interact, such that the rate by which performance instability becomes worse over time-on-task is increased under conditions of sleep loss and/or circadian misalignment (Wesensten et al., 2004).

Performance instability in particular makes fatigue problematic in operational settings, as it leads to human error at unpredictable moments in time. Under time pressure or otherwise high operational demand, and when the consequences of human error are severe, unexpected incidents and accidents may ensue (Van Dongen & Hursh, 2010). Even after a fatigue-related incident or accident has occurred, it tends to be difficult to conclusively demonstrate that fatigue was involved as a root cause. Indeed, in part because of the stochastic nature of cognitive performance impairment due to fatigue, the relationship between fatigue and accidents in the workplace tends to be intractable. To get a better handle on this critical issue for safety and productivity in 24/7 operational settings, there is a need for new discoveries and new insights regarding the neurobiology of sleep and fatigue (e.g., Krueger et al., 2008). In the interim, Figure 30.5 provides a practical heuristic, inspired by Reason's Swiss cheese model of accident causation (Reason, 1990), to understand the relationship between fatigue and operational errors, incidents and accidents.

MODEL-BASED MANAGEMENT OF FATIGUE AND FATIGUE RISK

Although the cognitive performance deficits resulting from fatigue are unpredictable from moment to moment (Figure 30.4), the overall magnitude of fatigue and, as such, the level of performance instability is driven by the underlying homeostatic and circadian processes (Figures 30.1–30.3) and therefore deterministic across hours and days. This means that fatigue and its impact on performance is not only predictable, but also potentially manageable, by considering these two neurobiological processes. The circadian process can be manipulated by carefully controlling light exposure (Minors, Waterhouse & Wirz-Justice, 1991) and other so-called "zeitgebers" (external time cues), but this is not easily achievable in practice. However, adjusting the homeostatic process is relatively straightforward, as it depends on the timing and duration of wakefulness and sleep (i.e., sleep/wake history). For this reason, managing fatigue in operational settings focuses primarily on controlling work times with the purpose of managing opportunities to sleep (see Tucker & Folkard, Chapter 28, this volume).

The normative way to control work times in order to manage fatigue involves promulgation of hours of service (HOS) regulations, which typically prescribe maximum durations of duty time and minimum durations of rest. However, because HOS regulations disregard sleep/wake history and circadian rhythmicity, they may at times be overly restrictive and inflexible, and at other times overly permissive and unsafe. In a seminal review paper, Dawson & McCulloch (2005) introduced an alternative, five-layer approach to fatigue risk management (FRM). The first layer is concerned with making sure that work schedules provide adequate sleep opportunities, either nominally or as ascertained with observations in the field. The second layer involves verifying that an operator has in fact obtained adequate sleep. The third layer is conceptualized as involving observation of fatigue symptoms (and invoking shift changes or other fatigue countermeasures as needed). The fourth layer deals with the recording and analysis of fatigue-related errors. The fifth and final layer focuses on the analysis (and subsequent prevention) of fatigue-related incidents and accidents. Dawson & McCulloch (2005) postulated that together, these FRM steps would provide a multi-layer defense in depth against fatigue risk.

Based on a literature review in the context of their first two layers of defense, Dawson and McCulloch (2005) derived that work schedules resulting in less than 5 hours of sleep in the prior 24 hours, or less than 12 hours of sleep in the prior 48 hours, or longer wakefulness than the total amount of sleep obtained in the prior 48 hours, would be inconsistent with safe operations. However, this is merely a rule of thumb, the validity of which depends

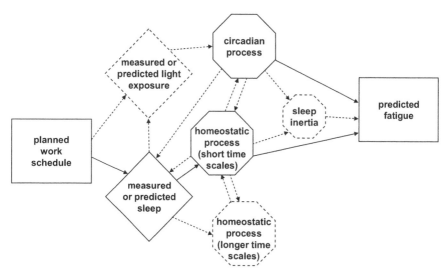

Figure 30.6   Schematic of the general structural components of mathematical models of fatigue as relevant for the evaluation of planned work schedules

Note: In the most basic form (solid boxes and arrows), sleep/wake times are determined from the work schedule under consideration, based on past (historic records) or online (e.g., actigraphic) observations, or predicted by a dedicated algorithm or procedure (e.g., Åkerstedt & Folkard, 1996; Darwent, Dawson & Roach, 2010). Based on these sleep/wake times, the dynamics of the homeostatic process are predicted. This is combined with the circadian process, which in the most basic form is a simple oscillator generating a steady 24-hour rhythm. Together, the two processes predict the temporal profile of fatigue associated with the planned work schedule. Note that transient internal and external states affecting fatigue are not accounted for (see Figure 30.1). There are additional structural components (dotted boxes and arrows) in some of the available mathematical models of fatigue (see Mallis et al., 2004). They may contain a more dynamic circadian process responsive to light exposure (e.g., Jewett & Kronauer, 1999), whether measured directly or predicted based on geographical location and/or work environment and schedule. The circadian process (like the homeostatic process) may be implemented to modulate the timing and duration of sleep, which in turn would control light exposure (which is largely blocked when the eyes are closed). The circadian process may also be made to interact nonlinearly with the homeostatic process (Van Dongen & Dinges, 2003; St. Hilaire et al., 2007). Some models predict sleep inertia, which is the transient state of fatigue that may be experienced immediately after awakening (Dinges, 1990). Sleep inertia appears to be exacerbated when fatigue from the homeostatic and circadian processes is high (Dinges, Orne & Orne, 1985), but it is dissipated rapidly (in the order of minutes), and may therefore have little operational importance depending on the setting. Finally, contemporary mathematical models of fatigue predict both the short-term and long-term dynamics of the homeostatic process (see Johnson et al., 2004; McCauley et al., 2009), so as to account for the fatiguing effects of chronic sleep restriction (Belenky, Wesensten et al., 2003; Van Dongen, Maislin et al., 2003).

Source: Figure based on Van Dongen, Avinash et al. (2006).

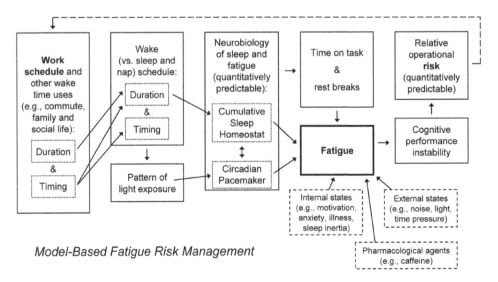

Figure 30.7   Schematic of the neurobiology of sleep and fatigue and its role
in model-based fatigue risk management

Note: See text for details.

on (prior) circumstances, most notably on circadian timing. A more scientifically valid and operationally optimal approach makes use of a mathematical model of fatigue to forecast the time-dependent fatiguing effects of work schedules under consideration. This is the basis of model-based FRM, which is gaining widespread acceptance in several transportation industries and beyond.

A number of mathematical models has been developed to predict fatigue and performance under conditions of sleep loss and/or circadian misalignment (e.g., Hursh, Balkin et al., 2004; Åkerstedt et al., 2008; McCauley et al., 2009). Figure 30.6 shows a schematic of how these models work in general. An in-depth discussion of their scientific foundations, implementation details, capabilities and limitations, strengths and weaknesses, and validation is beyond the scope of this chapter and can be found elsewhere (Friedl et al., 2004; Mallis et al., 2004; Van Dongen, 2004; Hursh & Van Dongen, 2010). Of importance here is that the most reputable models predict temporal profiles of fatigue that reflect the underlying neurobiology and, in that sense, should be generalizable across most work schedules and operational settings.

Figure 30.7 illustrates how a mathematical model of fatigue could be applied as the foundation for model-based FRM. The process begins with a current or planned work schedule, and an account of other requirements for wakefulness (e.g., for eating, commuting, training/education, and family/social life). The duration and timing of waking activities determine the wake schedule and thereby the opportunities for sleep. Opportunities for napping may be considered, if operationally feasible, for a split daily sleep schedule can be as effective as a consolidated daily sleep schedule (Mollicone et al., 2008). From the temporal sleep/wake profile, the short and long (cumulative) timescale components of the homeostatic process are tracked. From the pattern of light exposure measured or predicted to be associated with the sleep/wake profile, the circadian process is also tracked. A mathematical model of fatigue is used to trace these two neurobiological processes and make a quantitative prediction of the temporal profile of fatigue for the work schedule in question.

In a sophisticated implementation of model-based FRM, time-on-task and rest-break effects could be taken into account to predict fatigue with respect to the varying task requirements across the work period (if known in advance). Presumably transient internal states such as motivation, sleep inertia (see Figure 30.6, caption) or illness, and presumably transient external states such as ambient noise and light or time pressure, are typically not accounted for. The same is true for pharmacological agents such as caffeine and other stimulants, and hypnotics and other fatigue-inducing medications (Pagel, 2005). The reason is that it is virtually impossible to foresee or track these factors in the operational world. The predicted profile of fatigue should therefore be seen as nominal, applying to a healthy operator under normal circumstances. Although this implies that the fatigue predictions should be viewed with reference to the specific conditions of the operator and the work setting, this is not ultimately a significant methodological limitation (see below).

From the predicted fatigue profile, it is possible to predict the degree of cognitive performance instability – the main operational risk factor related to fatigue (see Figure 30.7). In practice, some correlate thereof is usually predicted. The specific outcome metric used, whether expressed on an abstract fatigue scale or calibrated on a subjective fatigue scale or objective performance output, is of limited importance within the framework of model-based FRM. The main requirement is that it should be reasonable to assume that greater or lesser prediction values, all else being equal, would consistently correspond with nominally better or worse performance on the job at hand. While this is a weak requirement, it has an important implication, namely that model-based FRM is inherently relative. That is, model-based FRM does not normally allow for absolute statements about the level of fatigue, risk or safety associated with a given work schedule – rather, work schedules in question should be evaluated by comparing them to appropriate benchmark schedules or to each other.

Whatever the outcome metric used, the mathematical model predictions are typically reduced to a summary statistic, such as the average or the maximum level of fatigue-related performance impairment during work times. This summary statistic then maps onto a level of operational risk for the work schedule in question. It should be recognized that this predicted level of risk is again a relative statistic, as many other factors contribute to the overall risk associated with the work schedule (see Figure 30.8 for an illustration). Nevertheless, the fatigue prediction is quantitative, in that it can be compared to fatigue predictions for other work schedules in order to select the schedule(s) with the smallest fatigue risk.[2] This information may be used to adjust the planned work schedule, and another pass through the whole process could be made to see if the adjusted schedule is associated with a reduction in predicted fatigue (see Figure 30.7). This is a powerful feature of model-based FRM not offered by other (non-quantitative) FRM schemes.

---

2   If the same operator(s) will be working the schedule, then this schedule selection process is fairly robust to quantitative inter-individual differences in vulnerability to cognitive impairment from fatigue. This is not the case if the assigned operator(s) may be different individuals depending on which of the work schedules considered ends up being selected. However, since the inter-individual differences are typically not known, the schedule selection process is still robust to inter-individual differences in terms of *a priori* fatigue risk.

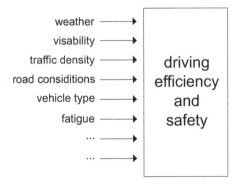

Figure 30.8 **Multiple factors modulate operational performance and risk simultaneously**

Note: This is illustrated here for driving efficiency and safety – each factor's effect, including that of fatigue, is relative to the others (whether or not there are interactions among the factors).

BRIGHT LINES FOR SAFETY

The decades-old question of where to draw the line of minimally acceptable fatigue risk remains unresolved. Model-based fatigue predictions do not map onto an absolute level of fatigue risk measurable on a scale of "acceptability" because no such scale exists. Yet, for regulatory, logistical, liability-driven and other reasons, it is often desirable to establish a "bright line" (i.e., a threshold level) to distinguish safe from unsafe (overly fatiguing) work schedules. Not surprisingly, research to establish such a threshold is frequently called for. A comparison has been made between fatigue and alcohol consumption (Dawson & Reid, 1997), as the effects on cognitive performance for selected doses have been shown to be similar in magnitude (although they are qualitatively different). This equivalence would seem to imply that fatigue in the workplace is unacceptably high if the ensuing performance impairment is greater than that of alcohol consumption to the limit of (legal) intoxication. However, this inference has not been validated, and the effectiveness of a fatigue safety threshold based on equivalence with alcohol consumption may vary considerably across occupations and job tasks. At present, the scientific basis for this approach is not strong enough to warrant its use.

An innovative strategy for deriving a fatigue safety threshold, with a high degree of face validity, was introduced in a study of railroad operations (Hursh et al., 2006; Raslear, Hursh & Van Dongen, 2011). Here, a "bright line" was derived from a data set of 400 human-factors accidents and 1,000 non-human-factors railroad accidents. A mathematical model of fatigue called SAFTE (Hursh et al., 2004) was used to get predicted estimates of operator performance at the times of the accidents, based on the work schedules and estimated sleep opportunities of the locomotive crew in the prior 30 days. SAFTE performance predictions are expressed on an effectiveness scale ranging from 0 (worst) to 100 (best). One of the striking findings of the study was a high correlation between reduced predicted crew effectiveness and the risk of a human-factors accident. This finding suggests that fatigue, on which the effectiveness predictions were based, played a major role in accident causation. At predicted effectiveness scores below 70, the risk of human-factors accidents was elevated above chance level, and greater than the risk of non-human-factors accidents (which were not significantly correlated with predicted effectiveness). On the basis of the available data, therefore, it could be argued that a fatigue safety threshold of 70 on the effectiveness scale of the SAFTE model constitutes a valid "bright line" for railroad operations.

The railroad study also revealed a significant relationship between reduced predicted crew effectiveness and increased accident damage costs (see Hursh & Van Dongen, 2010). As such, the study uncovered how fatigue in rail translates quantitatively to operational costs. Notably, human-factors accident costs were 2.5 times greater when predicted effectiveness was below approximately 77.5 as compared to above 90. Interestingly, 77.5 has been informally adopted as a "bright line" for evaluating the safety of work schedules in some non-rail operations including commercial aviation. However, it is unclear to what extent this threshold generalizes to other operations besides rail. Further research in other settings is needed to avoid the use of threshold levels that are either overly restrictive and therefore not cost-effective, or overly permissive and therefore potentially unsafe.

We recently showed that the general safety record of past operations (as opposed to specific accidents) can also be used to establish a plausible and inherently valid fatigue safety threshold for a particular work environment (see Belenky, Bowen & Van Dongen, 2009). This was illustrated with fatigue predictions from the SAFTE model applied to the landing phases of flight schedules for 4-pilot long-range (LR) and ultra-long-range (ULR) operations of a hypothetical commercial airline – see Figure 30.9. ULR flights (defined as longer than 16 hours of flight time in more than 10 percent of cases) have only recently

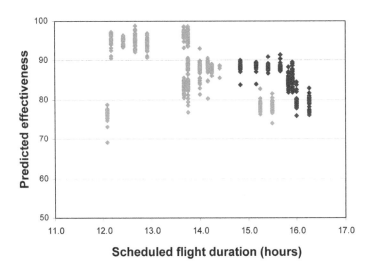

Scheduled flight duration (hours)

Figure 30.9   Fatigue predictions from the SAFTE model for a variety of hypothetical 4-pilot long-range flights (gray) and ultra-long-range flights (black), plotted as a function of scheduled flight duration

Note: Predictions are shown as a summary statistic, calculated as predicted effectiveness (expressed on a 0–100 scale) averaged across the landing phase (defined here as the hour prior to arrival). Note that the relationship between these predictions and flight duration is modulated not only by the flight duration but also by time of day (not indicated in the figure).

Source: Figure taken from Belenky, Bowen & Van Dongen (2009).

been commenced, and are as yet too few to have an extensive safety record. However, LR flights have a long safety track record, and the fatigue predictions for these routes (gray points in Figure 30.9) could therefore serve as a reference data set in order to establish a fatigue safety threshold for the ULR flights.

The key concept is that since the flight schedules in the reference data set have been historically safe, the corresponding fatigue predictions should be within the safety range. To be conservative in this judgment and account for potential outliers, the lowest ~5 percent fatigue predictions for the reference data set could be removed, and a "bright line" could then be drawn such that only schedules for which the fatigue scores are within the upper 95 percent of the reference data set are deemed acceptable. The threshold thus derived from the LR flights in Figure 30.9 is at 77 on the effectiveness scale. In our hypothetical scenario, most ULR flights (black points in Figure 30.9) are predicted by the SAFTE model to meet this safety criterion, but a few routes call for schedule adjustments (or other fatigue mitigation strategies – see, e.g. Caldwell et al., 2009).

When using historically safe operations as a reference, consideration should be given to defining the appropriate fatigue statistic. In flight operations, for instance, predicted fatigue during the take-off phase (not represented in Figure 30.9) and during the landing phase may be most relevant. It should also be carefully contemplated which work schedules are sufficiently representative to be included in the reference data set. Distinct kinds of operations may need to have distinct reference data sets, which could lead to different fatigue safety thresholds. For example, short-haul flight operations with multiple take-offs and landings might involve differential thresholds depending on the number of consecutive segments flown.[3] Different types of flight operations may also involve varying degrees of flight deck automation, more or less sophisticated avionics, and pilots with different levels of experience. These factors may interact with the safety consequences of fatigue, but this should be reflected in the reference data set for the operation at hand, and thus in the derived fatigue safety threshold. Re-evaluation of an established threshold in light of new operational developments (e.g., improved automation) may be appropriate from time to time. In a safety management system (SMS) with electronic record-keeping and a high degree of automation, this could be easily accomplished.

FATIGUE-FRIENDLY SCHEDULE OPTIMIZATION

Assessments of fatigue risk based on "bright lines" may be useful for the evaluation of work schedules on a case-by-case basis, and could also be appropriate for rule-making and legislation. However, fatigue and fatigue risk do not really vary categorically (safe vs. unsafe), but rather on a continuous scale. The interpretation of fatigue scores on a continuous scale is complicated by the lack of an anchor point on the high fatigue end of the scale – that is, it is not clear how to define maximum fatigue, making the scaling of fatigue scores intrinsically arbitrary. This problem can be overcome by systematically relating fatigue scores to a concrete, observable outcome variable. For instance, in large

---

3   There is a lack of scientific evidence regarding the effect of increased workload from multiple-segment flight operations on fatigue, and extant mathematical models do not account for it. Experimental research and model development to address this issue are ongoing (McCauley & Van Dongen, 2011). However, the U.S. Federal Aviation Administration (FAA) has not awaited the outcomes of these efforts and has promulgated a new rule in which, for the first time, flight duty period restrictions in unaugmented operations vary by time of day (a good development) as well as number of segments (an uncertain improvement). See Table B to Part 117 of 14 Code of Federal Regulations (CFR) Parts 117, 119, and 121 "Flightcrew Member Duty and Rest Requirements."

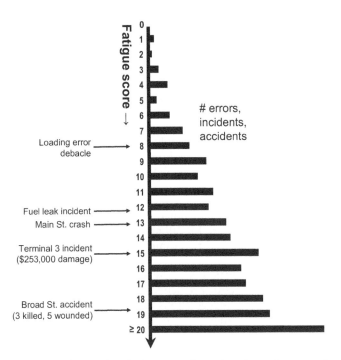

Figure 30.10   Illustration of a fatigue scale related to the number of errors, incidents and accidents in a large-scale operation

Note: In this example, a continuous fatigue scale (vertical line, with 0 indicating no fatigue) is put in context by relating it to the relative frequency of hypothetical errors, incidents and accidents (size of horizontal bars to the right) associated with (rounded) estimated operator fatigue scores of 0, 1, 2, etc. Some notable, hypothetical events (on the left) in the trucking division of the operation provide further context to help interpret scores on this fatigue scale.

operations, this could be the relative frequency of reported errors, incidents and accidents, as illustrated in Figure 30.10.

Fatigue scores expressed on a continuous scale provide a basis for an inherently valid, enhanced approach to implementing FRM enterprise-wide. This involves simultaneous comparison of a collection of schedules of interest versus another set of schedules used as benchmark. In this comparative framework, precisely defined safety criteria for the schedules of interest may be proposed, as follows:

1. *Shifting the overall distribution.* The mean of the fatigue predictions for the schedules of interest should be better (or at least no worse) than the mean of the fatigue predictions for the benchmark schedules. (This can be statistically tested with a two-sample t test.)
2. *Cutting off the tail of the distribution.* No single fatigue prediction among the schedules of interest should be worse than the worst prediction encountered among the benchmark schedules.
3. *Favorably skewing the distribution.* The variability (e.g., standard deviation) of the fatigue predictions for the schedules of interest should be smaller (or at least no

bigger) than the variability in the fatigue predictions for the benchmark schedules. (This can be statistically tested with a variance ratio F test.)

If these three criteria are met, then it can be claimed with reasonable certainty that the schedules of interest, as a set, are at least as fatigue-friendly as the benchmark set, and should therefore be at least as safe with regard to fatigue risk (cf. Huppert, 2009).

With reference to Figure 30.9, a good example of this concept would be the evaluation of the fatigue profile of an airline's ULR operations by comparison of the entire set of ULR flights (N=250) against the airline's comparable, historically safe set of LR flights, where "comparable" is defined as including only those flights with a scheduled duration of 14 hours or more (N=155). For the hypothetical flight schedules depicted in Figure 30.9, this comparison works out as follows:

1. The mean of the ULR fatigue predictions, which equals 85.4, is significantly better than the mean of the benchmark LR fatigue predictions, which equals 84.1 on the effectiveness scale. (A two-tailed two-sample t test yields t_{403}=2.95, p=0.003, i.e., a significant improvement.)
2. No single ULR fatigue prediction is worse than the worst prediction encountered among the benchmark LR fatigue predictions.
3. The variance of the ULR fatigue predictions, which equals 14.7, is significantly smaller than the variance of the benchmark fatigue predictions, which equals 24.0. (A variance ratio test yields $F_{154,249}$=1.63, p<0.001, i.e., a significant reduction in variability.)

Thus, the hypothetical ULR flights are at least as safe with respect to fatigue risk, collectively, as the comparable LR flights serving as a benchmark.

The procedure outlined above can be fruitfully applied to quantify the effects of proposed enterprise-wide scheduling improvements. Moreover, if other operational risk factors can also be quantified, then it becomes possible to compare different work schedule options on the basis of overall risk, and thus select the schedules that are predicted to be safest altogether. This latter possibility represents the quintessential aim of model-based FRM, that is, to improve operational safety in an optimal and evidence-based manner, while maintaining operational integrity. It can be achieved by incorporating fatigue modeling into rostering and scheduling software, so as to pursue – within the constraints of the operation (e.g., personnel and equipment availability, rules and regulations, and costs) – fatigue-friendly optimization of work schedules (Belenky & Van Dongen, 2009). See Figure 30.11.

A major advantage of fatigue-friendly schedule optimization through rostering and scheduling is that operational risk from fatigue and other sources can be considered along with productivity and other desirable outcomes, provided they can all be expressed on a common metric (such as operational costs) or otherwise weighed against each other. In this regard, it is pertinent that fatigue-friendly work schedules are not necessarily associated with reduced cost-effectiveness. In fact, relative to prescriptive HOS regulations, fatigue-friendly schedule optimization may come with increased productivity and reduced cost (Romig & Klemets, 2009), both directly by allowing greater flexibility in work schedules, and indirectly by promoting increased operator alertness, decreased absenteeism, reduced incident and accident costs, and smaller downstream expenses from fatigue-related decreases in fuel efficiency and increases in maintenance costs (e.g., Dorrian, Hussey & Dawson, 2007).

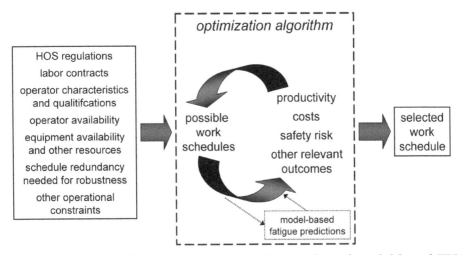

Figure 30.11   A simplified schematic of the integration of model-based FRM with rostering and scheduling software

Note: Given constraints on work scheduling associated with the operational setting (left box), a range of possible work schedules is considered. A computerized optimization algorithm iterates through these possible schedules until the schedule with the most favorable conditions is found. This typically involves consideration of productivity, costs and safety risk, but may also involve other relevant outcomes (e.g., meeting operators' preferences for specific work shifts). Model-based FRM can be inserted into the optimization process by mathematically predicting the safety risk from fatigue for each of the work schedules considered, and weighing the most fatigue-friendly schedules more favorably. The work schedule identified to have the most favorable conditions overall ends up being selected to run the operation.

Source: Figure based in part on Kohl & Karisch, 2004.

As a cautionary note on the schedule optimization process, it would be possible to assign greater relative weight to, say, operational costs than to fatigue-friendliness. This would result in work schedules being selected on the basis of cost reduction at the expense of safety. To use fatigue modeling properly as part of schedule optimization in the service of FRM, the following additional safety criteria are needed to supplement the three criteria listed above:

4.   The mean of the fatigue predictions for the work schedules selected after fatigue-based optimization should be better (or at least no worse) than the mean of the fatigue predictions for the schedules that would have been selected after optimization without consideration of fatigue.

5.   No single fatigue prediction for the work schedules selected after fatigue-based optimization should be worse than the worst predicted fatigue encountered among the schedules that would have been selected after optimization without consideration of fatigue.

6.   The variability in the fatigue predictions for the work schedules selected after fatigue-based optimization should be smaller (or at least no bigger) than the variability in the fatigue predictions for the schedules that would have been selected after optimization without consideration of fatigue.

All six criteria are relatively straightforward to implement in computer software. If they were to be combined with one additional straightforward criterion, namely that no schedules should be selected in which predicted fatigue exceeds the "bright line" threshold for the operation at hand (as discussed above), then the optimization process is practically guaranteed to yield fatigue-friendly work schedules that are both productive and increasingly safe.

CONCLUSIONS

Because of the stochastic nature of cognitive lapses due to fatigue, and the limited reliability of self-assessments of fatigue state, operators typically have insufficient insight into when and how fatigue may be putting them at risk. Model-based FRM is a science-based, quantitative tool readily available to help address this problem. Fatigue predictions are not absolute; they must be considered in the context of the individual operator and the work conditions he or she is in. Nevertheless, when comparing work schedules, all else being equivalent, the schedule predicted to be the least fatiguing may be assumed to be the safest. Model-based FRM cannot substitute for other operational safety measures or for common sense, experience and a good organizational safety culture. However, access to fatigue predictions – like, say, weather reports – can help significantly to make an operation safer.

In large-scale operations, model-based FRM can be integrated with rostering and scheduling software and embedded in the organizational SMS for additional effectiveness. Without losing sight of productivity and cost-effectiveness, this paves the way for evidence-based operational risk management, constituting a significant improvement over prescriptive HOS regulations (which it might ultimately replace) and allowing for increased work schedule flexibility. Although no formal cost/benefit assessments have as yet been published, reports from progressive industries in which model-based FRM has been implemented suggest it to be well worth the (comparatively small) investment. There are no turnkey solutions for incorporating model-based FRM yet, and significant expertise in mathematical modeling of fatigue as well as knowledge of the operational setting is required for successful implementation. Nonetheless, there is little doubt that model-based FRM will be an integral component of the future of work scheduling and operational safety management.

REFERENCES

Åkerstedt, T. (1995). Work hours, sleepiness and the underlying mechanisms. *Journal of Sleep Research*, 4(Suppl. 2), 15–22.

Åkerstedt, T. (1998). Shift work and disturbed sleep/wakefulness. *Sleep Medicine Reviews*, 2, 117–28.

Åkerstedt, T. & Folkard, S. (1996). Predicting duration of sleep from the three process model of regulation of alertness. *Occupational and Environmental Medicine*, 53, 136–41.

Åkerstedt, T., Ingre, M., Kecklund, G., Folkard, S. & Axelsson, J. (2008). Accounting for partial sleep deprivation and cumulative sleepiness in the three-process model of alertness regulation. *Chronobiology International*, 25, 309–19.

Axelsson, J., Kecklund, G., Åkerstedt, T., Donofrio, P., Lekander, M. & Ingre, M. (2008). Sleepiness and performance in response to repeated sleep restriction and subsequent recovery during semi-laboratory conditions. *Chronobiology International*, 25, 297–308.

Banks, S., Van Dongen, H.P.A., Maislin, G. & Dinges, D.F. (2010). Neurobehavioral dynamics following chronic sleep restriction: Dose–response effects of one night for recovery. *Sleep*, 33, 1013–26.

Belenky, G. & Van Dongen, H.P.A. (2009). Computer implemented scheduling systems and associated methods. Patent Application No. 12/253,703; Publication No. US 2009/0132332 A1.

Belenky, G., Bowen, A.K. & Van Dongen, H.P.A. (2009). White paper on fatigue risk management in commercial aviation: From bright lines to iterative safety improvements. Spokane, WA: Sleep and Performance Research Center, Washington State University. Copy available from the authors (hvd@wsu.edu).

Belenky, G., Wesensten, N.J., Thorne, D.R., Thomas, M.L., Sing, H.C., Redmond, D.P., Russo, M.B. & Balkin, T.J. (2003). Patterns of performance degradation and restoration during sleep restriction and subsequent recovery: A sleep dose–response study. *Journal of Sleep Research*, 12, 1–12.

Bergum, B.O. & Lehr, D.J. (1962). Vigilance performance as a function of interpolated rest. *Journal of Applied Psychology*, 46, 425–7.

Bills, A.G. (1931). Blocking: A new principle of mental fatigue. *American Journal of Psychology*, 43, 230–45.

Borbély, A.A. (1982). A two process model of sleep regulation. *Human Neurobiology*, 1, 195–204.

Caldwell, J.A., Mallis, M.M., Caldwell, J.L., Paul, M.A., Miller, J.C., Neri, D.F. & Aerospace Medical Association Fatigue Countermeasures Subcommittee of the Aerospace Human Factors Committee (2009). Fatigue countermeasures in aviation. *Aviation, Space, and Environmental Medicine*, 80, 29–59.

Darwent, D., Dawson, D. & Roach, G.D. (2010). Prediction of probabilistic sleep distributions following travel across multiple time zones. *Sleep*, 33, 185–95.

Davies, D.R. & Parasuraman, R. (1982). *The Psychology of Vigilance*. New York, NY: Academic Press.

Dawson, D. & McCulloch, K. (2005). Managing fatigue: It's about sleep. *Sleep Medicine Reviews*, 9, 365–80.

Dawson, D. & Reid, K. (1997). Fatigue, alcohol and performance impairment. *Nature*, 388, 235.

De Valck, E. & Cluydts, R. (2003). Sleepiness as a state–trait phenomenon, comprising both a sleep drive and a wake drive. *Medical Hypotheses*, 60, 509–12.

Dijk, D.-J. & Czeisler, C.A. (1994). Paradoxical timing of the circadian rhythm of sleep propensity serves to consolidate sleep and wakefulness in humans. *Neuroscience Letters*, 166, 63–8.

Dinges, D.F. (1990). Are you awake? Cognitive performance and reverie during the hypnopompic state, in R.R. Bootzin, J.F. Kihlstrom & D.L. Schacter (eds), *Sleep and Cognition*. Washington, DC: American Psychological Association, 159–75.

Dinges, D.F. & Powell, J.W. (1985). Microcomputer analyses of performance on a portable, simple visual RT task during sustained operations. *Behavioral Research Methods, Instruments, and Computers*, 17, 652–5.

Dinges, D.F., Orne, M.T. & Orne, E.C. (1985). Assessing performance upon abrupt awakening from naps during quasi-continuous operations. *Behavioral Research Methods, Instruments, and Computers*, 17, 37–45.

Doran, S.M., Van Dongen, H.P.A. & Dinges, D.F. (2001). Sustained attention performance during sleep deprivation: Evidence of state instability. *Archives of Italian Biology*, 139, 253–67.

Dorrian, J., Hussey, F. & Dawson, D. (2007). Train driving efficiency and safety: Examining the cost of fatigue. *Journal of Sleep Research*, 16, 1–11.

Folkard, S. & Åkerstedt, T. (2004). Trends in the risk of accidents and injuries and their implications for models of fatigue and performance. *Aviation, Space, and Environmental Medicine*, 75, A161–7.

Friedl, K.E., Mallis, M.M., Ahlers, S.T., Popkin, S.M. & Larkin, W. (2004). Research requirements for operational decision-making using models of fatigue and performance. *Aviation, Space, and Environmental Medicine*, 75, A192–9.

Hoddes, E., Zarcone, V., Smythe, H., Phillips, R. & Dement, W.C. (1973). Quantification of sleepiness: A new approach. *Psychophysiology*, 10, 431–6.

Huppert, F.A. (2009). A new approach to reducing disorder and improving well-being. *Perspectives on Psychological Science*, 4, 108–11.

Hursh, S.R. & Van Dongen, H.P.A. (2010). Fatigue and performance modeling, in M.H. Kryger, T. Roth & W.C. Dement (eds), *Principles and Practice of Sleep Medicine*. Fifth edition. St. Louis, MO: Elsevier Saunders, 745–52.

Hursh, S.R., Balkin, T.J., Miller, J.C. & Eddy, D.R. (2004). The fatigue avoidance scheduling tool: Modeling to minimize the effects of fatigue on cognitive performance. *SAE Transactions*, 113, 111–19.

Hursh, S.R., Raslear, T.G., Kaye, A.S. & Fanzone, J.F. Jr. (2006). Validation and calibration of a fatigue assessment tool for railroad work schedules, summary report. Washington, DC: Federal Railroad Administration.

Jackson, M.L. & Van Dongen, H.P.A. (2011). Cognitive effects of sleepiness, in M.J. Thorpy & M. Billiard (eds), *Sleepiness: Causes, Consequences and Treatment*. Cambridge, UK: Cambridge University Press, 72–81.

Jewett, M.E. & Kronauer, R.E. (1999). Interactive mathematical models of subjective alertness and cognitive throughput in humans. *Journal of Biological Rhythms*, 14, 588–97.

Johnson, M.L., Belenky, G., Redmond, D.P., Thorne, D.R., Williams, J.D., Hursh, S.R. & Balkin, T.J. (2004). Modulating the homeostatic process to predict performance during chronic sleep restriction. *Aviation, Space, and Environmental Medicine*, 75, A141–6.

Kaplan, K.A., Itoi, A. & Dement, W.C. (2007). Awareness of sleepiness and ability to predict sleep onset: Can drivers avoid falling asleep at the wheel? *Sleep Medicine*, 9, 71–9.

Kohl, N. & Karisch, S.E. (2004). Airline crew rostering: Problem types, modeling, and optimization. *Annals of Operations Research*, 127, 223–57.

Krueger, J.M., Rector, D.M., Roy, S., Van Dongen, H.P.A., Belenky, G. & Panksepp, J. (2008). Sleep as a fundamental property of neuronal assemblies. *Nature Reviews Neuroscience*, 9, 910–19.

Leproult, R., Colecchia, E.F., Berardi, A.M., Stickgold, R., Kosslyn, S.M. & Van Cauter, E. (2003). Individual differences in subjective and objective alertness during sleep deprivation are stable and unrelated. *American Journal of Physiology*, 284, R280–90.

Mallis, M.M., Mejdal, S., Nguyen, T.T. & Dinges, D.F. (2004). Summary of the key features of seven biomathematical models of human fatigue and performance. *Aviation, Space, and Environmental Medicine*, 75, A4–A14.

McCauley, P., Kalachev, L.V., Smith, A.D., Belenky, G., Dinges, D.F. & Van Dongen, H.P.A. (2009). A new mathematical model for the homeostatic effects of sleep loss on neurobehavioral performance. *Journal of Theoretical Biology*, 256, 227–39.

McCauley, P. & Van Dongen, H.P.A. (2010). Regional Airline Association fatigue and performance study – phase 1. Spokane, WA: Sleep and Performance Research Center, Washington State University. Copy available from the authors (hvd@wsu.edu).

McCormack, P.D. (1958). Performance in a vigilance task as a function of inter-stimulus interval and interpolated rest. *Canadian Journal of Psychology*, 12, 242–6.

Minors, D.S., Waterhouse, J.M. & Wirz-Justice, A. (1991). A human phase–response curve to light. *Neuroscience Letters*, 133, 36–40.

Mollicone, D.J., Van Dongen, H.P.A., Rogers, N.L. & Dinges, D.F. (2008). Response surface mapping of neurobehavioral performance: Testing the feasibility of split sleep schedules for space operations. *Acta Astronautica*, 63, 833–40.

Pagel, J.F. (2005). Medications and their effects on sleep. *Primary Care: Clinics in Office Practice*, 32, 491–509.

Raslear, T.G., Hursh, S.R. & Van Dongen, H.P.A. (2011). Predicting cognitive impairment and accident risk. *Progress in Brain Research*, 190, 155–67.

Reason, J. (1990). *Human Error*. Cambridge, UK: Cambridge University Press.

Ricci, J.A., Chee, E., Lorandeau, A.L. & Berger, J. (2007). Fatigue in the U.S. workforce: Prevalence and implications for lost productive work time. *Journal of Occupational and Environmental Medicine*, 49, 1–10.

Romig, E. & Klemets, T. (2009). Fatigue risk management in flight crew scheduling. *Aviation, Space, and Environmental Medicine*, 80, 1073–4.

Rosekind, M.R. (2005). Underestimating the societal costs of impaired alertness: Safety, health and productivity risks. *Sleep Medicine*, 6, S21–5.

St. Hilaire, M.A., Klerman, E.B., Khalsa, S.B.S., Wright, K.P. Jr., Czeisler, C.A. & Kronauer, R.E. (2007). Addition of a non-photic component to the light-based mathematical model of the human circadian pacemaker. *Journal of Theoretical Biology*, 247, 583–99.

Van Dongen, H.P.A. (2004). Comparison of mathematical model predictions to experimental data of fatigue and performance. *Aviation, Space, and Environmental Medicine*, 75, A15–36.

Van Dongen, H.P.A. (2006). Shift work and inter-individual differences in sleep and sleepiness. *Chronobiology International*, 23, 1139–47.

Van Dongen, H.P.A. & Belenky, G. (2009). Individual differences in vulnerability to sleep loss in the work environment. *Industrial Health*, 47, 518–26.

Van Dongen, H.P.A. & Dinges, D.F. (2003). Investigating the interaction between the homeostatic and circadian processes of sleep–wake regulation for the prediction of waking neurobehavioural performance. *Journal of Sleep Research*, 12, 181–7.

Van Dongen, H.P.A. & Dinges, D.F. (2005). Circadian rhythms in sleepiness, alertness, and performance, in M.H. Kryger, T. Roth & W.C. Dement (eds), *Principles and Practice of Sleep Medicine*. Fourth edition. Philadelphia, PA: Elsevier Saunders, 435–43.

Van Dongen, H.P.A. & Hursh, S.R. (2010). Fatigue, performance, errors and accidents, in M.H. Kryger, T. Roth & W.C. Dement (eds), *Principles and Practice of Sleep Medicine*. Fifth edition. St. Louis, MO: Elsevier Saunders, 753–59.

Van Dongen, H.P.A., Avinash, D., Robinson, B.M., Dinges, D.F. & Mallis, M.M. (2006). Development of an Astronaut Scheduling Assistant. *Habitation*, 10, 210–11.

Van Dongen, H.P.A., Baynard, M.D., Maislin, G. & Dinges, D.F. (2004). Systematic interindividual differences in neurobehavioral impairment from sleep loss: Evidence of trait-like differential vulnerability. *Sleep*, 27, 423–33.

Van Dongen, H.P.A., Belenky, G. & Krueger, J.M. (2011). Investigating the temporal dynamics and underlying mechanisms of cognitive fatigue, in P.L. Ackerman (ed), *Cognitive Fatigue*. Washington, DC: American Psychological Association, 127–47.

Van Dongen, H.P.A., Maislin, G., Mullington, J.M. & Dinges, D.F. (2003). The cumulative cost of additional wakefulness: Dose–response effects on neurobehavioral functions and sleep physiology from chronic sleep restriction and total sleep deprivation. *Sleep, 26*, 117–26.

Wesensten, N.J., Belenky, G., Thorne, D.R., Kautz, M.A. & Balkin, T.J. (2004). Modafinil vs. caffeine: Effects on fatigue during sleep deprivation. *Aviation, Space, and Environmental Medicine*, 75, 520–25.

Wever, R. (1970). Strength of a light–dark cycle as a zeitgeber for circadian rhythms in man. *Pflügers Archiv*, 321, 133–42.

Williams, H.L., Lubin, A. & Goodnow, J.J. (1959). Impaired performance with acute sleep loss. *Psychological Monographs: General and Applied*, 73, 1–26.

Index